Relativity

Relativity

SPECIAL, GENERAL, AND COSMOLOGICAL

Wolfgang Rindler

Professor of Physics
The University of Texas at Dallas

OXFORD

UNIVERSITY PRESS

OXFORD
UNIVERSITY PRESS

Great Clarendon Street, Oxford OX2 6DP

Oxford University Press is a department of the University of Oxford.
It furthers the University's objective of excellence in research, scholarship,
and education by publishing worldwide in

Oxford New York

Athens Auckland Bangkok Bogotá Buenos Aires Cape Town
Chennai Dar es Salaam Delhi Florence Hong Kong Istanbul
Karachi Kolkata Kuala Lumpur Madrid Melbourne Mexico City Mumbai
Nairobi Paris São Paulo Shanghai Singapore Taipei Tokyo Toronto Warsaw

with associated companies in Berlin Ibadan

Oxford is a trade mark of Oxford University Press
in the UK and in certain other countries

Published in the United States
by Oxford University Press, Inc., New York

A catalogue record for this book is available from the British Library

Library of Congress Cataloging in Publication Data
Rindler, Wolfgang, 1924–
Relativity : special, general, and cosmological/Wolfgang Rindler.
Includes bibliographical references and text.

1. Relativity (Physics) 2. Cosmology. I. Title
QC173.55 .R563 2001   530.11–dc21   00-067605

ISBN 0 19 850835 2 (Hbk)
ISBN 0 19 850836 0 (Pbk)

Typeset by Newgen Imaging Systems (P) Ltd., Chennai, India
Printed in Great Britain
on acid free paper by
Biddles Ltd., Guildford & King's Lynn.

To my wife Linda
the most generous person
I have ever known

Preface

My earlier book, *Essential Relativity*, aimed to provide a quick if thoughtful introduction to the subject at the level of advanced undergraduates and beginning graduate
students, while 'containing enough new material and simplifications of old arguments
so as not to bore the expert teacher.' But general relativity has by now robustly entered
the mainstream of physics, in particular astrophysics, new discoveries in cosmology
are routinely reported in the press, while 'wormholes' and time travel have made
it into popular TV. Students thus want to know more than the bare minimum. The
present book offers such an extension, in which the style, the general philosophy, and
the mathematical level of sophistication have nevertheless remained the same. Anyone who knows the calculus up to partial differentiation, ordinary vectors to the point
of differentiating them, and that most useful method of approximation, the binomial
theorem, should be able to read this book. But instead of the earlier nine chapters
there are now eighteen, and instead of 167 exercises, now there are more than 300;
above all, tensors are introduced without apology and then thoroughly used.

Einstein's special and general relativity, the theories of flat and curved spacetime
and of the physics therein, and relativistic cosmology, with its geometry and dynamics
for the entire universe, not only seem necessary for a scientist's balanced view of
the world, but also offer some of the greatest intellectual thrills of modern physics.
Perhaps the chief motivation in writing this book has been once more the desire to
convey that thrill, as well as some of the insights that long preoccupation with a subject
inevitably yields. It is true that many aspects of general relativity have still not been
tested experimentally. Nevertheless enough have been tested to justify the view that
all of relativity is by now well out of the tentative stage. That is also the reason why
the introductory chapter contains an overview of all of relativity and cosmology, so
that the student can appreciate from the very beginning the local character of special
relativity and how it fits into the general scheme. The three main parts that follow
deal extensively with special relativity, general relativity, and cosmology. In each I
have tried to report on the most important crucial experiments and observations, both
historical and modern, but stressing concepts rather than experimental detail. In fact,
the emphasis throughout is on understanding the concepts and making the ideas come
alive. But an equal value is put on developing the mathematical formalism rigorously,
and on guiding the student to use both concepts and mathematics in conjunction with
the tricks of the trade to become and expert problem solver. A vital part in this process
should be played by the exercises, which have been put together rather carefully, and
which are mostly of the 'thinking' variety. Though their full solution often requires

some ingenuity, they should at least be looked at, as a supplement to the text, for the extra information they contain.

No book ever has enough diagrams. That is one of the luxuries that classroom teaching has over a book. So readers should constantly draw their own, especially since relativity is a very geometric subject in which the facility to think geometrically is a great asset. Readers should also constantly make up their own problems, however trivial: what would happen if . . . ? In an initially paradoxical subject like relativity, it is often the most skeptical student who is the most successful.

Each of the three parts could well be cut short drastically so that the book might serve as a text for a one-semester course. To present it fully will take two semesters, probably with material to spare. But apart from its serving as an introductory text for a formal course, I also envisage the book as having some use for the general scientist who might wish to browse in it, and for the more advanced graduate student in search of greener pastures, as a change from the rocky pinnacles of more severe texts.

At the end of the book there is an Appendix on curvature components for diagonal metrics (in a little more generality than the old 'Dingle formulae'), which could be useful even to workers in the field who have not read the rest of the book. And finally a word of warning: in many sections, as is the custom in relativity, the units are chosen so as to make the speed of light unity, and later even to make Newton's constant of gravitation unity, which must be borne in mind when comparing formulae; where the dimensions seem wrong, c's or G's are missing.

I owe much to many modern authors (Sexl and Urbantke, Misner, Thorne and Wheeler, Ohanian and Ruffini, Woodhouse, etc.) though an exact assignment of debt would be difficult at this stage. I have also benefitted from the many searching questions of my students over the years, among whom I might perhaps single out James Gilson and Jack Denur. But the greatest debt I owe, as so often before, to my friend Jürgen Ehlers—discussion partner, scientific conscience, font of knowledge without peer.

Dallas, Texas   W.R.
January 2001

Contents

Introduction

1

From absolute space and time to influenceable spacetime: an overview

1.1 Definition of relativity

At their core, Einstein's relativity theories (both the special theory of 1905 and the general theory of 1915) are the modern physical theories of space and time, which have replaced Newton's concepts of *absolute* space and *absolute* time. We specifically call Einstein's theories 'physical' because they claim to describe real structures in the real world and are open to experimental disproof.

Since all (or, at least, all classical) physical processes play out on a background of space and time, the laws of physics must be compatible with the accepted theories of space and time. If one changes the latter, one must adapt the former. This process gave rise to 'relativistic physics', which from the outset made some startling predictions (like $E = mc^2$) but which has nevertheless been amply confirmed by experiment.

Originally, in physics, relativity meant the abolition of absolute space—a quest that had been recognized as desirable ever since Newton's days. And this is indeed what Einstein's two theories accomplished: *special relativity* (SR) abolished absolute space in its Maxwellian role as the 'ether' that carried electromagnetic fields and, in particular, light waves, while *general relativity* (GR) abolished absolute space also in its Newtonian role as the ubiquitous and uninfluenceable standard of rest or uniform motion. Surprisingly, and not by design but rather as an inevitable by-product, Einstein's theory also abolished Newton's concept of an absolute time.

Since these ideas are fundamental, we devote the first chapter to a brief discussion centered on the three questions: What *is* absolute space? Why *should* it be abolished? How *can* it be abolished?

A more modern and positive definition of relativity has evolved *ex post facto* from the actual relativity theories. According to this view, the relativity of any physical theory expresses itself in the group of transformations which leave the laws of the theory invariant and which therefore describe symmetries, for example of the space and time arenas of these theories. Thus, as we shall see, Newton's mechanics possesses the relativity of the so-called Galilean group, SR possesses the relativity of the Poincaré (or 'general' Lorentz) group, GR possesses the relativity of the full group of smooth one-to-one space and time transformations, and the various cosmologies possess the relativity of the various symmetries with which the large-scale universe is credited. Even a theory valid only in one absolute Euclidean space, provided *that* is physically homogeneous and isotropic, would possess a relativity, namely the group of rotations and translations.

1.2   Newton's laws and inertial frames

We recall Newton's three laws of mechanics, of which the first (Galileo's law of inertia) is really a special case of the second:

(i) Free particles move with constant vector-velocity (that is, with zero acceleration, or, in other words, with constant speed along straight lines).

(ii) The vector-force on a particle equals the product of its mass into its vector-acceleration: $\mathbf{f} = m\mathbf{a}$.

(iii) The forces of action and reaction are equal and opposite; for example, if a particle A exerts a force $\mathbf{f}$ on a particle B, then B exerts a force $-\mathbf{f}$ on A. (Newton's absolute time is needed here: If the particles are at a distance and the forces vary, action today will not equal reaction tomorrow; they must be measured simultaneously, and simultaneity must be unambiguous.)

Physical laws are usually stated relative to some *reference frame*, which allows physical quantities like velocity, electric field, etc., to be defined. Preferred among reference frames are *rigid* frames, and preferred among these are the *inertial* frames. Newton's laws apply in the latter.

A classical rigid reference frame is an imagined extension of a rigid body. For example, the earth determines a rigid frame throughout all space, consisting of all those points which remain 'rigidly' at rest relative to the earth and to each other (like 'geostationary' satellites). We can associate an orthogonal Cartesian coordinate system with such a frame in many ways, by choosing three mutually orthogonal planes within it and measuring x, y, z as distances from these planes. Of course, this presupposes that the geometry in such a frame is Euclidean, which was taken for granted until 1915! Also, a time t must be defined throughout the frame, since this enters into many of the laws. In Newton's theory there is no problem with that. Absolute time 'ticks' world-wide—its rate directly linked to Newton's first law (free particles cover equal distances in equal times)—and any particular frame just picks up this 'world-time'. Only the choice of units and the zero-setting remain free.

Newton's first law serves to single out inertial frames among rigid frames: a rigid frame is called inertial if free particles move without acceleration relative to it. And, as it turns out, Newton's laws apply equally in all inertial frames. However, Newton postulated the existence of a quasi-substantial *absolute space* (AS) in which he thought the center of mass of the solar system was at rest, and which, to him, was the primary arena for his mechanics. That the laws were equally valid in all other reference frames moving uniformly relative to AS (the inertial frames) was to him a profoundly interesting theorem, but it was AS that bore, as it were, the responsibility for it all.

He called it the *sensorium dei*—God's sensory organ—with which God 'felt' the world.

1.3 The Galilean transformation

Now consider *any* two rigid reference frames S and S′ in uniform relative motion with velocity v. Let identical units of length and time be used in both frames. And let their times t and $t′$ and their Cartesian coordinates x, y, z and $x′$, $y′$, $z′$ be adapted to their relative motion in the following way (cf. Fig. 1.1): The S′ origin moves with velocity v along the x-axis of S, the $x′$-axis coincides with the x-axis, while the y- and $y′$-axes remain parallel, as do the z- and $z′$-axes; and all clocks are set to zero when the two origins meet. The coordinate systems S: $\{x, y, z, t\}$ and S′: $\{x′, y′, z′, t′\}$ are then said to be in *standard configuration*.

Suppose an *event* (like the flashing of a light bulb, or the collision of two point-particles) has coordinates (x, y, z, t) relative to S and $(x′, y′, z′, t′)$ relative to S′. Then the classical (and 'common sense') relations between these two sets of coordinates are given by the *standard Galilean transformation* (GT):

$$x′ = x - vt, \qquad y′ = y, \qquad z′ = z, \qquad t′ = t, \qquad (1.1)$$

which can be read off from the diagram, since vt is the distance between the spatial origins. The last of these relations expresses the absoluteness (which is to say, observer-independence) of time.

Differentiating the LHSs of (1.1) with respect to $t′$ and the RHSs with respect to t immediately leads to the classical velocity transformation, which relates the velocity components of a moving particle in S with those in S′:

$$u′_1 = u_1 - v, \qquad u′_2 = u_2, \qquad u′_3 = u_3, \qquad (1.2)$$

where $(u_1, u_2, u_3) = (\mathrm{d}x/\mathrm{d}t, \mathrm{d}y/\mathrm{d}t, \mathrm{d}z/\mathrm{d}t)$ and $(u′_1, u′_2, u′_3) = (\mathrm{d}x′/\mathrm{d}t′, \mathrm{d}y′/\mathrm{d}t′, \mathrm{d}z′/\mathrm{d}t′)$. Thus if I walk forward at 2 mph ($u′_1$) in a bus traveling at 30 mph (v), my speed relative to the road (u_1) will be 32 mph. In special relativity this will no longer be true.

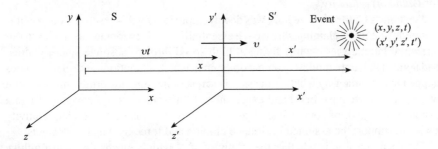

Fig. 1.1

A further differentiation yields (with $a_1' = \mathrm{d}u_1'/\mathrm{d}t'$, etc.)

$$a_1' = a_1, \qquad a_2' = a_2, \qquad a_3' = a_3, \tag{1.3}$$

that is, the invariance of acceleration.

In vector notation, these formulae can be written more concisely (and perhaps also more familiarly) in the form

$$\mathbf{r}' = \mathbf{r} - \mathbf{v}t, \qquad \mathbf{u}' = \mathbf{u} - \mathbf{v}, \qquad \mathbf{a}' = \mathbf{a}, \tag{1.4}$$

where $\mathbf{r}$, $\mathbf{u}$, $\mathbf{a}$ are the position-, velocity-, and acceleration-vectors, respectively, in S, and the primed symbols are similarly defined in S'; $\mathbf{v}$ denotes the (vector-)velocity of S' relative to S.

For future reference we note that inertial frames which are not in standard configuration are related by *general* GTs, which are simply compositions of standard GTs with rotations and spatial and temporal translations.

1.4   Newtonian relativity

Recall that an inertial frame is a rigid frame in which Newton's first law holds. Suppose the frame S of Fig. 1.1 is inertial. Since, by (1.2), constant velocities in S transform into constant velocities in S', we see that all particles recognized as free in S move uniformly in S', which is therefore also inertial. On the other hand, *only* frames moving uniformly relative to S can be inertial. For the fixed points in any inertial frame are pontential free particles, so all must move uniformly relative to S, and evidently no set of free particles can remain rigid unless all their velocities are identical. So the class of inertial frames consists precisely of all rigid frames that move uniformly relative to one known inertial frame; for example, absolute space.

Now, from the invariance of the acceleration, eqn (1.4) (iii), we see that all we need in order to have all three of Newton's laws invariant among inertial frames is (i) an axiom that the mass m is invariant, and (ii) an axiom that every force is invariant. Both these assumptions are indeed part of Newton's theory. The resulting property of Newtonian mechanics that it holds equally in all inertial frames is called *Newtonian* (or *Galilean*) *relativity*.

Newton, as Galileo before him, was deeply impressed by this result. Of course, it is a prerequisite for explaining why we see essentially pure Newtonian mechanics in our terrestrial laboratories—while flying at high speed through absolute space. Galileo had noticed the more modest example of a ship in which all motions and all mechanics happen in the same way whether the ship is at rest or moving uniformly through calm waters. Today we have first-hand experience of this sameness whenever we fly in a fast airplane. And, of course, each real-world manifestation of Newtonian relativity serves to support the assumed invariance of mass and force on which it depends.

The deep question is whether the relativity of Newton's mechanics is just a fluke or an integral part of nature, in which case it would probably go beyond mechanics.

There are some fascinating indications that Newton, at least during some periods of his life, might have thought the latter—in spite of his clinging to absolute space.[1]

1.5 Objections to absolute space; Mach's principle

Newton's concept of an absolute space has never lacked critics. From Huyghens and Leibniz and Bishop Berkeley, all near-contemporaries of Newton, to Mach in the nineteenth century and Einstein in the twentieth, cogent arguments have been brought against AS. There are two main objections:

(i) Absolute space cannot be distinguished by any intrinsic properties from all the other inertial frames. Differences that do not manifest themselves observationally should not be posited theoretically.

(ii) 'It conflicts with one's scientific understanding to conceive of a thing which acts but cannot be acted upon.' The words are Einstein's, but he attributes the thought to Mach.

It took a surprisingly long time, but by the late nineteenth century it gradually came to be appreciated that Newton's theory can logically very well dispense with absolute space; as an axiomatic basis, one can and should instead accept the existence of the infinite class of equivalent inertial frames (as Einstein still did in his special relativity). Then objection (i) is eliminated. But objection (ii) applies just as much to the entire class of inertial frames as it does to any one of them. Do the inertial frames really exist independently of the rest of the universe? This problem became the thorn in Einstein's consciousness that eventually spurred him on to general relativity.

But here we shall digress briefly to describe an earlier attempt to address this problem. It was made by the philosopher-scientist Mach, and it casts its shadow as far as the present day.[2] Mach's ideas on inertia, whose germ was already contained in the writings of Leibniz and Bishop Berkeley, are roughly these: (a) space is not a 'thing' in its own right; it is merely an abstraction from the totality of distance-relations between matter; (b) a particle's inertia is due to some (unfortunately unspecified) interaction of that particle with all the other masses in the universe; (c) the local standards of non-acceleration are determined by some average of the motions of all the masses in the universe; (d) all that matters in mechanics is the *relative* motion of *all* the masses. Thus Mach wrote: '... it does not matter if we think of the earth as turning round on its axis, or at rest while the fixed stars revolve around it .... The law of inertia must be so conceived that exactly the same thing results from the second supposition as from the first.' Mach called his view 'relativistic'. Had he found the sought-for law of inertia, *all* rigid frames would have become equivalent.

---

[1] See R. Penrose in *300 Years of Gravitation*, S. Hawking and W. Israel, eds, Cambridge University Press, 1987, especially Section 3.3 and p. 49.

[2] See, for example, *Mach's Principle*, J. Barbour and H. Pfister, eds, Birkhäuser, Boston, 1995.

A spinning elastic sphere bulges at its equator. To the question of how the sphere 'knows' that it is spinning and hence must bulge, Newton might have answered that it 'felt' the action of absolute space. Mach would have answered that the bulging sphere 'felt' the action of the cosmic masses rotating around it. To Newton, rotation with respect to AS produces centrifugal (inertial) forces, which are quite distinct from gravitational forces. To Mach, centrifugal forces *are* gravitational; that is, caused by the action of mass upon mass.

Einstein coined the term *Mach's principle* for this whole complex of ideas. Of course, with Mach these ideas were still embryonic in that a quantitative theory of the proposed effect of the *motion* of distant matter was totally lacking. One is reminded of Maxwell's theory, where the motion of the sources affects the field. Indeed, a Maxwell-type of gravitational theory has many Machian features.[3] But it violates special relativity. For example, whereas charge is necessarily invariant in Maxwell's theory, mass varies with speed in SR. Also, because of the relation $E = mc^2$, the gravitational binding energy of a body has (negative) mass; thus the total mass of a system cannot equal the sum of the masses of the parts, whereas in Maxwell's theory charge is strictly additive, as a direct consequence of the linearity of the theory.

Einstein's solution to the problem of inertia, GR, turned out to be much more complicated than Maxwell's theory. However, in 'first approximation' it reduces to Newton's theory, and in 'second approximation' it actually has Maxwellian features. (Cf. Section 15.5 below.) But in what sense GR is truly 'Machian' is still a matter of debate and, from a practical point of view, irrelevant. There certainly are GR solutions where the local standard of non-acceleration does *not* accord with the matter distribution. Thus, while in GR all matter, including its motion, undoubtedly *affects* local inertial behavior, it appears not entirely to cause it.

Mach's principle, nevertheless, continues a life of its own. One can perhaps appreciate this best from examples of its predictive successes—although there are also examples where its predictions are wrong.[4] The following instance of a potential success is due to Sciama. It is known today that our galaxy rotates differentially, somewhat like a huge planetary system, with a typical period of about 250 million years. Such a rotation was already postulated by Kant to account for the flattened shape of the galaxy, as evidenced by the Milky Way in the sky. Without orbiting, the stars would fall into the center of the galaxy in about 100 million years, which is much less than the age of the earth. Also it is known today that the best-fitting inertial frame for the solar system does not partake of this rotation, as indeed any Newtonian would expect. But had Mach been aware of this, he could have been led by his principle to postulate the existence of a whole vast extragalactic universe (which was not confirmed observationally until much later) simply in order to make the best-fitting inertial frame of the solar system come out right.

---

[3] See, for example, D. W. Sciama, *Mon. Not. R. Astron. Soc.* **113**, 34 (1953).

[4] See, for example, W. Rindler in *Mach's Principle, loc. cit.*, p. 439.

1.6   The ether

We now return to the first problem raised in Section 1.5—if and how one can distinguish absolute space among the inertial frames. It seems to have been Descartes (1596–1650) who introduced into science the idea of a space-filling material 'ether' as the transmitter of otherwise incomprehensible actions. Bodies in contact can push each other around, but it required an ether (today we call it a field!) to mediate between a magnet and the nail it attracts, or between the moon and the tides. A generation later, even Newton toyed with the idea of an ether with very strange elastic properties to 'explain' gravity. It is perhaps no wonder that he thought of absolute space as having substance. To Newton's contemporaries, like Hooke and Huyghens, the ether's main function was to carry light waves. Their 'luminiferous ether' evolved into a cornerstone of Maxwell's theory, and became a plausible marker for Newton's absolute space.

As is well known, in Maxwell's theory there occurs a constant c with the dimensions of a speed, which was originally defined as a ratio between electrostatic and electrodynamic units of charge, and which can be determined by simple laboratory experiments involving charges and currents. Moreover, Maxwell's theory predicted the propagation of disturbances of the electromagnetic field in vacuum with this speed c—in other words, the existence of electromagnetic waves. The surprising thing was that c coincided precisely with the known vacuum speed of light, which led Maxwell to conjecture that light must be an electromagnetic wave phenomenon. (At that time 'c' had not yet invaded the rest of physics; Maxwell would have been unlucky had light turned out to be *gravitational* waves!) To serve as a carrier for such waves, and for electromagnetic 'strains' in general, Maxwell resurrected the old idea of an ether. And it seemed reasonable to assume that the frame of 'still ether' coincided with the frame of the 'fixed stars'; that is, with Newton's absolute space. So absolute space is at least electromagnetically distinguishable from all other inertial frames. Or is it?

1.7   Michelson and Morley's search for the ether

The great success of Maxwell's theory since about 1860 made the ether as such a central object of study and debate in late nineteenth-century physics. There was considerable pressure on experimenters to 'observe' it directly. In particular, efforts were made to determine the speed of the orbiting earth through the ether, by measuring the 'ether wind' or 'ether drift' through the lab. The best known of all these experiments is that of Michelson and Morley of 1887. They split a beam of light and sent it along orthogonal paths of equal length and back again, whereupon interference fringes were produced between the returning beams. Different ether wind components along the two paths should have led to a difference in travel times. However, when the apparatus was rotated through 90°, so that this difference should be reversed, the expected displacement of the fringes did not occur.

Since the earth's orbital speed around the sun is 18 miles per second, one could expect the ether drift at *some* time during the year to be at least that much, no matter how the ether streamed past the solar system. And a drift of this magnitude was well within the capability of the apparatus to detect. The most obvious explanation, that the earth completely dragged the ether along with it in its neighborhood, could be ruled out because of various other optical effects like the aberration of starlight.

Thus electromagnetic theory was left with a serious puzzle: The average *to-and-fro* light speed in a *given* ether wind is direction-independent. (Modern laser versions have confirmed this experiment to an accuracy of one part in 10^{15}.[5]) The Michelson–Morley result is short of what we know today, namely that the *one-way* speed of light at *all* times is independent of any ether wind. This is nicely demonstrated by the workings of *international atomic time*, TAI (Temps Atomique International). TAI is determined by a large number of atomic clocks clustered in various national laboratories around the globe. Their readings are continuously checked against each other by the exchange of radio signals (no different from light except in having lower frequencies). Any interference with these signals by a variable ether wind of the expected magnitude would be detected by these super-accurate clocks. Needless to say, none has been detected: day or night, summer or winter, the signals from one clock to another always arrive with the same time delay. As another example, the incredible proven accuracy of some modern radio navigational systems (via satellites) hinges crucially on the speed of radio signals being independent of any ether wind.

1.8   Lorentz's ether theory

An ingenious 'explanation' of the Michelson–Morley null result was found by FitzGerald in 1889.[6] He suggested that the lengths of bodies moving through the ether at velocity v contract in the direction of their motion by a factor $(1 - v^2/c^2)^{1/2}$—which would just compensate for the ether drift in the Michelson–Morley apparatus. A few years later Lorentz—apparently independently—made the same hypothesis and incorporated it into his ever more comprehensive ether theory.[7] He was, moreover, able to justify it to some extent by appealing to the electromagnetic constitution of matter and to the known contraction of the field of moving charges (see Section 7.6 below). This 'Lorentz–FitzGerald contraction' then quickly diffused into the literature.

Let us see how it works. For simplicity, assume that one of the two paths or 'arms' of the Michelson–Morley apparatus, marked L_1 in Fig. 1.2, lies in the direction of an

---

[5]   A. Brillet and J. L. Hall, *Phys. Rev. Lett.* **42**, 549 (1979).

[6]   See S. G. Brush, *Isis*, **58**, 230 (1967).

[7]   See E. T. Whittaker, *A History of the Theories of Aether and Electricity*, Tomash/American Institute of Physics, reprint 1987, vol. 1, pp. 404, 405.

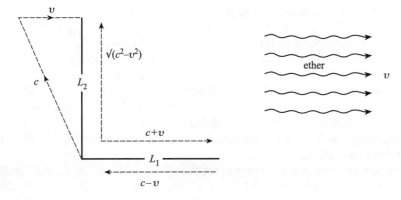

Fig. 1.2

ether drift of velocity v. Figure 1.2 should make it clear that the respective to-and-fro light travel times along the two arms would then be expected to be

$$T_1 = \frac{L_1}{c+v} + \frac{L_1}{c-v} = \frac{2L_1}{c(1-v^2/c^2)}, \tag{1.5}$$

$$T_2 = \frac{2L_2}{(c^2-v^2)^{1/2}} = \frac{2L_2}{c(1-v^2/c^2)^{1/2}}, \tag{1.6}$$

where L_1 and L_2 are the purportedly equal lengths of the two arms. The difference in these two times is at once eliminated if we assume that the arm *along* the ether drift undergoes Lorentz–FitzGerald contraction, so that $L_1 = L_2(1-v^2/c^2)^{1/2}$. A somewhat more complicated calculation (which must have lit the heart of FitzGerald) shows that under the same assumption the average to-and-fro speed of light, c', in *any* direction is the same,

$$c' = c(1-v^2/c^2)^{1/2}. \tag{1.7}$$

What the contraction hypothesis by itself does *not* achieve is to make the average to-and-fro speed of light independent of the ether drift—there is still a 'v' in (1.7)—nor does it make the *one-way* speed of light the same in all directions. Both these defects of the ether theory were eventually cured by insights taken over from Einstein's special relativity (for example, time dilation—the slowing down of moving clocks).

Lorentz—a giant among physicists and revered by Einstein ('I admire this man as no other')—could never free himself of the crutch of the ether, and when he died in 1928 he still believed in it. His ether theory came to include all of Einstein's basic findings and was, for calculational purposes, equivalent to special relativity, and less jolting to classical prejudices. But it was also infinitely less elegant and, above all, sterile in suggesting new results. Today it is best forgotten, except by historians.

1.9   Origins of special relativity

Einstein's solution of the ether puzzle was more drastic: it was like cutting the Gordian knot. In his famous *relativity principle* (RP) of 1905 he asserted that *all inertial frames are equivalent for the performance of all physical experiments.* That was the first postulate. For Einstein there is *no* ether, and *no* absolute space. All inertial frames are totally equivalent. In each IF the basic laws of *all* of physics are the same, and presumably simpler than in other rigid frames. In particular, every IF is as good for mechanics as Newton's absolute space, and as good for electromagnetism as Maxwell's ether frame.

This last remark almost forces another hypothesis on us, which Einstein, in fact, adopted as his second postulate: *light travels rectilinearly at speed c in every direction in every inertial frame.* For this property characterizes Maxwell's ether just as Newton's first law characterizes Newton's absolute space. Einstein knew, of course, that his second postulate clashed violently with our classical (Newtonian) ideas of space and time: no matter how fast I chase a light signal (by transferring myself to ever-faster IFs) it will always recede from me at speed c! Einstein's great achievement was to find a new framework of space and time, which in a natural and elegant way accommodates both his axioms. It depends on replacing the Galilean transformation by the Lorentz transformation as the link between IFs. This essentially leaves space and time unaltered within each IF, but it changes the view which each IF has of the others. Above all, it required a new concept of *relative* time, no different from the old within each IF, but different from frame to frame.

Einstein's two postulates 'explain' the failure of all the ether-drift experiments much as the principle of energy conservation explains a priori (that is, without the need for a detailed examination of the mechanism) the failure of all attempts to build a perpetual motion machine. Reciprocally, those experiments now served as empirical evidence for Einstein's two postulates. Einstein had turned the tables: predictions could be made. The situation can be compared to that obtaining in astronomy at the time when Ptolemy's intricate geocentric system (corresponding to Lorentz's 'etherocentric' theory) gave way to the ideas of Copernicus, Galileo, and Newton. In both cases the liberation from a time-honored but inconvenient reference frame ushered in a revolutionary clarification of physical thought, and consequently led to the discovery of a host of new and unexpected results.

Soon a whole theory based on Einstein's two postulates was in existence, and this theory is called *special relativity*. Its program was to modify all the laws of physics, where necessary, so as to make them equally valid in all inertial frames. For Einstein's relativity principle is really a *metaprinciple*: it puts constraints on *all* the laws of physics. The modifications suggested by the theory (especially in mechanics), though highly significant in many modern applications, have negligible effect in most classical problems, which is, of course, why they were not discovered earlier. However, they were not exactly needed empirically in 1905 either. This is a beautiful example of the power of pure thought to leap ahead of the empirical frontier—a feature of all good physical theories, though rarely on such a heroic scale.

Today, almost a century later, the enormous success of special relativity theory has made it impossible to doubt the wide validity of its basic premises. It has led, among other things, to a new theory of space and time in which the two mingle to form 'spacetime', to the existence of a maximum speed for all particles and signals, to a new mechanics in which mass increases with speed, to the formula $E = mc^2$, to a simple and successful macroscopic electrodynamics of moving bodies, to a new thermodynamics, to a kinetic gas theory that includes photons as well as particles, to de Broglie's association of waves with particles, to Sommerfeld's fine structure of atomic spectra, to Dirac's particle–antiparticle symmetry, and to Pauli's explanation of the connection between spin and statistics. Not least, it has paved the way for general-relativistic gravity and cosmology.

There is a touch of irony in the fact that Newton's theory, which had always been known to satisfy a relativity principle in the classical framework of space and time, now turned out to be in need of modification, whereas Maxwell's vacuum electrodynamics, with its apparent conceptual dependence on a preferred ether frame, came through with its formalism intact—in itself a powerful recommendation for special relativity. But on being freed from a material carrier, the electromagnetic field now became a non-substantial physical entity in its own right, an entity to which no state of rest and no velocity can be ascribed. Thus was born the modern field concept. Fields are not regarded as necessarily 'generated' by bodies, though influenced by them through field equations, and interacting with them by exchanging energy and momentum.

How original was Einstein in his special relativity? As Freud has stressed, most revolutionary ideas have at least been surmised or incompletely enunciated before. Like Copernicus, like Newton ('If I have seen further it is by standing on the shoulders of giants'), Einstein, too, had precursors, most notably Lorentz and Poincaré. Lorentz had actually found the 'Lorentz transformation' (LT) before Einstein, in 1903, as that which (in conjunction with a suitable transformation of the field) leaves Maxwell's equations invariant. But Lorentz neither penetrated the physical meaning of the LT nor ever renounced the ether. Poincaré, France's foremost mathematician of the day, and another strong participant in the hectic development of electromagnetic theory, occupies a position somewhat between Lorentz and Einstein. He used the LT equations to stress the need for a new mechanics in which c would be a limiting velocity, and yet, like Lorentz, he gave no indication of appreciating, in particular, the physicality of their time coordinate. He intuited as early as 1895 the impossibility of ever locating the ether frame, and even enunciated and *named* the 'relativity principle' in 1904, one year before Einstein. But, unlike Einstein, he did nothing *with* it. Einstein was the first to derive the LT from the relativity principle independently of Maxwell's theory, as that which connects *real* space and *real* time in various inertial frames. He was the first wholeheartedly to discard the ether and the old ideas of space and time (except as approximations) and to find equally symmetric and elegant substitutes for them. *That* was the vital and original breakthrough which made the subsequent rapid development of the theory possible. It took an extraordinarily agile and unprejudiced mind to do this, and Einstein fully deserves the credit for having changed our world view.

1.10   Further arguments for Einstein's two postulates

The relativity principle has become such a fundamental pillar of modern physics that it merits further discussion. Of course, as with all axioms, the proof of the pudding is in the eating: axioms are best justified by the success of the theory that follows from them, and this, in the case of special relativity, is overwhelming. But from a logical point of view, several arguments could and can be advanced for the RP a priori:

(i) The failure of all the ether-drift experiments—and there were others besides that of Michelson and Morley (see, for example, Exercise 7.18). Though Einstein made surprisingly little of these in his famous 1905 paper, they cried out for an explanation, which the relativity principle neatly provided.

(ii) The evident 'relativity' of Maxwell's theory, if not in spirit, yet in fact. This, to Einstein's mind, carried a great deal of weight. Take, for example, the interaction of a circular conducting loop and a bar magnet along its axis. If we move the loop, the Lorentz force due to the field of the stationary magnet drives the free electrons along the wire, thus producing a current. If, on the other hand we leave the loop stationary and move the magnet, the changing magnetic flux through the loop produces an identical current, by Faraday's law. So Maxwell's theory is as valid in the rest-frame of the magnet as it is in the rest-frame of the loop.

(iii) The unity of physics. This is an argument of more recent origin. But it has become increasingly obvious that physics cannot be separated into strictly independent branches; for example, no electromagnetic experiment can be performed without the use of mechanical parts, and no mechanical experiment is independent of the electromagnetic constitution of matter, etc. If there exists a *strict* relativity principle for mechanics, then a large part of electromagnetism must be relativistic also, namely that part which has to do with the constitution of matter. But if part, why not all? In short, if physics is indivisible, either all of it or none of it must satisfy the relativity principle. And since the RP is so strongly evident in mechanics, it is only reasonable to expect electromagnetism (and all the rest of physics) to obey it too.

(iv) The remarkableness of relativity. We are so utterly used to the relativity of all physical processes (always the same, in our terrestrial labs hurtling through the cosmos, in space capsules, in airplanes, etc.) that its remarkableness no longer strikes us. But recall how deeply Galileo was struck by his discovery that no force was necessary to keep a particle moving uniformly: he immediately suspected a law of nature behind it. It is much the same with relativity: if it holds approximately, that is so remarkable that it strongly suggests an exact law of nature.

As for the second postulate (the invariance of the speed of light) however essential, Einstein did not even devote a whole sentence to it in his original paper, nor did he deem it in need of a single word of justification! Here his instinct was sounder than that of many who followed him in the exposition of the theory. Much time and effort was spent wondering about such empirical questions as whether double-star systems rotating about a common center would *appear* to rotate uniformly, which would support the hypothesis that the velocity of light is independent of the velocity of the source; but then, maybe a cloud of gas around the system would absorb and re-emit the light and so mask the difference, etc., etc. But in fact (as Einstein very probably intuited), once we have accepted the RP, the second postulate is nothing but a two-way switch: As we shall see in Section 2.11, the RP (plus the assumption of causality invariance) implies that there *must* exist an invariant velocity—the only question is which. If that velocity is infinite (that is, an infinite speed in one inertial frame corresponds to an infinite speed in every other inertial frame), then the Galilean transformation group and, with it, Newtonian space and time result. If, on the other hand, the invariant velocity is finite, say c, then the Lorentz transformation group results, and with it the Einsteinian spacetime framework. So the only function of the second postulate is to fix the invariant velocity. And Maxwell's theory and the ether-drift experiments clearly suggest that it should be c.

From the above, it is also clear that to include 'rectilinearity' in the second postulate is superfluous—even without it one arrives at the LT. We have included it merely for convenience.

1.11   Cosmology and first doubts about inertial frames

We next turn our attention to the second problem of Section 1.5: How securely is the 'zeroth axiom' of both Newton's theory and Einstein's special relativity, namely the existence of the set of infinitely extended inertial frames, anchored in physical reality? It will be useful even at this early stage to review briefly the main features of the universe as they are known today. Our galaxy contains about 10^{11} stars— which account for most of the objects in the night sky that are visible to the naked eye. Beyond our galaxy there are other more or less similar galaxies, shaped and spaced roughly like coins three feet apart. The 'known' part of the universe, which stretches to a radius of about 10^{10} light-years, contains about 10^{11} such galaxies. It exhibits incredible large-scale regularity. Most cosmologists therefore accept the *cosmological principle* which asserts (in the absence of counter-indications) that all the galaxies are roughly on the same footing; that is, to say, the large-scale view of the universe from everywhere is the same. So there is no end to the galaxies, for in an 'island universe' there would have to be atypical edge-galaxies. But we do not know whether the universe is flat and infinite, or curved—in which case it could still be infinite (negative curvature), but it could also curve back on itself and be finite (positive curvature). If it is intrinsically curved, inertial frames are out anyway, since they are flat by hypothesis. So let us suppose the universe is flat and infinite, and uniformly filled with galaxies. Write 'stars' for 'galaxies' and add 'static'— and you have Newton's picture of the universe. How could Newton think that an infinite distribution of static matter could *remain* static in the face of all those mutual gravitational attractions?

The answer hinges on absolute space and symmetry: relative to absolute space, each galaxy would be pulled up as much as down, one way as much as the other, and so it would be in equilibrium and not move. But take away absolute space, and then this infinite array of galaxies could contract, everywhere at the same rate, without violating its *intrinsic* symmetry: each galaxy would see radial contraction onto itself. Today the universe is known not to contract but to expand, in the same symmetric way, as though it were the result of some primeval explosion (the 'big bang') billions of years ago. Gravity would slow the expansion and might eventually reverse it—or not. But the last thing such a universe would do is to expand at a *constant* rate, as though gravity were switched off. So how could this universe accommodate both Newton's infinite family of uniformly moving inertial frames *and* the cosmological principle? It could not. At most *one* galaxy could be at rest in an inertial frame and all the others would decelerate relative to that frame. Or *accelerate*: recent observations have lent some support to the presence of a cosmological 'lambda' force opposing gravity, the mathematical possibility of whose existence had already been noted by Einstein. In either case we conclude that extended inertial frames cannot exist in such a universe.

The cosmological principle suggests that under these conditions the center of each galaxy provides a basic *local* standard of non-acceleration, and the lines of sight from this center to the other galaxies (rather than to the stars of the galaxy itself, which may rotate) provide a local standard of non-rotation: together, a *local inertial frame*. Intertial frames would no longer be of infinite extent, and they would not all be in uniform relative motion. A frame which is locally inertial would cease to be so at a distance, if the universe expands non-uniformly. Nevertheless, *at each point* there would still be an infinite set of local inertial frames, all in uniform relative motion.

The extent of sufficient validity for Newtonian mechanics of such local inertial frames is clearly of practical importance in celestial mechanics. As a rule, they can be used to deal with gravitationally bound systems such as the solar system, a whole galaxy, and even clusters of galaxies small enough to have detached themselves from the cosmic expansion.

1.12   Inertial and gravitational mass

Much smaller 'local inertial frames' are used in general relativity. Einstein came upon them not via dynamic cosmology (he long thought the universe was static) but through his equivalence principle (EP) of 1907, which begins with a closer look at the concept of 'mass'. It is not always stressed that at least two quite distinct types of mass enter into Newton's theory of mechanics and gravitation. These are (i) the *inertial mass*, which occurs as the ratio between force and acceleration in Newton's second law and thus measures a particle's resistance to acceleration, and (ii) the *gravitational mass*, which may be regarded as the gravitational analog of electric charge, and which occurs in the equation

$$f = \frac{Gmm'}{r^2} \tag{1.8}$$

for the attractive force between two masses (*G* being the gravitational constant.)

One can further distinguish between *active* and *passive* gravitational mass, namely between that which causes and that which yields to a gravitational field, respectively. Because of the symmetry of eqn (1.8) (due to Newton's third law), no essential difference between active and passive gravitational mass exists in Newton's theory. In GR, on the other hand, the concept of passive mass does not arise, only that of active mass—the source of the field.

It so happens in nature that for *all* particles the inertial and gravitational masses are in the same proportion, and in fact they are usually made equal by a suitable choice of units; for example, by designating the same particle as unit for both. Newton took this proportionality as an axiom. He tested it to an accuracy of about one part in 1000 by observing (as Galileo had done before him) that the periods of pendulums were independent of the material of the bob. (The gravitational mass acts to shorten the period, the inertial mass acts to lengthen it.) Much more delicate verifications were performed by Eötvös, first in 1889, and finally in 1922 to an accuracy of five parts in 10^9. Eötvös suspended two equal weights of different material from the arms of a delicate torsion balance pointing west–east. Everywhere but at the poles and the equator the earth's rotation would produce a torque if the inertial masses of the weights were unequal—since centrifugal force acts on inertial mass. By an ingenious variation of Eötvös's experiment, using the earth's orbital centrifugal force which changes direction every 12 h and so lends itself to amplification by resonance, Roll, Krotkov, and Dicke (Princeton 1964) improved the accuracy to one part in 10^{11}, and Braginski and Panov (Moskow 1971) even to one part in 10^{12}. Plans are underway for an even more ambitious experiment called STEP (Satellite Test of the Equivalence Principle) which would test the free fall of different particles orbiting the earth in a drag-free space capsule, and which could yield an accuracy of one part in 10^{17}.

The question is sometimes asked whether antimatter might have negative gravitational mass; that is, whether it would be repelled by ordinary matter. Direct experiments to test the rate of falling of a beam of low-energy antiprotons are being planned at CERN (Holzscheiter *et al.*). However, quantum-mechanical calculations by Schiff have long ago shown that there are enough virtual positrons in ordinary matter to have upset the Eötvös–Dicke experiments *if* positrons fall up. And there seems to be astrophysical evidence that the gravitational mass of the meson K° and its antiparticle differ by at most a few parts in 10^{10}.[8] Thus all indications point to the universality of Newton's axiom.

This proportionality of gravitational and inertial mass is often called the *weak equivalence principle*. A fully equivalent property is that *all* free particles experience the same acceleration at a given point in a gravitational field. As in the case of the pendulum, gravitational mass tends to increase the acceleration, inertial mass tends to decrease it. More precisely, the field times passive mass gives the force, and the force divided by inertial mass gives the acceleration, so the acceleration equals the field, $\mathbf{a} = \mathbf{g}$, independently of the particle. (That, of course, is why $\mathbf{g}$ is often called

[8] For this and many other experimental data, see, for example, H. C. Ohanian and R. Ruffini, *Gravitation and Spacetime*, 2nd edn, Norton, New York, 1994.

the 'acceleration of gravity'.) It follows that free motion in a gravitational field is fully determined by the field and an initial velocity. Project a piano and a ping-pong ball side by side and with the same velocity anywhere into the solar system, and the two will travel side by side forever! This path unicity in a gravitational field is usually referred to as *Galileo's principle*, by a slight extension of Galileo's actual findings. (Recall his alleged experiment of dropping pairs of disparate particles from the Leaning Tower of Pisa—the most direct test of the weak equivalence principle.)

An important consequence of weak equivalence was already demonstrated by Newton in the *Principia*, namely: a cabin falling freely and without rotation in a parallel gravitational field is mechanically equivalent to an inertial frame without gravitation. (Recall the televised pictures of astronauts in their spacecraft being weightless and, if unrestrained, moving according to Newton's first law!) For proof, consider the motion of any particle in the cabin; let $\mathbf{f}$ and $\mathbf{f}_G$, respectively, be the total and the gravitational force on it, relative, say, to the earth (here treated as a Newtonian inertial frame), and m_I and m_G its inertial and gravitational mass. Then $\mathbf{f} = m_I\mathbf{a}$ and $\mathbf{f}_G = m_G\mathbf{g}$, where $\mathbf{a}$ is the acceleration of the particle, and $\mathbf{g}$ is the gravitational field and thus the acceleration of the cabin. The acceleration of the particle relative to the cabin is $\mathbf{a} - \mathbf{g}$ and so the force relative to the cabin is $(\mathbf{a} - \mathbf{g})m_I$. This equals the non-gravitational force $\mathbf{f} - \mathbf{f}_G$ *if* $m_I = m_G$; hence Newton's second law (including the first) holds in the cabin. And the same is true of the third law. Gravity has been 'transformed away' in the cabin.

1.13   Einstein's equivalence principle

The proportionality of inertial and gravitational mass is a profoundly mysterious fact. Why inertial mass (whose significance as 'resistance to acceleration' makes sense even in a world without gravity) should serve as gravitational charge when there *is* gravity, is totally unexplained in Newton's theory and seems purely fortuitous. Newton's theory would work perfectly well without it: it would then resemble a theory of motion of electrically charged particles under an attractive Coulomb law, where particles of the same (inertial) *mass* can carry different (gravitational) *charges*. GR, on the other hand, contains Galileo's principle as a primary ingredient and could not survive without it.

At the beginning of GR, in fact, stands Einstein's encounter with $m_I = m_G$. He dealt with it as he had dealt with relativity: boldly and universally. No need to wait for precision experiments. If it is even approximately known to be true, then that is such an astonishing fact that there must be an exact law of nature behind it. In what he later called 'the happiest thought of my life', he realized that inertia and gravity, in some deep sense, must really be the same thing. And this is how: You sit in a box from which you cannot look out. You feel a 'gravitational force' towards the floor, just as in your living room. But you have *no* way to exclude the possibility that the box is part of an accelerating rocket in free space, and that the force you feel is what in Newtonian theory is called an 'inertial force'. To Einstein, inertial and gravitational forces are identical.

All this is encapsuled in *Einstein's equivalence principle* (EP) which most usefully is expressed in terms of *freely-falling non-rotating* cabins. In a 'thought experiment' the walls of such a cabin could be made of bricks loosely stacked *without* mortar: if, in free fall, the bricks do not come apart, then the cabin is non-rotating and the gravitational field is uniform (parallel). The useful size of such a cabin in a specific case is determined by how much the actual field diverges from being parallel: the cabin has to be small enough for the field to be essentially parallel throughout its interior. Even so, if we use it for too long a time, the bricks may still come apart: so not only its size but also the duration of its use must be suitably restricted.

We can now state Einstein's equivalence principle as follows: *All freely-falling non-rotating cabins are equivalent for the performance of all physical experiments.* If true, then all such cabins will be equivalent, in particular, to cabins hovering motionless in an extended inertial frame in a world without gravity, and so the physics in all these cabins is SR. The cabins themselves are called *local inertial frames* (LIFs). Note how much smaller these are than the *Newtonian* local inertial frames discussed in Section 1.11, which can encompass whole clusters of galaxies. Note also that (just like extended inertial frames) the LIFs at one event form an infinite family, all in uniform relative motion; but LIFs at different events usually accelerate relative to each other.

Einstein's EP is an extension to all of physics of a principle that was previously well known to hold for mechanics (namely the one discussed at the end of the previous section). As in the case of the relativity principle, the *unity of physics* by itself would be a strong enough reason to justify this extension. But let us see how it also corresponds to Einstein's idea that inertia and gravity are the same thing.

Let C be a cabin freely falling with acceleration g near the earth's surface (see Fig. 1.3). Let C′ be another cabin *within* C and accelerating relative to C with acceleration g *upward*, and thus at rest in the earth's gravitational field. An observer

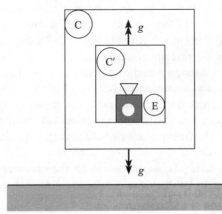

Fig. 1.3

performs a variety of experiments, mechanical and non-mechanical, in C'. Each such experiment E can be viewed as a compound experiment in C, namely, to accelerate upwards and *then* do E. Now, according to the EP, all these compound experiments have the same outcome as when C is freely floating in empty space. But then C' is an accelerating rocket and its internal field is purely 'inertial'. So no experiment can tell the difference between a gravitational and an inertial field: they are (locally) equivalent.

1.14   Preview of general relativity

Recall Einstein's philosophical objection to Newton's extended inertial frames: they are absolute structures that act but cannot be acted upon. In his equivalence principle Einstein saw a way to rid physics of these objectionable pre-existing structures. For special relativity he had still needed them in order to specify the arena of validity of the theory. But the EP changed all that. No longer is there a need to assume any absolute structure. What undoubtedly exists in the physical world is the totality of free-fall orbits—which in turn determine the LIFs everywhere and thus the local arena for SR. And, of course, those orbits are not absolute: they are influenced by the matter content of the universe. Only in the complete absence of gravity (as ideally assumed in SR) do the orbits straighten out, and then the LIFs join together to form extended IFs. Otherwise the concept of 'extended inertial frame' joins 'absolute space' and 'ether' into banishment.

In the following paragraphs (of which the first three are meant to be read only very lightly at this stage!) we sketch the road from the EP to GR.[9] Technically, Einstein's procedure was to introduce four physically deliberately meaningless coordinates x_1, x_2, x_3, x_4, whose sole purpose is to label the events (that is, the 'points' of spacetime) unambiguously and continuously. Inertia–gravity is then incorporated by an encoded prescription (the so-called 'metric') which allows us at every event to transform from x_1, x_2, x_3, x_4 to the LIFs with their physically *meaningful* coordinates x, y, z, t. This, in turn, allows us to predict the motion of 'free' particles. (Note that in GR a particle is called 'free' if subject only to inertia–gravity, like the planets in the solar system, but *not* the protons in a particle accelerator.) Locally, by the EP, each free particle moves rectilinearly and with constant speed in the LIF; that means going 'straight' in the local 4-dimensional spacetime. Now in GR the LIF families at the various events are patched together to form (in the presence of gravity) a *curved* spacetime: for if the big spacetime were flat, the various LIFs would all fit together to make an extended IF and all 'free' particles would move straight in *that*—which we know is not the case when there is gravitating matter around. The path of a particle or photon in spacetime is called its 'worldline'. In SR the worldlines of free particles (and photons) are straight. In GR they are 'locally straight'; that is, straight in every

---

[9] For a historical fantasy of how this could have happened earlier, see W. Rindler, *Am. J. Phys.* **62**, 887 (1994).

LIF along the way. This corresponds to being 'as straight as possible' in the big curved spacetime.

Such lines are called 'geodesics'. On the surface of the earth they are the great circles. There is just *one* geodesic in each direction. In a LIF, knowing a spacetime direction $dx : dy : dz : dt$ is as good as knowing a velocity dx/dt, dy/dt, dz/dt. So the initial velocity of a particle in a given LIF determines its initial direction in spacetime and thus the unique geodesic that will be its worldline. The piano and the ping-pong ball will follow the *same* worldline! Thus does GR 'explain' Galileo's principle. To Einstein, the law of geodesics is primary, and a natural extension of free motion in inertial frames.

Since the geometry of spacetime determines its geodesics and thus the motions of free particles, it must be the gravitating masses that determine the geometry. Newtonian *active* gravitational mass (the creator of the field) goes over into GR as the creator of curvature. Newtonian *passive* gravitational mass (that which is pulled by the field) goes into banishment along with the ether, etc. *Inertial* mass survives in non-gravitational contexts only, for example, as that which determines the outcome of collisions or the acceleration of charged particles in electromagnetic fields.

In sum, general-relativistic spacetime is curved. Its curvature is caused by active gravitational mass. The relation between curvature and mass is governed by Einstein's famous field equations. Finally, free particles (and photons) have geodesic worldlines in this curved spacetime, which accounts for Galileo's principle.

It seems almost miraculous that Newton's theory and GR—so different in spirit—are predictively almost equivalent in the classical applications of celestial mechanics, which are characterized by relatively weak fields and relatively slow motions (compared to the speed of light). With *one* exception: whereas Newtonian theory predicts a perfectly repetitive elliptical orbit for a (test-)planet in the field of a fixed sun, GR predicts a similar ellipse that *precesses*. Now such a precession in the case of the planet Mercury had been observed as early as 1859 by Leverrier. Although it amounts to only 43 *seconds* of arc per *century* (!), this result was so secure (the figure has hardly changed since 1882) that it constituted a notorious puzzle in Newton's theory. Einstein's discovery (late in 1915) that his theory gave *exactly* this result, was (and one can feel with him) 'by far the strongest emotional experience in his scientific life, perhaps in all his life. Nature had spoken to him. He had to be right.'[10]

In the very same paper Einstein gave another momentous result, one that was capable of early observational verification. Since light travels rectilinearly with speed c in every LIF, its worldline can be calculated in GR much like that of a particle. What the calculation yielded was a deflection of light from distant stars by the sun's gravity (for light just grazing the sun) through an angle of $1.7''$—just twice as much as the bending one gets in Newtonian theory by treating light corpuscularly. Such bending can be looked for during total eclipses of the sun, when the background stars become visible. And, indeed, the prediction was confirmed in 1919 by Eddington, who had

---

[10] A. Pais, *Subtle is the Lord...*, Oxford University Press, 1982, p. 253. This is one of the finest biographies of Einstein.

led an expedition for that purpose to Principe Island, off the coast of Spanish Guinea. A *British* expedition validating a *German* scientist, so soon after a terrible war that had pitted these two nations against each other—this somehow captured the popular imagination. It was this happening, not the precession of Mercury's orbit, not the relativity of time, nor even $E = mc^2$, that made Einstein into a popular hero and his name and his bushy countenance suddenly famous.

Subsequent theoretical developments and experimental tests have by now established GR as *the* modern theory of gravitation whose predictions are trusted. In principle, it has replaced Newton's inverse-square theory. In practice, of course, Newton's much simpler theory continues to be used whenever its known accuracy suffices. And that applies to most of celestial mechanics including, for example, the incredibly delicate operations of sending probes to the moon and the planets. GR here serves as a kind of superviser: Since it contains Newton's theory as a limit, it allows us to estimate the errors Newton's theory may incur, and shows them to be quite negligible in the above situations. But when the fields are strong, or quickly varying, or when large velocities are involved (as with photons), or large distances (as in cosmology), then GR diverges significantly from Newton's theory. Thus it predicts the existence of black holes and gravitational waves (both already 'almost' validated by indirect evidence) and it provides a consistent optics in the presence of gravity, which Newton's theory does not. The latter has already led to successes in connection with the study of 'gravitational lensing'. And in relativistic cosmology GR provides a consistent dynamics for the *whole* universe.

In GR two of Einstein's concerns merged: gravity as an aspect of inertia, and the elimination of the absolute (that is, uninfluenceable) set of extended IFs. The new inertial standard is spacetime, and this is directly influenced by active gravitational mass via the field equations. Yet in the total absence of mass and other disturbances like gravitational waves, spacetime would straighten itself out into the old family of extended inertial frames. This would seem to contradict Mach's idea that *all* inertia is caused by the cosmic masses. Einstein was eventually quite willing to drop that idea, and so shall we. The equality of inertial and *active* gravitational mass then remains as puzzling as ever. It would be nice if the inertial mass of an accelerating particle were simply a back-reaction to its own gravitational field, but that is not the case.

1.15   Caveats on the equivalence principle

Consider the following notorious paradox: an electric charge is at rest on the surface of the earth. By conservation of energy (or just by common sense!), it will not radiate. And yet, relative to an imagined freely falling cabin around it, that charge is accelerating. But charges that accelerate relative to an IF radiate. Why doesn't ours? Again, consider a charge that is fixed inside an earth-orbiting space capsule. Now, circularly moving charges *do* radiate, and one cannot imagine how the earth's gravitational field could change that. But relative to the freely falling space capsule the charge is at rest, and charges at rest in an inertial frame do not radiate. Where is the catch? Much has been written on these paradoxes, but the proper solution seems to have been first recognized by Ehlers: *It is necessary to restrict the class of experiments covered by the EP to those that are isolated from bodies or fields outside the cabin.* In the case of the charges discussed above, their electric field extends beyond the cabin and is, in fact, 'anchored' outside; since radiation is a property of that whole field, it follows that these 'experiments' lie outside the scope of the EP.

Beyond such restriction, there is a school of thought, represented most forcefully by the eminent Irish relativist Synge,[11] which holds that the EP is downright false and should be scrapped: Since every 'real' gravitational field **g** (as opposed to the 'fictitious' field in an accelerating rocket) is non-uniform, there will always be tidal forces present in the cabin, causing relative accelerations d**g** between neighboring free particles. And with perfect instruments these could be detected, no matter how small the cabin. Hence, the argument goes, we could *always* recognize a 'real' gravitational field and *never* mimic it with acceleration.

But consider this: the EP asserts a *limiting* property, like $\sin x / x \to 1$. True, $\sin x$ never *equals* x in any finite domain, but $\sin x \sim x$ is not useless information. The EP is, in fact, the *exact* 4-dimensional analog of the statement that in sufficiently small regions of a curved surface, plane (Euclidean) geometry applies. If the earth were a perfect sphere, surely the errors I commit by surveying my backyard using plane geometry would be miniscule. If, instead, I draw a large 'geodesic' triangle whose area is, say, one-*n*th of the surface of the earth, the sum of its internal angles is, in fact, given by

$$A + B + C = \pi\left(1 + \frac{4}{n}\right). \tag{1.9}$$

So for a triangle the size of France ($n \approx 1000$) the deviation from π would still only be 0.4 per cent. The critic insists that even if restricted to an area the size of a penny, he could, by use of (1.9), in principle determine that this area has a curvature equal to that of the earth and is decidedly not flat. We, on the other hand, find it useful to know that even on a scale the size of France, plane geometry will still only be out by some 0.4 per cent.

What does all this imply for SR? SR always was and always will be a self-consistent (and rather elegant) theory of an *ideal* physics in an *ideal* set of infinitely extended IFs. For comparison, Euclidean plane geometry always was and always will be a self-consistent and elegant geometry in an ideal infinite Euclidean plane. If the real universe is curved, it is possible that there may be *nowhere* embedded in it an infinite Euclidean plane, or even a portion of one. But that in no way invalidates plane geometry *per se*, nor does it make it useless for practical applications. It will apply (as it always has) with greater or lesser accuracy, according to circumstances, in limited regions. If the accuracy is orders of magnitude beyond what our instruments can measure or what our circumstances may require, what more could we want? And so it is with SR. Its internal logic is unaffected by the recognition that there are no

---

[11] See, for example, J. L. Synge, *Relativity: The General Theory*, North-Holland, Amsterdam, 1960, pp. IX, X.

extended IFs in the real world. Only the stage on which SR applies to the real world has shrunk. According to the EP, the *best* stage we can find for it is a freely falling cabin. And it is GR that will allow us to estimate the errors incurred when using SR in specific reference frames, such as our terrestrial labs—just as spherical geometry allowed us to estimate the errors incurred when using plane geometry on the sphere.

1.16   Gravitational frequency shift and light bending

Einstein's EP leads directly (that is, without the field equations of GR and also without the photon concept) to two interesting predictions about the behavior of light in the presence of gravity. The first is that, as light climbs up a gravitational gradient, its frequency decreases; and the other, that light is deflected 'ballistically' in a parallel gravitational field.

Of course, both these effects are intuitive once we know that light consists of photons: We 'only' need to know the Planck–Einstein formula $E = h\nu$ for the kinetic energy of a photon of frequency ν, Einstein's formula $E = m_I c^2$ relating energy to inertial mass, and the weak EP, $m_I = m_G$. For the work done by a gravitational field with potential Φ on a particle of gravitational mass m_G as it traverses a potential difference $d\Phi$ is $-m_G \, d\Phi$. This must equal dE, the gain in the particle's kinetic energy. For a photon, $dE = h \, d\nu$, and so

$$h \, d\nu = -m_G \, d\Phi = -m_I \, d\Phi = -\frac{E}{c^2} d\Phi = -\frac{h\nu}{c^2} d\Phi,$$

whence

$$\frac{d\nu}{\nu} = \frac{-d\Phi}{c^2}. \tag{1.10}$$

Integrating this equation over a finite path from A to B, we find

$$\frac{\nu_B}{\nu_A} = e^{-(\Phi_B - \Phi_A)/c^2} = \frac{e^{-\Phi_B/c^2}}{e^{-\Phi_A/c^2}}. \tag{1.11}$$

As for light bending, we can imagine a ray of light as a stream of photons; since these have inertial and gravitational mass, we might expect them to obey Galileo's principle and follow a curved path just like a Newtonian bullet traveling at velocity c. That would make, for example, the downward curvature of a horizontal beam in the earth's field (with x horizontal and y up) equal to

$$\frac{d^2 y}{dx^2} = \frac{d^2 y}{c^2 \, dt^2} = -\frac{g}{c^2}.$$

In units of years and light-years, $c = 1$, and it so happens that also $g \approx 1$; which shows that the radius of curvature of such a beam is approximately one light-year! Already in 1801 (unknown to Einstein) the German astronomer Soldner, concerned

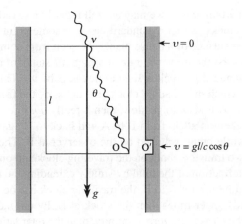

Fig. 1.4

whether bending of light might vitiate the accuracy of astronometrical measurements, had argued precisely along these lines. He wrote: 'No one would find it objectionable, I hope, that I treat a light ray as a heavy body . . .'. He calculated the entire orbit of a ray grazing the edge of the sun, and correctly found just one-half, 0.″84, of the later relativistic value for its total 'Newtonian' deflection.[12] But such ballistic treatment of light is inherently inconsistent: because of energy conservation, no Newtonian particle can travel at constant speed through a variable potential.

Einstein's EP allows us to obtain these results purely kinematically, without appeal to photons. Consider a freely falling cabin, say in an elevator shaft on earth (cf. Fig. 1.4). Suppose it is released from rest at the moment when a light ray enters its ceiling at an angle θ to the vertical and with frequency ν. If the height of the cabin is l, the ray arrives at the floor a time $l/c \cos \theta$ later, when the floor already moves with velocity $v = gl/c \cos \theta$. An observer O at rest on that floor sees the ray arrive with unaltered frequency ν, since the cabin is a LIF. But an observer O′ at rest in the shaft at O's level moves *into* the light relative to O with velocity v; by the classical Doppler argument that observer therefore sees a frequency shift given by

$$\frac{d\nu}{\nu} = \frac{v \cos \theta}{c} = \frac{gl}{c^2} = \frac{-d\Phi}{c^2},$$

thus confirming (1.10). Equation (1.11) follows as before.

That result (known as the *gravitational frequency shift*) has the important consequence that standard clocks fixed in a *stationary* gravitational field at low potential go slower than clocks fixed at higher potential. This can be seen as follows. Since the rates of standards atomic clocks (for example, cesium clocks) are directly linked to

---

[12] Pais, *loc. cit.*, p. 200.

the frequencies of certain atoms, we may as well regard these radiating atoms them-selves as standard clocks. So let a standard clock at some point A of low potential (for example, on the surface of a dense planet) be seen from some point B of higher potential, and let ν_A be the universal rate at which *all* standard clocks tick. Let the rate at which the standard A-clock *is seen* to tick at B be ν_B. This, by (1.11), is less than the rate ν_A at which the standard clock at B ticks, by the factor on the RHS.

But if the clock at A is *seen* to go slow, then it really *does* go slow. For suppose a standard clock is taken on a trip from B to A and back to B again after a *very* long stay at A, all the while being watched by the observer at B. The times observed to elapse during the two transfers, both on the traveling clock and on another fixed at B, can be dwarfed as a fraction of the total by simply extending the sojourn at A. Thus, in essence, by the time it returns to B, the traveling clock has been seen to tick (and therefore *has* ticked) fewer times than the clock at B, by the factor on the RHS of (1.11). If Φ is chosen so as to be zero at infinity, then the retardation factor relative to infinity at a point of (negative) potential Φ is $\exp(\Phi/c^2)$. One speaks of *gravitational time dilation*.

Owing to this effect, the US atomic standard clocks kept since 1969 at the National Bureau of Standards at Boulder, Colorado, at an altitude of 5400 ft as part of the International Atomic Time network, gain about five microseconds each year relative to similar clocks kept at the Royal Greenwich Observatory, England, at an altitude of only 80 ft. Since both sets of clocks are intrinsically accurate to one-tenth of a microsecond per year, the effect is observable and is one of several that must be corrected for.

We shall next derive a formula for the local bending of light by 'translating' the shape of a ray from a freely falling cabin S to a frame S' fixed to the earth. Let us choose Cartesian coordinates x and y in S with the x-axis horizontal and the y-axis straight up. In S' let similar coordinates x' and y' be chosen so that the corresponding axes of S and S' coincide at time $t = 0$, which is also the time when S is released from rest in S'. Then an argument analogous to that which led to the Galilean transformation (1.1) would now lead to the transformation

$$x' = x, \qquad y' = y - \tfrac{1}{2}gt^2, \tag{1.12}$$

if the spacetime were Newtonian. But, as we have seen, it may be curved. In that case, purely geometric arguments allow us to estimate the possible errors incurred in using (1.12): they are of the third or higher order in x, y, and t. (Distances in the tangent plane differ from corresponding distances in a curved surface by quantities of the third-order.)

Now in S, by Einstein's EP, the ray will travel straight and with velocity c. If it travels at an angle θ to the horizontal, its equation will be

$$x = ct \cos\theta + \mathrm{O}(t^3)$$
$$y = ct \sin\theta + \mathrm{O}(t^3). \tag{1.13}$$

Why the $O(t^3)$? Because there may be tidal forces in the cabin; only at the origin is $d^2x/dt^2 = d^2y/dt^2 = 0$ guaranteed. Now all that remains to be done is to translate (1.13) via (1.12) into S'. Thus we find for the path of the ray in S':

$$y' = x' \tan \theta - \tfrac{1}{2}c^{-2}gx'^2 \sec^2 \theta + O(x'^3), \qquad (1.14)$$

and consequently for its curvature κ at the origin:

$$\kappa = \frac{d^2y'/dx'^2}{[1 + (dy'/dx')^2]^{3/2}} = -\frac{1}{c^2}g \cos \theta, \qquad (1.15)$$

exactly. We conclude that the radius of curvature of the ray in the terrestrial frame is c^{-2} times the component of the gravitational field normal to the ray.

That, of course, is what one also finds in Newtonian theory for the path of a particle momentarily traveling at the speed of light. Yet earlier we mentioned that Einstein predicted a bending of light *twice* as large as what one gets in Newtonian theory. That, however, referred not to the above *local* bending (which is the same in all theories that accept the EP) but to the entire path past the sun, or, in other words, to the *integrated* ('global') bending; and here it is the general-relativistic curvature of spacetime itself that comes into play and doubles the result. (We shall deal with that in Chapter 11, where also the observations will be discussed.)

Einstein's proof of the local bending of light from the EP is really a most remarkable argument. From the mere fact that light travels at finite speed he deduces that 'light has weight.' He made absolutely no other assumption about light. *All* phenomena (gravitational waves, ESP?) that propagate with finite velocity in an inertial frame would thus be forced by gravity into a locally curved path. This suggests rather strongly that what has been discovered here is not so much a new property of light, but, instead, a new property of space in the presence of mass, namely curvature: if space itself (or spacetime) were curved, *all* naturally straight phenomena would thereby be forced onto curved 'rails'. And, indeed, the rails of curved spacetime are its geodesics, as we have seen.

Exercises 1
The order of the exercises (here and later) is roughly that in which the topics appear in the text, rather than that of ascending difficulty.

1.1. Verify that the Lorentz–FitzGerald length contraction hypothesis in the old ether theory indeed leads to the result that the two-way speed of light along *any* rod moving through the ether with uniform speed v (no matter at what inclination) is given by eqn (1.7). [*Hint*: only that component of the rod's length is shortened which is parallel to the motion.] Since this is a tricky problem, and merely of historical interest, it may well be omitted.

1.2. It is well known that a moving electric charge creates circular magnetic lines of force around its line of motion, so that a stationary magnet near it experiences a torque. Using the relativity principle, deduce that a magnet moving through a static electric field experiences a torque tending to point it in a direction orthogonal to both its line of motion and the electric field. Would this be easy to prove directly from the usual laws of electromagnetism?

1.3. An electric charge moving through a magnetic field experiences a (Lorentz-) force orthogonal to both its velocity and the field. From the relativity principle deduce that it must therefore be possible to set a stationary charge in motion by moving a magnet in its vicinity. Would this be easy to prove directly from the usual laws of electromagnetism?

1.4. Give some examples of the absurdities that would result if the inertial mass of some particles were negative. [For example, consider a negative-mass object sliding on (or under?) a rough table.] It is for reasons such as these that $m_I \geq 0$ is taken as an axiom.

1.5. A bob of gravitational mass m_G and inertial mass m_I is suspended by a weight-less string of length l in a gravitational field g. Prove that the period for small oscillations of this simple pendulum is given by $2\pi \sqrt{m_I/m_G}\sqrt{l/g}$. So if this is independent of what the bob is made of, m_I/m_G must be a universal constant.

1.6. On the grounds of Mach's principle, Einstein at one time conjectured that the inertial mass of a particle might increase in the presence of heavy bodies. This is not the case in GR. But if it were, then a cesium clock and an inertial clock (for example, one that counts the oscillations of a light spring connecting two equal masses) might stay in step when both are in a freely falling cabin in outer space, while in another such cabin near a heavy mass the inertial clock would fall behind. Would that violate the equivalence principle?

1.7. An 'elevator shaft' is drilled diametrically through an ideally homogeneous and spherical earth of radius $R = 6.37 \times 10^8$ cm and density $\rho = 5.52\,\mathrm{g\,cm^{-3}}$. An Einstein cabin is dropped into this shaft from rest at the surface. Two free particles, A and A', are initially at rest in the cabin, at the center of the ceiling and the floor, respectively. Two others, B and B', are initially at rest at the centers of two opposite sides. Prove that the whole cabin relative to the shaft, as well as each of these pairs of particles relative to the cabin, executes simple harmonic motion of period $\sqrt{3\pi/G\rho} = 1.41\,\mathrm{h}$, $G = 6.67 \times 10^{-8}\,\mathrm{cm^3\,g^{-1}\,s^{-2}}$ being the gravitational constant. Prove also that the tidal acceleration between each particle pair at separation dr is $(8\pi G\rho/3)\,dr$, equal to that which would be caused by the gravitation of a ball of density ρ that just fits between them, and thus equal to a fraction $1/R = 1.57 \times 10^{-9}$ of g (the acceleration of gravity on earth) per centimeter of separation. This shows how little the cabin diverges from being inertial. [*Hint*: Use Gauss's theorem, equating influx of the gravitational field through a given surface to $4\pi G$ times the enclosed mass.]

1.8. Consider two *identical* pendulum clocks placed at *different* potential levels in a given stationary gravitational field. Invent a situation where these clocks run at the *same* rate as judged by mutual viewing. [*Hint*: Consider the inverse-square field of a point mass.]

1.9. If a Mössbauer apparatus is capable of measuring the Doppler shift of a source moving with a velocity of as little as 10^{-5} cm/s, verify that it can detect the gravitational frequency shift down a 22-meter tower on earth. (As was done, in a famous experiment at Harvard, by Pound and Rebka in 1960.)

Part I
Special Relativity

2

Foundations of special relativity; The Lorentz transformation

2.1 On the nature of physical theories

It will be well to preface our detailed discussion of special relativity with some comments on the nature of physical laws and theories in general. According to modern thought (largely influenced by Einstein) even the best of physical theories do not claim to assert an absolute truth, but rather an approximation to the truth. Moreover, they are not mere summaries of experimental facts, available to any diligent seeker cleansed of prejudices, as was thought by Bacon in the sixteenth, and still by Mach in the twentieth century. Human invention necessarily enters into the systematization of these facts. As Einstein wrote in 1952: 'There is, of course, no logical way to the establishment of a theory ... '.[1] For example, the observations of planetary orbits, by themselves, do not imply the existence of a gravitational force. True, in Newton's theory the planets move *as if* pulled by the sun with an inverse-square force. But Newton *invented* this force. There is no such force in Einstein's general relativity; there the planets move as straight as possible in a spacetime curved in a specific manner by the sun.

A physical theory, in fact, is a man-made amalgam of concepts, definitions, and laws, constituting a mathematical *model* for a certain part of nature, and asserting not so much what nature *is*, but rather what it *is like*. Agreement with experiment is the most obvious requirement for the usefulness of such a theory. However, no amount of experimental agreement can ever 'prove' a theory, partly because no experiment (unless it involves counting only) can ever be infinitely accurate, and partly because we can evidently not test all relevant instances. Experimental disagreement, on the other hand, does not necessarily lead to the rejection of a physical theory, unless an equally appealing one can be found to replace it. Such disagreement may simply lead to a narrowing of the known 'domain of sufficient validity' of the model. We need only think of Newton's laws of particle mechanics, which today are known to fail in the case of very fast-moving particles, or Newton's gravitational theory, which today is known to fail for some of the finer details of planetary orbits. The 'truer' relativistic laws are also mathematically more complicated, and so Newton's laws continue to be used in areas where their known accuracy suffices.

---

[1] See p. 35 of R. S. Shankland, *Am. J. Phys.* **32**, 16 (1964). See also pp. 11, 12 of Einstein's Autobiographical Notes in *Albert Einstein: Philosopher-Scientist* (ed. P. A. Schilpp), Library of Living Philosophers, Evanston, Illinois, 1949.

Apart from the obvious requirements of a satisfactory *experimental fit*, of *internal consistency*, and of *compatibility* with other scientific concepts of the day, there are two more characteristics of a good theory. One is conceptual or mathematical *simplicity* or *elegance*. But it must be borne in mind that mathematical elegance often depends on which specific mathematical formalism is used (vectors, tensors, spinors, Clifford algebra, matrices, groups, etc.). What is elegant in one formalism may be a mess in another! And, lastly, there is the vital requirement of *falsifiability*, or the possibility of experimental disproof, as has been stressed by Popper. The better a theory, the more predictions it will make by which it could be disproved. And as a corollary, a theory should not allow indefinite *ad hoc* readjustments to take care of each new counter-result that may be found. In this way, theories are the engines of physics: it is the quest for their experimental falsification that drives physics on. When too many counter-results accumulate, it is time for a new theory to be invented.

2.2   Basic features of special relativity

We are now ready to begin our detailed study of special relativity, building on the historical introduction given in Sections 1.9 and 1.10. Special relativity (SR)—just like Newtonian mechanics—is a prime example of a physical theory that is a mathematical model. Already its initial fine fit with the real world and its inherent elegance (very much as in the case of Newton's theory) persuaded its early proponents of its enormous potential. Most of its development came out of the mathematics rather than out of the lab. Thus it very quickly produced striking predictions (time dilation, mass increase, $E = mc^2$, etc.) that went far beyond the testing capabilities of the day. Nevertheless, as the twentieth century wore on, one after another they were all validated.

Special relativity is crucially based on the concept of inertial frames (IFs), as is Newton's mechanics. One can picture an inertial frame as three mutually orthogonal straight wires (the x, y, and z axes) soldered together at the origin, with equal length scales etched along each axis. The geometry in each IF is Euclidean and in each IF Newton's first law holds; gravity is assumed to be absent in SR. Further, one can picture infinitely many such reference systems, with all possible orientations of their axes, moving with all possible uniform velocities (but without rotation) relative to each other—in Newton's theory. In SR there is a speed limit: all relative velocities between IFs are less than c. In Newton's theory all inertial reference systems share the same universal ('absolute') time and are necessarily related by Galilean transformations. The crucial *mathematical* discovery that made SR possible was that, if one is willing to give up the idea of absolute time (and Einstein had the courage to do this!), then a whole new family of transformation groups becomes possible, still allowing Newton's first law and Euclidean geometry to hold in each IF, and still respecting the relativity principle, namely the complete equivalence of all IFs. These are the various Lorentz transformation (LT) groups, each characterized by exactly one finite invariant speed; that is, a speed that transforms into the same speed in all IFs. The Galilean transformation turns out to be that limiting LT which transforms *infinite* speeds (in other words, linear sets of simultaneous events) in one IF into infinite speeds in every other IF, which proper LTs do not. Of course, Einstein picked that LT group which leaves the speed of light invariant—exactly the content of his second axiom.

The above 3-dimensional picture of IFs as reference triads flying through space received a new 4-dimensional interpretation by the mathematician Minkowski as early as 1907. Minkowski's view corresponds to a movie of the universe that has been cut up into its separate frames and then stacked to make a 4-dimensional *spacetime*, whose points are called *events*. Higher up in the stack corresponds to later in time. This spacetime has many interesting and useful properties, such as something akin to a distance between any two of its points. Many of these properties degenerate when there is an absolute time; that is, when the stack foliates uniquely into world-wide moments. For this reason a certain 4-dimensional 'metric' vector- and tensor-calculus in special-relativistic spacetime becomes essentially useless in Newton's theory. But in SR that is the formalism in terms of which the theory becomes most elegant. However, the pre-Minkowskian 3-D view of triads flying through space should not be disdained. It is the view we shall elaborate until Chapter 5. In order to become an efficient problem solver in SR one eventually has to learn to switch effortlessly from one point of view to the other and pragmatically pick what is best from either.

Special relativity is thus, to start with, the theory of space and time in a world filled with Lorentz-related IFs. This includes results like time dilation, length contraction, relativistic velocity addition, the existence of a speed limit, the relativistic kinematics of waves, etc. But, as we have remarked before, since classical physics plays out on a background of space and time, one cannot change that background (in the model) without adapting the rest of physics to it. And, of course, just like the spacetime background, the new physics must satisfy the relativity principle (RP). Most physical laws make reference not only to length and time, but also to non-kinematic quantities like forces, fields, masses, etc. Additional axioms must specify how these quantities transform from one IF to another. (Recall the axioms of force and mass invariance in Newton's theory.) To satisfy the RP, the mathematical statement of each law must then transform into itself under LTs.

So, altogether, SR is Lorentz-invariant physics. Newton's mechanics, for example, is Galileo-invariant but not Lorentz-invariant, and thus it is inconsistent with SR. It was the program of SR to review *all* existing laws of physics and to subject them to the test of the RP with the help of the LTs. Any law found to be lacking must be modified accordingly. It is rather remarkable that, in the mathematical formalism of SR, most of the new laws were neither very difficult to find nor in any way less elegant than their classical counterparts.[2]

But let us stress once more the *model* aspect that relativity shares with all other physical theories. Special relativity is the theory of an *ideal* physics in a *hypothetical* set of infinite Euclidean inertial frames free of gravity, each perfectly homogeneous and isotropic for all physical phenomena. The basic laws of this physics are assumed

---

[2] However, in some modern areas such as the quantum theory of interacting systems, there still remain fundamental difficulties with the relativistic formulation.

to hold in these frames in their simplest forms—idealizations of the laws observed in our imperfect terrestrial laboratories. As we have seen in Chapter 1, we do not expect to find such extended or perfect inertial frames in nature. It is the equivalence principle that provides the bridge between the ideal SR model and the real world. According to it, we can find at each event a set of *local inertial frames* (LIFs), which may be small or large depending on (i) the distribution of nearby masses, and (ii) the accuracy we require. It is to these frames (or other frames not differing too much from these frames) that SR is applied in practice, and with great success. As we mentioned earlier, there is a close analogy between SR as applied to LIFs and Euclidean plane geometry as applied to small portions of a curved surface. As abstract theories, SR and plane geometry are global and exact. But as applied to the real world or to a curved surface, respectively, both are local and approximative.

2.3   Relativistic problem solving

Apart from leading to new laws, SR leads to a useful technique of problem solving, namely the possibility of switching inertial reference frames. This often simplifies a problem. For although the totality of laws is the same, the configuration of the problem may be simpler, its symmetry enhanced, its unknowns fewer, and the applicable subset of laws more convenient, in a judiciously chosen inertial frame. In the present section we shall illustrate the flavor and the power of many of the relativistic arguments to follow. To make it simple and transparent, we consider two *Newtonian* examples. (An Einsteinian version of the first will be given later, in Section 6.5.)

First, then, we wish to prove from minimal assumptions the result, familiar to billiard players, that if a stationary and perfectly elastic ball is struck by another similar one, then the diverging paths of the two balls after collision will subtend a right angle. If the incident ball travels at velocity $2\mathbf{v}$, say, relative to the table, let us transfer ourselves to an inertial frame traveling in the same direction with velocity $\mathbf{v}$. In this frame the two balls approach each other symmetrically with velocities $\pm\mathbf{v}$ [see Fig. 2.1(a)], and the result of the collision is clear: simply by symmetry, the rebound velocities must be equal and opposite ($\pm\mathbf{u}$, say), and by the time-reversibility of Newton's laws, they must be numerically equal to $\mathbf{v}$ (that is, $u = v$). With this information, we can revert to the frame of the table, by adding $\mathbf{v}$ to all the velocities in Fig. 2.1(a), thus arriving at Fig. 2.1(b).

Here the rebound velocities are evidently $\mathbf{v} \pm \mathbf{u}$, and the simple expedient of drawing a semicircle centered at the tip of the arrow representing $\mathbf{v}$ makes the desired result self-evident, by elementary geometry. Alternatively, we have $(\mathbf{v}+\mathbf{u})\cdot(\mathbf{v}-\mathbf{u}) \equiv v^2 - u^2 = 0$, which also shows that the vectors $\mathbf{v} \pm \mathbf{u}$ are orthogonal. Of course, once we know about momentum and energy conservation, we do not need the relativistic detour.

As a second example, consider the familiar exercise machine called a treadmill. An endless belt connects two rollers at an incline θ (see Fig. 2.2). A motor drives the belt at speed v and an exerciser of mass m walks uphill, staying at constant height.

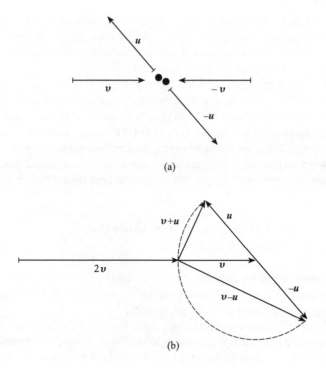

Fig. 2.1

At what rate is he working? The component of his weight along the belt is $mg \sin \theta$ and that is the force his driving leg applies at velocity v. Hence the rate of work is $mg \sin \theta \cdot v$. But, more simply, let us look at this problem in the IF attached to the top belt. Forces are invariant in Newton's theory, so there is still a vertical gravitational

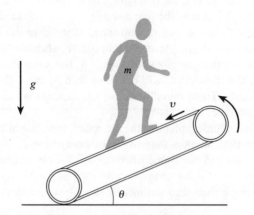

Fig. 2.2

field g. The man is now gaining height and therefore potential energy, the latter at the rate $mg \cdot v \sin \theta$, which must again equal the rate of work. Notice the coming into play of *different* parts of Newton's theory in the two inertial frames! (For further examples of Newtonian relativity, see Exercises 2.1–2.5.)

It was Einstein's recognition of the fact that arguments of a similar nature were apparently possible also in electromagnetism that significantly influenced his progress towards SR. At the beginning of his 1905 paper he discusses the apparent relativity of electromagnetic induction. And as late as 1952 (in a letter to a scientific congress), we find him writing: 'What led me more or less directly to the special theory of relativity was the conviction that the electromagnetic force acting on a [charged] body in motion in a magnetic field was nothing else but an electric field [in the body's rest-frame].'[3]

2.4   Relativity of simultaneity, time dilation and length contraction: a preview

Later we shall derive these three effects quantitatively by calculation from the Lorentz transformation. But here it will be our aim to see intuitively (from a simple thought experiment) why they *must* arise in special relativity.

According to Einstein's second axiom, *light in every inertial frame behaves like light in Maxwell's ether.* If light is sent from stationary clock A to stationary clock B over a distance L, the arrival time at B is L/c units later than the emission time at A. If light is emitted half-way between A and B, the arrival times at A and B are equal. Now suppose we are in an IF and a fast airplane flying overhead constitutes a second IF; let it be the top plane in Fig. 2.3(a). Suppose that a flash-bulb goes off in the exact middle of its cabin. Then the passengers at the front and back of the cabin will see the flash at the same time, say when their clocks or watches read '3'. But now consider the progress of this same flash in *our* IF. Here, too, the light travels with equal speed fore and aft. Here, too, the flash occurred exactly half-way between the front and back of the plane. (For, surely, the two halves of the plane will be considered to be of equal length by us.) But now the rear passengers, who travel into the signal, will receive it before the front passengers who travel away from the signal. The top of Fig. 2.3(a) is a snapshot of that plane taken in *our* IF when the signal hits the back of the cabin. We know the rear clock then reads 3. But since the signal has not yet reached the front, the front clock will read less than 3, say 1 (in units very much smaller than seconds!). These two different clock readings are simultaneous events in *our* IF. Thus simultaneity is relative!

Now add a second identical plane to the argument, traveling at the same speed but in the opposite direction. Suppose the two planes were just level with each other [as in Fig. 2.3(a)] at the instant in *our* frame when we we took our snapshot. By symmetry, the second plane's clocks in that snapshot will also read two units apart, say again 3 in back and 1 in front, if their zero-settings are suitably adjusted. But this implies that

---

[3] See p. 35 of R. S. Shankland, *Am. J. Phys.* **32**, 16 (1964).

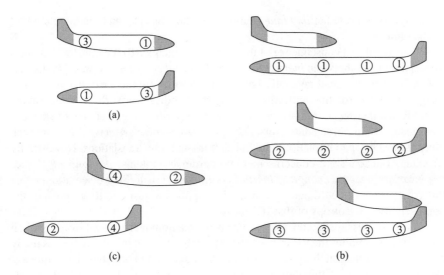

Fig. 2.3

the bottom plane sees the top plane fully alongside of it from time 1 to time 3, as in Fig. 2.3(b), which shows the situation from the point of view of the bottom plane. It therefore sees a plane one-third its own length fly by! Here we have the phenomenon of *length contraction*. Note its perfect symmetry: the top plane quite analogously considers the bottom plane to have only one-third of its own length.

Lastly, consider the instant in our IF when the rear ends of the two planes pass each other. We know from Fig. 2.3(b) that this will happen when the rear clock of the bottom plane reads '4', and clearly we expect the front and back clocks in each plane still to read two units apart. So, by symmetry, in both planes they now read 4 and 2, as in Fig. 2.3(c). Observe from Figs 2.3(a) and (c) that the passage of the rear clock of the top plane along the length of the bottom plane takes only one unit of time by its own reading, but three units (from time 1 to time 4) by the reckoning of the bottom plane. So the bottom plane considers this moving clock to go slow by a factor 3—the same factor that applies to length contraction. This is the phenomenon of *time dilation*. Note again its perfect symmetry relative to the two planes.

2.5 The relativity principle and the homogeneity and isotropy of inertial frames

Each ideal inertial frame is perfectly symmetric. By that we mean that each IF is spatially homogeneous and isotropic not only in its Euclidean geometry, but for all of physics, and that it is temporally homogeneous, too. In other words, a given physical experiment can be set up *anywhere* in an IF (homogeneity), face in *any direction* (isotropy), be repeated at *any time* (temporal homogeneity)—and the outcome will be the same.

All this is a direct consequence of the relativity principle. To see this, let us make a logical distinction between *inertial frames* and *inertial coordinate systems* (a distinction which later we shall generally ignore). The former are mere extensions (real or imagined) of non-rotating uniformly moving rigid bodies in a world without gravity, as in SR. An inertial coordinate system is such an IF *plus*, in it, a choice of standard coordinates *x, y, z, t* in standard units. For a given *frame* these *systems* differ from each other at most by spatial rotations and translations and time translations. Now, strictly speaking, Einstein's RP concerns inertial coordinate systems: *the laws of physics are invariant under a change of inertial coordinate system.* So we see at once that this equivalence of coordinate systems, when applied to just one IF, guarantees the homogeneity and isotropy of that IF.

It is perhaps less well known that, conversely, the homogeneity and isotropy of all inertial frames implies the RP, as has been especially stressed by Dixon. So it is really a question of taste which of the two is taken as axiom and which as consequence. One demonstration of this depends on the simple *midframe lemma* which asserts that 'between' any two inertial frames S and S′ there exists an inertial frame S″ relative to which S and S′ have equal and opposite velocities. For proof, consider a one-parameter family of inertial frames moving collinearly with S and S′, the parameter being the velocity relative to S. It is then obvious from continuity that there must be one member of this family with the required property (see Fig. 2.4). Now imagine two intrinsically identical experiments E and E′ being performed in S and S′, respectively. We can transform E′, by a spatial translation and rotation and a temporal translation, in S′, into a position where it differs from E only by a 180° rotation in S″. Thus, by the assumed homogeneity and isotropy of S′ and S″, the outcome of E and E′ must be the same, which establishes Einstein's RP in its following alternative form: *the outcome of any physical experiment is the same when performed with identical initial conditions relative to any inertial coordinate system.*

We may note that temporal homogeneity implies (at least in special relativity) that all methods of time-keeping based on repetitive processes are equivalent, and it denies such possibilities (envisaged by Milne) as that inertial time (relative to which free

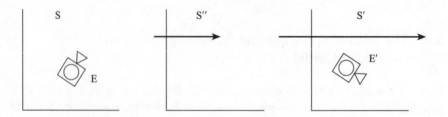

Fig. 2.4

particles move uniformly) falls out of step over the centuries with atomic time; for example, that indicated by a cesium clock.

As just another application of the midframe lemma, we end this section with a proof of the obvious-seeming '*reciprocity theorem*' which asserts of any pair S and S' of inertial frames, using identical standards of length and time, that each ascribes the same velocity to the other. For the manipulation performed in S to determine the velocity of S' can be regarded as an experiment in the midframe S''. By a suitable 180°-rotation in S'' this experiment is transformed into a manipulation in S' for determining the velocity of S. And by the assumed isotropy of S'', the two outcomes must be the same.

2.6   The coordinate lattice; Definitions of simultaneity

The main formal task of the present chapter is to derive the Lorentz transformation equations. These equations constitute the mathematical core of SR. But before we *transform* coordinates from one frame to another, it will be well to clarify how they are assigned in a *single* IF, at least conceptually.

First of all, we need universal units of time and of length. In this age of atoms it makes good sense to fall back on atomic frequencies and wavelengths to provide these units. Thus in 1967 the (international) General Conference of Weights and Measures (CGPM-1967) defined the second as follows: 'The second is the duration of 9 192 631 770 periods of the radiation corresponding to the transition between the two hyperfine levels of the ground state of the cesium-133 atom.' The international standard of *length* had been defined back in 1960 in terms of the wavelength of a certain line in the spectrum of krypton-86. More recently, however, it has become clear that the precision available from the krypton-86 line is surpassed by the precision with which, on the one hand, the second, and, on the other hand, the speed of light are determinable. Thus, demonstrating its complete confidence in special relativity, CGPM-1983 re-defined the meter as the distance traveled by light in vacuum in a time-interval of 1/299 792 458 of a second. Note that, consequently, the speed of light is and remains *precisely* 299 792 458 meters per second; improvements in experimental accuracy will modify the meter relative to atomic wavelengths, but not the value of the speed of light! Clearly, the above units can be reproduced in each IF.

The standard spatial coordinates for inertial frames are orthonormal Cartesian coordinates x, y, z. To assign these to events, the 'presiding' observer at the origin of an inertial frame needs to be equipped only with a standard clock (for example, one based on the vibrations of the cesium atom), a theodolite, and a means of emitting and receiving light signals. The observer can then measure the distance of any particle (at which an event may be occurring) by the radar method of bouncing at light-echo off that particle and multiplying the elapsed time by $\frac{1}{2}c$. Angle measurements with the theodolite on the returning light signal will serve to determine the relevant (x, y, z) once a set of coordinate directions has been chosen. The same signal can be used to determine the time t of the reflection event at the particle as the average of the time of emission and the time of reception.

But conceptually it is preferable to *pre*coordinatize the frame and to read off the coordinates of all events *locally*. For this purpose we imagine minute standard clocks placed at rest at the vertices $(m\varepsilon, n\varepsilon, p\varepsilon)$ of an arbitrarily fine orthogonal lattice of thin rods, where m, n, p run over the integers and ε is arbitrarily small. The spatial coordinates of these clocks can be engraved upon them. To synchronize the clocks it is sufficient to emit a single light signal from the origin, say at time t_0: each lattice clock is set to read $t_0 + r/c$ as the signal passes it, where r is its distance from the origin. Once this calibration is in place, we can read off the coordinates of any event by simply looking at the nearest clock.

In Maxwell's ether frame the above procedure for synchronizing clocks with a single control signal from the origin would clearly be satisfactory; in SR *every* IF is as good as Maxwell's ether frame, so here too the method is satisfactory. But what is satisfactory? What do we require of a 'good' time coordinate? In general relativity, for example, directly meaningful time (and space) coordinates generally do not exist, and the coordinates are just arbitrary labels for events. But in SR, as in Newton's theory, the symmetry of the inertial frames *allows* us to choose meaningful coordinates (our 'standard coordinates') and it pays us to do so. In particular, the time coordinate t can be chosen so that the mathematical expression of the physical laws reflects their inherent symmetries. Already Newton's first law then fixes the time *rate* up to a constant multiplier (that is, up to a unit) to be such that equal spatial increments along a free path correspond to equal time increments. A non-linear scale change away from t, like $t \mapsto t' = \sinh t$, would destroy this correspondence, and the *appearance* of temporal homogeneity in general. While adhering to this preferred *rate* of time, we could still make a non-universal change of zero-point, like $t \mapsto t' = t + kx$ ($k = \text{const} > 0$). But this would destroy the appearance of spatial isotropy. For example, any given rifle would then shoot bullets faster in the negative x-direction than in the positive x-direction (that is, with greater *coordinate* velocity dx/dt). From this point of view, a 'good' standard time is unique, except for changes of rate and zero-point by *constants* only.

In Newton's theory the clocks of any IF can simply 'take over' the time from absolute space, and a time that is satisfactory in the above sense will then result in each IF ('universal time'). Not so in SR, where what is a satisfactory time for one IF turns out to be, if taken over directly, isotropy-violating *and* unit-violating in another IF. Hence the need to synchronize the clocks *independently* in each IF, for example, by the light-signaling method described above.

But in spite of the traditional and conceptually very convenient use of light signals in the usual presentations of SR, SR is logically quite independent of the existence of light signals, or indeed of any real-world effect that travels at the speed of light. If light were banished from the world, SR would survive. Its success in high-speed mechanics alone would justify it, and with it the LTs. The latter leave invariant the velocity c, which at the same time acts (as we shall see) as a kind of speed limit. But whether anything physical actually travels *at* speed c is really irrelevant for the logic of the theory. Such theoretical arguments as those of our Section 2.4 could be pushed through with imagined geometrical points traveling with velocity c. And for the synchronization of clocks in each IF we could use alternative 'signals'. For example, the observer at the origin could shoot standard cannon balls from standard cannons in all directions at time t_0. When one of these balls passes a lattice point at distance r, its clock is set to read $t_0 + r/u$, u being the muzzle velocity of the cannons. Obviously this synchronization is equivalent to that using light, since it would be so in absolute space, and since in SR every inertial frame is as good as absolute space.

2.7   Derivation of the Lorentz transformation

In the last section we described methods for assigning coordinates (x, y, z, t) to events in any inertial frame, based on universal units of length and time. Such coordinates are called *standard coordinates* for an IF. We shall now consider the transformation $(x, y, z, t) \mapsto (x', y', z', t')$ of the standard coordinates of a given event from one inertial frame S to another, S'. To start with, the transformation must be linear, as can be proved in many ways—for example, from Newton's first law and temporal and spatial homogeneity: Consider a standard clock C freely moving through S, its motion being given by $x_i = x_i(t)$, where $x_i (i = 1, 2, 3)$ stands for (x, y, z). Then $dx_i/dt = $ const. If τ is the time indicated by C itself, homogeneity requires the constancy of $dt/d\tau$. (Equal outcomes here and there, now and later, of the experiment that consists of timing the ticks of a standard clock moving at constant speed.) Together these results imply $dx_\mu/d\tau = $ const and thus $d^2x_\mu/d\tau^2 = 0$, where we have written $x_\mu (\mu = 1, 2, 3, 4)$ for (x, y, z, t). In S' the same argument yields $d^2x'_\mu/d\tau^2 = 0$. But we have

$$\frac{dx'_\mu}{d\tau} = \sum \frac{\partial x'_\mu}{\partial x_\nu} \frac{dx_\nu}{d\tau}, \qquad \frac{d^2x'_\mu}{d\tau^2} = \sum \frac{\partial x'_\mu}{\partial x_\nu} \frac{d^2x_\nu}{d\tau^2} + \sum \frac{\partial^2 x'_\mu}{\partial x_\nu \delta x_\sigma} \frac{dx_\nu}{d\tau} \frac{dx_\sigma}{d\tau}.$$

Thus for *any* free motion of such a clock the last term in the above line of equations must vanish. This can only happen if $\partial^2 x'_\mu/\partial x_\nu \partial x_\sigma = 0$; that is, if the transformation is linear.

An immediate consequence of linearity is that all the defining particles (that is, those at rest in the lattice) of any inertial frame S' move with identical, constant velocity through any other inertial frame S. For suppose that the coordinates of S and S' are related by

$$x_\mu = \left(\sum A_{\mu\nu} x'_\nu \right) + B_\mu.$$

Then setting $x'_i = $ const $(i = 1, 2, 3)$ for a particle fixed in S', we get $dt = A_{44} dt'$, $dx_i = A_{i4} dt'$, and thus $dx_i/dt = A_{i4}/A_{44} = $ const, as asserted. The defining particles of S' thus constitute, as judged in S, a rigid lattice whose motion is fully determined by the velocity of any one of its particles.

Another consequence of linearity (plus symmetry) is that the standard coordinates in two arbitrary inertial frames S and S' can always be chosen so as to be in *standard configuration* with each other (as in Fig. 1.1). It is clearly always possible (i) to choose the line of motion of the spatial origin of S′ as the x-axis of S, and (ii) to choose the zero points of time in S and S′ so that the two origin clocks both read zero when they pass each other. Next, any two orthogonal planes intersecting along the x-axis can serve as the coordinate planes $y = 0$ and $z = 0$ of S. By the result of the preceding paragraph, these two planes, fixed in S, plus the moving plane $x = vt$ (v being the velocity of S′ relative to S) correspond to plane sets of particles fixed in S′. Moreover, the planes $y = 0$ and $z = 0$ must also be regarded as orthogonal in S′, otherwise the isotropy of S (in particular, its axial symmetry about the x-axis) would be violated. So we can take these planes as the coordinate planes $y′ = 0$ and $z′ = 0$, respectively, of S′. Similarly $x = vt$ must be regarded as orthogonal to the $x′$-axis in S′, otherwise the projection of that axis on to that plane would violate the isotropy of S. Hence we can take $x = vt$ as $x′ = 0$. In what follows, therefore, we assume S and S′ to be in standard configuration.

The RP implies that the transformation between any pair of inertial frames in standard configuration, with the same v, must be the same. Suppose we reverse the x- and z-axes of both S and S′. Referring to Fig. 1.1, and recalling the reciprocity theorem established at the end of Section 2.5, we see that this operation produces an identical pair or IFs with the roles of the 'first' and 'second' interchanged. So if we then interchange primed and unprimed coordinates, the transformation equations must be unchanged. In other words, the transformation must be invariant under what we shall call an *xz reversal*:

$$x \leftrightarrow -x′, \qquad y \leftrightarrow y′, \qquad z \leftrightarrow -z′, \qquad t \leftrightarrow t′. \tag{2.1}$$

Obviously, the same holds for an *xy* reversal.

Now, by linearity, $y′ = Ax + By + Cz + Dt + E$, where the coefficients are constants, possibly depending on v. Since, by our choice of coordinates, $y = 0$ must entail $y′ = 0$, we have $y′ = By$. Applying an *xz* reversal yields $y = By′$ and so $B = \pm 1$. But $v \to 0$ must continuously lead to the identity transformation and thus to $y′ = y$, whence $B = 1$. The argument for z is similar, and so we arrive at the two 'trivial' members of the transformation,

$$y′ = y, \qquad z′ = z \tag{2.2}$$

just as in the Newtonian case, and for the same reasons.

Next, suppose $x′ = \gamma x + Fy + Gz + Ht + J$, where for traditional reasons we have denoted the first coefficient by γ. By our choice of coordinates, $x = vt$ must imply $x′ = 0$, so $\gamma v + H, F, G, J$ all vanish and

$$x′ = \gamma(x - vt). \tag{2.3}$$

An *xz* reversal then yields

$$x = \gamma(x′ + vt′). \tag{2.4}$$

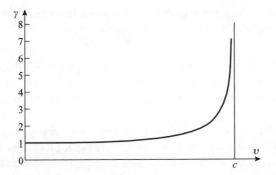

Fig. 2.5

At this stage Newton's axiom $t' = t$ would lead from (2.3) and (2.4) to $\gamma = 1$ and $x' = x - vt$; that is, to the GT (1.1). Instead, we now appeal to Einstein's law of light propagation. According to it, $x = ct$ and $x' = ct'$ are valid simultaneously, being descriptions of the same light signal in S and S'. Substituting these expressions into (2.3) and (2.4) we get the equations $ct' = \gamma t(c - v)$ and $ct = \gamma t'(c + v)$, whose product, divided by tt', yields

$$\gamma = \gamma(v) = \frac{1}{(1 - v^2/c^2)^{1/2}}. \tag{2.5}$$

Since $v \to 0$ must lead to $x' = x$ continuously, we see from (2.3) that we must choose the positive root in (2.5). This particular function of v is the famous '*Lorentz factor*', which plays an important role in the theory. Its graph is shown in Fig. 2.5. Its slow initial increase and its asymptote at $v = c$ should be noted.

The elimination of x' between (2.3) and (2.4) finally leads to the most revolutionary of the four equations,

$$t' = \gamma(t - vx/c^2).$$

Thus, collecting our results, we have found the *standard Lorentz transformation equations*

$$x' = \gamma(x - vt), \qquad y' = y, \qquad z' = z, \qquad t' = \gamma(t - vx/c^2), \tag{2.6}$$

with γ as given by (2.5).

Since in the above derivation we have used Einstein's light postulate only along the x-axis, we must still check whether the transformation (2.6) respects it generally. First, the linearity of the transformation implies that *any* uniformly moving point transforms into a uniformly moving point. This, incidentally, recovers the invariance of Newton's first law, but, of course, it also applies to light signals. Next, one easily derives from (2.6) the enormously important *fundamental identity* [see also after eqns (2.16) below]

$$c^2\,dt'^2 - dx'^2 - dy'^2 - dz'^2 = c^2\,dt^2 - dx^2 - dy^2 - dz^2. \tag{2.7}$$

Now, the distance dr between neighboring points in a Euclidean frame S is given by the 'Euclidean metric'

$$dr^2 = dx^2 + dy^2 + dz^2. \tag{2.8}$$

From identity (2.7) it then follows that $dr^2 = c^2 dt^2$ (which is characteristic of any effect traveling at the speed of light) implies $dr'^2 = c^2 dt'^2$ and vice versa. So the Euclidicity of the metric and the invariance of the speed of light are *jointly* respected by the LT.

Again, in our derivation of the LT, we have used the relativity principle only minimally. To show its full incorporation, we must verify that the LTs form a group—which is postponed to (viii) of the following section.

If a law of physics is invariant under the standard LT (2.6), *and* under spatial rotations, spatial translations and time translations, then it is invariant between *any* two inertial coordinate systems. For the general transformation between two inertial frames S and S', whose coordinates are standard but not in standard configuration with each other, can be broken down into a product of such transformations: Rotate the x-axis of S to be parallel to the velocity $\mathbf{v}$ of S', thus arriving at a frame $\tilde{S}$; next apply a standard LT with velocity v to $\tilde{S}$ to arrive at $\tilde{\tilde{S}}$, whose defining particles coincide with those of S'; a spatial rotation, and a spatial and a temporal translation (at most) will finally bring $\tilde{\tilde{S}}$ into S'. The resultant transformation is called a *general Lorentz transformation*, or a *Poincaré transformation*. It is, of course, linear, since each link in this chain of transformations is linear.

In today's terminology (which we may not always strictly follow) it has become customary to include reflections ($x \longmapsto -x$, etc.) in the definition of Poincaré and general Lorentz transformations, and to restrict the term Lorentz transformation to Poincaré transformations without translations.

In our derivation of the transformation equations allowed by the RP we saw that with Newton's 'second axiom' $t' = t$ one arrives at the GT, while with Einstein's second axiom $c = $ invariant one arrives at the LT. We shall postpone (but not for long, only to Section 2.11) the proof that these two 'second axioms' exhaust the possibilities.

Lastly, a remark about the xy and xz reversals we used in the derivation of the LT. Their advantage was that they do not involve v when one has no a priori knowledge of the velocity dependence of the coefficients in the LT. But in future we shall find another symmetry more useful: Any relativistic transformation formula relating unprimed and primed quantities (from S and S' respectively) remains valid when v is replaced by $-v$ and primed and unprimed quantities are interchanged. We call this a v *reversal*. The reason for its validity is that just as reversing the x- and z-axes of a pair of frames in standard configuration reverses their roles, so also does the mere reversing of v. For S then moves with velocity v along the x'-axis of S'. So whatever formula was originally true for primed in terms of unprimed quantities is now true for unprimed in terms of primed. As an example, let us apply a v reversal to the transformation (2.6), thus obtaining the inverse transformation

$$x = \gamma(x' + vt'), \qquad y = y', \qquad z = z', \qquad t = \gamma(t' + vx'/c^2), \tag{2.9}$$

which, of course, can also be found directly by algebra (or, alternatively, by an xy or xz reversal). A v reversal in general can often save a great deal of algebra.

2.8 Properties of the Lorentz transformation

(i) *Relativity of simultaneity*: The most striking new feature of the Lorentz transformation is the transformation of time, which exhibits the relativity of simultaneity; events with equal t do not necessarily correspond to events with equal t'. (For a view of t' clocks from another frame, see Fig. 3.2, Section 3.5.)

(ii) *Symmetry in x and ct*: Equations (2.6) are symmetric not only in y and z but also in x and ct. (The reader can verify this by writing T/c for t and T'/c for t' in (2.6) and multiplying the first equation by c.) In what follows we shall often find ct a more convenient variable than t itself.

(iii) *Lorentz factor*: For $v \neq 0$ the Lorentz factor γ is always greater than unity, though not much so when v is small. For example, as long as $v/c \lesssim 1/7$ (at which speed the earth is circled in one second), γ is less than 1.01; when $v/c = \sqrt{3}/2 = 0.866$, $\gamma = 2$; and when $v/c = 0.99 \ldots 995$ ($2n$ nines), γ is approximately 10^n. The following are frequently useful identities satisfied by the Lorentz factor:

$$\gamma v = c(\gamma^2 - 1)^{1/2}, \qquad c^2 \, d\gamma = \gamma^3 v \, dv = \gamma^3 \mathbf{v} \cdot d\mathbf{v},$$

$$d(\gamma v) = \gamma^3 \, dv. \tag{2.10}$$

They are particularly needed later when $\gamma(v)$ becomes associated with a particle moving arbitrarily at velocity $\mathbf{v}$. The proofs are left as an exercise to the reader.

(iv) *Newtonian limit*: The Lorentz transformation replaces the older Galilean transformation, to which it nevertheless approximates when v/c is small. This accounts for the high accuracy of Newtonian mechanics (invariant under the Galilean transformation) in describing a large domain of nature. Note also that the two transformations become identical if we let c formally tend to infinity. This occasionally provides a useful check on relativistic formulae.

(v) *Only one invariant speed*: *Any* effect whose speed in vacuum is always the same could have been used to derive the Lorentz transformation, as light was used in our derivation. Since only one transformation can be valid, it follows that all such effects (weak gravitational waves, ESP?) must propagate at the speed of light.

(vi) *Difference and differential versions*: If Δx, Δy, etc., denote the finite coordinate differences $x_2 - x_1$, $y_2 - y_1$, etc., corresponding to two events $\mathcal{P}_1$ and $\mathcal{P}_2$, then by substituting the coordinates of $\mathcal{P}_1$ and $\mathcal{P}_2$ successively into (2.6) and subtracting, we get the following transformation:

$$\Delta x' = \gamma(\Delta x - v\Delta t), \qquad \Delta y' = \Delta y, \qquad \Delta z' = \Delta z,$$

$$\Delta t' = \gamma(\Delta t - v\Delta x/c^2). \tag{2.11}$$

If, instead of forming differences, we take differentials in (2.6), we obtain equations identical with the above but in the differentials:

$$dx' = \gamma(dx - v\,dt), \qquad dy' = dy, \qquad dz' = dz,$$
$$dt' = \gamma(dt - v\,dx/c^2).$$
(2.12)

Analogous formulae arise from the inverse transformation (2.9). Thus the finite coordinate differences, as well as the differentials, satisfy the same transformation equations as the coordinates themselves. This, of course, is always the case with linear homogeneous transformations. Each form has its uses. The original form serves to transform single events and also whole families of events. The delta form is surprisingly often useful, but one must be very clear in one's mind as to precisely which two events are being considered. And the differential form is useful for problems of motion.

(vii) *Squared displacement*: It follows from (vi) that together with (2.7) we must also have

$$c^2\Delta t'^2 - \Delta x'^2 - \Delta y'^2 - \Delta z'^2 = c^2\Delta t^2 - \Delta x^2 - \Delta y^2 - \Delta z^2$$
(2.13)

and

$$c^2 t'^2 - x'^2 - y'^2 - z'^2 = c^2 t^2 - x^2 - y^2 - z^2$$
(2.14)

under a standard Lorentz transformation. But then, by our remarks on the decomposition of Poincaré transformations in the last section, and since time and space translations leave *all* the Δ- and d-terms unchanged while rotations preserve the spatial and temporal parts of (2.7) and (2.13) separately, these two identities hold even under Poincaré transformations; that is, between *any* two inertial coordinate systems. [Identity (2.14) holds only in the absence of translations.] Conversely, it can be shown that Poincaré transformations are the most general transformations which satisfy (2.7).[4] That identity, therefore, concisely characterizes Poincaré transformations, just as the invariance of the differential form (2.8) characterizes the rotations translations and reflections of Euclidean 3-space.

The common value of the two quadratic forms in (2.13) is defined as the *squared displacement* $\Delta\mathbf{s}^2$ between the two events in question:

$$\Delta\mathbf{s}^2 := c^2\Delta t^2 - \Delta x^2 - \Delta y^2 - \Delta z^2.$$
(2.15)

It can evidently be positive, negative, or zero, and so it must not be thought of as the square of an ordinary number. It is, in fact, as we shall see later, the square of a 4-vector, hence the bold type. The square root of its absolute value, written Δs, is often called the *interval*.

(viii) *Group properties*: The standard Lorentz transformation (with ct as the fourth variable) has unit determinant, as can easily be verified, and it possesses the two so-called group properties, *symmetry* and *transitivity*, without which it could not

---

[4] cf. W. Rindler, *Special Relativity*, Oliver & Boyd, Edinburgh 1966, p. 17.

consistently apply to *all* pairs of IFs. We have already seen that the inverse of a Lorentz transformation is another Lorentz transformation ('symmetry'), with parameter $-v$ instead of v. Also, it is found that the resultant of two Lorentz transformations with parameters v_1 and v_2, respectively, is another Lorentz transformation ('transitivity') with parameter $v = (v_1 + v_2)/(1 + v_1 v_2/c^2)$. [The direct verification of this is a little tedious; a more transparent proof is indicated after eqns (2.16).] Thus the standard Lorentz transformations constitute a group. The same is true of the Poincaré transformations, as can be seen at once if we accept here without proof that they are fully characterized by (2.7). As a result, when two frames are each related to a third by Poincaré transformations, they are so related to each other (by transitivity through the third). In Lorentz's ether theory, where each IF is primarily Poincaré related to the ether frame, the IFs are thus necessarily also Poincaré related to each other.

2.9   Graphical representation of the Lorentz transformation

In this section we concern ourselves solely with the transformation behavior of x and t under standard LTs, ignoring y and z which in any case are unchanged. What chiefly distinguishes the LT from the classical GT is the fact that space and time coordinates *both* transform, and, moreover, transform partly into each other: they get 'mixed,' rather as do x and y under a rotation of axes in the Cartesian x, y plane. We have already remarked on the formal similarity of the invariance of (2.15), which characterizes general LTs, to that of (2.8), which characterizes rotations. But in spite of the similarities, the character of a LT differs significantly from that of a rotation. This is brought out well by the graphical representation.

Recall first that there are two ways of regarding any transformation of coordinates (x, t) into (x', t'). *Either* we think of the point (x, t) as moving to a new position (x', t') on the same set of axes; that is, we regard the transformation as a motion in x, t space; this is the 'active' view. *Or* we regard (x', t') as merely a new label of the old point (x, t); this is the 'passive' view, whose graphical representation we shall discuss first. (See Fig. 2.6.) The events, once marked relative to a set of x, t axes, remain fixed; only the coordinate axes change. *For convenience we choose units in which $c = 1$* (such as years and light-years, or seconds and light-seconds). We draw the x- and t-axes corresponding to the frame S orthogonal, with t taking the place of the usual y. This orthogonality is just another convenience without physical significance. Our 2-dimensional diagram, with one dimension taken up by time, has room only to map whatever is going on along the *spatial* x-axis of S (that is, one of the three mutually orthogonal 'wires' of the reference triad, not to be confused with the *spacetime* x-axis, like the one in Fig. 2.6). Under the standard LT here considered, S$'$ shares its spatial x-axis with S. In fact, the diagram can describe whatever happens along this common spatial axis as seen by any number of frames in standard configuration with S.

Any curve representing a continuous one-valued function $x = f(t)$ in the x, t plane corresponds to the motion of some geometric point along the spatial x-axis. It is called the *worldline* of the moving point. The slope of such a line *relative to the*

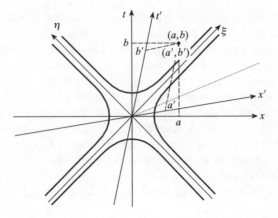

Fig. 2.6

t-axis, dx/dt, measures the velocity of the point in S. Not all such lines are possible histories of real particles, since the latter must obey the relativistic speed limit (as we shall see in Section 2.10). So the inclination to the vertical of a *particle* worldline must nowhere exceed 45°.

'Moments' in S have equation $t = $ const and correspond to horizontal lines, while the worldline of each fixed point on the spatial x-axis of S corresponds to a vertical line, $x = $ const. Similarly, moments in S' have equation $t' = $ const and thus, by (2.6), $t - vx = $ const, so in our diagram they correspond to lines with slope v. In particular, the x' axis ($t' = 0$) corresponds to $t = vx$. Again, worldlines of fixed points on the spatial x' axis have equation $x' = $ const and thus, by (2.6), $x - vt = $ const. In our diagram they are lines with slope v relative to the t axis. In particular, the t' axis ($x' = 0$) corresponds to $x = vt$. Thus the axes of S' subtend equal angles with their counterparts in S; but whereas in rotations these angles have the same sense, in LTs they have opposite sense. S' can have any velocity between $-c$ and c relatives to S; the corresponding x'- and t'-axes in the diagram are like scissors pointing NE (in the direction marked ξ), fully open for $v \rightarrow -c$, closed for $v \rightarrow c$.

We have already tacitly assumed the x- and t-axes in the diagram to be equally calibrated; for example, 1 cm corresponding to 1 s on the t-axis and to 1 lt s on the x-axis. (In fact, the vertical axis is often taken to be ct, so that the units are naturally the same.) For calibrating the primed axes (not quite as straightforward as in the case of rotations) we observe that for standard LTs (where $y' = y, z' = z$) and with $c = 1$, (2.14) reduces to $t'^2 - x'^2 = t^2 - x^2$. So if we draw the calibrating hyperbolae $t^2 - x^2 = \pm 1$, they will cut all four of the axes at the relevant unit time or unit distance from the origin. They are, in fact, the loci of events whose interval from the origin is unity [cf. after (2.15)]. The units can then be repeated along the axes, by linearity. The diagram shows how to read off the coordinates (a', b') of a given event relative to S': we must go along lines of constant x' or t' from the event to the axes.

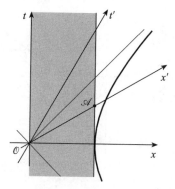

Fig. 2.7

Diagrams like Fig. 2.6 are often called *Minkowski diagrams*. They can be extremely helpful and illuminating in certain types of relativistic problems. For example, one can sometimes use them to get a rough preliminary idea of the answer. But one should beware of trying to use them for *everything*, for their utility is limited. Analytic or algebraic arguments are generally much more powerful.

As a first example, look at the dotted line in Fig. 2.6. It represents the uniform motion of a 'superluminally' moving point in S. Now imagine the x'- and t'-axes scissored towards each other until the x'-axis first coincides with and then surpasses the dotted line; in the first case the signal has *infinite* velocity in S', in the second case it moves in the opposite spatial sense, namely from greater to lesser x'! We shall discuss superluminal motions in the next section.

Length contraction and time dilation can also be read off, qualitatively, from the diagram. In Fig. 2.7 the shaded area shows the 'world-tube' (bundle of worldlines) corresponding to a unit rod at rest on the spatial x-axis between 0 and 1. In S' this rod moves at velocity $-v$. At $t' = 0$ it occupies the segment $\mathcal{OA}$ of the x'-axis, which, by reference to the calibrating hyperbola, is seen to be less than unity: the moving rod is short.

In Fig. 2.8 the t'-axis is the worldline of a standard clock fixed at the spatial origin of S' and therefore moving with velocity v through S. At $\mathcal{B}$, where its worldline intersects the calibrating hyperbola, it reads 1. However, the corresponding time t in S is evidently greater than 1: the moving clock goes slow.

These examples should convince the reader of the utility of such 'passive' Minkowski diagrams. We next turn to the graphing of *active* LTs. Viewed actively, the standard LT moves each point (x, t) to a new position (x', t'). *Motion* is the key concept. Students good at computer graphics can usually manage to write a program displaying active LTs. But it is really quite sufficient just to *imagine* a computer display of the x, t plane with one's inner eye. Suppose it is covered with lots of bright dots like stars on the night sky. As we 'press the velocity button', we see the speedometer indicate bigger and bigger velocities, and watch the points move: They do not rotate

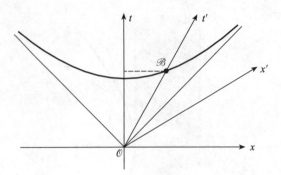

Fig. 2.8

about an origin, as does the night sky, but rather move along the set of hyperbolae $t^2 - x^2 = $ const, as indicated in Fig. 2.9. That they must *stay* on these hyperbolae is clear from the invariance of the interval. The four quadrants defined by the $\pm 45°$ lines through the origin thus separately transform into themselves, as do their boundaries. Of course, *all* straight lines transform into straight lines, by linearity.

The details of the motion become clearer when we cast the LT into an alternative form. Adding and subtracting the x and t members of (2.6) (after multiplying the latter by c), we find

$$ct' + x' = e^{-\phi}(ct + x)$$
$$ct' - x' = e^{\phi}(ct - x)$$

(2.16)

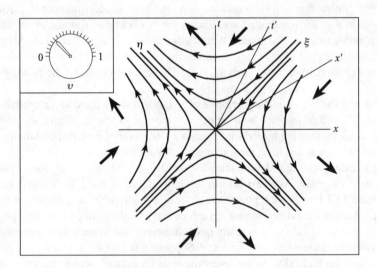

Fig. 2.9

with

$$e^\phi := \gamma\left(1 + \frac{v}{c}\right) = \left(\frac{1 + v/c}{1 - v/c}\right)^{1/2}. \qquad (2.17)$$

The ϕ here introduced is a useful alternative to v as parameter for the Lorentz group (it is its 'canonical' parameter), often called the 'rapidity' or 'hyperbolic parameter.' (See also Exercise 2.13.)

To digress briefly: from eqns (2.16) we can read off without effort the group properties discussed at the end of the last section: the inverse of a Lorentz transformation with ϕ is a Lorentz transformation with $-\phi$, and the composition of two Lorentz transformations with ϕ_1 and ϕ_2, respectively, is a Lorentz transformation with $\phi_1 + \phi_2$. And it is equally easy to read off from these equations (by multiplying them together) the fundamental invariance $c^2t^2 - x^2 = c^2t'^2 - x'^2$.

Returning to Fig. 2.9 (now again with $c = 1$), note that the (signed) distances, ξ and η, of a point (x, t) from the lines $t + x = 0$, $t - x = 0$, respectively (in other words, the Cartesian coordinates relative to the asymptotes of the hyperbolae), are given by

$$\xi = (t + x)/\sqrt{2}, \qquad \eta = (t - x)/\sqrt{2}. \qquad (2.18)$$

The corresponding coordinate axes are marked with ξ and η in Figs 2.6 and 2.9. The LT (2.16) now reads very simply

$$\xi' = e^{-\phi}\xi, \qquad \eta' = e^\phi\eta. \qquad (2.19)$$

This shows that, as v (and with it ϕ) increases from zero to positive values, all ξ coordinates decrease in absolute value while all η coordinates increase. This leads to the motion pattern shown in Fig. 2.9. The opposite happens when v increases negatively. The fundamental invariance now reads $\xi'\eta' = \xi\eta$.

Recall that (x', t') are the coordinates in S' of some event $\mathscr{A}$ whose coordinates in S are (x, t). If we mark (x', t') as a point in the x, t plane, we see where $\mathscr{A}$ would be mapped if S' made the map. In other words, after applying an active LT through the appropriate value of v (or ϕ), we see the world as mapped by S' on a standard set of orthogonal axes—which is, of course, exactly what has by then become of the original x'- and t'-axes (shown in Fig. 2.9). This is one of the chief uses of the active transformation.

As a simple but important example, consider a particle whose worldline, relative to a given frame S, is the right-hand branch of the hyperbola

$$x^2 - t^2 = x_0^2. \qquad (2.20)$$

(Cf. Fig. 2.10.) A worldline is, of course, always traversed into the future (t increasing); so this particle comes from positive infinity along the spatial x-axis of S, moves in as far as $x = x_0$ (at $t = 0$), where it momentarily comes to a halt (at the event marked $\mathscr{A}$) and then goes back to infinity. But what is interesting about this motion is that the particle has constant *proper acceleration*; that is, its acceleration relative to the IF instantaneously comoving with it (its 'rest-frame') is always the same. To

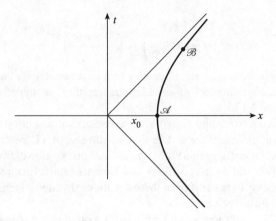

Fig. 2.10

see this, consider an arbitrary event $\mathcal{B}$ on the worldline and apply an active LT so as to bring $\mathcal{B}$ onto the horizontal axis: the hyperbola has not changed, $\mathcal{B}$ is now where $\mathcal{A}$ was before, and we are evidently in the rest-frame of the particle at $\mathcal{B}$ (its worldline at $\mathcal{B}$ being vertical). Whatever its proper acceleration was at $\mathcal{A}$ is therefore its proper acceleration also at $\mathcal{B}$ and our result is established. (We discuss this 'hyperbolic motion' analytically in Section 3.7.) Note also the interesting fact that a photon dispatched to 'chase' the particle at time $t = 0$ from the spatial origin $x = 0$ (its worldline is the top asymptote of the hyperbola) will never catch up with it; in fact, in any rest-frame of the particle the photon's distance from the particle is *always* precisely x_0, as the same active LT reveals.

2.10   The relativistic speed limit

When $v = c$ the γ factor (2.5) becomes infinite, and $v > c$ leads to imaginary values of γ. This shows that the relative velocity of two inertial frames must be less than the speed of light, since finite real coordinates in one frame must correspond to finite real coordinates in any other frame. This is a first indication that no particle can move superluminally relative to an IF, for a set of such particles moving parallelly would constitute an IF moving superluminally relative to the first. But there are many other indications that the speed of particles, and more generally, of all physical 'signals', is limited by c. Consider, for example, any signal or process whereby an event $\mathcal{P}$ causes an event $\mathcal{Q}$ (or whereby information is sent from $\mathcal{P}$ to $\mathcal{Q}$) at superluminal speed $U > c$ relative to some frame S. Choose coordinates in S so that these events both occur on the x-axis, and let their time and distance separations be $\Delta t > 0$ and $\Delta x > 0$. Then in the usual second frame S' we have, from (2.11),

$$\Delta t' = \gamma\left(\Delta t - \frac{v\Delta x}{c^2}\right) = \gamma\Delta t\left(1 - \frac{vU}{c^2}\right). \tag{2.21}$$

For a v that satisfies

$$c^2/U < v < c, \tag{2.22}$$

we would then have $\Delta t' < 0$. Hence there would exist inertial frames in which $\mathcal{Q}$ precedes $\mathcal{P}$, in which cause and effect are thus reversed and in which the signal is considered to travel in the opposite spatial direction (as we have already seen graphically in connection with Fig. 2.6).

So if $\mathcal{Q}$ were, for example, the breaking of a glass somehow caused in S by the signal from $\mathcal{P}$, then in S' the glass would break spontaneously and at the same time *emit* a signal *to* $\mathcal{P}$. Since in macro-physics no such uncaused events are observed, nature must have a way to prevent superluminal signals.

Other paradoxical consequences would result. In Einstein's phrase, we could 'telegraph into our past', and so tamper with it. Suppose we have a gun that can shoot 'telegrams' at speed $U > c$ in its rest-frame S. Let us run it *backwards* at speed v satisfying (2.22) relative to 'our' rest-frame S', and have it shoot our telegram to a distant relay station fixed in S', where it arrives, say, 20 years *before* it was emitted. From there it is returned at once, by a similarly backwards moving gun, reaching our location 40 years before we sent it. The telegram could be addressed to our parents and read: 'Have no children.' Thus we could foil events which have already happened and get into deep logical trouble. Or, if the first shot sent a question and the second an answer, we would have our answer before we asked the question!

Or again, let us look graphically at the superluminal signal discussed in eqn (2.21). In Fig. 2.11 the t- and t'-axes are the worldlines of two observers O and O', respectively. The segment $\mathcal{P}\mathcal{Q}$ represents a superluminal 'message' sent by O to O'. But for O', $\mathcal{Q}$ precedes $\mathcal{P}$. (Some lines of constant t and constant t' are indicated in the diagram.) There is symmetry between the two observers. *Each* can think to be the one who sent the message. What sort of a message is that? Suppose that at an event

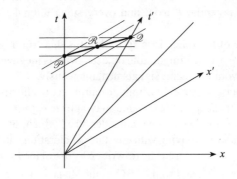

Fig. 2.11

$\mathcal{R}$ midway between $\mathcal{P}$ and $\mathcal{Q}$, the message is hit by lightening and destroyed. Who will then *not* know what is in the message?

Since the whole concept of superluminal signaling is thus seen to be fraught with paradox, we accept the *axiom* that *no superluminal signals can exist*: c is an upper bound to the speed of macroscopic information-conveying signals. In particular, this speed limit must apply to particles, since they can convey messages. We shall see in Chapter 6 how relativistic mechanics provides a speed governor by having the mass of particles increase beyond all bounds as their speed approaches the speed of light. Note also from Fig. 2.10 how a particle that constantly accelerates in its successive rest-frames, no matter how strongly, approaches but never attains the speed of light.

It will be well to check that the speed limit c actually *does* guarantee the invariance of causality. If two events happen on a line making an angle θ with the x-axis in S (thus not restricting their generality relative to S and S$'$), and are connectible by a signal with speed $u \le c$ in S, we see on replacing U by $u \cos \theta$ in (2.21), that for all v between $\pm c$, Δt and $\Delta t'$ do indeed have the same sign.

Arbitrarily large velocities are possible for moving points that carry no information, for example, the sweep of the light spot on the moon where a movable laser beam from earth impinges, or the intersection point of two rulers that cross each other at an arbitrarily small angle.

At this point we should also note the important fact that superluminal speeds always transform to superluminal speeds and subluminal to subluminal speeds. For if u and u' are the speeds of a point (or a signal, or a particle) in S and S$'$, respectively, and the differentials refer to its worldline, we can cast eqn (2.7) into the form

$$\mathrm{d}t'^2 \, (c^2 - u'^2) = \mathrm{d}t^2 \, (c^2 - u^2), \tag{2.23}$$

which allows the stated result to be read off directly.

One last item of interest to be extracted from eqn (2.21) (now regarded as referring to some superluminally moving *point*) is that in the particular frame S$'$ having $v = c^2/U$, the point has zero transit time and thus infinite velocity! So in SR 'infinity' is not an invariant speed, as it is in Newtonian theory: *every* superluminally moving point has infinite velocity in *some* IF, just as every subluminally moving point has *zero* velocity in some IF. This is also clear from Fig. 2.6, where every subluminal signal is a potential t'-axis, and every superluminal signal is a potential x'-axis.

One consequence of the relativistic speed limit is that 'rigid bodies' and 'incompressible fluids' have become impossible objects, even as idealizations or limits. For, by definition, they would transmit signals instantaneously.

An interesting fact about rigidity, though unrelated to the speed limit, is that a body which retains its shape in one frame may appear *deformed* in another frame, if it accelerates. As a simple example, consider a rod which, in a frame S$'$, remains parallel to the x'-axis while moving with constant acceleration a in the y'-direction. Its equation of motion, $y' = \frac{1}{2}at'^2$, can be re-written by use of the Lorentz transformation as $y = \frac{1}{2}a\gamma^2(t - vx/c^2)^2$, and so the rod has the shape of part of a parabola at each instant $t = $ const in the usual second frame S. (For another example, see Exercise 3.6.) The reason for this phenomenon is the relativity of simultaneity: the S-observer picks a different set of events at the rod to form an instantaneous view of it. We draw the reader's attention to the technique used here of 'translating' a whole set of events, characterized by some equation, from one inertial frame to another. It has many applications.

2.11 Which transformations are allowed by the relativity principle?

We shall now establish the result stated in the penultimate paragraph of Section 2.7 that, given the RP, Newton's and Einstein's 'second axioms' exhaust the possibilities, or, in other words, that the RP by itself necessarily leads *either* to the GT *or* to the LT. Logically we could have given the proof there and then, but it will make more sense now that we have discussed superluminal signals.

We shall assume that the future sense along any particle-worldline is the same in all inertial frames. Without this or an equivalent restriction ('causality invariance') another but wholly unphysical group of transformations becomes possible.[5]

Now, either there is, or there is not, an upper bound to the possible speeds of particles. Suppose, first, that there is. Then, mathematically speaking, there must be a *least* upper bound, which we will call c. This speed c, whether attained or not by actual particles, must be invariant. For suppose some velocity of magnitude c in an inertial frame S corresponds to one of magnitude $c' > c$ in another inertial frame S'. By continuity there will then exist a velocity of magnitude slightly less than c in S (that is, a possible particle velocity) that still corresponds to one of magnitude greater than c in S', a contradiction. Similarly $c' < c$ can be ruled out. So there exists an invariant speed, which is essentially Einstein's postulate.

On the other hand, suppose that particles can travel at all speeds. Let S and S' be in standard configuration. Then any event in S with $t > 0$ can be reached by a particle from the origin-crossing event, and must therefore correspond to $t' > 0$. Similarly $t < 0$ must always correspond to $t' < 0$. So $t = 0$ must correspond to $t' = 0$, and then, as in our argument preceding eqn (2.2), it follows that $t' = t$. Hence there exists a universal time, which is Newton's postulate.

Thus the relativity principle *together with causality invariance* necessarily implies that all inertial frames are related either by Galilean transformations, or by Lorentz transformations with some universal 'c'. The role of the second axiom is to separate these two possibilities, and (in the second case) to fix the value of c. In fact, fixing c is the *only* role of the second axiom: $c = \infty$ corresponds to the Galilean transformation.

---

[5] Cf. W. Rindler, *Essential Relativity*, 2nd edn, Springer-Verlag, 1977, p. 51.

Exercises 2

Note: Problems in special relativity should generally be worked in units which make $c = 1$. The 'missing' cs can either be inserted later throughout, or simply at the answer stage by using cs to balance the dimensions. However, the first five exercises below are on *Newtonian* relativity.

2.1. Steady rain comes down vertically at velocity u. You drive into it at velocity v. At what angle to the horizontal do you see the rain come down? [*Hint*: eqn (1.2).]

2.2. On a perfectly windless day you find yourself in a sailboat on a broad smoothly flowing river. You need to get as quickly as possible to a given place some miles down-river. Should you or should you not put up your sail? [*Hint*: change of reference frame.]

2.3. Use Newtonian relativity (as in Section 2.3) and symmetry considerations to prove that if one billiard ball hits a second stationary one head-on, and no energy is dissipated, the second assumes the velocity of the first while the first comes to a total stop.

2.4. A heavy plane slab moves with uniform speed v in the direction of its normal through an inertial frame. A ball is thrown at it with velocity u, from a direction making an angle θ with its normal. Assuming that the slab has essentially infinite mass (no recoil) and that there is no dissipation of energy, use Newtonian relativity to show that the ball with leave the slab in a direction making an angle ϕ with its normal, and with a velocity w, such that

$$\frac{u}{w} = \frac{\sin \phi}{\sin \theta}, \qquad \frac{u \cos \theta + 2v}{u \sin \theta} = \cot \phi.$$

2.5. In Newtonian mechanics the mass of each particle is *invariant*; that is, it has the same measure in all inertial frames. Moreover, in any collision, mass is *conserved*; that is, the total mass of all the particles going into the collision equals the total mass of all the particles (possibly different ones) coming out of the collision. Establish this law of mass conservation as a *consequence* of mass invariance, momentum conservation, and Newtonian relativity. [*Hint*: Let $\sum^*$ denotes a summation which assigns positive signs to terms measured *before* a certain collision and negative signs to terms measured *after* the collision. Then momentum conservation is expressed by $\sum^* m\mathbf{u} = 0$. Also, if primed quantities refer to a second inertial frame moving with velocity $\mathbf{v}$ relative to the first, we have $\mathbf{u} = \mathbf{u}' + \mathbf{v}$ for all $\mathbf{u}$.] Prove similarly that if in any collision the kinetic energy $\frac{1}{2} \sum m u^2$ is conserved in *all* inertial frames, then mass *and* momentum must also be conserved.

2.6. Consider the usual two inertial frames S and S' in standard configuration. In S' the standard lattice clocks all emit a 'flash' at noon. Prove that in S this flash occurs on a plane orthogonal to the x-axis and traveling in the positive x-direction at (de Broglie-) speed c^2/v.

2.7. Prove that an any instant there is just one plane in S on which the clocks of S agree with the clocks of S', and that this plane moves with velocity $(c^2/v)(1 - 1/\gamma)$. How is this plane related to the frame S'' of the 'midframe' lemma? [*Hint*: Fig. 2.3.]

2.8. Establish the approximation $\gamma(v) \approx 10^n$ when $v/c = 0.99 \ldots 995$ ($2n$ nines). [*Hint*: $1 - v^2/c^2 = (1 - v/c)(1 + v/c)$.]

2.9. Establish the identities (2.10).

2.10. Consider two events whose coordinates (x, y, z, t) relative to some inertial frame S are $(0, 0, 0, 0)$ and $(2, 0, 0, 1)$ in units which make $c = 1$. Find the speeds of frames in standard configuration with S in which (i) the events are simultaneous, (ii) the second event precedes the first by one unit of time. Is there a frame in which the two events occur at the same point? [*Answers*: $\frac{1}{2}c$, $\frac{4}{5}c$. *Hint*: (2.11).]

2.11. Looking at Fig. 2.3, and taking the proper length of each plane to be L, deduce that the relative velocity between the two planes is given by $v = L/3$ in the time units indicated by the clocks. If these units are 10^{-7} s (100 nanoseconds), prove $v = (2\sqrt{2}/3)c$ and L = 85 m. What is the velocity of each plane relative to the ground? [*Answer*: $(1/\sqrt{2})c$. *Hint*: Consider the two events where the top rear clock reads 3 and 4, and apply (2.11).]

2.12. Illustrate on a Minkowski diagram a situation similar to that of Exercise 2.6 above: A flash occurs everywhere at once on the spatial x'-axis of the frame S' at some instant $t' = t_0'$. Show that in the usual second frame S this flash is a bright spot traveling forward along the spatial x-axis at superluminal speed c^2/v, and note particularly the last:

2.13. Prove the following additional relations between the 'hyperbolic parameter' (or rapidity) ϕ defined in (2.17) and the velocity v, and note particularly the last:

$$\cosh \phi = \gamma, \qquad \sinh \phi = \frac{v}{c}\gamma, \qquad \tanh \phi = \frac{v}{c}.$$

2.14. Three inertial frames S, S', S'' are in standard configuration with each other; S' has velocity v_1 relative to S, and S'' has velocity v_2 relative to S'. Prove that S'' has velocity $(v_1 + v_2)/(1 + v_1 v_2/c^2)$ relative to S. [*Hint*: Utilize the additivity of the hyperbolic parameter discussed after eqn (2.17) and the last result of the preceding exercise. Recall that trigonometric identities can be converted to hyperbolic-function identities by the rule $\cos x \mapsto \cosh x$, $\sin x \mapsto i \sinh x$; so, in particular, $\tanh(\phi_1 + \phi_2) = (\tanh \phi_1 + \tanh \phi_2)/(1 + \tanh \phi_1 \tanh \phi_2)$.]

2.15. Prove that, under an active Lorentz transformation of the xt plane, straight lines transform into straight lines and parallel straight lines transform into parallel straight lines. Also prove that a tangent line to a curve transforms into a tangent line of the transformed curve. Hence prove that the tangent to any hyperbola $t^2 - x^2 = $ const at the point where the t'-axis (x'-axis) intersects it (see Fig. 2.6) is parallel to the x'-axis (t'-axis).

2.16. In an inertial frame S a train of plane light waves of wavelength λ travels in the negative x-direction towards the observer at the origin. The loci of the wavecrests then satisfy an equation of the form $x = -ct + n\lambda$ (n = integer) . Sketch some such loci on a Minkowski diagram and then apply an active LT (2.19) to deduce that in the usual second frame S′ the wavelength will be given by

$$\lambda' = \sqrt{\frac{c-v}{c+v}}\lambda.$$

This is a relativistic *Doppler* formula.

2.17. What is the diagram analogous to Fig. 2.6 for the Galilean transformation? If the GT is a limiting case of the LT, why are the diagrams different in character? [*Hint*: consider ordinary units for which $c \neq 1$.]

2.18. Prove that the proper acceleration of the particle whose worldline is given by eqn. (2.20) is $1/x_0$ (or c^2/x_0 in full units).

2.19. S and S′ are in standard configuration. In S′ a straight rod parallel to the x' axis moves in the y' direction with velocity u. Show that in S the rod is inclined to the x-axis at an angle $-\tan^{-1}(\gamma uv/c^2)$. [*Hint*: End of Section 2.10.]

2.20. In a frame S′ a straight rod in the x', y' plane rotates anticlockwise with uniform angular velocity ω about its center, which is fixed at the origin. It lies along the x' axis when $t' = 0$. Find the exact shape of the rod in the usual second frame S at the instant $t = 0$, and draw a diagram to illustrate this shape in the immediate neighborhood of the origin. Also show that when the rod is orthogonal to the x'-axis in S′ it appears straight in S.

3

Relativistic kinematics

3.1 Introduction

We have seen how Einstein's second postulate (the invariance of the speed of light) seems to violate common sense and certainly violates Newtonian kinematics. In this chapter we meet the new 'relativistic' kinematics that accommodates both of Einstein's postulates. Its main ingredients are length contraction, time dilation, and the relativistic velocity addition law. This latter has the strange property that one can 'add' any subluminal velocity to the velocity of light and still get the velocity of light, and one can add any number of subluminal velocities and will always get another subluminal velocity. And we have already seen in Section 2.4 that relativistic length contraction and time dilation are symmetric phenomena between two inertial frames, conceptually very different from the similarly named effects in the old ether theory. There, the ether flow was regarded as actually 'doing' something to the moving clock or rod. In relativity, I (at rest in any inertial frame) can observe my stationary clock and my stationary yardstick in their perfect undisturbed state, which is quite unaffected by the motions of all other inertial frames and their observers; if to them my yardstick is short and my clock goes slow, well, that's *their* business. But all these effects play together to make a consistent kinematics, and all are direct consequences of the Lorentz transformation.

3.2 World-picture and world-map

In relativity it is especially important to distinguish between the set of events that an observer *sees* at one instant and the set of events that the observer considers to have *occurred* at that instant. What an observer actually sees or can photograph at one instant is called a *world-picture*. It is a composite of events that occurred progressively earlier as they occurred farther away. For our present purposes it is irrelevant. But it assumes importance in cosmology, where it constitutes essentially our entire data set, and in studies of causality, where it shows all particles at the latest position from which they could have influenced us *now*.

The concept that plays a pervasive role in special relativity is that of the *world-map*. As the name implies, this may be thought of as a map of events, namely those constituting an observer's instantaneous 3-space $t = t_0$. It could be produced by having auxiliary observers at the coordinate lattice-points all map their immediate neighborhoods at a pre-determined time $t = t_0$, and then joining all these local maps into a single global map. Alternatively the world-map can be regarded as a 3-dimensional life-sized photograph exposed everywhere simultaneously, or a *frozen instant in the observer's inertial frame*.

When we speak of 'a snapshot taken in S' (as, for example, we did in connection with Fig. 2.3) or of 'the length of a moving object in S' or of 'the shape of an accelerating object in S', etc., we invariably think of the world-map. The world-map is generally what matters. These remarks are already relevant in the next section, where we show how moving bodies shrink. The shrinkage refers to the world-map. How the eye actually *sees* a moving body is rather different, and in itself not very significant, except that in relativity some of the facts of vision are a little surprising, as we shall see in Section 4.5.

3.3   Length contraction

Consider two inertial frames S and S' in standard configuration. In S' let a rigid rod of length $\Delta x'$ be placed at rest along the x'-axis. We wish to find its length in S, relative to which it moves with velocity v. To measure the rod's length in any inertial frame in which it moves, its end-points must be observed simultaneously. No such precaution is needed in its rest-frame S'. Consider, therefore, two events occurring simultaneously at the extremities of the rod in S, and use (2.11)(i). Since $\Delta t = 0$, we have $\Delta x' = \gamma \Delta x$, or, writing for Δx, $\Delta x'$ the more specific symbols L, L_0 respectively,

$$L = \frac{L_0}{\gamma} = \left(1 - \frac{v^2}{c^2}\right)^{1/2} L_0. \tag{3.1}$$

This shows, quite generally, that *the length L of a body in the direction of its motion with uniform velocity v is reduced by a factor* $(1 - v^2/c^2)^{1/2}$.

Evidently the greatest length is ascribed to a uniformly moving body in its *rest-frame*, namely the frame in which its velocity is zero. This length, L_0, is called the *rest length* or *proper length* of the body. (In general a 'proper' measure of a quantity is that taken in the relevant instantaneous rest-frame.) On the other hand, in a frame in which the body moves with a velocity approaching that of light, its length approaches zero.

This *length contraction* is no illusion, no mere accident of measurement or convention. It is real in every sense. A moving rod is *really* short! It could *really* be pushed into a hole at rest in the lab into which it would not fit if it were not moving and shrunk. (See Section 3.4.) Whatever physical forces of attraction and repulsion are responsible for holding together the constituent elementary particles of the body when the body is at rest, will all change in accordance with the laws of relativistic mechanics when the body is in motion relative to an inertial frame, in such a way as to produce the shortening *in that frame*. We cannot and need not know the details of all this, but we know a priori that it must be so. As a trivial example, we shall see in

Chapter 7 that the Coulomb attraction between a stationary electron and a stationary proton is judged to increase when viewed from an inertial frame that moves in the direction of their separation.

The contraction effect follows inevitably from the Lorentz transformations, which themselves follow inevitably from Einstein's two postulates. So if a consistent mechanics can be constructed within special relativity (and it can) then there *must* be a detailed mechanical explanation of the shortening. It is much like the total and the detailed Newtonian energy balance of some engine. We know a priori that as much energy goes in (in the form of heat, electricity, etc.) as comes out (in the form of kinetic energy, etc.). Although we may not know or understand every minute process that occurs within (friction, vibration, heat exchange, etc.), we *know* that when you add it all up it must give the above result.

Unfortunately there is little prospect of ever being able to test the length contraction of a macroscopic object directly, since we have no means of accelerating macroscopic bodies to relativistically significant speeds. But in Chapter 7 we shall see a measurable effect that length contraction has on the line density of charge in a current-carrying wire, even though the speeds involved are quite small.

3.4   Length contraction paradox

Consider the admittedly unrealistic situation of a man carrying horizontally a 20-ft pole and wanting to get it into a 10-ft garage. He will run at speed $v = 0.866c$ to make $\gamma = 2$, so that the pole contracts to 10 ft. It will be well to insist on having a sufficiently massive block of concrete at the back of the garage, so that there is no question of whether the pole finally stops in the inertial frame of the garage, or vice versa. So the man runs with his (now contracted) pole into the garage and a friend quickly closes the door. In principle we do not doubt the feasibility of this experiment; that is, the reality of length contraction. When the pole stops in the rest-frame of the garage, it will tend to assume, if it can, its original length relative to the garage. Thus, if it survived the impact, it must now either bend, or burst the door, or remain compressed.

At this point a paradox might occur to the reader:[1] What about the symmetry of the phenomenon? Relative to the runner, won't the garage be only 5 ft long? And, if so, how can the 20-ft pole get into the 5-ft garage? Very well, let us consider what happens in the rest-frame of the pole. The open 5-ft garage now comes towards the stationary pole. Because of the concrete block, it keeps on going even after the impact, taking the front end of the pole with it (see Fig. 3.1). But the back end of the pole is still at rest: it cannot yet 'know' that the front end has been struck, because of the finite speed of propagation of *all* signals. Even if the 'signal' (in this case the elastic shock wave) travels along the pole with the speed of light, that signal has 20 ft to

---

[1] It is perhaps surprising that no such paradox seems to have been noted in the literature before 1960. See W. Rindler, *Special Relativity*, Oliver & Boyd, Edinburgh, 1960, p. 37; also W. Rindler, *Am. J. Phys.* **29**, 365 (1961) and E. M. Dewan, *Am. J. Phys.* **31**, 383 (1963).

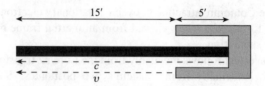

Fig. 3.1

travel against the garage front's 15 ft, before reaching the back end of the pole. This race would be a dead heat if v were $0.75c$. But v is $0.866c$! So the pole *more* than just gets in. It could even get into a garage whose length is as little as 5.4 ft at rest and thus 2.7 ft in motion: the garage front would then have to travel 17.3 ft against the shock wave's 20 ft, requiring speeds in the ratio 17.3 to 20 or 0.865 to 1 for a dead heat.

There is one important moral to this story: whatever result we get by correct reasoning in any one inertial frame, must be true; in particular, it must be true when viewed from any other inertial frame. As long as the set of physical laws we are using is self-consistent and Lorentz-invariant, there *must* be an explanation of the result in every other inertial frame, although it may be quite a different explanation from that in the first frame.

3.5   Time dilation; The twin paradox

Let us again consider two inertial frames S and S' in standard configuration. Let a standard clock be fixed in S' and consider two events at that clock when it indicates times differing by $\Delta t'$. We enquire what time interval Δt is ascribed to these events in S. From the Δ-form of (2.9)(iv) we see at once, since $\Delta x' = 0$, that $\Delta t = \gamma \Delta t'$, or, replacing Δt and $\Delta t'$ by the more specific symbols T and T_0, respectively,

$$T = \gamma T_0 = \frac{T_0}{(1 - v^2/c^2)^{1/2}}. \tag{3.2}$$

We can deduce from this quite generally that *a clock moving uniformly with velocity v through an inertial frame S goes slow by a factor $(1 - v^2/c^2)^{1/2}$ relative to the synchronized standard clocks at rest in S.* Clearly, then, the fastest rate is ascribed to a clock in its rest-frame, and this is called its *proper rate*. On the other hand, at speeds close to the speed of light, the rate of the clock would be close to zero.

This *time dilation*, like length contraction, is no accident of convention but a *real* effect. Moving clocks really do go slow. If a standard clock is taken at uniform speed v through an inertial frame S along a straight line from point A to point B and back again, the elapsed time T_0 indicated on the moving clock will be related to the elapsed time T on the clock fixed at A by the eqn (3.2), except for possible errors introduced when the clock is accelerated to initiate, reverse, and terminate its journey. But whatever these errors are, their contribution can be dwarfed by simply extending the periods of uniform motion. So, at least in theory, the effect is tangible. But it has by now also been amply observed in the real world, as we shall presently recount.

For any particular uniformly moving clock, the relativistic laws of mechanics (or electromagnetism, or whatever) *must* in principle be responsible for the details that make this clock go slow by exactly the Lorentz factor. (See, for example, Exercises 3.8 and 7.17.) In the case of *accelerated clocks*, we have no such shortcut through the details governing their behavior, and no general predictions can be made. Certain types of clock are unaffected by acceleration to the extent that the internal driving forces which act between parts of the clock exceed the *derivative* of the external force field (for example, the tidal field if the acceleration is gravitational). For the force itself affects all the parts equally and thus not their relative motion. For other types of clock it is their size that determines their susceptibility to acceleration (cf. Exercise 3.9).

As we shall see, certain natural clocks, like vibrating atoms, decaying muons, etc., seem to satisfy criteria of this sort to high accuracy under the accelerations to which they have so far been subjected. We define an *ideal* clock as one that is *completely* unaffected by acceleration; that is, as one whose instantaneous rate depends only on its instantaneous speed in accordance with (3.2). As has been stressed by Sexl, the absoluteness of acceleration implies that ideal clocks are possible objects, at least in principle. We need only take any good clock, observe (or calculate) how it reacts to given accelerations, derivatives of accelerations, etc., and then equip it with accelerometers from which a properly programmed computer can always correct for the errors incurred and thus display 'ideal' time.

The image to keep in mind is that of a lattice of synchronized standard clocks filling a given inertial frame S and an *ideal* clock moving arbitrarily through this lattice and losing time steadily against the fixed clocks according to (3.2). Its total time elapsed, if it starts at time t_1 and ends at time t_2, will be given by

$$\Delta\tau = \int_{t_1}^{t_2} (1 - v^2/c^2)^{1/2}\, dt =: \int_{\tau_1}^{\tau_2} d\tau, \qquad (3.3)$$

where t is the time and v the velocity in S, and where we have introduced a quantity of great importance in the sequel, namely the *proper time* τ that is indicated by an arbitrarily moving ideal clock.

We shall assume that accelerated observers use ideal clocks. One can similarly define ideal infinitesimal measuring rods, but it is more straightforward (in principle) to use clocks and radar to determine distances. It is often stated that an accelerating observer under these circumstances makes the same observations as an instanta-neously comoving inertial observer. While this is true in certain cases, it is false in others, as when the two observers observe their accelerometers. So each such observation must be discussed on its own merits. (See, for example, after eqn (4.5) below.) There is relevance in this question, since, according to the equivalence prin-ciple, we are all accelerating observers in our terrestrial labs, relative to the inertial observers in free fall in their LIFs.

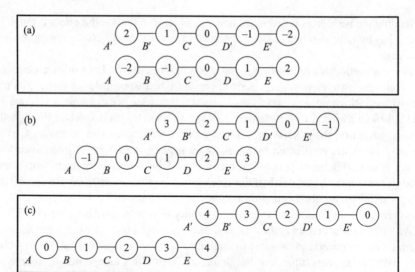

Fig. 3.2

Time dilation, like length contraction, must a priori be symmetric: if one inertial observer considers the clocks of a second inertial observer to run slow, the second must also consider the clocks of the first to run slow. Figure 3.2—which is really an extension of Figs 2.3(a) and (c)—shows in detail how this happens. Synchronized standard clocks A, B, C, ... and A′, B′, C′, ... are fixed at certain equal intervals along the x-axes of two frames S and S′ in standard configuration. The figure shows three world-maps made, at convenient equal time intervals, in the 'midframe' S″ relative to which S and S′ have equal and opposite velocities. In each world-map the clocks of S and S′ are all seen to indicate different times, since simultaneity is relative. [Write t'' for t in (2.6) (iv) and set $t'' = $ const.] Suppose the clocks in the diagram indicate seconds. As can be seen, A′ reads 4 seconds ahead of A in Fig. 3.2(a), only 2 seconds ahead of C in Fig. 3.2(b), and equal with E in Fig. 3.2(c). Thus A′ loses steadily relative to the clocks in S. Similarly E loses steadily relative to the clocks in S′, and indeed all clocks in the diagram lose at the same rate relative to the clocks of the other frame.

Unlike length contraction, time dilation has been amply confirmed experimentally. For example, certain mesons (muons) reaching us from the top of the atmosphere (where they are produced by incoming cosmic rays), are so short-lived that, even had they traveled at the speed of light, their travel time in the absence of time dilation would exceed their lifetime by factors of the order of 10. Rossi and Hall in 1941 timed such muons between the summit and foot of Mt Washington, and found their lifetimes were indeed dilated in accordance with eqn (3.2). Equivalent experiments with muons circling 'storage rings' (with velocities corresponding to $\gamma \approx 29$) at the CERN laboratory in 1975 and thereafter have refined these results to an impressive accuracy of $\sim 2 \times 10^{-3}$ and have additionally shown that, to such accuracy, proper accelerations of up to $10^{18}g$ (!) do not contribute to the muons' time dilation.[2] It may be objected that muons are not clocks—but time dilation applies to *any* temporal process and therefore also to muon decay and even human lifetimes (as, for example, that of astronauts on fast space journeys). To see this, at least for uniform motion, we need only imagine a standard clock to travel *with* the muon or the space traveler in their respective inertial rest-frames.

Another striking instance of time dilation is provided by 'relativistic focusing' of electrically charged particles, which plays a role in the operation of high-energy particle accelerators. Any stationary cluster of electrons (or protons, etc.) tends to expand at a characteristic rate because of mutual electrostatic repulsion. But the corresponding particles in a fast-moving beam are observed to spread at a much slower rate. If we regard the stationary cluster as a kind of clock, we have here an almost visible manifestation of the slowing down of a moving clock. (This can alternatively be explained by the same mechanism as that which causes parallel currents to attract each other.) Yet another such manifestation, the so-called transverse Doppler effect, will be discussed in Section 4.3. (It, too, has led to a demonstration of acceleration-independence for certain natural clocks.)

Nowadays, amazingly, time dilation can even be observed in macroscopic clocks. This was first done in 1971 (though only to an accuracy of about 10 per cent) by Hafele and Keating,[3] who simply took some very accurate cesium clocks around the world on commercial airliners! Eventually Vessot *et al.*[4] made a very precise determination of the gravitational time dilation effect of general relativity *jointly* with the time dilation of special relativity by sending rocket-launched hydrogen-maser 'clocks' up to 10 000 km in nearly vertical trajectories. The validity of the two effects jointly was established to an accuracy of $\sim 10^{-4}$.

No account of special relativity would be complete without at least a mention of the notorious *clock* or *twin paradox* dating back as far as 1911. Reams of literature were written on it unnecessarily for more than six decades. At its root apparently lay a deep psychological barrier to accepting time dilation as real. From a modern point of view it is difficult to understand the earlier fascination with this problem, or even to recognize it as a problem. The 'paradox' concerns the situation we already discussed above (in the second paragraph of this section) of transporting a clock from A to B and back again, and then finding that the traveling clock upon its return indicates a lesser time than the stationary one. The story is usually embellished by replacing the two clocks by two twins, of which the traveler upon returning is younger than the stay-at-home. The claim now is that all motion is relative. So the traveler can with equal right maintain that it was the stay-at-home who did the traveling and should therefore be the younger when they reunite! But, whereas uniform motion indeed is relative, acceleration is not, and accelerometers attached to the twins will easily settle

2 J. Bailey *et al.*, *Nature* **268**, 301 (1977).
3 J. C. Hafele and R. Keating, *Science* **177**, 166 (1972).
4 R. F. C. Vessot *et al.*, *Phys. Rev. Lett.* **45**, 2081 (1980).

the dispute: one remained fixed in an inertial frame, and the other did not. The single most worthwhile remark on this 'paradox' was made by Sciama: it has, he said, the same status as Newton's experiment with the two buckets of water—one, rotating, suspended below the other, at rest. If these were the whole content of the universe, it would indeed be paradoxical that the water surface in the one should be curved and that in the other flat. But inertial frames have a real existence too, and relative to the inertial frames there is no symmetry between the buckets, and no symmetry between the twins, either.

From a practical point of view, the reader might well ask in what *real-world* reference frame the twins could play out their experiment. Surely the small freely falling Einstein cabins are of no interest in this context. And any large 'Newtonian' local inertial frame, say one containing our whole galaxy, will be distorted by curvature, according to GR. Still, provided the traveler doesn't pass too close to any highly concentrated mass, that curvature will be quite negligible in the present context, as can be seen, for example, from our discussion in Section 1.16 of the influence of the gravitational potential (curvature!) on clock rates. So the experiment will work.

3.6   Velocity transformation; Relative and mutual velocity

Once again, let us consider two inertial frames S and S' in standard configuration. Let **u** be the instantaneous vector velocity in S of a particle or simply of a geometrical point (so as not to exclude the possibility $u \geq c$). We wish to find the velocity **u'** of this point in S'. As in classical kinematics, we define

$$\mathbf{u} = (u_1, u_2, u_3) = (dx/dt, dy/dt, dz/dt), \tag{3.4}$$

$$\mathbf{u}' = (u'_1, u'_2, u'_3) = (dx'/dt', dy'/dt', dz'/dt'). \tag{3.5}$$

Substituting from (2.12) into (3.5), dividing each numerator and denominator by dt, and comparing with (3.4), now immediately yields the velocity transformation formulae:

$$u'_1 = \frac{u_1 - v}{1 - u_1 v/c^2}, \qquad u'_2 = \frac{u_2}{\gamma(1 - u_1 v/c^2)}, \qquad u'_3 = \frac{u_3}{\gamma(1 - u_1 v/c^2)}. \tag{3.6}$$

No assumption as to the uniformity of **u** was made, and these formulae apply equally to the instantaneous velocity in a non-uniform motion. Note also how they reduce to the classical formulae (1.2) when either $v \ll c$, or $c \to \infty$ formally.

We can obtain the inverse of (3.6) without further effort by applying a 'v-reversal transformation' (see last paragraph of Section 2.7) to (3.6):

$$u_1 = \frac{u'_1 + v}{1 + u'_1 v/c^2}, \qquad u_2 = \frac{u'_2}{\gamma(1 + u'_1 v/c^2)}, \qquad u_3 = \frac{u'_3}{\gamma(1 + u'_1 v/c^2)}. \tag{3.7}$$

These last equations can be regarded alternatively as giving the resultant, **u**, of *first* imparting to a particle a velocity $\mathbf{v} = (v, 0, 0)$ and then, relative to its *new* rest-frame, another velocity $\mathbf{u}'$. They are therefore occasionally referred to as the relativistic *velocity addition* formulae. In particular, the first member gives the resultant of two collinear velocities v and u_1' and is therefore of the same form as the velocity parameter for the resultant of two successive Lorentz transformations [see property (viii) of Section 2.8]. It is occasionally convenient to write $\mathbf{u} = \mathbf{v} \dotplus \mathbf{u}'$ and $\mathbf{u}' = -\mathbf{v} \dotplus \mathbf{u}$. But note that $\mathbf{v} \dotplus \mathbf{w} \neq \mathbf{w} \dotplus \mathbf{v}$ except for 1-dimensional motion (cf. Exercise 3.12).

Writing $u = (u_1^2 + u_2^2 + u_3^2)^{1/2}$ and $u' = (u_1'^2 + u_2'^2 + u_3'^2)^{1/2}$ for the magnitudes of corresponding velocities in S and S' we have, by first factoring out dt'^2 and dt^2, respectively, from the left- and right-hand sides of (2.7), and then using (2.12)(iv),

$$dt^2(c^2 - u^2) = dt'^2(c^2 - u'^2) = dt^2 \gamma^2(v)(1 - u_1 v/c^2)^2(c^2 - u'^2). \qquad (3.8)$$

The first of these equations we have already noted in Chapter 2 [in Δ-form: (2.23)], as well as its consequence: $u \gtreqless c$ implies $u' \gtreqless c$ and vice versa. This also shows that any 'sum' $\mathbf{u} = \mathbf{v} \dotplus \mathbf{u}'$ of two velocities less than c is itself a velocity less than c. So, however many velocity increments (less than c) a particle receives in its successive rest-frames (that is, the sequence of inertial frames in which the particle is momentarily at rest), it can never attain the velocity of light. The velocity of light thus plays the role in relativity of an infinite velocity, inasmuch as no sum of lesser velocities can ever equal it.

If we now cancel dt^2 from the extremities of eqn (3.8) and rearrange terms, we find the following transformation of u^2, the squared magnitude of a particle's velocity:

$$c^2 - u'^2 = \frac{c^2(c^2 - u^2)(c^2 - v^2)}{(c^2 - u_1 v)^2}. \qquad (3.9)$$

Note that $u_1 v = \mathbf{u} \cdot \mathbf{v}$, so that the RHS is actually symmetric in $\mathbf{u}$ and $\mathbf{v}$. This means that the *magnitudes* of $(-\mathbf{v}) \dotplus \mathbf{u}$ and $\mathbf{u} \dotplus (-\mathbf{v})$ are the same, and this evidently holds for any two subluminal 3-velocities.

Rewriting (3.9) in terms of $\gamma(u)$, $\gamma(u')$, $\gamma(v)$, we get an equation which, on taking square roots, yields the first of the following two useful relations,

$$\frac{\gamma(u')}{\gamma(u)} = \gamma(v)\left(1 - \frac{u_1 v}{c^2}\right), \qquad \frac{\gamma(u)}{\gamma(u')} = \gamma(v)\left(1 + \frac{u_1' v}{c^2}\right), \qquad (3.10)$$

while the second again results from a v-reversal transformation. These relations show how the γ-factor of a moving particle transforms.

We end this section by remarking on one useful velocity concept (and formula) which applies equally in Newtonian and relativistic kinematics. This concerns the rate of change, in *one* given inertial frame S, of the connecting vector between two particles, whose position vectors and velocities are (let us say) $\mathbf{r}_1$, $\mathbf{u}_1$ and $\mathbf{r}_2$, $\mathbf{u}_2$, respectively: $(d/dt)(\mathbf{r}_2 - \mathbf{r}_1) = \mathbf{u}_2 - \mathbf{u}_1$. We call this, for lack of a better name, the *mutual velocity* between the particles in S, to distinguish it from the *relative velocity*, which is what one particle ascribes to the other. It can be as big as $2c$, as when two photons collide head-on. If I want to know how much time will elapse in my own frame before two collinearly moving particles collide, I simply divide their present distance apart by their mutual velocity, in relativistic just as in Newtonian kinematics.

3.7   Acceleration transformation; Hyperbolic motion

When a particle moves non-uniformly, it is useful to know how to transform not only its velocity, but also its acceleration. *In this section we restrict our discussion to 1-dimensional motion*; the general case will be discussed, with more powerful mathematical tools, in Section 5.5.

Suppose, then, a particle moves along the common x-axis of the usual two inertial frames S and S' in standard configuration. Let u and u' denote its velocities in S and S', respectively. What we seek is the relation between du'/dt' and du/dt. It will simplify the calculation if we work in terms of the rapidity function (cf. Exercise 2.13)

$$\phi(u) = \tanh^{-1}(u/c), \tag{3.11}$$

which allows the velocity addition formula (3.7)(i) to be re-expressed in the strikingly simple form

$$\phi(u) = \phi(v) + \phi(u'), \tag{3.12}$$

as we have already seen in Exercise 2.14. Differentiating eqn (3.12) with respect to t yields

$$\frac{d}{dt}\phi(u) = \frac{d}{dt'}\phi(u')\frac{dt'}{dt}. \tag{3.13}$$

But from (3.11) we have

$$\frac{d}{dt}\phi(u) = \frac{1}{c}\gamma^2(u)\frac{du}{dt}, \tag{3.14}$$

and from (3.8)(i) [or, more directly, from (3.2)],

$$\frac{dt'}{dt} = \frac{\gamma(u')}{\gamma(u)}. \tag{3.15}$$

When we substitute these last two equations [and the primed version of (3.14)] into (3.13), we get the desired acceleration transformation formula

$$\gamma^3(u')\frac{du'}{dt'} = \gamma^3(u)\frac{du}{dt}. \tag{3.16}$$

Using (3.10)(i) we now could, if we wished, express du'/dt' entirely in terms of unprimed quantities.

Under the Galilean transformation, as we saw earlier, the acceleration is invariant: $du'/dt' = du/dt$. Eqn (3.16) shows that in relativity this is no longer true.

If we define the *proper acceleration* α of a particle as that which is measured in its instantaneous rest-frame, say S$'$, we find, on setting $u' = 0$ and $du'/dt' = \alpha$ in (3.16), that [cf. (2.10)]

$$\alpha = \gamma^3(u)\frac{du}{dt} = \frac{d}{dt}[\gamma(u)u]. \tag{3.17}$$

Proper acceleration is precisely the push we feel when sitting in an accelerating rocket. Also, by the equivalence principle, the gravitational field in our terrestrial lab is the negative of our proper acceleration, our instantaneous rest-frame being an imagined Einstein cabin falling with acceleration g.

A case of particular interest is that of rectilinear motion with *constant* proper acceleration α. We can then integrate (3.17) at once, choosing $t = 0$ when $u = 0$: $\alpha t = \gamma(u)u$. Squaring, solving for u, integrating once more and setting the constant of integration equal to zero once more, yields the following equation for the motion:

$$x^2 - c^2 t^2 = c^4/\alpha^2. \tag{3.18}$$

Thus, for obvious reasons, rectilinear motion with constant proper acceleration is called *hyperbolic motion*. We have already seen this kinematic significance of a hyperbolic worldline from the perspective of an active Lorentz transformation (cf. Fig. 2.10). The corresponding Newtonian calculation gives the familiar result $x = \frac{1}{2}\alpha t^2$; that is, 'parabolic' motion. There is no limit on the proper acceleration that a particle can have; by (3.18), $\alpha = \infty$ implies $x = \pm ct$, hence the proper acceleration of a photon can be taken to be infinite. A sequence of hyperbolic worldlines for various values of α is shown in Fig. 3.3 (where we have written X for c^2/α). As $\alpha \to \infty$, they fit ever more snugly into the limiting photon paths $x = \pm ct$.

3.8   Rigid motion and the uniformly accelerated rod

Look again at Fig. 3.3 and consider, in fact, the equation

$$x^2 - c^2 t^2 = X^2 \tag{3.19}$$

for a continuous range of positive values of the parameter X. For each fixed X it represents a particle moving with constant proper acceleration c^2/X in the x-direction. Altogether it represents, as we shall presently show, a 'rigidly moving' rod.

By the *rigid motion* of a body one understands a motion during which every small volume element of the body shrinks always in the direction of its motion in inverse proportion to its instantaneous Lorentz factor relative to a given inertial frame. Thus every small volume element preserves its dimensions in its own successive instantaneous rest-frames, which shows that the definition is intrinsic, that is, frame-independent. It also shows that during the rigid motion of an initially unstressed body no elastic stresses arise. A body moving rigidly cannot start to rotate, since circumferences of

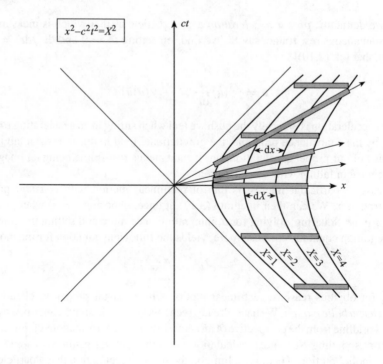

Fig. 3.3

circles described by points of the body would have to shrink, while their radii would have to remain constant, which is impossible.[5] In general, therefore, the motion of one point of a rigidly moving body determines that of all others. Note that the transversely accelerating rod discussed at the end of Section 2.10 moves rigidly in the present technical sense, even though it continually changes its shape in the frame S.

Clearly, if the front-end of a rigidly moving rod moves forward with constant proper acceleration, the back-end must move with greater acceleration, because of the ever-increasing contraction of the rod. (In Fig. 3.3 the horizontal bars show the rod at various times in S and thus at various stages of contraction.) Since the same is true of each portion of the rod, the acceleration must increase steadily towards the rear. But that all points move with *constant* proper acceleration, and that the total picture is given by the neat equation (3.19), comes as a pleasant surprise. And yet it is 'obvious': Because of the invariance (2.14), eqn (3.19) translates into itself (that is,

[5] For this reason, a 'rigidly moving' disk that is set to rotate would have to bend. In the early days of relativity this was regarded as paradoxical ('Ehrenfest's paradox,' 1909). Of course a flat unstressed *uniformly* rotating disk can be built in motion (of light material, to avoid centrifugal stress). The geometry on it, as measured by meter sticks at rest relative to it, would be non-Euclidean: for example, its circumference would exceed $2\pi r$. This played a role in suggesting to Einstein one of the key ideas of general relativity, namely the relevance of non-Euclidean geometry (curvature) in the presence of gravity—here mimicked by the centrifugal force.

into $x'^2 - c^2 t'^2 = X^2$) in any frame S' in standard configuration with the original frame S; in other words, the worldline pattern of Fig. 3.3 is invariant under an active LT. So in *any* such frame S' the entire rod comes to rest momentarily (vertical worldlines) at $t' = 0$, and between the same x values as in S; that is, with the same length. And this is equally true for each portion of it. The rod thus goes from instantaneous rest in one inertial frame to instantaneous rest in the next, and the next, and so on, each element always having the same rest length. But that is precisely what characterizes 'rigid motion'. The slanting bars in Fig. 3.3 show the rod in its successive rest-frames. The final geometric result established in Exercise 2.15 allows us to 'see' that the slopes of the worldlines (and thus the velocities) along each slant are equal—as they must be, since the entire rod is then at rest in a single inertial frame.

From the successive rest-frames it is clear that X measures ruler distance along the rod; that is, distance measured with ideal infinitesimal rulers at rest relative to the rod (since an ideal accelerating infinitesimal ruler and a momentarily comoving inertial ruler coincide.) Note that the rod cannot be extended to $X \leq 0$, since $X = 0$ already corresponds to a photon. A bundle of such rods can be regarded as a rocketship, possibly quite a long one. Life in such a rocketship would be similar to life in a static skyscraper, in which the gravitational field, here mimicked by the proper acceleration c^2/X has parallel field lines but varies in strength inversely with 'height' X. (A *constant* gravitational vacuum field can *not* be generated in this way.) We shall have occasion to make significant use of the above analysis of the uniformly accelerating rocket in Chapter 12.

Exercises 3

3.1. Invent a more realistic arrangement than the garage and pole described in Section 3.4 whereby length contraction could be verified experimentally, if only our instruments were delicate enough. [For example, consider two meter sticks moving with equal and opposite velocities through the lab, like the airplanes in Fig. 2.3(a).]

3.2. Suppose the rod discussed at the end of Section 2.10 is at rest on the x'-axis, between $x' = 0$ and $x' = l$ for all $t' \leq 0$, and that the acceleration begins at $t' = 0$ Indicate on an x, y diagram a sequence of snapshots (world-maps) of this rod in S, beginning from its being straight and on the x-axis, and paying particular attention to the various phases of lift-off.

3.3. Use a Minkowski diagram to establish the following result: Given two rods of rest lengths l_1 and l_2 ($l_2 < l_1$), moving along a common line with relative velocity v, there exists a unique inertial frame S' moving along the same line with velocity $[l_1 - l_2/\gamma(v)]/l_1(v/c)$ relative to the longer rod, in which the two rods have equal lengths, provided $l_1^2(c - v) < l_2^2(c + v)$.

3.4. In the pole-and-garage problem of Section 3.4, what is the longest pole that can be run into a 12-ft garage at a speed v making $\gamma(v) = 3$, assuming that the elastic shock wave travels at the speed of light? [*Answer*: 69.941 ft.]

3.5. An 18-ft pole, while remaining parallel to the x-axis, moves with velocity $(v, -w, 0)$ relative to frame S, with $\gamma(v) = 3$, and v, w positive. The centre of the pole passes the centre of a 9-ft hole in a plate that coincides with the plane $y = 0$. Explain, from the point of view of the usual second frame S' moving with velocity v relative to S, how the pole gets through the (now 3-ft) hole.

3.6. It was pointed out by M. von Laue that a cylinder rotating uniformly about the x'-axis of S' will seem *twisted* when observed instantaneously in S, where it not only rotates but also travels forward. If the angular speed of the cylinder in S' is ω', prove that in S the twist per unit length is $\gamma\omega'v/c^2$ in the opposite direction. (This effect is to be expected, since we can regard the cylinder as composed of a stack of circular disks, each disk by its rotation serving as a clock, with arbitrary radii which are all parallel in S' designated as hands; in S these radii are not parallel, since simultaneity is relative, cf. Fig. 3.2.)

3.7. Ideal clocks are taken from event $\mathcal{A}$ to event $\mathcal{B}$ along various worldlines. Prove that the longest proper time for the trip will be indicated by that clock which follows the *straight* worldline—the clock that moves rectilinearly and with constant speed. [*Hint*: Pick the most convenient inertial frame from which to look at this situation.]

3.8. A 'light-clock' consists of two mirrors at opposite ends of a rod with a photon bouncing between them. Assuming only length contraction and the invariance of c, prove that such a clock will go slow by the expected Lorentz factor as it travels (i) longitudinally, (ii) transversely, through an inertial frame. [*Hint*: for (i) use 'mutual' velocity.]

3.9. A light-clock, as in Exercise 3.8, has proper length l and moves longitudinally through an inertial frame with proper acceleration α. (Ignore any variation of α along the rod.) By looking at the time it takes the photon to make one to-and-fro bounce in the instantaneous rest-frame, show that the frequency and proper frequency are related, in lowest approximation, by $v = v_0\gamma^{-1}(1 + \frac{1}{2}\alpha l/c^2)$. So the deviation from idealness is proportional to α and to l. What is it when $\alpha = 10^{17}g$ ($g = 980\,\text{cm/s}^2$) and $l = 1\,\text{cm}$?

3.10. (i) Two particles move along the x-axis of S at velocities $0.8c$ and $0.9c$, respectively, the faster one momentarily $1m$ behind the slower one. How many seconds elapse before collision? (ii) A rod of proper length $10\,\text{cm}$ moves longitudinally along the x-axis of S at speed $\frac{1}{2}c$. How long (in S) does it take a particle, moving oppositely at the same speed, to pass the rod?

3.11. In a given inertial frame, two particles are shot out simultaneously from a given point, with equal speeds v, in orthogonal directions. What is the speed of each particle relative to the other? [*Answer*: $v(2 - (v^2/c^2))^{1/2}$.]

3.12. Show that the result of relativistically 'adding' a velocity $\mathbf{u}'$ to a velocity $\mathbf{v}$ in the sense of (3.7) is not, in general, the same as that of 'adding' a velocity $\mathbf{v}$ to a velocity $\mathbf{u}'$. [*Hint*: consider $\mathbf{u}' = (0, u', 0)$ and $\mathbf{v} = (v, 0, 0)$.] Note, however, that it makes sense to add two vector velocities like $\mathbf{u}'$ and $\mathbf{v}$, which are defined in different frames, only if the axes of these frames are parallel in some well-defined way.

3.13. Two inertial frames S and S' are in standard configuration while a third, S'', moves with velocity v' along the y'-axis, its axes parallel to those of S'. If the line of relative motion of S and S'' makes angles θ and θ'' with the x-and x''-axes, respectively, prove that $\tan\theta = v'/v\gamma(v)$, $\tan\theta'' = v'\gamma(v')/v$. [*Hint*: use (3.7).] The inclination $\delta\theta$ of S'' relative to S is defined as $\theta'' - \theta$. If $v, v' \ll c$, prove that $\delta\theta \approx vv'/2c^2$. If a particle describes a circular path at uniform speed $v \ll c$ in a given frame S, and consecutive instantaneous rest-frames, say S' and S'', always have zero relative inclination, prove that after a complete revolution the instantaneous rest-frame is tilted through an angle $\pi v^2/c^2$ in the sense opposite to that of the motion. ['*Thomas precession.*' *Hint*: Let a tangent and radius of the circle coincide with the x'- and y'-axes. Then $v' \approx v^2\,dt/a$, $a = $ radius.]

3.14. How many successive velocity increments of $\frac{1}{2}c$ are needed to produce a resultant velocity of (i) 0.99c, (ii) 0.999c? [*Answer*: 5, 7. *Hint*: $\tanh 0.55 = 0.5$, $\tanh 2.65 = 0.99$, $\tanh 3.8 = 0.999$.]

3.15. If $\phi = \tanh^{-1}(u/c)$, and $e^{2\phi} = z$, prove that n consecutive velocity increments u from the instantaneous rest-frame produce a velocity $c(z^n - 1)/(z^n + 1)$.

3.16. In the notation of the preceding exercise, show that for $u \approx c$, $\gamma(u) \approx \frac{1}{2}e^\phi$. Deduce that the γ-factor of the resultant of n consecutive velocity increments $u \approx c$ is $\sim 2^{n-1}\gamma^n(u)$.

3.17. A rod moves along the x-axis with speed u and has length L relative to S. What is its length L' relative to the usual second frame S'? [*Answer*: $L/\gamma(v)(1 - uv/c^2)$.]

3.18. A particle moves 1-dimensionally. If τ denotes *proper time*, prove $d\phi(u)/d\tau = \alpha/c$. [*Hint*: eqns (3.14) and (3.17); or, more directly, differentiate (3.12) at $u' = 0$.]

3.19. Derive the Newtonian equation for motion with constant acceleration, $x = \frac{1}{2}\alpha t^2$, as the $c \to \infty$ limit of hyperbolic motion. [*Hint*: Shift the origin to the vertex of the hyperbola, and then write eqn (3.18) in the form $x + c^2/\alpha = (c^2/\alpha)(1 + \alpha^2 t^2/c^2)^{1/2}$.]

3.20. Consider a particle in hyperbolic motion according to eqn (3.18). If $u = $ velocity, $\phi = $ rapidity, $\tau = $ proper time, and t and τ vanish together, prove the following formulae:

$$u = c^2 t/x, \qquad \gamma(u) = \alpha x/c^2, \qquad \phi(u) = \alpha\tau/c,$$

$$u/c = \tanh(\alpha\tau/c), \qquad \gamma(u) = \cosh(\alpha\tau/c),$$

$$\alpha t/c = \sinh(\alpha\tau/c). \qquad \alpha x/c^2 = \cosh(\alpha\tau/c).$$

3.21. A certain piece of elastic breaks when it is stretched to twice its unstretched length. At time $t = 0$, all points of it are accelerated longitudinally with constant proper acceleration α, from rest in the unstretched state. Prove that the elastic breaks at $t = \sqrt{3}c/\alpha$.

3.22. Given that g, the acceleration of gravity at the earth's surface, is $\sim 9.8\,\mathrm{ms}^{-2}$, and that a year has $\sim 3.2 \times 10^7$ s, verify that, in units of years and light-years, $g \approx 1$. A short rocket moves from rest in an inertial frame S with constant proper acceleration g (thus giving maximum comfort to its passengers). Find its Lorentz factor relative to S when its own clock indicates times $\tau = 1$ day, 1 year, 10 years. Find also the corresponding distances and times traveled in S. If the rocket accelerates for 10 years of its own time, then decelerates for 10 years, and then repeats the whole manoeuvre in the reverse direction, what is the total time elapsed in S during the rocket's absence? [*Answers*: $\gamma = 1.000\,0038, 1.5431, 11\,013$; $x = 0.000\,0038, 0.5431, 11\,012$ light-years; $t = 0.0027, 1.1752, 11\,013$ years; $t = 44\,052$ years. To obtain some of these answers you will have to consult tables of $\sinh x$ and $\cosh x$. At small values of their arguments a Taylor expansion suffices.]

3.23. Show that the equation $x\,\mathrm{d}x = X\,\mathrm{d}X$, which results from differentiating eqn (3.19) for constant t, in conjunction with one of the results of Exercise 3.20 above, directly establishes the 'rigid motion' of the rod described by (3.19).

3.24. Consider a long uniformly accelerating 'rocketship' analogous to the uniformly accelerating rod with eqn (3.19). Prove that the 'radar distance' (the proper time of a 'light-echo' multiplied by $c/2$) of a point on the rocket at parameter X_2, from an observer riding on the rocket at parameter X_1 ($<X_2$) is always $X_1 \sinh^{-1}[(X_2^2 - X_1^2)/2X_1X_2$. [*Hint*: refer to Fig. 3.3, and work in the rest-frame of the rocket at the reflection event.]

3.25. In the rocketship of the preceding exercise, a standard clock is dropped from rest at level X_2 and allowed to fall freely. How much time elapses at that clock before it passes level X_1?

4

Relativistic optics

4.1   Introduction

Optics is one of the fields to which relativity brought considerable simplification. In the old ether theory, light behaved simply *only* in the ether frame, and it made a difference how the observer and the source separately move through the ether. Relativity eliminated this middleman: every inertial frame is now known to be as 'primary' as the ether was once thought to be.

In this chapter we take a pragmatic approach, indifferently treating light as particles (photons, rays) or waves. The kinematic equivalence of these concepts will be justified in the next chapter. When discussing frequency, we shall consider a source to emit a regular series of 'pulses' that correspond to the wavecrests of the diverging wave pattern.

The reader should bear in mind that with *optical* formulae we can, in general, *not* expect to retrieve the classical equivalents by simply letting $c \to \infty$, since even the classical formulae must contain cs. Only *some* of the cs in the relativistic formulae come from the new kinematics, while the others refer to the actual light.

4.2   The drag effect

To begin with, we discuss a problem for which relativity provided an ideally simple solution. Prior to that it had challenged the ingenuity of ether theoreticians. The question is to what extent a flowing transparent liquid (for example, in a long straight tube) will 'drag' light along with it. Flowing air, of course, drags sound along totally, but the optical situation is different: on the basis of an ether theory, it would be conceivable that there is no drag at all, since light is a disturbance of the ether and not of the liquid. Fizeau's experiment of 1851 indicated that there *was* a drag: the liquid seemed to force the ether along with it, but only partially. If the speed of light in the liquid *at rest* is u', and the liquid is set to move with velocity v, then the speed of light relative to the outside was found to be of the form

$$u = u' + kv, \qquad k = 1 - 1/n^2, \tag{4.1}$$

where k is the 'drag coefficient', a number between zero and one indicating what fraction of its own velocity the liquid imparts to the ether within, and n is the refractive index c/u' of the liquid. (For water, $k \simeq 0.44$.) Decades earlier (~ 1820) Fresnel had developed an ether-based theory of light propagation in transparent media, which predicted this result, and which now seemed vindicated. From the point of view of special relativity, however, (4.1) is nothing but the relativistic velocity addition formula! The light travels relative to the liquid with velocity u', and the liquid travels relative to the observer with velocity v, and therefore [cf. (3.7)(i)]

$$u = \frac{u' + v}{1 + u'v/c^2} \approx (u' + v)\left(1 - \frac{u'v}{c^2}\right) \approx u' + v\left(1 - \frac{u'^2}{c^2}\right) = u' + kv, \quad (4.2)$$

neglecting terms of order v^2/c^2 in the middle steps. Einstein already gave the velocity addition formula in his 1905 paper, but it took two more years before Laue made this beautiful application of it.

4.3   The Doppler effect

Waves from an approaching light-source have higher frequency than waves from a stationary source. In the frame of the source, this is because the observer moves *into* the wave train, and in the frame of the observer it is because the source, chasing its waves, bunches them up. The opposite happens when the source recedes. Similar effects exist for sound and other wave phenomena; all are named after the Austrian physicist Doppler.

In the analysis of the optical Doppler effect, relativity not only eliminated the complicating presence of an ether, but added the new element of time dilation of the moving source or observer, as the case may be. The following derivation of the relativistic formula, though not the most elegant (see, for example, Section 5.7), is nevertheless instructive.

Let a light-source P traveling through an inertial frame S have instantaneous velocity u, and radial velocity component u_r relative to the origin-observer O [see Fig. 4.1(a)]. Let the time between successive pulses be dt_0 as measured by a comoving clock *at*

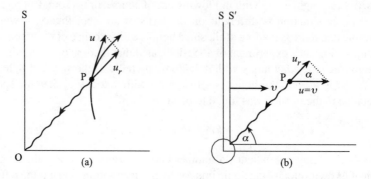

Fig. 4.1

the source, and therefore $dt_0 \gamma(u)$ in S (by time dilation). In that time, the source has increased its distance from O by $dt_0 \gamma(u)u_r$. Consequently these pulses arrive at O a time $dt = dt_0 \gamma + dt_0 \gamma u_r/c$ apart. But dt_0 and dt are inversely proportional, respectively, to the proper frequency v_0 of the source and its frequency v as observed by O. So

$$\frac{v_0}{v} = \frac{1 + u_r/c}{(1 - u^2/c^2)^{1/2}} = 1 + \frac{u_r}{c} + \frac{1}{2}\frac{u^2}{c^2} + O\left(\frac{u^3}{c^3}\right). \tag{4.3}$$

Our series expansion separates the 'pure' Doppler effect $1 + u_r/c$ from the contribution $\frac{1}{2}u^2/c^2$ of time dilation, to the order shown. The corresponding pre-relativistic formula had the Lorentz factor missing, but it was considered valid only for an observer at rest in the ether.

Note that the above argument and formula apply equally well to the visually observed frequency v of a moving clock of proper frequency v_0.

When the motion of the source is purely radial, $u_r = u$ and eqn (4.3) reduces to

$$\frac{v_0}{v} = \left(\frac{1 + u/c}{1 - u/c}\right)^{1/2}. \tag{4.4}$$

It is also useful to have a formula relating the frequencies v and v' ascribed by *two* observers O and O' at the *same* event to an incoming ray of unspecified origin. Let O and O' be associated with the usual frames S and S', respectively. Let α be the angle which the *negative* direction of the ray makes with the x-axis of S. The trick here is to assume, obviously without loss of generality, that the ray originated from a source P at rest in S' [see Fig. 4.1(b)]. Then (4.3) applies with the following specializations: $v_0 = v', u = v, u_r = v \cos \alpha$, and so

$$\frac{v'}{v} = \frac{1 + (v/c) \cos \alpha}{(1 - v^2/c^2)^{1/2}}. \tag{4.5}$$

This formula allows us, when convenient, to evaluate the Doppler ratio in one inertial frame and then transform it to the frame of interest. This is just what we would do if, for example, the source were at rest in an inertial frame S through which the *observer* moves non-uniformly. At the observer's location we would simply transform the frequency $v = v_0$ seen in S to S', the observer's instantaneous rest-frame. In this and similar cases, however, we need a criterion that tells us how much acceleration can be tolerated before the process of replacing source or observer by momentarily comoving *inertial* ones becomes invalid.

To start with, assume that both observer and source can be regarded as ideal clocks. If not, then the tolerable deviation from idealness sets first limits on the acceleration. Next note that in the time $t = 1/v$ which it takes for a complete wave to pass a point in the observer's instantaneous rest-frame, the observer, if accelerating with acceleration α, will have moved a distance $\frac{1}{2}\alpha t^2 = \frac{1}{2}\alpha/v^2$. This must be negligible compared to the wavelength $\lambda = c/v$. So we need $\alpha \ll 2cv$ in order to legitimately replace the actual observer by the momentarily comoving inertial one. The same restriction clearly holds for a possibly accelerating source. In the case of yellow light, for example, this amounts to $\alpha \ll 3 \times 10^{29}$ cm/s^2; that is, $\alpha \ll 3 \times 10^{26} g$.

A simple case in point is the frequency shift between a source at the center of a rapidly turning rotor, and an 'observer' (a piece of apparatus) at the rim, which moves, let us say, with linear velocity v. Setting $\alpha = 90°$ and $v = v_0$ in (4.5), we find $v' = v_0 \gamma(v)$ for the observed frequency of a source of proper frequency v_0. This, of course, is entirely due to the time dilation of the moving observer. The experiment was performed (with a view to demonstrating such time dilation) by Hay, Schiffer, Cranshaw, and Engelstaff in 1960, using Mössbauer resonance. Agreement with the theoretical predictions was obtained to within an expected experimental error of a few per cent. Of course, the accelerations produced on any conceivable laboratory rotor fall within the limit discussed in the last paragraph, by a huge margin. They were $\sim 6 \times 10^4 g$ in the 1960 experiment.

A longer calculation along the same lines would give us the frequency shift between two *arbitrary* points on the rotor. But we can get this more easily by using a trick. Consider a large disk which rotates at uniform angular velocity ω in an inertial frame S, and which has affixed to it a light source and receiver, at points P_0 and P_1, at distances r_0 and r_1 from the center, respectively. Since each signal from P_0 to P_1 clearly takes the same time in S, two successive signals are emitted and received with the same time difference in S, say Δt. These differences correspond to *proper* time differences $\Delta t / \gamma(\omega r_0)$ and $\Delta t / \gamma(\omega r_1)$ at P_0 and P_1, respectively. But those are inversely proportional to the frequencies v_0 and v_1, respectively, and so

$$\frac{v_0}{v_1} = \frac{\gamma(\omega r_0)}{\gamma(\omega r_1)}. \tag{4.6}$$

Time dilation is the *only* cause of the frequency shift whenever there is no radial motion between source and observer. This is the so-called *transverse Doppler effect*, and has long been considered as a possible basis for time dilation experiments. Prior to the rotor experiments, however, it was difficult to ensure exact transverseness in the motion of the sources (for example, fast-moving hydrogen ions). The slightest radial component would swamp the transverse effect. Ives and Stilwell in 1938 cleverly used a to-and-fro motion of ions whereby the first-order Doppler effect canceled out, and only the contribution of time dilation remained, which they were able to measure to an accuracy of a few per cent. Theirs was a historic experiment, being the first to demonstrate the reality of time dilation. (Yet by a curious irony of fate, Ives and Stilwell were lone hold-outs for Lorentz's ether theory and rejected special relativity!) A conceptually similar experiment, using lasers and 'two-photon spectroscopy' on a beam of fast-moving neon atoms as source,[1] has more recently verified time dilation to an accuracy of 4×10^{-5}. Even so, a 1000-fold improvement in the possible accuracy of this type of experiment is envisioned.

A canceling of the first-order contribution also occurs in the so-called *thermal Doppler effect*. Radioactive nuclei bound in a hot crystal move thermally in a rapid and

---

[1]   M. Kaivolo *et al.*, *Phys. Rev. Lett.* **54**, 255 (1985).

random way. Because of this randomness, their first-order (classical) Doppler effects average out, but not the second-order (relativistic) time dilation effects. The former cause a mere broadening of the spectral lines, the latter a shift of the entire spectrum. This shift was observed, once again by use of Mössbauer resonance, in 1960 by Rebka and Pound. The accuracy of that experiment was only about 10 per cent. However, it also yielded some evidence for the existence of approximately ideal clocks: in spite of proper accelerations up to $10^{16}g$, these nuclear 'clocks' were slowed only by the velocity factor $(1 - v^2/c^2)^{1/2}$.

4.4 Aberration

Anyone who has driven into vertically falling rain or snow knows that it seems to come at one obliquely. For similar reasons, if two observers measure the angle which an incoming ray of light makes with their relative line of motion, their measurements will generally not agree. This phenomenon is called aberration, and, of course, it was well known long before relativity.

Already in 1728 Bradley had discovered the aberration of starlight, as manifested by the apparent motion relative to the ecliptic (along a small ellipse) of each fixed star in the course of a year. He had thereby provided the first *empirical* proof of Copernicus's claim (~ 1530) that the earth orbited the sun and hence moved relative to the fixed stars.

As in the case of the Doppler effect, the relativistic formula contains a kinematic 'correction', and it applies to all pairs of observers, whereas the prerelativistic formula was simple only if one of the observers was at rest in the ether frame.

To obtain the basic aberration formulae, consider an incoming light signal whose negative direction makes angles α and α' with the x-axes of the usual two frames S and S', respectively [as in Fig. 4.1(b)]. The velocity transformation formula (3.6)(i) can evidently be applied to this signal, with $u_1 = -c \cos \alpha$ and $u_1' = -c \cos \alpha'$, yielding

$$\cos \alpha' = \frac{\cos \alpha + v/c}{1 + (v/c) \cos \alpha}. \tag{4.7}$$

Similarly, from (3.6)(ii) we obtain the alternative formula (assuming temporarily, without loss of generality, that the signal lies in the xy-plane)

$$\sin \alpha' = \frac{\sin \alpha}{\gamma[1 + (v/c) \cos \alpha]}. \tag{4.8}$$

But the most interesting version of the aberration formula is obtained by substituting from eqns (4.7) and (4.8) into the trigonometric identity

$$\tan \frac{1}{2}\alpha' = \frac{\sin \alpha'}{1 + \cos \alpha'},$$

which gives

$$\tan \frac{1}{2}\alpha' = \left(\frac{c-v}{c+v}\right)^{1/2} \tan \frac{1}{2}\alpha. \tag{4.9}$$

Since $\tan \frac{1}{2}\alpha$ is a monotonic function between $\alpha = 0$ and $180°$, it follows that α' is less than or greater than α according as v is positive or negative, respectively.

In some situations it is the aberration of *outgoing* rays that is of interest. *To obtain the corresponding formulae we simply replace c by −c in* (4.7)–(4.9), as is clear when we examine the way we obtained these formulae. Interestingly, this is equivalent to an interchange of α and α', as can be seen particularly easily from (4.9). Thus for *outgoing* rays α is smaller than α' when v is positive, and significantly so when $v \approx c$. A well-known consequence of this is the so-called *headlight effect*: a source that radiates isotropically in its rest-frame throws almost all of its radiation into a narrow forward cone at high speed (cf. Exercise 4.11). Much the same is true even if the source does *not* radiate isotropically in its rest-frame. This effect is very pronounced, for example, in the *synchrotron radiation* emitted by highly accelerated charged particles in circular orbit.

Consider an infinitesimally thin forward cone of radiation emanating from a point source; that is, one whose bounding angles α in S and α_0 in S′ (the rest-frame) are *both* infinitesimal. Then from (4.9) (with $c \mapsto -c$) we have

$$\frac{\alpha_0}{\alpha} = \left(\frac{c+v}{c-v}\right)^{1/2} =: D, \tag{4.10}$$

where we have written D for the usual Doppler ratio [cf. (4.4)]. Hence the forward ray density in S is increased by a factor D^2 over that in the rest-frame. But, additionally, the energy $h\nu$ of each photon as well as the number of photons arriving per unit time is increased by a Doppler factor ν/ν_0, which eqn (4.4) (with $-v$ instead of u for an approaching source) determines to be D. So we see that the energy current ε due to a directly approaching point source is increased by *four* Doppler factors over that of a stationary source:

$$\varepsilon = D^4\varepsilon_0 = \left(\frac{c+v}{c-v}\right)^2 \varepsilon_0. \tag{4.11}$$

4.5   The visual appearance of moving objects

Aberration causes (at least, theoretically) certain distortions in the visual appearance of extended uniformly moving objects. For, from the viewpoint of the rest-frame of the object, as the observer moves past the conical pattern of rays converging from the object to the observer's eye, rays from its different points are unequally aberrated. Alternatively, from the viewpoint of the observer's rest-frame, the light from different parts of the moving object has taken different times to reach the eye at a given instant, and thus it was emitted at different past times; the more distant points of the object consequently appear displaced relative to the nearer points in the direction opposite to the motion.

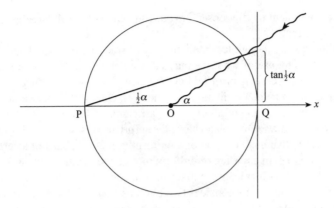

Fig. 4.2

It is in connection with the visual appearance of uniformly moving objects that the relativistic results are a little unexpected. Consider the two observers O, O′ at the origins of the usual pair of frames S and S′. Following an ingenious argument of Penrose, let us draw a sphere of *unit* diameter around each observer (see Fig. 4.2), cutting the negative and positive x-axis at points P and Q, respectively. All that the observer sees at any instant can be mapped onto this sphere (the 'sky'). Let it further be mapped from this sphere onto the tangent plane at Q (the 'screen') by stereographic projection from P. We recall that the angle subtended by an arc of a circle at the circumference is half of the angle subtended at the center, and we have accordingly labeled the diagram (for a single incident ray). From (4.9) it therefore follows that whatever the two momentarily coincident observers in S and S′ see at that moment, the corresponding images on their imaginary screens will be identical except for size ('geometrically similar').

Consider, in particular, a sphere Z at rest anywhere in the frame of observer O′. O′ sees a circular outline of Z on the sky and projects this outline as a circle or straight line onto the screen (for under stereographic projection, circles on the sphere correspond to circles or straight lines on the plane). Relative to the second observer O, of course, Z moves. Nevertheless, according to this analysis, the screen image of O will differ from that of O′ only in size, and must therefore also be circular or straight; consequently, O's 'sky' image of Z must be circular too. This shows that a moving sphere presents a circular outline to *all* observers *in spite* of length contraction! (Or rather: *because* of length contraction; for without length contraction the outline would be distorted.) By the same argument, all uniformly moving straight lines or rods will be seen as arcs of circles on the sky.

Stereographic projection is *conformal*; that is, angle-preserving. Now small conformal mappings of a small triangle are clearly similar to the original, so the same must be true of all small conformal mappings of small figures (by triangulation). It then follows from the Penrose construction that when the images of some object as seen by two coincident inertial observers *both* subtend small solid angles, then these images are similar.

Another interesting though less realistic way of studying the visual appearance of moving objects is by use of what we may call 'supersnapshots'. These are life-size snapshots made by receiving *parallel* light from an object and catching it directly on a photographic plate held at right angles to the rays. One could, for example, make a supersnapshot of the *outline* of an object by arranging to have the sun behind it and letting it cast its shadow onto a plate. Now, the surprising result (due to Terrell) is this: all supersnapshots that can be made of a uniformly moving object at a certain place and time by observers in any state of uniform motion are identical. In particular, they are all identical to the supersnapshot that can be made in the rest-frame of the object.

To prove this result, let us consider two photons P and Q traveling abreast along parallel straight paths a distance Δr apart, relative to some frame S. Let us consider two arbitrary events $\mathscr{P}$ and $\mathscr{Q}$ at P and Q, respectively. If $\mathscr{Q}$ occurs a time Δt after $\mathscr{P}$, then the space separation between $\mathscr{P}$ and $\mathscr{Q}$ is evidently $(\Delta r^2 + c^2\Delta t^2)^{1/2}$, and thus, by (2.15), the squared displacement between $\mathscr{P}$ and $\mathscr{Q}$ is $-\Delta r^2$, independently of their time separation. But if, instead of traveling abreast, Q leads P by a distance Δl, then the space separation between $\mathscr{P}$ and $\mathscr{Q}$ is $[\Delta r^2 + (c\Delta t + \Delta l)^2]^{1/2}$, and the squared displacement is *not* independent of Δt. Now, since squared displacement is an invariant and since parallel rays transform into parallel rays, it follows that two photons traveling abreast along parallel paths a distance Δr apart in one frame do precisely the same in all other frames. But a supersnapshot results from catching an array of photons traveling abreast along parallel paths on a plate orthogonal to those paths. By our present result, these photons travel abreast along parallel lines with the same space separation in *all* inertial frames, and so the equality of supersnapshots is established.

Now, *real* snapshots, or images seen by the eye, of objects that subtend sufficiently small solid angles at the eye (so that all the arriving light essentially left the object at one instant), are geometrically similar to supersnapshots taken at the same place. For consider, as in Fig. 4.3, any ray AE from the object to the eye, and a corresponding ray AP from the object to the supersnapshot. Then the (small) angle α between the normal and the incoming ray to the eye is proportional to the distance EP of the superpicture P of A, which proves our assertion. That, in turn, proves once more that all such images are similar to each other. But there will *always* be images that are *not* similar to the others, since there will always be observers for whom the object's solid angle is large: If I move sufficiently fast away from an object, I can increase its apparent angular radius, α', beyond all bounds, as eqn (4.9) (with $v < 0$) shows.

As an example, suppose the origin-observer O in S sees at $t = 0$ a small object, apparently on the y-axis ($\alpha = 90°$). Suppose this object is at rest in the usual second frame S'. The origin-observer O' in S' will see the object at an angle $\alpha < 90°$, given by (4.9). If the object is a cube with its edges parallel to the coordinate axes of S and S', O' of course sees the cube not face-on but rotated. Thus O, who might have expected to see a contracted cube face-on, also sees an uncontracted rotated cube!

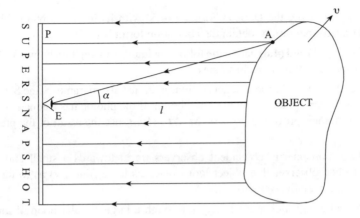

Fig. 4.3

(However, the basic rotation effect is not specifically relativistic; classically there would also be rotation, though with distortion.)

Exercises 4

4.1. If the twins $\tilde{A}$ (the stay-at-home) and $\tilde{B}$ (the traveler) in the twin-paradox 'experiment' discussed in Section 3.5 visually observe the regular ticking of each other's standard clocks, describe exactly what each sees as $\tilde{B}$ moves uniformly to a distant point B and back. [*Hint*: draw a spacetime diagram, treating $\tilde{B}$'s velocity changes as instantaneous.] Note that $\tilde{B}$ receives slow ticks for half the time and fast ticks for the other half, whereas $\tilde{A}$ receives slow ticks for *more* than half the time: hence $\tilde{A}$ receives fewer ticks, hence $\tilde{B}$ is younger when they meet again. This is one of the arguments often used to illustrate the 'non-paradoxicality' of the paradox. Where does relativity come into it—that is, why does it not lead to an age difference classically?

4.2. A source of monochromatic light of frequency v_0 is fixed at the origin of a frame S. An observer travels through S with instantaneous velocity u and radial velocity u_r relative to the origin, when observing the source. What frequency does the observer see? [*Answer*: $\gamma(u)(1 - u_r/c)v_0$.]

4.3. Relative to a given inertial frame, a source of proper frequency v_0 and an observer move along coplanar straight lines and with the same speed u. A signal travels from source to observer along a line l. At emission the source crosses l at 30° 'to the left' while at reception the observer crosses l at 30° 'to the right' (for someone looking along l). What is the frequency observed? What is the observed frequency if the second angle is 60° instead of 30°?

4.4. In the hyperbolically moving rocketship discussed in Section 3.8, a light-signal is sent from a source at rest at $X = X_1$ to an observer at rest at $X = X_2$. Prove that the Doppler shift v_1/v_2 in the light is given by X_2/X_1. [*Hint*: refer to Fig. 3.3, and transform to a frame in which the observer is at rest at reception.]

4.5. By combining the Doppler formula (4.5) with its inverse (obtained by a v-reversal transformation), re-obtain the aberration formula (4.7).

4.6. From (4.5) and (4.8) derive the following interesting relation between Doppler shift and aberration: $v'/v = \sin\alpha/\sin\alpha'$.

4.7. Let Δt and $\Delta t'$ be the time separations in the usual two frames S and S' between two events occurring at a freely moving photon. If the photon has frequencies v and v' in these frames, prove that $v/v' = \Delta t/\Delta t'$. [*Hint*: use the result of the preceding exercise.]

4.8. Two momentarily coincident observers travel towards a small and distant object. To one observer that object looks twice as large (linearly) as to the other. Prove that their relative velocity is $3c/5$.

4.9. A rocketship flies at a velocity v through a large circular hoop of radius a, along its axis. How far *beyond* the hoop is the rocketship when the hoop appears exactly lateral to the pilot? [*Note*: this '*hindsight*' effect *is not* specifically relativistic (although the exact result *is*); and it is quite analogous to hearing a high-flying aeroplane overhead long after it has actually passed.]

4.10. Writing $\alpha' = \alpha + d\alpha$, and assuming $v \ll c$, derive the approximate aberration formula $d\alpha = -(v/c)\sin\alpha$. From it, and given that the earth's orbital speed is ~ 30 km/s, prove that each fixed star's apparent position relative to the ecliptic (that is, the earth's orbit) changes in the course of one year by $\sim 41''$ (seconds of arc) in the direction parallel to the ecliptic, and by $\sim (41'')\sin\alpha$ in the orthogonal direction, α being the star's elevation relative to the ecliptic; that is, its 'ecliptic latitude'.

4.11. A source of light is fixed in S' and in that frame it emits light uniformly in all directions. Show that for large v, the light in S is mostly concentrated in a narrow forward cone; in particular, half the photons are emitted into a cone whose semi-angle is given by $\sin\theta = 1/\gamma$. This is called the '*headlight effect*'. Is the situation essentially different in the classical theory?

4.12. A plane mirror moves in the direction of its normal with uniform velocity v in a frame S. A ray of light of frequency v_1 strikes the mirror at an angle of incidence θ, and is reflected with frequency v_2 at an angle of reflection ϕ. Prove that $\tan\frac{1}{2}\theta / \tan\frac{1}{2}\phi = (c+v)/(c-v)$ and

$$\frac{v_2}{v_1} = \frac{\sin\theta}{\sin\phi} = \frac{c\cos\theta + v}{c\cos\phi - v} = \frac{c + v\cos\theta}{c - v\cos\phi}.$$

These results are of some importance in thermodynamics. [*Hint*: let the mirror be fixed in S' and write down the obvious relations in that frame; then transform to S.]

4.13. A particle moves uniformly in a frame S with velocity $\mathbf{u}$ making an angle α with the positive x-axis. If α' is the corresponding angle in the usual second frame S', prove the '*particle aberration formula*'

$$\tan\alpha' = \frac{\sin\alpha}{\gamma(v)(\cos\alpha - v/u)},$$

and compare this with (4.7) and (4.8). [*Hint*: use the velocity transformation formula as for (4.7) and (4.8).]

4.14. In a frame S, consider the equation $x \cos \alpha + y \sin \alpha = wt$. For fixed α it represents a plane propagating in the direction of its normal with speed w, that direction being parallel to the xy-plane and making an angle α with the positive x-axis. We can evidently regard this plane as a wave front. Now transform x, y, and t directly to the usual frame S$'$. From the resulting equation deduce the following aberration formula for the wave normal:

$$\tan \alpha' = \frac{\sin \alpha}{\gamma(v)(\cos \alpha - wv/c^2)}.$$

By comparison with the result of the preceding exercise, note that waves and particles traveling with the same velocity aberrate differently, unless the velocity is c. But waves with velocity $w = c^2/u$ aberrate like particles with velocity u—a result that will be of interest to us later.

4.15. A plane slab of width λ travels at uniform speed w away from the origin in a frame S, its normal making an angle α with the positive x-axis. Prove that its width λ' in the usual second frame S$'$ is given by

$$\lambda' = \lambda[\gamma^2(v)(\cos^2 \alpha + vw/c^2) + \sin^2 \alpha]^{1/2}.$$

Note that this formula also gives the transformation of the wavelength of a plane wave train traveling in the direction α with velocity w. But it does *not* simply follow from the length contraction of the width of the slab : a rod orthogonal to its bounding planes in S is *not* orthogonal to them in S$'$! [*Hint*: use the method of the preceding exercise.]

4.16. Repeat Exercise 4.14 under the assumption of the Galilean transformation. Note that there is then no aberration of the wave normal at all! To explain aberration classically, one needed to use either the concept of light-corpuscles (photons) or of 'rays' whose directions do not coincide with the wave normal except in still ether.

4.17. (i) The center of a long straight rod moves at high uniform speed along the x-axis of some inertial frame, while the rod remains parallel to the y-axis. A (real) snapshot taken by a camera on the z-axis pointing towards the origin shows the center of the rod *at* the origin. How does the rest of the picture look? [*Hint*: When did the light leave the top and the bottom? What general information do we have from the Penrose construction?] (ii) The same camera also photographs a large circular disk in the xy-plane whose center moves at high speed along the x-axis. If the snapshot shows this center at the origin, how does the rest of the picture look? [*Answer*: like a boomerang. *Hint*: The world-*map* of the disk will be a very narrow ellipse—almost a rod!]

4.18. Show that the ratio of the solid angles subtended in S and S$'$ by a thin pencil of light-rays converging on the coincident origins of these frames, its negative direction making angles α, α' with the respective x-axes, is given by

$$\frac{d\Omega}{d\Omega'} = \left(\frac{d\alpha}{d\alpha'}\right)^2 = \gamma^2(v)\left(1 + \frac{v}{c}\cos\alpha\right)^2 = \frac{v'^2}{v^2} = \frac{\sin^2\alpha}{\sin^2\alpha'},$$

[*Hint*: without loss of generality, consider a solid angle with circular normal cross section and recall the argument associated with Fig. 4.2.]

4.19. A cube with its edges parallel to the coordinate axes moves with Lorentz factor 3 along the x-axis of an inertial frame S. A 'supersnapshot' of this cube is made in a plane $z = $ const by means of light-rays parallel to the z-axis. Make an exact scale drawing of this supersnapshot.

4.20. Uniform parallel light is observed in two arbitrary inertial frames, say the usual frames S and S', with the light *not* necessarily parallel to the x-axes. If v and v' are the respective frequencies of the light in S and S', prove that the ratio $\rho : \rho'$ of the respective photon densities (number of photons per unit volume) is $v : v'$. [*Hint*: supersnapshots.]

5

Spacetime and four-vectors

5.1 The discovery of Minkowski space

We have now reached a point at which further progress can be greatly clarified *and* simplified by the introduction of a new entity combining space and time and a corresponding mathematical technique that is tailor-made for special relativity. The new entity is Minkowski's 4-dimensional 'spacetime' (now usually called *Minkowski space*) and the mathematical technique is the '4-vectors' calculus *in* that spacetime. (Later we shall recognize this as only a part of a more comprehensive '4-*tensor*' calculus.)

We have already briefly mentioned in Section 2.2 how (in 1907) Minkowski replaced Einstein's (and Newton's) picture of inertial frames moving uniformly through space by a *static* picture in four dimensions. It was Einstein who taught us to focus our attention on *events* (x, y, z, t)—the 'atoms' of (physical) history. Minkowski now pictured them as *points* in a 4-dimensional space he called spacetime—a stack of 3-dimensional world-maps. This much can certainly be done just as well in Newton's theory. But what is new in special relativity is the existence of an invariant 4-dimensional *differential form* [cf. eqn (2.7)]

$$\mathbf{ds}^2 := c^2 \, dt^2 - dx^2 - dy^2 - dz^2. \tag{5.1}$$

Apart from the strange mix of plus and minus signs this is reminiscent of the Euclidean 'metric' form

$$\mathbf{dr}^2 = dx^2 + dy^2 + dz^2, \tag{5.2}$$

which is entirely responsible for the structure of Euclidean space and thus for *all* of Euclidean geometry! Minkowski now realized that the $\mathbf{ds}^2$ of (5.1) provides a natural metric structure for spacetime, with its resultant rich geometry, and, above all, with its resultant vector and tensor calculus, which he quickly proceeded to develop. Minkowski was so struck by his discovery that the proclaimed: 'Henceforth space by itself and time by itself are doomed to fade away into mere shadows, and only a kind of union of the two will preserve an independent reality'.

Minkowski's spacetime is far more than a mere mathematical artifice. Relativity made the older concepts of absolute space and absolute time untenable, and yet they had served as a deeply ingrained framework in our minds on which to 'hang' the rest of physics. Spacetime is their modern successor: it is the new absolute framework for exact physical thought. This has great heuristic value. Moreover, by automatically combining not only space and time, but also, as we shall see, momentum and energy, force and power, electric and magnetic fields, etc., Minkowski's mathematical formalism illuminates some profound physical interconnections.

But, once again, Freud's maxim (few completely new ideas!—cf. Section 1.9) was confirmed. The same Poincaré who had in some ways anticipated Einstein with the relativity principle, later also anticipated Minkowski with 4-dimensional arguments, yet never got as far as metric spacetime. Still, Poincaré's ideas (consciously or not) may well have served those bolder spirits as seeds from which sprang their own breakthroughs.

In the mathematical model that is SR, smooth, flat, completely uniform Minkowski space fills the whole world—past, present, and future—and constitutes an *absolute* structure that determines all inertial frames, including their times. As such, it is subject to the same main criticism that Einstein brought against absolute space: it acts, but cannot be acted on. Only in the further development of the theory, in 'general' relativity, did Einstein show that gravitating matter actually *does* act on spacetime: it distorts it, giving it curvature.

5.2   Three-dimensional Minkowski diagrams

Minkowski geometrized special relativity. One of the advantages of geometry is that it allows one to visualize relations. But in order to develop a good visual intuition, one needs to play with pictures. The pictures relevant to Minkowskian geometry are *Minkowski diagrams*. We have, in fact, already come across examples of such diagrams (in two dimensions) in Section 2.9. A slight generalization will take us to three dimensions.

The pictures we can draw on a piece of paper of 2-dimensional Euclidean figures like triangles, circles, etc., are direct; that is, they are the actual objects. On the other hand, the best we can do for figures in Minkowski space is to *map* them onto Euclidean space, as did Mercator with his flat map of the curved surface of the earth. Such maps necessarily *distort* metric relations and one has to learn to compensate for this distortion.

With Minkowski diagrams, even 3-dimensional ones, one also has to learn to compensate for the unavoidable absence of at least one spatial dimension (usually the z-dimension), reading circles as spheres, planes as Euclidean 3-spaces, etc. A 3-dimensional Minkowski diagram like Fig. 5.1 is a perspective rendering onto a plane piece of paper of structures in Euclidean 3-space that are *mappings* of structures in Minkowskian 3-space (x, y, t). For example, Fig. 5.1 shows the calibrating hyperboloids of revolution

$$c^2 t^2 - x^2 - y^2 = \pm 1. \tag{5.3}$$

These are obtained by rotating the calibrating hyperbolae of Fig. 2.6 about the t-axis. The inner (2-sheeted) hyperboloid is the locus of events satisfying $\Delta s^2 = 1$, while the outer (1-sheeted) hyperboloid represents all events satisfying $\Delta s^2 = -1$, relative

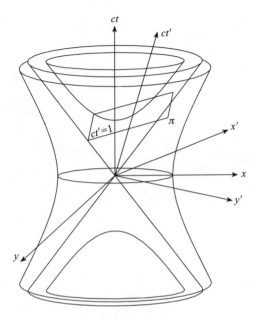

Fig. 5.1

to the origin. Note that we now work in ordinary units (not making $c = 1$) and label the vertical axis accordingly ct instead of t. Then particle worldlines, as before, must subtend angles less than $45°$ with the vertical, since the slope relative to the t-axis, as before, measures v/c. Of course, any particle moving uniformly in any inertial frame has a straight worldline in the diagram. Straight lines at $45°$ to the vertical represent photons.

The present diagram can accommodate all inertial frames with relative motions parallel to the single spatial (x, y) plane it describes. Without loss of generality we can have the spatial origins of all these frames coincide at time zero. The worldline of the spatial origin $x' = y' = 0$ of any such frame S' is also the time axis $t' = \text{var}$ corresponding to S' in Minkowski space. The simultaneities $t' = \text{const}$ are all the planes parallel to the tangent plane π to the hyperboloid where the t'-axis pierces it—by an argument quite analogous to that of Exercise 2.15. And the spacetime x'- and y'-axes are parallel to these simultaneities and pass through the spacetime origin.

5.3   Light cones and intervals

The most fundamental invariant structure in Minkowski space is the set of so-called *light cones* or *null cones*—one at each point (event) $\mathcal{P}$. These are the bundles of worldlines of all the photons passing through $\mathcal{P}$, or, equivalently, the loci of events that can send light to or receive light from $\mathcal{P}$, or yet, equivalently, the loci of events

Fig. 5.2

at zero interval from $\mathscr{P}$:

$$c^2 \Delta t^2 - \Delta x^2 - \Delta y^2 - \Delta z^2 = 0. \tag{5.4}$$

At the origin, and under suppression of the z-dimension, this equation becomes $c^2 t^2 - x^2 - y^2 = 0$ and is recognized as the 45°-cone asymptotic to the calibrating hyperboloids in Fig. 5.1. Regarded sequentially in time t in the frame S, this cone represents a circular light-front converging onto the origin and diverging away from it again. Because of the invariance of the defining equation (5.4) under an active LT (and also directly from the physics), the light cone presents this same aspect in *any* inertial frame. In full dimensions, of course, it is not a circular but a *spherical* light-front that converges onto and then diverges from the event, but 'light cone' is still the term used for its locus.

The light cones at each event imprint a 'grain' onto spacetime, which has no analog in isotropic Euclidean space, but is somewhat reminiscent of crystal structure. Light travels along the grain, and particle worldlines have to be within the light cone at each of their points (cf. Fig. 5.2).

It is of some interest to see what happens to the light cones as we formally let $c \to \infty$. For this purpose we must use t rather than ct as the time variable. The light cones (in normal units) will then have slope c relative to the time axis and consequently flatten out completely as $c \to \infty$: they have been transformed into the fundamental absolute structure of *Newtonian* spacetime, namely the absolute simultaneities. Thus while Newtonian spacetime splits naturally into a stack of 3-spaces and absolute time, Minkowski spacetime is *essentially* 4-dimensional, as is already indicated by the intertwining of x and t in the Lorentz transformation.

Next, we investigate the physical meaning of the fundamental invariant form (5.1) of Minkowski space, or, better, of its finite version

$$\mathbf{\Delta s}^2 = c^2 \Delta t^2 - \Delta x^2 - \Delta y^2 - \Delta z^2 = \Delta t^2 \left(c^2 - \frac{\Delta r^2}{\Delta t^2} \right), \qquad (5.5)$$

where the Δ terms refer to two not necessarily neighboring events $\mathcal{P}$ and $\mathcal{Q}$. The meaning of $\mathbf{\Delta s}^2$ depends on its sign. The simplest case is $\mathbf{\Delta s}^2 = 0$, which means precisely that $\mathcal{P}$ and $\mathcal{Q}$ are connectible by a light signal.

When $\mathbf{\Delta s}^2 > 0$, then, in any inertial frame, $\Delta r^2 / \Delta t^2 < c^2$, so a clock can be sent at uniform speed from $\mathcal{P}$ to $\mathcal{Q}$ or vice versa. In *its* rest-frame S', $\mathcal{P}$ and $\mathcal{Q}$ occur at the same location, so $\mathbf{\Delta s}^2 = c^2 \Delta t'^2$. Consequently the *interval* $\Delta s = |\mathbf{\Delta s}^2|^{1/2}$ between $\mathcal{P}$ and $\mathcal{Q}$ in that case is c times the time elapsed ('proper time') on a clock that freely falls from one to the other.

Note, by reference to our earlier result (3.3), that the time elapsed on an *ideal* clock moving *arbitrarily* between two events $\mathcal{P}$ and $\mathcal{Q}$ is given by

$$\Delta \tau = \int_{\mathcal{P}}^{\mathcal{Q}} (dt^2 - dr^2/c^2)^{1/2} = \frac{1}{c} \int_{\mathcal{P}}^{\mathcal{Q}} ds, \qquad (5.6)$$

which reduces to our present result in the case of uniform motion, since then $\Delta s = \int ds$ (cf. Exercise 5.1). Equation (5.6) interprets ds in its most important practical context.

Lastly, suppose $\mathbf{\Delta s}^2 < 0$; this corresponds to $\Delta r^2 / \Delta t^2 > c^2$ (that is, to two events on a superluminal signal) for which, as we have seen after eqn (2.2), there always exists an inertial frame in which the events are simultaneous. In that frame, say S', $\mathbf{\Delta s}^2 = -\Delta r'^2$. So Δs is the spatial separation between the two events in the frame in which they are simultaneous. From eqn (5.5) it is clear that this is also the *shortest* spatial separation assigned to $\mathcal{P}$ and $\mathcal{Q}$ in *any* inertial frame.

The light cone at each event $\mathcal{P}$ effects a very important, causal partition of all other events relative to $\mathcal{P}$ (see Fig. 5.3). All events on and within the *future cone* (that is, the top half of it) can be influenced by $\mathcal{P}$; for they can receive signals from $\mathcal{P}$, since they are all reachable at speeds $u \leq c$. Moreover, as we have already seen in Section 2.10 (fourth paragraph from end), along any such signal all observes agree that events in this region happen *after* $\mathcal{P}$: those events are therefore said to constitute the *absolute* (or *causal*) *future* of $\mathcal{P}$. Similar remarks apply to the events on or within $\mathcal{P}$'s *past cone* (the bottom half of the light cone): they absolutely precede and can influence $\mathcal{P}$ and thus constitute $\mathcal{P}$'s *absolute* (or *causal*) *past*. These two classes of events that are causally related to $\mathcal{P}$ are characterized by $\mathbf{\Delta s}^2 \geq 0$. On the other hand, no event in the region *outside* the light cone, characterized by $\mathbf{\Delta s}^2 < 0$, can influence $\mathcal{P}$ or be influenced by it, since that would require superluminal communication. But, as we have seen, each event in this region is simultaneous with $\mathcal{P}$ in *some* inertial frame, and accordingly we call it the *causal present* of $\mathcal{P}$.

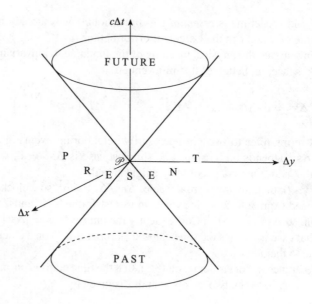

Fig. 5.3

5.4   Three-vectors

Before introducing 4-vectors, it will be well to review the salient features of 3-vectors; that is, of 'ordinary' vectors. Anyone who has done 3-dimensional geometry or mechanics will be aware of the power of the vector calculus. Just what *is* that power? First, of course, there is power simply in abbreviation. A comparison of Newton's second law in scalar and in vector form makes this clear:

$$\left.\begin{array}{l} f_1 = m(\mathrm{d}^2x/\mathrm{d}t^2) \\ f_2 = m(\mathrm{d}^2y/\mathrm{d}t^2) \\ f_3 = m(\mathrm{d}^2z/\mathrm{d}t^2) \end{array}\right\} \mathbf{f} = m\mathbf{a}.$$

This is only a very mild example. In looking through older books on physics or geometry, one often wonders how anyone could have seen the underlying physical reality through the triple maze of coordinate-dependent scalar equations. Yet abbreviation, though in itself often profoundly fruitful, is only one aspect of the matter. The other is the abolition of the coordinate-dependence just mentioned: vectors are absolute.

In studying the geometry and physics of 3-dimensional Euclidean space, each 'observer' can set up standard coordinates x, y, z (right-handed orthogonal Cartesians) with any point as origin and with any orientation. Does this mean that there are as many spaces as there are coordinate systems? No: underlying all subjective 'observations' there is a single space with *absolute* elements and properties, namely those on which all observers agree, such as specific points and specific straight lines, distances between specific points and lines, angles between lines, etc. Vector calculus treats these absolutes in a coordinate-free way that makes their absoluteness evident. All relations that can be expressed vectorially, such as $\mathbf{a} = \mathbf{b} + \mathbf{c}$, or $\mathbf{a} \cdot \mathbf{b} = 5$, are necessarily absolute. On the other hand, an observer's statement like $f_1 = f_2$ (about a force), which has no vector formulation, is of subjective interest only.

A (Cartesian) 3-vector $\mathbf{a}$ can be defined as a number-triple (a_1, a_2, a_3) which depends on the choice of a Cartesian reference frame $\{x, y, z\}$. Technically, $\mathbf{a}$ is a *mapping* from the set of Cartesian coordinate systems $\{x, y, z\}$ to the space $\mathbf{R}^3$ of number triples, $\mathbf{a} : \{x, y, z\} \rightarrow (a_1, a_2, a_3)$. The various vector operations can be defined via these 'components'; for example, $\mathbf{a} + \mathbf{b} = (a_1 + b_1, a_2 + b_2, a_3 + b_3)$, $\mathbf{a} \cdot \mathbf{b} = a_1 b_1 + a_2 b_2 + a_3 b_3$. But they can be *interpreted* absolutely (that is, coordinate-independently): for example, $\mathbf{a}$ itself as having a certain length and direction, $\mathbf{a} + \mathbf{b}$ by the parallelogram rule, etc. Only operations that have absolute significance are admissible in vector calculus. To check a vector equation, observers could proceed *directly* by measuring absolutes like lengths and angles, but they would then really be 'superobservers'. The observers we have in mind simply possess a standard coordinate lattice, and in fact they can be identified with such a lattice. Thus they can only read off components of all relevant vectors. To check a relation like $\mathbf{a} = \mathbf{b} + \mathbf{c}$, they would each obtain a set of three scalar equations $a_i = b_i + c_i$ $(i = 1, 2, 3)$ which differ from observer to observer; but either all sets are false, or all are true. *A vector (component) equation that is true in one coordinate system is true in all*: this is the most basic feature of the vector calculus. Speaking technically, vector (component) equations are form-invariant under the rotations about the origin and the translations of axes (and combinations thereof) which relate the different 'observers' in Euclidean space and which in fact constitute the 'relativity group' of Euclidean geometry. The reason for this form-invariance will appear presently.

The prototype of a 3-vector is the displacement vector $\mathbf{\Delta r} = (\Delta x, \Delta y, \Delta z)$ joining two points in Euclidean space. Under a translation of axes its components remain unchanged, and under a rotation about the origin they suffer the same (linear-homogeneous) transformation as the coordinates themselves [cf. Section 2.8(iv)], say

$$\Delta x' = \alpha_{11}\Delta x + \alpha_{12}\Delta y + \alpha_{13}\Delta z$$
$$\Delta y' = \alpha_{21}\Delta x + \alpha_{22}\Delta y + \alpha_{23}\Delta z \qquad (5.7)$$
$$\Delta z' = \alpha_{31}\Delta x + \alpha_{32}\Delta y + \alpha_{33}\Delta z,$$

where the αs are certain functions of the angles specifying the rotation. *Any quantity having three components (a_1, a_2, a_3) which undergo exactly the same transformation (5.7) as $(\Delta x, \Delta y, \Delta z)$ under the contemplated changes of coordinates (rotations and translations) is said to constitute a 3-vector.* This property of 3-vectors is often not stated explicitly in texts; but it is implicit in the usual assumption that each 3-vector quantity can be *represented* by a displacement vector (a 'directed line segment') in Euclidean space. Note that the position 'vector' $\mathbf{r} = (x, y, z)$ of a point relative to the origin is a vector *only* under rotations and not under translations! The *zero-vector* defined in all coordinate systems alike as $\mathbf{0} = (0, 0, 0)$ is a 3-vector according to (5.7); it is usually (if incorrectly) written as 0, as in $\mathbf{a} = 0$.

From (5.7) it follows that if the components of two 3-vectors are equal in one coordinate system, they are equal in all coordinate systems; for both sets of new components are the same linear combination of the old components. So if both sides of a suspected equation are known to be vectors, the equation will be established if it is shown to be true in *one* coordinate system.

If $\mathbf{a} = (a_1, a_2, a_3)$ and $\mathbf{b} = (b_1, b_2, b_3)$ separately transform like $(\Delta x, \Delta y, \Delta z)$, then so does $\mathbf{a} + \mathbf{b} = (a_1 + b_1, a_2 + b_2, a_3 + b_3)$, because of the linearity of (5.7); hence sums of vectors are vectors. Similarly, if k is a *scalar invariant* (often shortened to just 'scalar' or 'invariant'—that is, a real number independent of the coordinate system), then clearly $k\mathbf{a}$, defined as (ka_1, ka_2, ka_3), is a vector, again from (5.7).

If (x, y, z) is the current point of a curve in space, and each of the three coordinates is expressed as a function of the arc l, then the 'unit tangent'

$$\mathbf{t} = \left(\frac{\mathrm{d}x}{\mathrm{d}l}, \frac{\mathrm{d}y}{\mathrm{d}l}, \frac{\mathrm{d}z}{\mathrm{d}l} \right) \tag{5.8}$$

is a vector, usually written as $\mathrm{d}\mathbf{r}/\mathrm{d}l$. This can be seen by considering the transformation of the coordinates x, y, z themselves (which differs from the pattern (5.7) only by the possible presence of additive constants at the end of each line) and differentiating both sides with respect to l. Passing from geometry to Newtonian mechanics, consider a particle moving along this curve; it will now be convenient to express x, y, z as functions of the time t, but by the same argument it follows that the velocity $\mathbf{u} = \mathrm{d}\mathbf{r}/\mathrm{d}t$ is a vector. One easily sees, by differentiating the transformation pattern (5.7) satisfied by *any* vector, that the derivative $\mathrm{d}\mathbf{a}/\mathrm{d}\theta := (\mathrm{d}a_1/\mathrm{d}\theta, \mathrm{d}a_2/\mathrm{d}\theta, \mathrm{d}a_3/\mathrm{d}\theta)$ of a vector with respect to any scalar invariant θ is a vector. Thus the acceleration $\mathbf{a} = \mathrm{d}\mathbf{u}/\mathrm{d}t = (\mathrm{d}u_1/\mathrm{d}t, \mathrm{d}u_2/\mathrm{d}t, \mathrm{d}u_3/\mathrm{d}t)$ is a vector. Multiplying $\mathbf{u}$ and $\mathbf{a}$ by the mass m (a scalar) yields two more vectors, the momentum $\mathbf{p} = m\mathbf{u}$ and the force $\mathbf{f} = m\mathbf{a}$. Note how the five basic vectors $\mathbf{t}, \mathbf{u}, \mathbf{a}, \mathbf{p}, \mathbf{f}$ all arise by differentiation and scalar multiplication from the coordinates themselves.

Associated with each 3-vector $\mathbf{a} = (a_1, a_2, a_3)$ there is a very important scalar, its *magnitude*, written $|\mathbf{a}|$ or simply a, and defined by

$$a^2 = a_1^2 + a_2^2 + a_3^2, \quad a \geq 0. \tag{5.9}$$

That this is on invariant follows at once from the invariance of the metric $\Delta x^2 + \Delta y^2 + \Delta z^2$ of Euclidean space under rotations and translations, since (a_1, a_2, a_3) transforms just like $(\Delta x, \Delta y, \Delta z)$. If $\mathbf{a}$ and $\mathbf{b}$ are vectors, then $\mathbf{a} + \mathbf{b}$ is a vector, whose magnitude must be invariant; but

$$|\mathbf{a} + \mathbf{b}|^2 = (a_1 + b_1)^2 + (a_2 + b_2)^2 + (a_3 + b_3)^2$$
$$= a^2 + b^2 + 2(a_1 b_1 + a_2 b_2 + a_3 b_3),$$

and since a^2 and b^2 are invariant, it follows that the 'scalar product'

$$\mathbf{a} \cdot \mathbf{b} = a_1 b_1 + a_2 b_2 + a_3 b_3 \qquad (5.10)$$

is invariant; that is, coordinate-independent. If this were our first encounter with vectors, we would now look for the absolute (that is, coordinate-independent) significance of $\mathbf{a} \cdot \mathbf{b}$, which a priori must exist; and by going to a specific coordinate system (for example, that in which $a_2 = a_3 = 0$), we would soon discover it. The product $\mathbf{a} \cdot \mathbf{b}$, as defined by (5.10), is easily seen to obey the commutative law $\mathbf{a} \cdot \mathbf{b} = \mathbf{b} \cdot \mathbf{a}$, the distributive law $\mathbf{a} \cdot (\mathbf{b} + \mathbf{c}) = \mathbf{a} \cdot \mathbf{b} + \mathbf{a} \cdot \mathbf{c}$, and the Leibniz differentiation law $\mathrm{d}(\mathbf{a} \cdot \mathbf{b}) = \mathrm{d}\mathbf{a} \cdot \mathbf{b} + \mathbf{a} \cdot \mathrm{d}\mathbf{b}$. Also note that $\mathbf{a} \cdot \mathbf{a} = a^2$ (which we may write as $\mathbf{a}^2 = a^2$, so that $\mathbf{a} \cdot \mathrm{d}\mathbf{a} = a\,\mathrm{d}a$.

5.5 Four-vectors

We are now ready to develop the calculus of 4-vectors by close analogy with 3-vectors. And it is easy to see what we are going to get: We get a calculus of vectors in Minkowski space, the equations of which, when 'projected' into the various inertial frames (that is, when written in component form) are form-invariant under *general* Lorentz transformations (cf. Section 2.7, third paragraph from end). So they automatically possess the property required by the relativity principle of all physical laws! This often enables us to recognize by its vector form alone that a given or proposed law is Lorentz-invariant, and so assists us greatly in the construction of relativistic physics. However, let it be said at once that not *all* Lorentz-invariant laws are expressible as relations between 4-vectors and scalars; some require 4-*tensors*, and, in quantum mechanics, even *spinors*.

The prototye of a 4-*vector* $\mathbf{A} = (A_1, A_2, A_3, A_4)$ *is the displacement* 4-*vector* $\mathbf{\Delta R} = (\Delta x, \Delta y, \Delta z, \Delta ct)$ between two events, and the defining property of a 4-vector is that it transforms like $\mathbf{\Delta R}$. Note that we here take as our fourth coordinate ct rather than t. The advantages of this are threefold: (i) all four components of a 4-vector then have the same physical dimensions; (ii) in Minkowski space the choice ct is preferable since it gives light signals the standard slope unity; and (iii) it is the convention adopted by most other authors—many of whom, however, take ct as the *first* rather than the last coordinate. The small disadvantage of this convention is that it forces us to carry along some extra cs, but in one's private calculations this is usually irrelevant anyway, since one uses 'relativistic' units making $c = 1$.

Each 4-vector, then, can be *represented* by a directed line segment in Minkowski space. The admissible coordinate systems are the 'standard' coordinates of inertial observers, and hence the relevant transformations are the general Lorentz transformations (compounded of translations, rotations about the spatial origin, and standard LTs). These give rise to a 4-dimensional analog of (5.7) for any pair of inertial frames, with 16 constant αs instead of nine. *We shall consistently use lower-case boldface letters to denote 3-vectors, and boldface capitals to denote 4-vectors.* Under spatial and temporal translations, the components of $\mathbf{\Delta R}$ (and thus of *any* 4-vector) are unchanged; under spatial rotations about the origin, the first three components of $\mathbf{\Delta R}$ (and thus of *any* 4-vector) transform like a 3-vector, while the last component is unchanged; and under a standard LT, the components of any 4-vector $\mathbf{A}$ transform as do those of the prototype $(\Delta x, \Delta y, \Delta z, \Delta ct)$—which differs from the scheme (2.11) only by some c-factors:

$$\Delta x' = \gamma\left(\Delta x - \frac{v}{c}\Delta ct\right), \qquad \Delta y' = \Delta y,$$

$$\Delta z' = \Delta z, \qquad \Delta ct' = \gamma\left(\Delta ct - \frac{v}{c}\Delta x\right). \tag{5.11}$$

Thus $A_1' = \gamma(A_1 - (v/c)A_4)$, $A_2' = A_2$, etc. The position 4-'vector' $\mathbf{R} = (x, y, z, ct)$ is a 4-vector *only* under homogeneous general LTs; that is, those that leave the coordinates of the event $(0, 0, 0, 0)$ unchanged. The *zero-vector* $\mathbf{0} = (0, 0, 0, 0)$ is a true 4-vector. Sums, scalar multiples, and scalar derivatives of 4-vectors are defined by analogy with 3-vectors and are recognized as 4-vectors.

The *square* of any 4-vector $\mathbf{A} = (A_1, A_2, A_3, A_4)$ is defined by

$$\mathbf{A}^2 = A_4^2 - A_1^2 - A_2^2 - A_3^2, \tag{5.12}$$

and its invariance follows from that of the square of the prototype $\mathbf{\Delta R} = (\Delta x, \Delta y, \Delta z, \Delta ct)$. Denoting the latter alternatively by $\mathbf{\Delta s}$ now at last justifies the notation (5.5) that we have used all along. The *magnitude* or 'length' of a 4-vector $\mathbf{A}$ is written as $|\mathbf{A}|$ or A and is defined by

$$A = |\mathbf{A}^2|^{1/2} \geq 0. \tag{5.13}$$

Precisely as for (5.10) we can now deduce from the invariance of $|\mathbf{A}|$, $|\mathbf{B}|$, and $|\mathbf{A}+\mathbf{B}|$ that the *scalar product* $\mathbf{A} \cdot \mathbf{B}$ defined by

$$\mathbf{A} \cdot \mathbf{B} = A_4 B_4 - A_1 B_1 - A_2 B_2 - A_3 B_3 \tag{5.14}$$

is also invariant. And if $\mathbf{A}, \mathbf{B}, \mathbf{C}$ are arbitrary 4-vectors, one then verifies at once from the definitions that

$$\mathbf{A} \cdot \mathbf{B} = \mathbf{B} \cdot \mathbf{A}, \qquad \mathbf{A} \cdot (\mathbf{B} + \mathbf{C}) = \mathbf{A} \cdot \mathbf{B} + \mathbf{A} \cdot \mathbf{C}, \qquad \mathbf{A} \cdot \mathbf{A} = \mathbf{A}^2, \tag{5.15}$$

$$d(\mathbf{A} \cdot \mathbf{B}) = d\mathbf{A} \cdot \mathbf{B} + \mathbf{A} \cdot d\mathbf{B}. \tag{5.16}$$

We are now ready to construct our first two real-life 4-vectors in analogy to the velocity $\mathbf{u}$ and the acceleration $\mathbf{a}$ of 3-vectors calculus. An important scalar along the worldline of a particle is the differential interval ds. However, it is often convenient to work instead with the corresponding *proper time* interval dτ, defined by

$$d\tau^2 = \frac{ds^2}{c^2} = dt^2 - \frac{dx^2 + dy^2 + dz^2}{c^2}. \tag{5.17}$$

This gets its name from the fact that $d\tau$ coincides with the time indicated by an ideal clock attached to the particle, as we have already seen in eqn (5.6). We shall not be surprised, therefore, to find $d\tau$ appearing in many relativistic formulae where in the classical analog there is a dt. For example, if (x, y, z, ct) are the coordinates of a moving particle, we find by differentiating with respect to τ the 4-dimensional analog of (5.7) for the coordinates themselves, that

$$U = \frac{dR}{d\tau} = \left(\frac{dx}{d\tau}, \frac{dy}{d\tau}, \frac{dz}{d\tau}, \frac{dct}{d\tau}\right) \tag{5.18}$$

is a 4-vector. This is called the 4-*velocity* of the particle. It really is the analog of *both* $t = dr/dl$ and $u = dr/dt$ and so U can also be regarded as the tangent vector of the worldline of the particle in spacetime. Now from (5.17),

$$\frac{d\tau^2}{dt^2} = 1 - \frac{u^2}{c^2},$$

u being the speed of the particle, whence (not surprisingly)

$$\frac{dt}{d\tau} = \gamma(u). \tag{5.19}$$

Since $dx/d\tau = (dx/dt)(dt/d\tau) = u_1\gamma(u)$, etc., we see that

$$U = \gamma(u)(u_1, u_2, u_3, c) = \gamma(u)(\mathbf{u}, c). \tag{5.20}$$

We shall often recognize in the first three components of a 4-vector the components of a familiar 3-vector (or a multiple thereof), and in such cases we adopt the notation exemplified by (5.20).

As in 3-vector theory, scalar derivatives of 4-vectors (defined by differentiating the components) are themselves 4-vectors. Thus, in particular,

$$A = \frac{dU}{d\tau} = \frac{d^2R}{d\tau^2} \tag{5.21}$$

is a 4-vector, called the *4-acceleration*. Its relation to the 3-acceleration $\mathbf{a}$ is not quite as simple as that of U to $\mathbf{u}$; by (5.19), (5.20) and (5.21) we have

$$A = \gamma\frac{dU}{dt} = \gamma\frac{d}{dt}(\gamma\mathbf{u}, \gamma c) = \gamma\left(\frac{d\gamma}{dt}\mathbf{u} + \gamma\mathbf{a}, \frac{d\gamma}{dt}c\right) \tag{5.22}$$

and it is seen from this that the components of A *in the instantaneous rest-frame* of the particle ($u = 0$) are given by

$$A = (\mathbf{a}, 0), \tag{5.23}$$

since the derivative of γ contains a factor u [cf. (2.10)(ii)]. Thus $A = 0$ if and only if the 3-acceleration in the rest-frame vanishes. The 4-velocity U, on the other hand, never vanishes.

In fact, from (5.20) and (5.12) (in which **A** is *any* 4-vector) we find

$$\mathbf{U}^2 = c^2. \tag{5.24}$$

But it is even easier to get this result by first putting $\mathbf{u} = 0$ in (5.20)! Why does this work? Because $\mathbf{U}^2$ is an invariant and can thus be evaluated in *any* IF—in particular, therefore, in the particle's rest-frame where $\mathbf{u} = 0$. The same trick yields

$$\mathbf{A}^2 = -\mathbf{a}_0^2 =: -\alpha^2, \tag{5.25}$$

where we have written $\mathbf{a}_0$ for the 3-acceleration in the rest-frame and α for the magnitude of $\mathbf{a}_0$, namely, the proper acceleration.

Once again using the rest-frame, we find

$$\mathbf{U} \cdot \mathbf{A} = 0; \tag{5.26}$$

that is, the 4-acceleration is always 'orthogonal' to the 4-velocity. Alternatively, we could have differentiated eqn (5.24) with respect to proper time, applying the Leibniz rule (5.16) to its LHS $\mathbf{U} \cdot \mathbf{U}$:

$$\mathbf{A} \cdot \mathbf{U} + \mathbf{U} \cdot \mathbf{A} = 2\mathbf{U} \cdot \mathbf{A} = 0.$$

As in Euclidean space, we would expect the general scalar product to have some absolute significance. But the anisotropy of Minkowski space complicates matters (cf. Exercises 5.15, 5.16). As an important special case, however, consider the scalar product of the 4-velocities $\mathbf{U}, \mathbf{V}$ of two particles which either move uniformly or whose worldlines just cross. Evaluating this product in the rest-frame of the **U**-particle, relative to which the **V**-particle has velocity v, say, we find

$$\mathbf{U} \cdot \mathbf{V} = c^2 \gamma(v); \tag{5.27}$$

that is, $\mathbf{U} \cdot \mathbf{V}$ is c^2 times the Lorentz factor of the *relative* velocity of the corresponding particles.

We have left a somewhat more lengthy but nevertheless rewarding calculation to the end. By working out $\mathbf{A}^2$ in the *general* frame and comparing with eqn (5.25), we shall obtain a formula for α that generalizes our previous 1-dimensional result (3.17). From (5.22) we have

$$\mathbf{A}^2 = \gamma^2[\dot{\gamma}^2 c^2 - (\dot{\gamma}\mathbf{u} + \gamma\mathbf{a})^2], \tag{5.28}$$

where the dots denote differentiation with respect to t. Using the relation $\dot{\gamma} = \gamma^3 u\dot{u}/c^2$ [cf. (2.10)(ii)] and the familiar 3-vector results $\mathbf{u}^2 = u^2$, $\mathbf{u} \cdot \dot{\mathbf{u}} = u\dot{u}$, and $(\mathbf{u} \times \mathbf{a})^2 = u^2 a^2 - (\mathbf{u} \cdot \mathbf{a})^2$, we find, from (5.28), the desired formula in two alternative versions:

$$\alpha^2 = -\mathbf{A}^2 = \gamma^2[\dot{\gamma}^2 u^2 + 2\gamma\dot{\gamma}u\dot{u} + \gamma^2 a^2 - \dot{\gamma}^2 c^2]$$
$$= \gamma^6 u^2 \dot{u}^2/c^2 + \gamma^4 a^2 = \gamma^6[a^2 - c^{-2}(\mathbf{u} \times \mathbf{a})^2]. \tag{5.29}$$

When **a** is parallel to **u**, this yields our previous result $\alpha = \gamma^3 a$ and when **a** is orthogonal to **u** (that is, whenever $\dot{u}$ vanishes), it yields $\alpha = \gamma^2 a$. This, in particular, is interesting for fast circular motion, for example, in proton storage rings: it shows how the proper acceleration exceeds the lab acceleration $u^2/r \approx c^2/r$ (r = radius) by the possibly quite large factor γ^2.

5.6 The geometry of four-vectors

We have seen (in Section 5.3) how the light cone partitions events relative to its vertex into three classes, according to the sign of Δs^2. The same partitioning carries over to the displacement vector Δs itself and, by extension, to *all* 4-vectors, since they can be represented by displacements. We call a 4-vector **A** *timelike* if $\mathbf{A}^2 > 0$, *null* if $A^2 = 0$, and *spacelike* if $A^2 < 0$. In the first case, **A** points *into* the light cone, in the second case it points *along* the light cone, and in the third case it points *outside* the light cone.

Timelike and null vectors share certain properties and are sometimes classed together as *causal* vectors. In particular, the sign of their fourth component is invariant, since all observers agree on the time sequence along the corresponding displacements. One can therefore invariantly subdivide causal vectors **A** into those that are 'future-pointing' and those that are 'past-pointing', according as $A_4 \gtrless 0$. For example, the 4-velocity **U** of a particle is timelike and future-pointing. (The 4-acceleration **A** is spacelike.)

In our discussion of **U** we have already seen how a specific choice of reference frame (namely, the rest-frame) simplifies its component representation to $(0, 0, 0, c)$. Similar simplifications can be achieved for all 4-vectors. This is analogous to reducing a 3-vector **a** to the form $(a, 0, 0)$ by laying the x-axis along it. For a 4-vector **A** whose components in some inertial frame S are (A_1, A_2, A_3, A_4) we can similarly eliminate A_2 and A_3 by rotating the spatial axes so that the spatial x-axis lies along (A_1, A_2, A_3) which behaves as a 3-vector under rotations. If **A** is null, we have now reduced it to the form $(N, 0, 0, N)$ (if the x- and t-components turn out to have opposite signs, reverse the x- and y-axes), and this is as far as we can go. But if **A** is *not* null, then, as Fig. 2.6 shows, we can next find a standard LT to an inertial frame S' such that either the x'-axis or the t'-axis lies along **A**, depending on whether it is spacelike or timelike. In this way we arrive at the special representations $\mathbf{A} = (A, 0, 0, 0)$ or $\mathbf{A} = (0, 0, 0, \pm A)$, A now being the magnitude of **A**.

The above simplifications are useful in the proofs of certain geometrical results. For example, that two null vectors **A** and **B** cannot be orthogonal without being also parallel: choose axes so that $\mathbf{A} = (N, 0, 0, N)$; orthogonality to **B** then implies $B_4 = B_1$, but $B_4^2 - B_1^2 - B_2^2 - B_3^2 = 0$ and consequently $B_2 = B_3 = 0$, which establishes the result.

Another such result is that all 4-vectors orthogonal to a given causal vector are spacelike, except for the causal vector itself if that is null. (The algebraic proof is left to the reader.) A spacetime diagram illustrates this result nicely: Consider any timelike vector **A** and pick the ct'-axis in Fig. 5.1 along **A**. All vectors in the tangent plane π

and also in the plane parallel to π through the origin (the $x'y'$-plane) are orthogonal to A (as is clear in primed coordinates.) Now let A tilt towards the light cone. In the limit, the $x'y'$-plane will have moved to touch the light cone along what is now the generator A; all vectors in this plane are orthogonal to A, and all are spacelike except A itself.

We shall later require also the following result: the scalar product of two future-pointing causal vectors A and B is positive or zero, and it is zero only if A and B are null and parallel. *Proof*: If one of the vectors is timelike, take it to be $A = (0, 0, 0, A_4)$; then $A \cdot B = A_4 B_4$, which is positive by the hypothesis. If both vectors are null, let $A = (A_4, 0, 0, A_4)$; then $A \cdot B = A_4(B_4 - B_1)$ and it is easily seen that this is positive unless $B_1 = B_4$ and consequently $B_2 = B_3 = 0$, which establishes the assertion. A most important corollary is that *the sum of any number of future-pointing causal vectors* A, B, ... *is a future-pointing timelike vector, unless all the summands are null and parallel, in which case the sum is clearly null also.* For timelikeness we must show that the square of the sum is positive—the positivity of the fourth component is obvious. Now

$$(A + B + \cdots) \cdot (A + B + \cdots) = A^2 + B^2 + \cdots + 2(A \cdot B + \cdots).$$

The terms on the RHS are either positive or zero; and *all* of them are zero if and only if all the vectors are null and parallel, as claimed.

What we have just proved can also be seen (or, at least, surmised) from a spacetime diagram. We add vectors head to tail. The order of summation is immaterial. Start at the origin with a timelike vector if there is one. This takes us to point $\mathscr{P}$ inside the origin's future cone. Draw the future cone at $\mathscr{P}$: none of the remaining additions can take us out of that and so the sum is timelike. If there is no timelike summand but at least one pair of non-parallel null vectors, start with them; their sum also lies *inside* the origin's future cone, and the rest of the argument goes as before. That does it.

In regard to graphical arguments like the above, one must bear in mind that the Minkowski diagram, being only a map, distorts lengths and angles. In Fig. 5.1 *all* vectors that start at the origin and end on one of the calibrating hyperboloids have equal 'Minkowski length' $|A|$. Vectors that appear orthogonal in the diagram are not necessarily 'Minkowski-orthogonal' in the sense $A \cdot B = 0$, as exemplified by the axes of ξ and η in Fig. 2.6. Conversely, the axes of x' and t' *are* Minkowski-orthogonal but do not appear orthogonal in that diagram. On the other hand, it is clear that vector sums in the diagram correspond to 'Minkowski sums' $A + B$, that parallel vectors in the diagram correspond to 'Minkowski-parallel' vectors (in the sense $A = kB$, k real), and that the 'Minkowski ratio' k of such vectors is also the apparent ratio in the diagram.

We end this section with an important result, namely the *zero-component lemma*, whose analog in 3-vector calculus is geometrically obvious: *If a 4-vector has a partic-ular one of its four components zero in all inertial frames, then the entire vector must be zero.* Here it is best to proceed algebraically. Let the vector be $V = (V_1, V_2, V_3, V_4)$ and suppose, typically, that V_2 were always zero. Consider a LT in the y-direction: $V_2' = \gamma(V_2 - (v/c)V_4)$. This shows that also V_4 must vanish in all frames. But then a LT in the x-direction, $V_4' = \gamma(V_4 - (v/c)V_1)$, shows that V_1 vanishes in all frames, and so does V_3 by an analogous argument.

5.7 Plane waves

In this section we shall make the acquaintance of a third fundamental kinematical 4-vector, the *wave 4-vector* L. While U and A refer to the motion of a particle, L refers to the motion of a wave.

Consider a series of plane 'disturbances' or 'wavecrests', a wavelength λ apart, and progressing in a unit direction $\mathbf{n} = (l, m, n)$ at speed w relative to an inertial frame S. A single *stationary* plane in S with normal $\mathbf{n}$ and at distance p from a point $P_0 = (x_0, y_0, z_0)$ satisfies the equation $\mathbf{n} \cdot \mathbf{r} = p$, or, in full,

$$l(x - x_0) + m(y - y_0) + n(z - z_0) = p.$$

If this plane propagates with speed w the equation becomes, in delta-notation,

$$l\Delta x + m\Delta y + n\Delta z = w\Delta t = \frac{w}{c}\Delta ct,$$

where Δt measures the time from when the plane crosses P_0. A whole set of such traveling planes, at distances λ apart, has the same equation with $N\lambda$ added to the RHS, N being any (positive or negative) integer. And this can be written, after absorbing a minus sign into N and dividing by λ,

$$\mathbf{L} \cdot \Delta \mathbf{s} = N, \tag{5.30}$$

where

$$\mathbf{L} = \frac{1}{\lambda}\left(\mathbf{n}, \frac{w}{c}\right) = \nu\left(\frac{\mathbf{n}}{w}, \frac{1}{c}\right), \tag{5.31}$$

$\nu = w/\lambda$ being the frequency. Conversely, any equation of form (5.30) will be recognized in S as representing a moving set of equidistant planes, with λ, ν, w, and $\mathbf{n}$ determined by (5.31). We have written L *like* a 4-vector, though we do not yet know that it is one. In fact, let us more specifically write $\mathbf{L}(S)$ for just the set of components (5.31) in S, and similarly define $\Delta \mathbf{s}(S)$. Then eqn. (5.30) reads $\mathbf{L}(S) \cdot \Delta \mathbf{s}(S) = N$. In this equation, when written out, replace Δx, Δy, Δz, Δct by their appropriate Lorentz transforms in an arbitrary second frame S', each of which will be a linear homogeneous combination of $\Delta x'$, $\Delta y'$, $\Delta z'$, $\Delta ct'$. Collecting all the $\Delta x'$ terms, etc., we can clearly rewrite this translated equation in the form

$$\mathbf{L}(S') \cdot \Delta \mathbf{s}(S') = N, \tag{5.32}$$

where $\mathbf{L}(S')$ stands for the quadruplet of the coefficients of $-\Delta x'$, $-\Delta y'$, $-\Delta z'$, $\Delta ct'$. Thus in S', too, one has a moving wave train, whose kinematic characteristics λ', ν', w', and $\mathbf{n}'$ are related to $\mathbf{L}(S')$ analogously to (5.31). The only question is whether this $\mathbf{L}(S')$ is related to $\mathbf{L}(S)$ by the appropriate transformation to make it a

4-vector. Suppose $\mathbf{V}(S')$ stands for the vector transforms into S' of the components $\mathbf{L}(S)$. Then, by the invariance of the scalar product, we shall have, from (5.30),

$$\mathbf{V}(S') \cdot \Delta\mathbf{s}(S') = \mathbf{L}(S) \cdot \Delta\mathbf{s}(S) = N. \tag{5.33}$$

Each value of N determines a specific one of the set of traveling planes; hence, forming the difference of eqns (5.32) and (5.33), we get

$$\{\mathbf{L}(S') - \mathbf{V}(S')\} \cdot \Delta\mathbf{s}(S') = 0 \tag{5.34}$$

for any event on any one of the planes. We can certainly find four linearly independent vectors $\Delta\mathbf{s}(S')$ corresponding to events on the planes, and thus satisfying (5.34). Hence $\{\ \} = 0$, that is, $\mathbf{L}(S') = \mathbf{V}(S')$, and so $\mathbf{L}$ *does* transform as a vector and therefore *is* a vector.

We note, incidentally, that our analysis of an extended *plane* wave train applies *locally* to arbitrary wave trains, provided only that the wavelength is then much smaller than the radius of curvature of the wave-fronts, so that a sufficiently small portion of the train has the appearance of parallel planes.

We shall now use the transformation properties of the wave-vector $\mathbf{L}$ to derive formulae for the Doppler effect, for aberration, and for the velocity transformation for waves of arbitrary speed w. Consider the usual two frames S and S' in standard configuration, and in S a train of plane waves with frequency v and velocity w in a direction $\mathbf{n} = -(\cos\alpha, \sin\alpha, 0)$. The components of the wave-vector is S are then given by

$$\mathbf{L} = \left(\frac{-v\cos\alpha}{w}, \frac{-v\sin\alpha}{w}, 0, \frac{v}{c} \right). \tag{5.35}$$

Transforming these components by the scheme (5.11), we find for the components in S':

$$\frac{v'\cos\alpha'}{w'} = \frac{\gamma v(\cos\alpha + vw/c^2)}{w}, \tag{5.36}$$

$$\frac{v'\sin\alpha'}{w'} = \frac{v\sin\alpha}{w}, \tag{5.37}$$

$$v' = v\gamma\left(1 + \frac{v}{w}\cos\alpha\right). \tag{5.38}$$

The last equation expresses the Doppler effect for waves of all velocities, and, in particular, for light waves ($w = c$). In the latter case, it is seen to be equivalent to our previous formula (4.5).

From (5.36) and (5.37), we obtain the general wave aberration formula

$$\tan\alpha' = \frac{\sin\alpha}{\gamma(\cos\alpha + vw/c^2)}. \tag{5.39}$$

In the particular case when $w = c$, this is seen to be equivalent to our previous formulae (4.7) and (4.8)—in fact, it corresponds to their quotient.

Finally, to get the transformation of w, we *could* eliminate the irrelevant quantities from eqns (5.36)–(5.38), but it is simpler to make use of the invariance of L^2. Writing this out in S and S', we obtain

$$v^2\left(1 - \frac{c^2}{w^2}\right) = v'^2\left(1 - \frac{c^2}{w'^2}\right),$$ (5.40)

whence, by use of (5.38),

$$1 - \frac{c^2}{w'^2} = \frac{(1 - c^2/w^2)(1 - v^2/c^2)}{(1 + v\cos\alpha/w)^2}.$$ (5.41)

Neither this velocity transformation formula nor the aberration formula (5.39) are equivalent to the corresponding formulae for particles [cf. Exercise 4.13—also 4.14 for an alternative derivation of eqn (5.39)]. The reason is that a particle riding the crest of a wave in the direction of the wave normal in one frame does not, in general, do so in another frame: there it rides the crest of the wave also, but not in the normal direction. The one exception is when $w = c$.

Exercises 5

5.1. Prove that, for any straight-line segment in spacetime, $\Delta s = \int ds$, where Δs is the magnitude of $\Delta \mathbf{s}$ [cf. eqn (2.15)] and ds that of the corresponding $d\mathbf{s}$. [*Hint*: let the segment be defined by $x = x_0 + A\theta, \ldots, ct = ct_0 + D\theta, 0 \le \theta \le 1$.]

5.2. An inertial observer O bounces a radar signal off an arbitrary event $\mathscr{P}$. If the signal is emitted and received by O at times τ_1 and τ_2, respectively, as indicated by O's standard clock, prove that the squared interval $\Delta \mathbf{s}^2$ between O's origin-event $\tau = 0$ and $\mathscr{P}$ is $c^2\tau_1\tau_2$. This, in fact, constitutes a uniform method (apparently due to A. A. Robb) for assigning $\Delta \mathbf{s}^2$ to any pair of events.

5.3. A 4-vector has components (V_1, V_2, V_3, V_4) in an inertial frame S. Write down its components (i) in a frame which coincides with S except that the directions of the x- and z-axes are reversed; (ii) in a frame which coincides with S except for a 45° rotation of the xy-plane about the origin followed by a translation in the z-direction by 3 units; (iii) in any frame which has its axes parallel to those of S and moves with velocity v in they y-direction.

5.4. We can define the 4-velocity of superluminally moving particles (or just geometric points) in analogy to that of normal particles: $\mathbf{U} = c\, d\mathbf{R}/ds$. Prove that then $\mathbf{U} = (u^2/c^2 - 1)^{-1/2}(\mathbf{u}, c)$, and $\mathbf{U}^2 = -c^2$.

5.5. Use the fact that $\mathbf{U} = \gamma(\mathbf{u}, c)$ transforms as a 4-vector to re-derive the transformation equations (3.6) and (3.10). [Our earlier derivation of (3.6) has the advantage of applying to *all* velocities, including the velocity of light.]

5.6. Solve the relative-motion problem of Exercise 3.11 by using formula (5.27).

5.7. An inertial observer O has 4-velocity $\mathbf{U}_0$ and a particle P has (variable) 4-acceleration $\mathbf{A}$. If $\mathbf{U}_0 \cdot \mathbf{A} = 0$, what can you conclude about the speed of P in O's rest-frame?

5.8. Prove that, in any inertial frame where **a** is orthogonal to **u**, $\mathbf{A} = \gamma^2(\mathbf{a}, 0)$. Deduce that $\mathbf{a} \perp \mathbf{u}$ is an invariant property among all inertial frames with relative motion parallel to **u**.

5.9. Muons circle a storage ring of radius $10\,\text{m}$ at a speed that makes $\gamma = 30$. Verify that their proper accelerations are $\sim 0.8 \times 10^{18} g$.

5.10. A particle moves in an inertial frame (for a while) according to the equations $x = wt$, $y = \frac{1}{2} gt^2$, $z = 0$ ($w, g = $ const). Find its proper acceleration as a function of time.

5.11. Prove: (i) all 4-vectors orthogonal to a given causal vector are spacelike except for the causal vector itself if it is null; (ii) the sum or difference of any two orthogonal spacelike vectors is spacelike; (iii) every 4-vector can be expressed as the sum of two null vectors. [*Hint*: The component specializations of Section 5.6 may help.]

5.12. (i) Find four linearly independent timelike vectors.
      (ii) Find four linearly independent spacelike vectors.
      (iii) Find four linearly independent null vectors.

5.13. Consider two inertial observes with non-intersecting and non-parallel world-lines. Prove that there exists a unique pair of events, one on each worldline and simultaneous to both observers. [*Hint*: let one of the worldlines be the t-axis.]

5.14. We shall say that three particles move *codirectionally* if their 3-velocities are parallel in *some* inertial frame. Prove that the necessary and sufficient condition for this to be the case is that the 4-velocities **U**, **V**, **W** of these particles be linearly dependent. [*Hint*: component specialization.]

5.15. For any two timelike vectors $\mathbf{V}_1$ and $\mathbf{V}_2$ which are *isochronous* (that is, both pointing into the future or both into the past), prove that $\mathbf{V}_1 \cdot \mathbf{V}_2 = V_1 V_2 \cosh \phi_{12}$, where ϕ_{12}, the 'hyperbolic angle' between $\mathbf{V}_1$ and $\mathbf{V}_2$, equals the relative rapidity of two particles having worldlines parallel to $\mathbf{V}_1$ and $\mathbf{V}_2$. [*Hint*: (5.27).] Moreover, prove that ϕ is additive; that is, for any three *coplanar* such vectors $\mathbf{V}_1, \mathbf{V}_2, \mathbf{V}_3$ (corresponding to codirectional particles), $\phi_{13} = \phi_{12} + \phi_{23}$.

5.16. Consider the two-plane π in spacetime spanned by two non-parallel spacelike vectors **A** and **B**. Prove: π cuts any light cone whose vertex lies on it if $|\mathbf{A} \cdot \mathbf{B}| > AB$, and touches it if $|\mathbf{A} \cdot \mathbf{B}| = AB$; if $|\mathbf{A} \cdot \mathbf{B}| < AB$, say $\mathbf{A} \cdot \mathbf{B} = -AB \cos \theta$, then π can be chosen as the xy plane of some inertial coordinate system S and θ is the ordinary angle between **A** and **B** in S.

5.17. A 3-dimensional *hyperplane* in Minkowski space is the locus of events satisfying a linear equation of the form $Ax + By + Cz + Dct = E$. For any displacement in it, we have $A\Delta x + B\Delta y + C\Delta z + D\Delta ct = 0$. The *normal* $(-A, -B, -C, D)$—when standardized to unit length unless it is null—is seen to be a 4-vector by the same kind of reasoning as we used for **L** in eqn (5.30). A hyperplane is called spacelike, timelike, or null according as its normal is timelike, spacelike, or null, respectively. Prove: (i) a spacelike hyperplane is a simultaneity for all inertial observers whose worldlines are orthogonal to it; (ii) all displacements in a spacelike hyperplane are spacelike; (iii) all displacements in a null hyperplane are spacelike except for those in one particular direction, which are null; (iv) a triad of mutually orthogonal 4-vectors in a hyperplane H necessarily consists of three spacelike vectors if H is spacelike, of two spacelike and one timelike vector if H is timelike, and of two spacelike and one null vector if H is null.

5.18. (i) Prove that the necessary and sufficient condition for a particle to move along a straight line in *some* inertial frame is that its worldline should be 'flat'; that is, lie in a 2-plane spanned by one timelike and one spacelike vector. (ii) Prove that if in *some* inertial frame a particle's 3-acceleration is always parallel to a constant 3-vector, the flatness condition on the worldline is satisfied. Conversely, prove that the flatness condition implies this property of the 3-acceleration in *every* inertial frame. It is thus an invariant property.

5.19. (i) Prove that the 4-vector equation $(\mathrm{d}/\mathrm{d}\tau)\mathbf{A} = \phi\mathbf{U}$, where ϕ is some scalar, implies $\alpha = c\sqrt{\phi} = \text{const}$. [*Hint*: Differentiate the equations $\mathbf{A} \cdot \mathbf{U} = 0$ and $\mathbf{A} \cdot \mathbf{A} = -\alpha^2$.] (ii) Prove that the above equation is equivalent to the 3-vector equation $(\mathrm{d}/\mathrm{d}t)(\gamma^3\mathbf{a}) = 0$. [*Hint*: eqns (5.20), (5.22), and (2.10) (ii).] Consequently $\gamma^3\mathbf{a} = \text{const}$ is an invariant property. By reference to the preceding exercise, prove that it characterizes hyperbolic motion in *some* inertial frame.

5.20. Perform in detail the transformation of eqn (5.30) that is outlined in the text above eqn (5.32), taking S′ to be related to S by a *standard* LT and simply writing L_1, L_2, L_3, L_4 for the components of $\mathbf{L}(\mathrm{S})$. Then observe directly that $\mathbf{L}(\mathrm{S}')$ is the vector transform of $\mathbf{L}(\mathrm{S})$. Complete this finding into an alternative proof that $\mathbf{L}$ transforms as a 4-vector under *general* LTs.

5.21. From eqn (5.37) deduce the relation $\lambda'/\sin\alpha' = \lambda/\sin\alpha$ for plane waves of arbitrary speed. Then use eqn (5.39) to rederive the result of Exercise 4.15.

5.22. (i) If a wave train with wave-vector $\mathbf{L}$ is observed by an observer with 4-velocity $\mathbf{U}_1$ to have frequency ν_1, prove $\nu_1 = \mathbf{L} \cdot \mathbf{U}_1$. Similarly, if the wave train is emitted by a source which has proper frequency ν_0 and 4-velocity $\mathbf{U}_0$, prove $\nu_0 = \mathbf{L} \cdot \mathbf{U}_0$. (ii) Use the first of these results to re-derive the Doppler formula (5.38). (iii) In some inertial frame S light travels along a straight line which the source crosses at $30°$ at emission and the observer crosses at $60°$ at reception; find the frequency shift ν_1/ν_0 by evaluating $\mathbf{L} \cdot \mathbf{U}_1/\mathbf{L} \cdot \mathbf{U}_0$ in S, and note how the frequency ν of the waves in S conveniently cancels out: *any* null vector proportional to $\mathbf{L}$ can be used in place of $\mathbf{L}$. (iv) Re-derive formula (4.3) and the result of Exercise 4.3.

5.23. Consider a plane wave train propagating along the x-axis of some inertial frame S. Prove that for all inertial frames in standard configuration with S the wave velocity w transforms just like a particle velocity; that is, according to formula (3.6)(i). [*Hint*: eqns (5.36) and (5.38).]

6

Relativistic particle mechanics

6.1 Domain of sufficient validity of Newtonian mechanics

What we have done so far has been essentially an elaboration of Einstein's two basic axioms, without the addition of further hypotheses. We have seen how Newton's concepts of space and time could be replaced by a somewhat more complicated but still elegant and harmonious spacetime structure that accommodates Einstein's axioms. But now we have arrived at the next point in the program of special relativity, namely the scrutiny of the existing laws of physics and the modification of those that are found to be not Lorentz-invariant. Chief among these is Newton's basic law $\mathbf{f} = m\mathbf{a}$ and his treatment of the force $\mathbf{f}$ and the mass m as invariants. Clearly this is at odds with the new kinematics, where $\mathbf{a}$ is no longer an invariant. And altogether, Newton's theory is Galileo-invariant and not Lorentz-invariant. Thus it was logically necessary to construct a new mechanics—long before any serious empirical deficiencies of the old mechanics had become apparent. The new mechanics is known as 'relativistic' mechanics. This is not really a good name, since, as we have seen, Newton's mechanics, too, is relativistic, but under the 'wrong' (Galilean) transformation group. Newton's theory has excellently served astronomy (for example, in foretelling eclipses and orbital motions in general), it has been used as the basic theory in the incredibly delicate operations of sending probes to the moon and some of the planets, and it has proved itself reliable in countless terrestrial applications. Thus it cannot be *entirely* wrong. Before the twentieth century, in fact, only a single case of irreducible failure was known, namely the excessive advance of the perihelion of the planet Mercury, by about 43 seconds (!) of arc per century. Since the advent of modern particle accelerators, however, vast discrepancies with Newton's laws have been uncovered, whereas the new mechanics consistently gives correct descriptions. (Of course, Newton's mechanics has undergone *two* 'corrections,' one due to relativity and one due to quantum theory. We are here concerned exclusively with the former.) The new mechanics practically overlaps with the old in a large domain of applications (dealing with motions that are slow compared to the speed of light) and, in fact, it delineates the domain of sufficient validity of the old mechanics as a function of the desired accuracy. Roughly speaking, the old mechanics is in error to the extent that the γ-factors of the various motions involved exceed unity. In laboratory collisions of elementary particles γ-factors of the order of 10^4 are not unusual, and γ-factors as high as 10^{11} have been calculated for some cosmic-ray protons incident in the upper atmosphere. Applied to such situations, Newtonian mechanics is not just *slightly* wrong: it is totally wrong. Yet within its known slow-motion domain, Newton's theory will undoubtedly continue to be used for reasons of conceptual and technical convenience. And as a logical construct it will remain as perfect and inviolate as Euclid's geometry. Only as a model of nature it must not be stretched unduly.

6.2 The axioms of the new mechanics

The mechanics for which we now seek a Lorentz-invariant substitute is the non-gravitational part of Newton's mechanics; that is to say, that which is covered by Newton's basic three laws (cf. Section 1.2) and which primarily concerns itself with particle collisions, particle systems and particles in external fields. We specifically assume the absence of gravity since, according to GR, gravity distorts the Minkowskian spacetime of SR which here plays a key role.

Though there are many approaches to this new mechanics, the result is always the same.[1] If Newton's well-tested theory is to hold in the 'slow-motion limit', and unnecessary complications are to be avoided, then only *one* Lorentz-invariant mechanics appears to be possible. Moreover, it is persuasively elegant, and has been uncannily successful in matching nature perfectly in modern high-speed interactions where Newton's theory is out by many orders of magnitude.

We must stress that what is required here is the judicious *invention* of axioms to be placed at the head of the new mechanics; there is no logically binding way to *derive* them.

Force, which is the central concept of Newton's theory, is somewhat more peripheral in relativity, where its chief manifestation is the Lorentz force of electromagnetism. Gravitational force, of course, has been replaced by spacetime curvature. So it is more convenient to take an analog of momentum conservation (which is a *derived* result in Newton's theory) as primary now.

We begin by assuming what we already know, that associated with each particle there is an intrinsic positive scalar, m_0, namely its Newtonian or *proper* or *rest-mass*. This allows us to define the 4-*momentum* $\mathbf{P}$ of a particle in analogy to its 3-momentum,

$$\mathbf{P} = m_0 \mathbf{U}, \tag{6.1}$$

$\mathbf{U}$ being the 4-velocity. Like $\mathbf{U}$, $\mathbf{P}$ is timelike and future pointing. And we take as the basic axiom of collision mechanics the conservation of this 4-vector quantity: The sum of the 4-momenta of all the particles going into a point-collision is the same as the sum of the 4-momenta of all those coming out. (The collision may or may not be elastic, and there may be more, or fewer, or other particles coming out than going in.) We can write this in the form

$$\sum{}^* \mathbf{P}_n = 0, \tag{6.2}$$

---

[1] The first development of special-relativistic mechanics was given by Planck in 1906, whose starting point was the relativization of Newton's law of motion, $\mathbf{f} = m\mathbf{a}$. A second soon followed (in 1909) by Lewis and Tolman, who chose as their starting point the relativization of Newton's law of momentum conservation in particle collisions. A development from energy conservation can be found in J. Ehlers, R. Penrose and W. Rindler, *Am. J. Phys.* **33**, 995 (1965).

where a different value of $n = 1, 2, \ldots$ is assigned to *each* particle going in and to *each* particle coming out, and $\sum^*$ is a sum that counts pre-collision terms positively and post-collision terms negatively. The LHS of (6.2) is thus a 4-vector, which makes our axiom automatically Lorentz-invariant.

Using the component form (5.15) of **U**, we find the following components for **P**:

$$\mathbf{P} = m_0 \mathbf{U} = m_0 \gamma(u)(\mathbf{u}, c) =: (\mathbf{p}, mc), \tag{6.3}$$

where, in the last equation, we have introduced the symbols

$$m = \gamma(u)m_0, \tag{6.4}$$

$$\mathbf{p} = m\mathbf{u}. \tag{6.5}$$

The formalism thus leads us naturally to this quantity m, which we shall call the *relativistic inertial mass* (or usually just 'mass'), and to **p**, which we shall call the *relativistic momentum* (or usually just 'momentum'). Observe that m increases with speed; when $u = 0$ it is least, namely m_0, which is why we call m_0 the *rest-mass* of the particle. On the other hand, m becomes infinite as u approaches c—which is nature's way to avoid superluminal velocities.

In terms of these quantities the original conservation law (6.2) splits into two separate laws, the conservation of relativistic momentum,

$$\sum{}^* \mathbf{p} = 0; \quad \text{that is,} \quad \sum{}^* m\mathbf{u} = 0, \tag{6.6}$$

and the conservation of relativistic mass,

$$\sum{}^* m = 0, \tag{6.7}$$

where, for brevity, we have omitted the summation index n. Evidently in the slow-motion limit ($u \ll c$) these are the corresponding Newtonian conservation laws, and so our proposed relativistic law passes the three basic tests: Lorentz-invariance, simplicity, and Newton-conformity. Also in the formal $c \to \infty$ limit do the relativistic laws become Newtonian, which can occasionally be used as a check on our equations.

As in Newton's theory, eqns (6.6) can be regarded as implicit definitions of the mass m_0. They show that from a sufficient number of collision experiments we can determine at least the *ratios* of the rest-masses of all particles.

Note from (6.3) that the quantities on the LHSs of eqns (6.6) and (6.7) are the spatial and temporal components of the 4-vector $\sum^* \mathbf{P}$, respectively, and so (by the zero-component lemma of the end of Section 5.6) the vanishing of *either* in all inertial frames implies the vanishing of the other. Logically, therefore, each of these two conservation laws *by itself* has the same implications as the full law (6.2).

For reassurance that we are doing the right thing, note that every 'slow-motion' collision, proceeding along Newtonian lines in some inertial frame S [that is, satisfying (6.6) and (6.7) with $m = m_0$] *necessarily* satisfies, when regarded from a fast-moving inertial frame S′, the same eqns (6.6) and (6.7), but now with $m = \gamma(u)m_0$. This is not a consequence of *assuming* (6.2), but an unavoidable fact, since a 4-vector equation [namely (6.2)] if valid in S must be valid also in S$'$.

For further reassurance, consider two identical particles moving along the respective z-axes of the usual two inertial frames S and S$'$ in opposite directions but with equal speeds u. Then, by (3.6)(iii), each has a z-velocity numerically equal to $u\gamma^{-1}(v)$ as judged from the other frame. Let the particles collide and fuse at the momentarily coincident origins of S and S$'$. By symmetry, the new compound particle can have no z-velocity in either frame. Thus conservation of z-momentum in S in accordance with a formula of type (6.6) requires

$$m(u)u = \frac{m(w)u}{\gamma(v)},$$

where $m(u)$ denotes the mass of one of the original particles at speed u, and w is the total speed in S of the particle moving on the z$'$-axis. Canceling u and then letting $u \to 0$ (that is, considering a sequence of experiments with ever smaller u) forces us to conclude

$$m(0) = \frac{m(v)}{\gamma(v)}.$$

This very directly shows that if we wish to salvage the *form* (6.6) of Newton's law of momentum conservation, we must allow m to vary precisely as in (6.4). And that, of course, is the crucial difference between relativistic and pre-relativistic collision mechanics.

6.3   The equivalence of mass and energy

Let us now take a closer look at eqn (6.7), $\sum^* m = 0$, the conservation of relativistic mass. At first sight this appears to be just an analog of the Newtonian law of mass conservation—but it is not! Newton defined mass as 'quantity of matter', and asserting its conservation was tantamount to asserting that matter can neither be created nor destroyed. But we now know that matter can be transmuted into radiation, as when an electron and a positron annihilate each other. So it is just as well that $\sum^* m = 0$ is not, in fact, an analog of the Newtonian law, except in a purely formal sense. What is conserved here is a quantity, γm_0, which varies with speed. In classical mechanics we know of only one such conserved quantity, namely the kinetic energy of particles in an elastic collision. Of course, (6.7) must hold in *all* collisions, elastic or not, if our approach is right; and we already know for a fact that it holds in all those fast collisions that are slow in *some* inertial frame. Could it be that m (or a multiple of it) is a measure of *total* energy? The answer turns out to be 'yes' and was regarded by Einstein, who found it in 1905 (but see the last paragraph of this section), as the most significant result of his special theory of relativity. Nevertheless, Einstein's assertion of the full equivalence of mass and energy, according to the famous formula

$$E = mc^2, \tag{6.8}$$

was in part a hypothesis, as we shall see. It cannot be uniquely deduced from the basic laws.

Consider the following expansion for the mass (6.4):

$$m = m_0 \left(1 - \frac{u^2}{c^2} \right)^{-1/2} = m_0 + \frac{1}{c^2} \left(\frac{1}{2} m_0 u^2 \right) + \cdots \tag{6.9}$$

This shows that the relativistic mass of a slowly moving particle exceeds its rest-mass by $1/c^2$ times its kinetic energy (assuming the approximate validity of the Newtonian expression for the latter). So kinetic energy *contributes* to the mass in a way that is consistent with (6.8). In fact, it is eqn (6.9) that supplies the constant of proportionality between E and m. And it is the enormity of this constant that explains why the mass-increases correponding to the easily measurable kinetic energies of particles in classical collisions had never been observed.

We can next show that, since kinetic energy 'has mass', all energy must have mass in the same proportion. For one of the characteristics of energy is its transmutability from one form to another. When two oppositely moving identical particles collide and fuse and remain at rest (we are here thinking of putty balls rather than protons) $\sum m$ remains constant throughout, so that whatever mass was contributed before impact by the kinetic energy is thereafter contributed by the equal amount of thermal energy into which it changes. But then *all* forms of energy must have mass in the same proportion. For inside the now stationary compound particle the extra heat can be transmuted arbitrarily into other forms of energy without setting the particle in motion; for each such transmutation-event can be regarded as a 'collision', in which the total momentum is conserved; but also the total mass is conserved, which proves our assertion.

Yet it is still logically possible that energy only *contributes* to mass, without causing *all* of it. Especially in Einstein's time it would have been perfectly reasonable to suppose that the elementary particles are indestructible, so that the *available* energy of a macroscopic particle would be $c^2(m - q)$, where q is the total rest-mass of its constituent elementary particles. To equate *all* mass with energy required an act of aesthetic faith, very characteristic of Einstein. Of course, today we know how amply nature has confirmed that faith.

Einstein's mass–energy equivalence is not restricted to mechanics. It has been found applicable in many other branches of physics, from electromagnetism to general relativity. It is truly a new fundamental principle of Nature.

Observe also how Einstein's principle determines a *zero-point* of energy. In Newton's theory, for example, one could theoretically extract an unlimited amount of energy from a macroscopic body by letting it collapse indefinitely under its own gravity. According to Einstein, on the other hand, Nature must find a mechanism to prevent the extraction of more than mc^2 units of energy. That mechanism is the general-relativistic 'black hole'!

The relativistic *kinetic energy* T of a particle is naturally defined as the difference between its total and its internal or rest energy:

$$mc^2 = m_0 c^2 + T, \qquad T = m_0 c^2 (\gamma - 1). \tag{6.10}$$

The leading term in the power-series expansion of T is, of course, the Newtonian $\frac{1}{2}m_0 u^2$, as in eqn (6.9); the rest is the relativistic 'correction'. Note that in an *elastic* collision, where by definition each particle's rest-mass is preserved (so that $\sum^* m_0 = 0$), the conservation law $\sum^* m = 0$ implies the conservation of kinetic energy, $\sum^* T = 0$.

Einstein's mass–energy equivalence allows us to include even particles of zero rest-mass (photons, ...) into the scheme of collision mechanics. If such a particle has finite energy E (all of it being kinetic!), it has finite mass $m = E/c^2$ and thus, because of (6.4), it *must* move at the speed of light. Formally we can regard its mass as the limit of a product, γm_0, of which the first factor has gone to infinity and the second to zero. According to (6.5), it then has a perfectly normal 3-momentum $\mathbf{p}$ and thus also a 4-momentum $\mathbf{P}$ given by the last member of eqn (6.3): $\mathbf{P} = (\mathbf{p}, E/c)$. In this case, however, $\mathbf{P}$ is null ($\mathbf{P}^2 = 0$).

In fact, the 4-momentum (6.3) of *any* particle can now be written in the form

$$\mathbf{P} = (\mathbf{p}, E/c). \tag{6.11}$$

Particle physicists, whose basic unit is the electrovolt rather than the gram, prefer this form and tend to discard the concept of relativistic mass altogether. And, of course, they are the main consumers and therefore trend-setters of relativistic mechanics. On the other hand, it is trivial to switch back and forth between m and E and we prefer to keep our options open.

So much for the formalism. What about the applications? For a macroscopic 'particle' the internal energy $m_0 c^2$ is vast: in each gram of mass there are 9×10^{20} ergs of energy, roughly the energy of the Hiroshima bomb (20 kilotons). A very small part of this energy resides in the thermal motions of the molecules constituting the particle, and can be given up as heat; a part resides in the intermolecular and interatomic cohesion forces, and some of that can be given up in chemical explosions; another part may reside in excited atoms and escape in the form of radiation; much more resides in nuclear bonds and can also sometimes be set free, as in the atomic bomb. But by far the largest part of the energy (about 99 per cent) resides simply in the mass of the ultimate particles, and cannot be further explained. Nevertheless, it too can be liberated under suitable conditions; for example, when matter and antimatter annihilate each other.

One kind of energy that does *not* contribute to mass is *potential* energy of position. In classical mechanics, a particle moving in an electromagnetic (or gravitational) field is often said to possess potential energy, so that the sum of its kinetic and potential energies remains constant. This is a useful 'book-keeping' device, but energy conservation can also be satisfied by debiting the *field* with an energy loss equal to the kinetic energy gained by the particle. In relativity there are good reasons for adopting the second alternative, though the first can be used as an occasional shortcut: the 'real' location of any part of the energy is no longer a mere convention, since energy (as mass) gravitates; that is, it contributes measurably to the curvature of spacetime *at its location*.

According to Einstein's hypothesis, *every* form of energy has a mass equivalent: (i) If all mass exerts and suffers gravity, we would expect even (the energy of) an electromagnetic field to exert a gravitational attraction, and, conversely, light to bend under gravity (this we have already anticipated by a different line of reasoning). (ii) We shall expect a gravitational field *itself* to gravitate. (iii) The radiation which the sun pours into space is equivalent to more than four million tons of mass per second! Radiation, having mass and velocity, must also have momentum; accordingly, the radiation from the sun is a (small) contributing factor in the observed deflection of the tails of comets away from the sun. (The major factor is 'solar wind.') (iv) An electric motor (with battery) at one end of a raft, driving a heavy flywheel at the other end by means of a belt, transfers energy and thus mass to the flywheel; in accordance with the law of momentum conservation, the raft must therefore accelerate in the opposite direction. (v) Stretched or compressed objects have (minutely) more mass by virtue of the stored elastic energy. (vi) The total mass of the separate components of a stable atomic nucleus always exceeds the mass of the nucleus itself, since energy (that is, mass) would have to be supplied in order to decompose the nucleus against the nuclear binding forces. This is the reason for the well-known 'mass defect'. Nevertheless, if a nucleus is split into two new nuclei, these parts may have greater *or* lesser mass than the whole. With the lighter atoms, the parts usually exceed the whole, whereas with the heavier atoms the whole can exceed the parts, owing mainly to the electrostatic repulsion of the protons. In the first case, energy can be released by 'fusion', in the second, by 'fission'.

From a historical perspective, Einstein's recognition of $E = mc^2$ did not quite come 'out of the blue'. There had been foreshadowings along those lines for almost a quarter of a century. Already in 1881, J. J. Thomson had calculated that a charged sphere behaves as if it had an *additional* mass of amount $\frac{4}{3}c^{-2}$ times the energy of its Coulomb field. That set off a quest for the 'electromagnetic mass' of the electron—an effort to explain its inertia purely in terms of the field energy. (This effort was beset by the 'wrong' factor $\frac{4}{3}$ due to the as yet unknown mass-equivalent of the stresses needed to hold the 'electron' together.) In 1900, Poincaré made the simpler observation that, since the electromagnetic momentum of radiation is $1/c^2$ times the Poynting flux of energy, radiation seems to possess a mass density $1/c^2$ times its energy density. And then in 1905 came Einstein. What truly sets him apart once more is the universality of his proposal.

6.4   Four-momentum identities

It will be useful to have a number of often-used identities collected here for future reference. We have already discussed the following alternative expressions for the 4-momentum **P**,

$$\mathbf{P} = m_0 \mathbf{U} = (\mathbf{p}, mc) = (\mathbf{p}, E/c), \qquad (6.12)$$

which lead to the following alternative expressions for its square:

$$\mathbf{P}^2 = m_0^2 c^2 = m^2 c^2 - p^2 = E^2/c^2 - p^2. \tag{6.13}$$

Note that for zero-rest-mass particles, and only for those, $\mathbf{P}$ becomes a null vector. From eqns (6.13),

$$E^2 = p^2 c^2 + m_0^2 c^4, \qquad p^2 c^2 = E^2 - E_0^2 = c^4(m^2 - m_0^2). \tag{6.14}$$

When two particles with respective 4-momenta $\mathbf{P}_1$ and $\mathbf{P}_2$ are involved, and v_{12} is their relative speed (that is, the speed of one in the rest-frame of the other) we have

$$\mathbf{P}_1 \cdot \mathbf{P}_2 = m_{01} E_2 = m_{02} E_1 = c^2 \gamma(v_{12}) m_{01} m_{02}, \tag{6.15}$$

where, typically, m_{01} is the rest-mass of the first particle and E_2 is the energy of the second particle *in the rest-frame of the first*. For proof we need only evaluate $\mathbf{P}_1 \cdot \mathbf{P}_2$ in the rest-frame of either particle. The first equation holds even if the second particle is a photon.

In the particular case of an *elastic* (that is, rest-mass preserving) collision of two particles with pre-collision momenta $\mathbf{P}, \mathbf{Q}$ and post-collision momenta $\mathbf{P}', \mathbf{Q}'$, we find, on squaring the conservation equation $\mathbf{P} + \mathbf{Q} = \mathbf{P}' + \mathbf{Q}'$, that

$$\mathbf{P} \cdot \mathbf{Q} = \mathbf{P}' \cdot \mathbf{Q}'; \tag{6.16}$$

that is, that the relative velocity between the particles is conserved. Since this result is independent of the value of c, it must hold in Newton's theory as well!

6.5 Relativistic billiards

As a first example on the new mechanics we shall consider the relativistic analog of a billiard ball collision—namely, an elastic collision of two particles of equal rest-mass, one of which is originally at rest. This analysis has many applications. Agreement with it provided the first direct confirmation of the relativistic collision laws when Champion in 1932 bombarded stationary electrons in a cloud chamber with fast electrons from a radioactive source. The much later bubble-chamber experiments on elastic proton–antiproton scattering fitted into the same framework, and it is still relevant with modern particle accelerators.

If we approach this problem naïvely, setting up the conservation equations for mass and momentum in the lab frame, we quickly get into a bad tangle of different γ-factors. So we look for a 'trick'. One would be to find an elegant 4-vector argument, but here none presents itself naturally, in spite of the availability of eqn (6.16). The method we shall use instead is also of very general utility: we go to a frame where the problem is simpler, or even trivial, and then transform back to the frame of interest, exactly as we did for the corresponding Newtonian problem in Section 2.3. Let S'

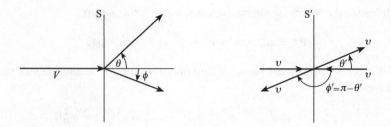

Fig. 6.1

be the frame in which the two particles originally approach each other with equal and opposite constant velocities,[2] say $\pm v$ along the x'-axis. The only way to satisfy momentum and energy conservation in S' is for the post-collision velocities also to be $\pm v$, but possibly along some other line, say one making an angle θ' with the x'-axis (see Fig. 6.1). Now let S be in standard configuration with S' at velocity v. The 'right' particle is then originally at rest in S, and so S is the required 'lab' frame. All we need to do now is to transform the remaining three velocities from S' to S. But in fact we are interested only in the *directions* of the post-collision velocities. So we can make use of the 'particle aberration' formula of Exercise 4.13—here needed in its inverse form $\tan \alpha = \sin \alpha' / \gamma(v)(\cos \alpha' + v/u')$. For the post-collision angles θ and ϕ in S, corresponding to θ' and $\phi' = \pi - \theta'$ in S', we then have

$$\tan \theta = \frac{\sin \theta'}{\gamma(v)(\cos \theta' + 1)}, \qquad \tan \phi = \frac{\sin \theta'}{\gamma(v)(-\cos \theta' + 1)},$$

where we measure θ anticlockwise and ϕ clockwise. Multiplying these expressions together gives the first of the following equations:

$$\tan \theta \, \tan \phi = \frac{1}{\gamma^2(v)} = \frac{2}{\gamma(V) + 1}. \tag{6.17}$$

The second results when we apply (3.10)(ii) to the 'bullet', setting $u' = u_1' = v$ and $u = V$, V being the incident velocity in the lab. It is of course clear from momentum conservation in S that the angles θ and ϕ are coplanar.

Equation (6.17) gives the required relation between the incident velocity V and the post-collision angles θ and ϕ. The Newtonian result $\theta + \phi = 90°$ corresponds to $\tan \theta \, \tan \phi = \tan \theta \, \cot \theta = 1$ and can be recovered from (6.17) by letting $c \to \infty$ (cf. penultimate paragraph of Section 6.2). In relativity, for any given θ, the corresponding ϕ is less than the Newtonian ϕ, so that $\theta + \phi$ is always less than 90°; and for $\theta \approx \phi$, this total angle can be very small indeed.

---

[2] The possible existence of a short-range electric force between the particles does not affect the argument: details of what happens in the 'collision zone' are irrelevant.

6.6 The zero-momentum frame

We have seen in the preceding section how useful an inertial frame can be in which the total momentum vanishes. Such a frame exists uniquely for every particle system. We call it the *zero-momentum frame*, S_{ZM}, and it corresponds to the classical center-of-mass frame (cf. Exercise 6.5).

Consider an arbitrary inertial frame S, and in it a system of occasionally colliding particles subject to no forces other than very short-range forces *during* collisions (cf. footnote 2) and thus moving uniformly *between* collisions. We define the total mass $\bar{m}$, total momentum $\bar{\mathbf{p}}$, and total 4-momentum $\bar{\mathbf{P}}$ of the system in S as the *instantaneous* sums of the respective quantities belonging to the individual particles:

$$\bar{m} = \sum m, \qquad \bar{\mathbf{p}} = \sum \mathbf{p}, \qquad \bar{\mathbf{P}} = \sum \mathbf{P} = \sum (\mathbf{p}, mc) = (\bar{\mathbf{p}}, \bar{m}c) \qquad (6.18)$$

[cf. (6.12).] Because of the conservation laws, each of the barred quantities remains constant in time.

The quantity $\bar{\mathbf{P}}$, being a sum of 4-vectors, seems assured of 4-vector status itself. But, in fact, it is not quite as simple as that. If all observers agreed on which Ps make up the sum $\sum \mathbf{P}$, then $\sum \mathbf{P}$ would clearly be a vector. But in each frame the sum is taken at one instant, which may result in different Ps making up the $\sum \mathbf{P}$ of different observers. A spacetime diagram, even an imagined one, such as Fig. 5.1, is useful in proving that $\sum \mathbf{P}$ is nevertheless a vector. A simultaneity in S corresponds to a 'horizontal' plane π in the diagram and a simultaneity in a second frame S' corresponds to a 'tilted' plane π'. In S, $\sum \mathbf{P}$ is summed over planes like π, and in S' over planes like π'. However, we now assert that in S' the same $\sum \mathbf{P}$ results whether summed over π' or π. For imagine a continuous motion of π' into π. As π' is tilted, each individual $\mathbf{P}$, located at a particle on π', remains constant (since the particles move uniformly between collisions) except when π' sweeps over a collision; but then the sub-sum of $\sum \mathbf{P}$ which enters the collision remains constant, by 4-momentum conservation. Thus, without affecting the value, all observers *could* sum their Ps over the same plane π, and thus $\bar{\mathbf{P}}$ is indeed a 4-vector.

Now, even if we allow some or all of the particles of the system to have zero rest-mass and thus null 4-momentum (as long as not *all* of them do *and* move in parallel), the sum $\bar{\mathbf{P}}$ will be timelike and future-pointing—by the italicized 'corollary' of Section 5.6. As we saw in that same section, we can therefore find a frame in which the spatial components of $\bar{\mathbf{P}}$ vanish; that is, in which $\bar{\mathbf{p}} = 0$. That frame is evidently the required S_{ZM}. *In S_{ZM} the 4-velocity $\mathbf{U}_{ZM}$ of S_{ZM} is $(0,0,0,c)$*, so that, from (6.18),

$$\bar{\mathbf{P}} = (0,0,0,\bar{m}_{ZM}c) = \bar{m}_{ZM}\mathbf{U}_{ZM}, \qquad (6.19)$$

where

$$\bar{m}_{ZM} = \sum m \quad \text{in } S_{ZM}, \qquad (6.20)$$

obviously an invariant.

The extremities of (6.19) constitute an important 4-vector relation. Comparing it with (6.1), we see that $\bar{m}_{ZM}$ and $\mathbf{U}_{ZM}$ are for the system what m_0 and $\mathbf{U}$ are for a single particle. They are the quantities that would be recognized as the rest-mass and 4-velocity of the system if its composite nature were *not* recognized (as in the case of an 'ordinary' particle—which is made up of possibly moving molecules).

Let us write out eqn (6.19) in component form in the general frame S, relative to which S_{ZM} has 3-velocity $\mathbf{u}_{ZM}$, say:

$$\mathbf{P} = (\bar{\mathbf{p}}, \bar{m}c) = \bar{m}_{ZM}\gamma(u_{ZM})(\mathbf{u}_{ZM}, c).$$

From this, we can read off the following useful relations:

$$\bar{m} = \gamma(u_{ZM})\bar{m}_{ZM} \tag{6.21}$$

and

$$\bar{\mathbf{p}} = \bar{m}\mathbf{u}_{ZM}, \quad \text{or} \quad \mathbf{u}_{ZM} = \bar{\mathbf{p}}/\bar{m}. \tag{6.22}$$

6.7   Threshold energies

An important application of relativistic mechanics occurs in so-called threshold problems. Consider, for example, the case of a free stationary proton (of rest-mass M) being struck by a moving proton, whereupon not only the two protons but also a pion (of rest-mass m) emerge. (Such reactions are often written in the form $p + p \rightarrow p + p + \pi^0$.) We shall ignore the electric interaction between the protons which is confined to a small collision zone. The question is, what is the minimum ('threshold') energy of the incident proton for this reaction to be possible? It is *not* simply $Mc^2 + mc^2$; that is, it is *not* enough for the proton's kinetic energy to equal the rest energy of the newly created particle. For, by momentum conservation, the post-collision particles cannot be at rest, and so a part of the incident kinetic energy must *remain* kinetic energy and thus be 'wasted'. (It is a little like trying to smash ping-pong balls floating in space with a hammer.)

In all such cases the theoretical minimum expenditure of energy occurs when all the end-products are *mutually* at rest. For consider, quite generally, two colliding particles: a 'bullet' and a stationary 'target', with respective 4-momenta $\mathbf{P}_B$ and $\mathbf{P}_T$. If the emergent particles have 4-momenta $\mathbf{P}_i$ ($i = 1, 2, \ldots$), then

$$\mathbf{P}_B + \mathbf{P}_T = \sum \mathbf{P}_i.$$

Squaring this equation, using (6.13) and (6.15), and adopting a self-explanatory notation, we then find

$$m_{0B}^2 + m_{0T}^2 + 2c^{-2}m_{0T}E_B = \sum m_{0i}^2 + 2\sum_{(i<j)} m_{0i}\,m_{0j}\,\gamma(v_{ij}). \tag{6.23}$$

All the m_0 in this equation are fixed by the problem. The only variable on the LHS is E_B, the energy of the bullet relative to the rest-frame of the target, and therefore relative to the lab. The minimum of the RHS evidently occurs when *all* the γ-factors are unity; that is, when there is no mutual motion between any of the outgoing particles; its value is then $(\sum m_{0i})^2$. So the *threshold energy* of the bullet is given by

$$E_B = (c^2/2m_{0T})\left[\left(\sum m_{0i}\right)^2 - m_{0B}^2 - m_{0T}^2\right]. \tag{6.24}$$

This formula applies even when the bullet is a photon that gets absorbed in the collision (cf. Exercise 6.16). For our original example of an extra meson coming out of a proton–proton collision, it yields

$$E_B = c^2\left(M + 2m + \frac{m^2}{2M}\right). \tag{6.25}$$

The *efficiency* of this and all analogous processes can be defined as the ratio k of the rest energy mc^2 of the new particle to the kinetic energy $E_B - m_{0B}c^2$ of the bullet. When eqn (6.25) applies, we find

$$k = m\left(2m + \frac{m^2}{2M}\right)^{-1} = \frac{2}{4 + (m/M)}. \tag{6.26}$$

The efficiency is thus always less than 50 per cent. In our particular example, $m/M \approx 0.14$ and $k \approx 48$ per cent. But when m greatly *exceeds* the rest-masses of all other particles involved, the details of the collision do not matter and eqn (6.24) yields the very unfavorable efficiency

$$k \approx \frac{2m_{0T}}{m}. \tag{6.27}$$

For example, when Richter and Ting created the 'psi' particle by colliding electrons with positrons, k would have been $\sim 1/1850$! The way out of the difficulty was to use a method that is almost 100 per cent efficient: the method of *colliding beams*. Here both target and bullet particles are first accelerated to high energy (for example, electrons and positrons can be accelerated in the same sychrotron, in opposite senses), then accumulated in magnetic 'storage rings', before being loosed at each other. No 'waste' kinetic energy need be present after the collision, since there was no net momentum going in.

6.8 Light quanta and de Broglie waves

One of the most striking discoveries of the twentieth century, and one that is crucial for our quantum-mechanical understanding of the structure of matter, was that all micro-particles (photons, electrons, nuclei, atoms, etc.) have both particle-like and wavelike properties. Not only does this wave-particle duality fit naturally into the framework of special relativity, it was actually suggested by it—very much as in the case of the equally important principle $E = mc^2$. All this speaks well for SR as correctly modeling some very fundamental structures in the real world.

As a desperate last resort to avoid the notorious infinity ('ultraviolet catastrophe') in the classical theoretical blackbody spectrum, Planck in 1900 had made the truly unprecedented suggestion that radiation of frequency v might be emitted only in definite 'quanta' of energy

$$E = hv, \tag{6.28}$$

where h is a new universal constant of nature, now known as Planck's constant.

Perhaps the only person who looked on Planck's formula as a chink in the curtain of nature rather than as an embarrassment (as did Planck himself!), was Einstein. He found that the entropy of blackbody radiation had a mathematical form analogous to that of a gas, and thus of particles. They could be made to coincide if the particles satisfied Planck's relation. Einstein also showed that quanta could explain certain strange effects connected with fluorescence, and others connected with the passage of ultraviolet radiation through a gas. But above all, he realized that Maxwell's continuous radiation theory could never explain the discreteness properties of the photoelectric effect recently brought to light by Lenard (1902); quanta, on the other hand, provided the perfect solution. In 1905 he therefore boldly suggested that not only is radiation *emitted* in quanta, but that it also travels and is absorbed in quanta, which were later called photons. According to Einstein, a photon of frequency v has energy hv, and thus (as he only came to realize several years later) a finite mass hv/c^2 and a finite momentum hv/c.

The generalization of Einstein's photon-wave duality to other particles (in particular, to electrons) could have been made almost at once. Of course, there was no experimental mandate for such a generalization in 1905—but then, neither was there one in 1923 when de Broglie finally made it. (Except mildly: it correctly explained the permissible electron orbits in the old Bohr model of the hydrogen atom as those containing a whole number of 'electron waves'.) What de Broglie now proposed was to extend Planck's and Einstein's relation $E = hv$ to *all* particles: a particle of total energy E would have associated with it a wave of frequency E/h traveling in the same direction. Of course, de Broglie knew that this wave cannot travel at the same speed as the particle (unless that speed is c), for that would not be a Lorentz-invariant association, as we have seen in Section 5.7. According to de Broglie, any particle of 4-momentum **P** has associated with it a wave of wave-vector **L** determined by what is now called *de Broglie's equation*:

$$\mathbf{P} = h\mathbf{L}; \quad \text{that is,} \quad (\mathbf{p}, E/c) = hv\left(\frac{\mathbf{n}}{w}, \frac{1}{c}\right) = h\left(\frac{\mathbf{n}}{\lambda}, \frac{v}{c}\right) \tag{6.29}$$

[cf. (5.31) and (6.11)]. In fact, this equation is inevitable (by the *zero-component lemma*) once we accept the universal validity of $E = hv$; for $(E - hv)/c$ is the fourth component of the 4-vector $\mathbf{P} - h\mathbf{L}$. Setting $p = Eu/c^2 = hvu/c^2$ and comparing spatial components in (6.29), we now find the following relation between the velocity u of the particle and the velocity w of its wave:

$$uw = c^2. \tag{6.30}$$

For particles of non-zero rest-mass, $u < c$ and thus $w > c$. The de Broglie wave speed then has an interesting and simple interpretation. Suppose a whole swarm of identical particles travel with equal velocity, and something happens to all of them simultaneously in their rest-frame: suppose, for example, they all 'flash'. Then this flash sweeps over the particles at the de Broglie velocity in any other frame. To see this, suppose the particles are at rest in the usual frame S$'$, traveling at speed v relative to a frame S. Setting $t' = 0$ for the flash, we find from (2.6)(iv) that $x/t = c^2/v$, and this is evidently the speed at which the plane 'flash wave' travels in S. Thus de Broglie waves can be regarded as 'waves of simultaneity'. It can also be shown quite easily that the *group velocity* of the de Broglie waves (the velocity of a *wave group* or *wave packet*), coincides with the velocity of the associated particle. (Cf. Exercise 6.11.)

Note from (6.28) that v can neither vanish nor be infinite, even if the particle is at rest (when $u = 0$ and $w = \infty$). But the wavelength, $\lambda = w/v$, is infinite for a particle at rest.

As a way to test his hypothesis, de Broglie suggested that one might look for diffraction effects when a beam of electrons traverses an aperture that is small compared to the wavelength. In a historic experiment in 1927 Davisson and Germer indeed discovered electron diffraction, though by a crystal rather than an 'aperture'. And, of course, the superiority of the modern electron microscope hinges on de Broglie's relation, according to which fast electrons allow us to 'see' with greater resolving power than photons since they have very much shorter wavelengths. But the richest development that came out of de Broglie's idea was undoubtedly Schrödinger's invention of wave mechanics in 1926, and its eventual merger with quantum mechanics.

6.9   The Compton effect

An extraordinary validation of Einstein's idea that photons can behave like little billiard balls with (relativistic) mass and momentum was provided by Compton's famous scattering experiment of 1922, in which photons were the 'bullets' and electrons the 'targets'; the results were found to conform precisely to the laws of relativistic collision mechanics.

To help analyze this situation, we first establish two general results. Our earlier momentum-product formula (6.15)(i) is still valid and useful when *one* of the two momenta is that of a photon: if $\mathbf{P}_1$ refers to a particle of finite rest-mass m_{01} and $\mathbf{P}_2$ to a photon of frequency v_2, eqn (6.15)(i) in conjunction with $E = hv$ now yields

$$\mathbf{P}_1 \cdot \mathbf{P}_2 = hm_{01}v_2. \tag{6.31}$$

But when both $\mathbf{P}_1$ and $\mathbf{P}_2$ refer to photons, we need to bring into play eqn (6.29) (with $w = c$): if the paths of the two photons subtend an angle θ with each other, that equation yields

$$\mathbf{P}_1 \cdot \mathbf{P}_2 = c^{-2}h^2v_1v_2(1 - \cos\theta). \tag{6.32}$$

Returning to Compton's experiment, we can use the left half of Fig. 6.1 to illustrate the situation: a photon of frequency ν is incident along the x-axis, strikes a stationary electron, and scatters at an angle θ with diminished frequency ν' while the electron recoils at an angle ϕ. We wish to relate ν, ν', and θ with m, the rest-mass of the electron. (No observations were made, at first, on the recoiling electron, but see Exercise 6.20.) If $\mathbf{P}$, $\mathbf{P}'$ are the pre- and post-collision 4-momenta of the photon and $\mathbf{Q}$, $\mathbf{Q}'$ those of the electron, then $\mathbf{P} + \mathbf{Q} = \mathbf{P}' + \mathbf{Q}'$, by conservation of 4-momentum. Next we isolate the unwanted vector $\mathbf{Q}'$ on one side of the equation and square to get rid of it:

$$(\mathbf{P} + \mathbf{Q} - \mathbf{P}')^2 = \mathbf{Q}'^2.$$

Since $\mathbf{Q}^2 = \mathbf{Q}'^2$ and $\mathbf{P}^2 = \mathbf{P}'^2 = 0$, we are left with $\mathbf{P} \cdot \mathbf{Q} - \mathbf{P}' \cdot \mathbf{Q} - \mathbf{P} \cdot \mathbf{P}' = 0$, that is,

$$\mathbf{P} \cdot \mathbf{P}' = \mathbf{Q} \cdot (\mathbf{P} - \mathbf{P}'), \tag{6.33}$$

from which we find at once, by reference to (6.31) and (6.32), the desired relation

$$hc^{-2}\nu\nu'(1 - \cos\theta) = m(\nu - \nu'). \tag{6.34}$$

In terms of the corresponding wavelengths λ, λ', and the half-angle $\theta/2$, this may be rewritten in the more standard form

$$\lambda' - \lambda = (2h/cm)\sin^2(\theta/2) = 2l\sin^2(\theta/2), \tag{6.35}$$

where l is the 'Compton wavelength' h/cm, gained by the photon when it scatters at $90°$.

Scattering of photons by stationary electrons is called *Compton scattering*, and clearly always results in a loss of energy to the photon. The opposite is true for *inverse Compton scattering*, in which a photon collides with a fast ('relativistic') electron or other charged particle and often experiences a spectacular gain in energy. This process is an important source of intergalactic X-rays, but it also has applications in the laboratory.

To obtain the result of an inverse Compton scattering, we simply apply (6.33) in the new 'lab' frame, where the electron now moves at high speed u. For simplicity let us consider only the extreme case of a head-on collision, which will result in the maximum energy gain by the photon. Assuming the collision paths are along the x-axis, we may write

$$\mathbf{Q} = \gamma m(-u, 0, 0, c), \mathbf{P} = (E/c)(1, 0, 0, 1), \mathbf{P}' = (E'/c)(-1, 0, 0, 1), \tag{6.36}$$

E and E' being the pre- and post-collision energies of the photon. Then eqn (6.33) reads

$$2EE'/c^2 = E\gamma m(1 + u/c) - E'\gamma m(1 - u/c). \tag{6.37}$$

If we now set $1 + u/c \approx 2$ and $1 - u/c \approx 1/2\gamma^2$ (since the product is γ^{-2}), we can recast (6.37) into the form

$$\frac{E'}{E} = \frac{4\gamma^2}{1 + \frac{4\gamma E}{mc^2}}. \tag{6.38}$$

The denominator contains the ratio E/mc^2 which is now quite small, even when multiplied by γ. So the photon energy can be amplified by a factor of the order of γ^2. For example, when a photon of the cosmic microwave background ($h\nu \approx 10^{-3}\,\text{eV}$) collides with a high-energy proton, say one with $\gamma = 10^{11}$ (the rest-energy of a proton being $\sim 10^{10}\,\text{eV}$), its energy can be amplified to $\sim 10^{19}\,\text{eV}$!

6.10   Four-force and three-force

The only influences on the motion of particles that we have so far considered were collisions, and we have come quite a long way without recourse to the concept of force. But that, too, plays an important role in relativistic mechanics, as when fast charged particles move through an electromagnetic field. Even without such practical need it would be desirable to have a relativistic version of the force concept, so that relativistic mechanics might 'contain' all of Newtonian mechanics in a suitable limit.

There are at hand essentially only two reasonable definitions for the *4-force* **F** on a particle of rest-mass m_0, 4-acceleration **A** and 4-momentum **P** : $\mathbf{F} = m_0\mathbf{A}$ or $\mathbf{F} = (d/d\tau)\mathbf{P}$. But **P** is a more fundamental quantity than **A**; so we choose as the more promising definition

$$\mathbf{F} = \frac{d}{d\tau}\mathbf{P} = \frac{d}{d\tau}(m_0\mathbf{U}) = m_0\mathbf{A} + \frac{dm_0}{d\tau}\mathbf{U}. \tag{6.39}$$

As long as we have no knowledge of specific 4-forces, eqn (6.39) must indeed be regarded as a mere definition. However, it certainly satisfies the desideratum that in the absence of a force, **P** remains constant. And when we later find that the Lorentz force of Maxwell's theory fits into this pattern, it will become a law.

In Newton's theory, the conservation of momentum is a consequence of Newton's second and third laws. In our scheme, by contrast, a limited (though 4-dimensional) analog of Newton's third law is a consequence of momentum conservation and the definition (6.39): During the contact phase of a collision of two compound particles, their proper times are the same, and, of course, the sum of their 4-momenta $\mathbf{P}_1$, $\mathbf{P}_2$ remains constant. So if $\mathbf{F}_1$ and $\mathbf{F}_2$ are the respective contact 4-forces on the two particles, we have

$$\mathbf{F}_1 + \mathbf{F}_2 = \frac{d}{d\tau}(\mathbf{P}_1 + \mathbf{P}_2) = 0, \tag{6.40}$$

whence, in analogy to Newton's third law, $\mathbf{F}_1 = -\mathbf{F}_2$. (Note how we could *not* prove this from the definition $\mathbf{F} = m_0\mathbf{A}$.)

From (6.3) and (5.19) we find

$$\mathbf{F} = \frac{d}{d\tau}\mathbf{P} = \gamma(u)\frac{d}{dt}(\mathbf{p}, mc) = \gamma(u)\left(\mathbf{f}, \frac{1}{c}\frac{dE}{dt}\right), \tag{6.41}$$

where we have introduced the *relativistic 3-force* **f** defined by

$$\mathbf{f} = \frac{d\mathbf{p}}{dt} = \frac{d(m\mathbf{u})}{dt}. \tag{6.42}$$

In the slow-motion limit, **f** becomes the Newtonian force. Note how the *power* dE/dt is the complement of the 3-force in making up the 4-vector **F**, just as the energy itself is the complement of the 3-momentum in the 4-vector **P**.

From (6.39) we find, by use of (5.24) and (5.26), the first of the following equations; the second results from multiplying the rightmost term of (6.41) by $\mathbf{U} = \gamma(u)(\mathbf{u}, c)$:

$$\mathbf{F} \cdot \mathbf{U} = c^2 \frac{dm_0}{d\tau} = \gamma^2(u) \left(\frac{dE}{dt} - \mathbf{f} \cdot \mathbf{u} \right). \tag{6.43}$$

(The middle term can also be obtained by specializing the RHS to the particle's rest-frame.) This shows that $\mathbf{F} \cdot \mathbf{U}$ is the proper rate at which the particle's *internal* energy is being increased. A *rest-mass preserving* force is therefore characterized by

$$\mathbf{F} \cdot \mathbf{U} = 0, \quad \text{or} \quad \mathbf{f} \cdot \mathbf{u} = \frac{dE}{dt}, \quad \text{or} \quad \mathbf{F} = \gamma(u)(\mathbf{f}, \mathbf{f} \cdot \mathbf{u}/c). \tag{6.44}$$

Collision forces on compound particles during contact will *not* satisfy this condition. However, if the force *is* rest-mass preserving, we also see from (6.44)(ii) that the Newtonian relation

$$\mathbf{f} \cdot d\mathbf{r} = dE \tag{6.45}$$

will be satisfied, and, of course, the energy increment will be purely kinetic, as in the Newtonian case. If rest-mass is not preserved, $\mathbf{f} \cdot d\mathbf{r}$ has no such simple significance.

By applying the standard transformation pattern (5.11) to the 4-vector **F** ($F'_1 = \gamma(v)(F_1 - v/cF_4)$, etc.), we can read off—with the help of our earlier eqn (3.10)—the transformation of the components (6.41) of **F** under a standard LT. Writing Q for the power dE/dt, we thus find

$$f'_1 = \frac{f_1 - vQ/c^2}{1 - u_1 v/c^2} \quad \left(f'_1 = \frac{f_1 - v\mathbf{f} \cdot \mathbf{u}/c^2}{1 - u_1 v/c^2}, \quad \text{if } m_0 = \text{const} \right)$$

$$f'_2 = \frac{f_2}{\gamma(v)(1 - u_1 v/c^2)}, \quad f'_3 = \frac{f_3}{\gamma(v)(1 - u_1 v/c^2)} \tag{6.46}$$

and

$$Q' = \frac{Q - vf_1}{1 - u_1 v/c^2}, \tag{6.47}$$

where we have used (6.44) to get the alternative formula for f'_1. In the formal limit $c \to \infty$, the above formulae reduce to the Newtonian equations $\mathbf{f}' = \mathbf{f}$ and $Q' = Q - \mathbf{f} \cdot \mathbf{v}$.

In relativity, **f** is no longer invariant. Indeed, the transformation of **f** generally involves the velocity **u** of the particle on which the force acts. Thus a velocity-independent 3-force is not a Lorentz-invariant concept. (The 'Lorentz force' of electromagnetism, which depends on the velocity of the charged particle on which it acts, is typical of a relativistic force.) However, even in relativity we have $\mathbf{f}' = \mathbf{f}$ for the special case of a rest-mass preserving force, and among those inertial frames in standard configuration with each other whose relative velocities are parallel to the force: Let the x-direction coincide with that of the force f, so that $f_2 = f_3 = 0$. Then $\mathbf{f} \cdot \mathbf{u} = f_1 u_1$, and with that the parenthesized formula in (6.46) yields $f_1' = f_1$, while $f_2' = f_3' = 0$ is immediate from the remaining two formulae. Thus if in a given inertial frame there is, for example, a *constant* 1-directional rest-mass preserving field $\mathbf{f}$ (such as a uniform electric field), then a particle moving along the field lines 'feels' a constant force also in its rest-frame, to which, by (6.42), it responds with constant proper acceleration f/m_0. The resulting motion is therefore hyperbolic.

Note the formal similarity of the transformation equations (6.46) for $\mathbf{f}$ to the transformation equations (3.6) for $\mathbf{u}$, especially in the y- and z-directions. This is due to the similarity of the component patterns on the RHSs of eqns (6.41) and (5.20). (Cf. Exercise 5.5.)

For a rest-mass preserving force $\mathbf{f}$ we have, from (6.42) and (6.44),

$$\mathbf{f} = m\mathbf{a} + \frac{dm}{dt}\mathbf{u} = \gamma m_0 \mathbf{a} + \frac{\mathbf{f} \cdot \mathbf{u}}{c^2}\mathbf{u}. \tag{6.48}$$

This shows that while $\mathbf{a}$ is necessarily coplanar with $\mathbf{f}$ and $\mathbf{u}$, it is in general *not* parallel to $\mathbf{f}$. Unless $\mathbf{u} = 0$, there are only two other cases when it *is*, namely when the last term in (6.48) is a multiple of $\mathbf{f}$, for which $\mathbf{f}$ must be either parallel or orthogonal to $\mathbf{u}$. In the first case we write $f = f_{\parallel}$, $a = a_{\parallel}$ and find, from (6.48),

$$f_{\parallel} = \gamma m_0 a_{\parallel} + f_{\parallel} u^2/c^2; \quad \text{that is,} \quad f_{\parallel} = \gamma^3 m_0 a_{\parallel}. \tag{6.49}$$

When $\mathbf{f}$ is orthogonal to $\mathbf{u}$, we write $f = f_{\perp}$, $a = a_{\perp}$ and immediately find from (6.48) that

$$f_{\perp} = \gamma m_0 a_{\perp}. \tag{6.50}$$

It appears, therefore, that a moving particle offers different inertial resistances to the same force, according to whether it is subjected to that force longitudinally or transversely. In this way there arose the (now somewhat dated but still useful) concepts of 'longitudinal mass' $m_{\parallel} = \gamma^3 m_0$ and 'transverse mass' $m_{\perp} = \gamma m_0$. In the general case, we can split both $\mathbf{f}$ and $\mathbf{a}$ into their components along and orthogonal to $\mathbf{u}$, and accordingly rewrite eqn (6.48) in the form

$$\mathbf{f}_{\parallel} + \mathbf{f}_{\perp} = \left(\gamma m_0 \mathbf{a}_{\parallel} + \frac{\mathbf{f} \cdot \mathbf{u}}{c^2}\mathbf{u} \right) + \gamma m_0 \mathbf{a}_{\perp}, \tag{6.51}$$

from which we deduce

$$\mathbf{a} = \mathbf{f}_{\parallel}/m_{\parallel} + \mathbf{f}_{\perp}/m_{\perp}. \tag{6.52}$$

Another formula which can occasionally be very useful (cf. Exercise 6.24)—and which also applies only if m_0 is constant—can be read off from the comparison of the fourth member of (6.41) with the third member of (6.39):

$$\gamma(u)\mathbf{f} = m_0 \frac{d^2\mathbf{r}}{d\tau^2}. \tag{6.53}$$

Exercises 6

6.1. How fast must a particle move before its *kinetic* energy equals its *rest* energy? [0.866 c.]

6.2. How fast must a 1-kg cannon ball move to have the same kinetic energy as a cosmic-ray proton moving with γ factor 10^{11}? [$\sim$ 5.5 m/s.]

6.3. The mass of a hydrogen atom is 1.00814 amu, that of a neutron is 1.00898 amu, and that of a helium atom (two hydrogen atoms and two neutrons) is 4.00388 amu. Find the binding energy as a fraction of the total energy of a helium atom. [*Answer*: $\sim$0.76%.]

6.4. In the 'relativistic-billiards' collision discussed in Section 6.5, prove that for a given incident velocity of the bullet, the angle $\theta + \phi$ between the tracks of the particles after collision is least when $\theta = \phi$. [*Hint*: Consider $\tan(\theta + \phi)$.]

6.5. The position vector of the center of mass (CM) of a system of particles in any inertial frame is defined by $\mathbf{r}_{\text{CM}} = \sum m\mathbf{r} / \sum m$, the ms being the *relativistic* masses. By considering two equal particles traveling in opposite directions along parallel lines, show that the CM of a system in one IF does not necessarily coincide with its CM in another IF. Prove that, nevertheless, if the particles of the system suffer collision forces only, the CM in *every* IF moves with the velocity of the ZM frame. [*Hint*: $\sum m$, $\sum m\dot{\mathbf{r}}$ are constant; $\sum \dot{m}\mathbf{r}$ is zero between collisions, and *at* any collision we can factor out the $\mathbf{r}$ of the participating particles: $\mathbf{r} \sum \dot{m} = 0$.]

6.6. A rocket propels itself rectilinearly by giving portions of its mass a constant (backward) velocity U relative to its instantaneous rest-frame. It continues to do so until it attains a velocity V relative to its initial rest-frame. Prove that the ratio of the initial to the final rest-mass of the rocket is given by

$$\frac{M_i}{M_f} = \left(\frac{c+V}{c-V}\right)^{c/2U}.$$

Note that the least expenditure of mass needed to attain a given velocity occurs when $U = c$; that is, when the rocket propels itself with a jet of photons. [*Hint*: $(-dM)U = Mdu'$, where M is the rest-mass of the rocket, and u' is its velocity relative to its instantaneous rest-frame.

6.7. Consider a *head-on* elastic collision of a 'bullet' of rest-mass M with a stationary 'target' of rest-mass m. Prove that the post-collision γ-factor of the bullet cannot exceed $(m^2 + M^2)/2mM$. This means that for large bullet energies (with γ-factors much larger than this critical value), the relative transfer of energy from bullet to target is almost total. [*Hint*: if $\mathbf{P}$, $\mathbf{P}'$ are the pre- and post-collision 4-momenta of the bullet, and $\mathbf{Q}$, $\mathbf{Q}'$ those of the target, show, by going to the ZM frame, that $(\mathbf{P}' - \mathbf{Q})^2 \geq 0$; in fact, in the ZM frame $\mathbf{P}' - \mathbf{Q}$ has no spatial components.] The situation is radically different in Newtonian mechanics, where the pre- and post-collision velocities of the bullet are related by $u/u' = (M + m)/(M - m)$. Prove this.

6.8. Generalize eqn (6.26), for the efficiency of an elastic bombardment, to the case where the target has a different rest-mass N from that of the bullet, M. Then note

Exercises 6   127

that (as intuition would suggest) for sufficiently large N, k can be arbitrarily close to unity. [Answer: $k^{-1} = 1 + (m + 2M)/2N$.]

6.9. If $\Delta s = (\Delta r, c\Delta t)$ is the 4-vector join of two events on the worldline of a uniformly moving particle (or photon), prove that the wave-vector of its de Broglie wave is given by $L = c^{-2}v\Delta s/\Delta t$, and deduce that $v/\Delta t$ is invariant. (Compare with Exercise 4.8.)

6.10. (i) If $l = h/cm_0$ is a particle's Compton wavelength and λ is its de Broglie wavelength, prove that $\lambda = l$ when $u/c = \sqrt{2}/2$. (ii) Verify that for de Broglie waves, the frequency transformation (5.38) is equivalent to (3.10)(i), when allowance is made for α in (5.38) measuring the *incoming* angle.

6.11. If $\omega = 2\pi\nu$ is the angular frequency and $k = 2\pi/\lambda$ the wave number of the de Broglie wave belonging to a particle moving with velocity u, establish from (6.29) the important identity

$$\omega^2 - k^2c^2 = (2\pi m_0c^2/h)^2 = c^2/l^2.$$

Then deduce that $\partial\omega/\partial k = u$, where we envisage particles of the same m_0 moving with speeds differing infinitesimally from u. Since $\partial\omega/\partial k$ is the usual formula for the group velocity, we see that de Broglie wave packets travel with the speed of their particle.

6.12. A rocket propels itself rectilinearly by emitting *radiation* in the direction opposite to its motion. If V is its final velocity relative to its initial rest-frame, prove *ab initio* that the ratio of the initial to the final rest-mass of the rocket is given by

$$\frac{M_i}{M_f} = \left(\frac{c + V}{c - V}\right)^{1/2},$$

and compare this with the result of Exercise 6.6 above. [*Hint*: equate energies and momenta at the beginning and at the end of the acceleration, writing $\sum h\nu$ and $\sum h\nu/c$ for the total energy and momentum, respectively, of the emitted photons.]

6.13. In an inertial frame S, two photons of frequencies ν_1 and ν_2 travel in the positive and negative x-directions respectively. Find the velocity of the ZM frame of these photons. [*Answer*: $v/c = (\nu_1 - \nu_2)/(\nu_1 + \nu_2)$.]

6.14. Radiation energy from the sun is received on earth at the rate of 1.94 calories per minute per square centimeter. Given the distance of the sun (150 000 000 km), and that one calorie $= 4.18 \times 10^7$ ergs, find the total mass lost by the sun per second, and also the force exerted by solar radiation on a black disk of the same diameter as the earth (12 800 km) at the location of the earth. [*Answers*: 4.3×10^9 kg, 5.8×10^{13} dyne. *Hint*: Force equals momentum absorbed per unit time.]

6.15. Show that a photon cannot spontaneously disintegrate into an electron–positron pair. [*Hint*: 4-momentum conservation.] But in the presence of a stationary nucleus (acting as a kind of catalyst) it can. If the rest-mass of the nucleus is N, and that of the electron (and positron) is m, what is the threshold frequency of the photon?

Verify that for large N the efficiency is ~ 100 per cent (cf. Exercise 6.8 above), so that the nucleus then comes close to being a pure catalyst.

6.16. If one neutron and one pi-meson are to emerge from the collision of a photon with a stationary proton, find the threshold frequency of the photon in terms of the rest-mass n of a proton or neutron (here assumed equal) and that, m, of a pi-meson. [*Answer*: $c^2(m^2 + 2mn)/2hn$.]

6.17. A particle of rest-mass m decays from rest into a particle of rest-mass m' and a photon. Find the separate energies of these end products. [*Answer*: $c^2(m^2 \pm m'^2)/2m$. *Hint*: use a 4-vector argument.]

6.18. A fast electron having rest-mass m and velocity u decelerates in a collision with a heavy stationary nucleus of rest-mass M and emits a bremsstrahlung (German for 'brake-radiation') photon of frequency v. Prove that the maximum energy of the photon is given by

$$hv = \frac{c^2 m M(\gamma(u) - 1)}{M + m(1 - u/c)}$$

and occurs when the photon is emitted in the forward direction and the electron and the nucleus travel 'as a lump' after the collision. [*Hint*: Square the equation $\mathbf{P} + \mathbf{Q} - \mathbf{N} = \mathbf{P}' + \mathbf{Q}'$, where $\mathbf{P}, \mathbf{Q}, \mathbf{P}', \mathbf{Q}'$ are the pre- and post-collision 4-momenta of the electron and nucleus and $\mathbf{N}$ is the 4-momentum of the photon.]

6.19. Uniform parallel radiation is observed in two arbitrary inertial frames S and S' in which it has frequencies v and v' respectively. If p, g, σ denote, respectively, the radiation pressure, momentum density, and energy density of the radiation in S, and primed symbols denote corresponding quantities in S', prove $p'/p = g'/g = \sigma'/\sigma = v'^2/v^2$. [*Hint*: Exercise 4.20.]

6.20. For the 'Compton collision' discussed in Section 6.9 prove the relation

$$\tan \phi = (1 + hv/mc^2)^{-1} \cot\tfrac{1}{2}\theta.$$

6.21. For motion under a rest-mass preserving inverse square force $\mathbf{f} = -k\mathbf{r}/r^3$ (k = constant), derive the energy equation $m_0 \gamma c^2 - k/r =$ const. [*Hint*: eqn (6.45).]

6.22. As we have seen in (6.48), there is an acceleration component orthogonal to $\mathbf{f}$ unless $\mathbf{u}$ is either parallel or orthogonal to $\mathbf{f}$. Find a physical explanation of this fact in terms of the conservation of momentum.

6.23. For a rest-mass preserving 4-force $\mathbf{F}$, prove that $\mathbf{F}^2 = -m_0^2 \alpha^2$, m_0 being the rest-mass and α the proper acceleration of the particle on which $\mathbf{F}$ acts. Deduce that if a 3-force $\mathbf{f}$ acts on a particle of constant rest-mass m_0 moving at velocity $\mathbf{u}$ making an angle θ with $\mathbf{f}$, then $\alpha = m_0^{-1} f \gamma(u)(1 - u^2/c^2 \cos^2 \theta)$.

6.24. In an inertial frame S there is a uniform electric field E in the direction of the positive x-axis. (Electromagnetic fields are always rest-mass preserving) A particle of rest-mass m_0 and charge q is projected into this field in the y-direction with initial velocity u_i and corresponding γ-factor γ_i. Prove that its trajectory is a catenary whose equation relative to a suitably chosen origin is

$$x = \frac{c^2 m_0 \gamma_i}{qE} \cosh \frac{qEy}{cm_0 u_i \gamma_i}.$$

[*Hint*: from (6.53),

$$\frac{d^2 y}{d\tau^2} = 0, \qquad m_0 \frac{d^2 x}{d\tau^2} = \gamma qE, \qquad \gamma = \frac{1}{c}\sqrt{\left(\frac{dx}{d\tau}\right)^2 + c^2 \gamma_i^2}. \ ]$$

In spite of the occurrence of the hyperbolic cosine, the motion in the x-direction is *not* 'hyperbolic'. To see this, go to the inertial frame moving in the x-direction in which the particle momentarily has no x-velocity, only a y-velocity u_2. Show that in a succession of such x-rest-frames the acceleration of the particle in the x-direction is $qE/m_0\gamma(u_2)$, and thus not constant.

7

Four-tensors; Electromagnetism in vacuum

7.1 Tensors: Preliminary ideas and notations

As we saw in the last chapter, the mathematical formalism of 4-vectors is ideally suited for the discussion of particle mechanics, where the chief quantities involved—velocity, acceleration, momentum, and force—are indeed 4-vectors. But when it comes to electromagnetism, it turns out that 4-vectors, though important, are not enough. One might perhaps have expected that the electric and magnetic field 3-vectors $\mathbf{e}$ and $\mathbf{b}$ should give rise to corresponding field 4-vectors $\mathbf{E}$ and $\mathbf{B}$, but this is not so. Instead, they give rise to a single electromagnetic field 4-*tensor*!

Three-tensors are often encountered in elementary physics without necessarily being identified as such. For example, the components of the inertia tensor are self-effacingly called 'coefficients of inertia'. The reason is that the property which primarily characterizes tensors (namely, the way they transform when the reference coordinates are changed) rarely comes into play in elementary physics, where a single reference system is usually all one needs. So here tensors indeed mainly arise as *coefficients* which combine with vectors to form other vectors or scalars. This actually is another basic property of tensors. Thus the inertia tensor I_{ij} gives the (scalar) moment of inertia of a body about a given point and a given unit direction $\mathbf{n} = (n_i)$ as $\sum_{i,j=1}^{3} I_{ij} n_i n_j$; the stress tensor t_{ij} gives the elastic force $\mathbf{f}$ inside a body on a unit area with unit normal $\mathbf{n}$ as $f_i = \sum_{j=1}^{3} t_{ij} n_j$; and so on. But since, for a given point in a given body, and a given direction $\mathbf{n}$, the value of $\sum I_{ij} n_i n_j$ must be independent of the choice of x, y, z axes, and since we already know how vectors like n_i transform, this invariance implies a certain transformation law for the coefficients I_{ij}—which turns out to be the *tensor* transformation law! And the same is true of t_{ij} and other such coefficients.

But let us begin at the beginning. Tensors exist in spaces of all dimensions. The tensors of special relativity are associated with Minkowski space and are 4-dimensional. But since there is nothing very special about four dimensions, we shall at first discuss tensors in an N-dimensional space V^N. Indices occur naturally in tensor theory and, once we have decided on a dimension N, all our indices, like i, j, k, etc., will range over the values $1, 2, \ldots, N$. All tensors can be denoted by a 'kernel' symbol like A, B, etc., adorned with indices (both subscripts and superscripts) to indicate their type: $A_{ij}, B_i{}^{jk}$ etc. Evidently 1-index tensors (these are called *vectors*!) have N components, 2-index tensors have N^2 components, etc. But not all arrays of components that *look* like tensors *are* tensors. To constitute a tensor, the array must obey the tensor transformation law. Before we write that down, however, it will pay us to introduce some streamlined notation.

To start with, we shall adopt Einstein's *summation convention*, namely: if any index appears twice in a given term, once as a subscript and once as a superscript, a summation over the range of that index is implied. Thus, for example,

$$A_i B^i = \sum_{i=1}^{N} A_i B^i,$$

$$A_{ijk} B^{jk} = \sum_{j=1}^{N} \sum_{k=1}^{N} A_{ijk} B^{jk},$$

and so on. By a slight extension of the rule we shall also understand summation in such expressions as

$$\frac{\partial u^i}{\partial x^i}, \qquad \frac{\partial q}{\partial x^i} \frac{\mathrm{d}x^i}{\mathrm{d}t}, \text{ etc.}$$

In certain manipulations the reader may at first find it helpful to imagine the summation signs in front of the relevant terms. Since the summations are all finite, all elementary rules, such as interchanging the order of summation, differentiating under the summation sign, etc., apply. The repeated indices signaling summation are called *dummy indices*, while a non-repeated index is called a *free index*. The same free index (or indices) must occur in each term of an equation, but can be replaced throughout by another: $f_i = g_i$ says the same as $f_j = g_j$. (It is understood that such equations hold for all values of the index.) And a dummy index pair in any term can be replaced by any other; for example, $A_i B^i = A_j B^j$; this is often necessary in order to avoid the triple occurrence of an index which would lead to ambiguities.

Our next convention is the *primed index notation*. This consists in the use of primed and multiply primed indices to distinguish between various coordinate systems for the underlying space V^N. All range from 1 to N. Thus:

$$i, j, k, \ldots; \quad i', j', k', \ldots; \quad i'', j'', k'', \ldots; \quad \ldots = 1, 2, \ldots, N.$$

No special relation is implied between, say, i and i': they are as independent as i and j'. A first system of coordinates can then be denoted by $\{x^i\} = \{x^1, x^2, \ldots, x^N\}$, a second by $\{x^{i'}\} = \{x^{1'}, x^{2'}, \ldots, x^{N'}\}$, etc. Similarly the components of a given tensor in different coordinate systems are distinguished by the primes on their indices. Thus, for example, the components of some 3-index tensor may be denoted by A_{ijk} in the $\{x^i\}$ system, by $A_{i'j'k'}$ in the $\{x^{i'}\}$ system, etc. When primed indices take particular numerical values, we can prime *these*, so as not to lose sight of the relevant coordinate system. Thus, for example, when $i' = 2, j' = 3, k' = 5$, $A_{i'j'k'}$ becomes $A_{2'3'5'}$. This will already have been noted for the case of the coordinates above. (However, sometimes we adopt the simpler device of priming the kernel: A'_{235}.)

When we make a coordinate transformation from one set of coordinates x^i to another $x^{i'}$ (we often drop the braces), it will be assumed that the transformation is non-singular; that is, that the equations which express the $x^{i'}$ in terms of the x^i can be solved uniquely for the x^i in terms of the $x^{i'}$. We also assume that the functions speci-fying a transformation are differentiable as often as may be required. For convenience we write

$$\frac{\partial x^{i'}}{\partial x^i} = p_i^{i'}, \qquad \frac{\partial x^i}{\partial x^{i'}} = p_{i'}^i, \qquad \frac{\partial^2 x^{i'}}{\partial x^i \partial x^j} = p_{ij}^{i'} \tag{7.1}$$

(p for 'partial derivative'), and use a similar notation for other such derivatives.

We observe that, by the chain rule of differentiation,

$$p_{i'}^i p_{i''}^{i'} = p_{i''}^i, \qquad p_{i'}^i p_j^{i'} = \delta_j^i, \tag{7.2}$$

where δ_j^i (the *Kronecker delta*) equals 1 or 0 according as $i = j$ or $i \neq j$. It is important to note the 'index-substitution' action of δ_j^i exemplified by $A_{ikl}\delta_j^i = A_{jkl}$.

Finally note how ps can be 'flipped' from one side of an equation to the other, for example,

$$A_i p_{i'}^i = B_{i'} \Rightarrow A_i = B_{i'} p_i^{i'}. \tag{7.3}$$

For proof, multiply the original equation by $p_j^{i'}$.

7.2   Tensors: Definition and properties

A. Definition of tensors

(i) An object having components $A^{ij\cdots n}$ in the x^i system of coordinates and $A^{i'j'\cdots n'}$ in the $x^{i'}$ system is said to behave as a *contravariant* tensor under the transformation $\{x^i\} \to \{x^{i'}\}$ if

$$A^{i'j'\cdots n'} = A^{ij\cdots n} p_i^{i'} p_j^{j'} \cdots p_n^{n'}. \tag{7.4}$$

(ii) Similarly, $A_{ij\cdots n}$ is said to behave as a *covariant* tensor under $\{x^i\} \to \{x^{i'}\}$ if

$$A_{i'j'\cdots n'} = A_{ij\cdots n} p_{i'}^i p_{j'}^j \cdots p_{n'}^n. \tag{7.5}$$

(iii) Lastly, $A_{l\cdots n}^{i\cdots k}$ is said to behave as a *mixed* tensor (contravariant in $i\cdots k$ and covariant in $l\cdots n$) under $\{x^i\} \to \{x^{i'}\}$ if

$$A_{l'\cdots n'}^{i'\cdots k'} = A_{l\cdots n}^{i\cdots k} p_i^{i'} \cdots p_k^{k'} p_{l'}^l \cdots p_{n'}^n. \tag{7.6}$$

Note that (7.6) evidently subsumes both (7.4) and (7.5) as special cases. The reader should perhaps be reminded that there are r separate summations going on in the above formulae if the tensors have r indices.

At a given point in V^N the ps are pure numbers. Thus the tensor transforma-tions (7.4)–(7.6) are linear: the components in the new coordinate system are linear functions of the components in the old system, the coefficients being products of the ps. Contravariant tensors involve derivatives of the new coordinates $x^{i'}$ with respect to the old, x^i, covariant tensors involve the derivatives of the old coordinates with respect to the new, and mixed tensors involve both types of derivatives. The convention of using subscripts for covariance and superscripts for contravariance, together with the requirement that the free indices on both sides of the equations must balance, serve as a good mnemonic for reproducing eqns (7.4)–(7.6).

If we simply say an object *is* a tensor it is understood that the object *behaves* as a tensor under *all* non-singular differentiable transformations of the coordinates of V^N. An object which behaves as a tensor only under a certain subgroup of non-singular differentiable coordinate transformations, like the Lorentz transformations, may be called a 'qualified tensor', and its name should be qualified by an adjective recalling the subgroup in question, as in 'Lorentz tensor', more commonly called '4-tensor'. These tensors are, as a matter of fact, the (qualified) tensors used in special relativity. The so-called 'Cartesian tensors' of classical physics behave tensorially under orthogonal transformations (rotations) of the Cartesian coordinates x, y, z.

The above definitions, when applied to a tensor with *no* indices (a *scalar*) imply $A' = A$ (no ps!), whence a scalar is just a function of position in V^N, independent of the coordinate system. A scalar is therefore often called an *invariant*.

The *zero tensor* of any type $A^{i\cdots k}_{l\cdots n}$ is defined as having all its components zero in all coordinate systems. It is clear from (7.6) that it *is* a tensor. For brevity it is usually written as 0, with the indices omitted.

Evidently we must call two tensors *equal* if they have the same components in all relevant coordinate systems. Now the *main theorem* of the tensor calculus (trivial in its proof, profound in its implications) is this: if two tensors of the same type (that is, having the same number of subscripts and superscripts) have equal components in *any one* coordinate system then they have equal components in *all* coordinate systems. This is an immediate consequence of the definition (7.6). This 'theorem' implies that tensor-(component) equations always express physical or geometrical facts, namely facts transcending the coordinate system used to describe them.

B. Three basic tensors

The most basic *contravariant* tensor is the coordinate differential dx^i. For, by the chain rule of differentiation, we have

$$dx^{i'} = p^{i'}_i \, dx^i.$$

Under *linear* transformations the coordinate differences Δx^i transform like the dx^i and are themselves tensors. The '4-vectors' we introduced in Chapter 5 are now seen to be the contravariant one-index Lorentz tensors in Minkowski space.

The most basic *covariant* tensor is the gradient of a function of position $\phi(x^1, x^2, \ldots, x^N)$. For, if we write

$$\phi_{,i} = \frac{\partial \phi}{\partial x^i}, \tag{7.7}$$

we have, again by the chain rule,

$$\phi_{,i'} = \phi_{,i} \, p^i_{i'}.$$

The Kronecker delta introduced in (7.2) is a mixed tensor. To see this, we first use its 'index substitution' property and then (7.2)(ii):

$$\delta^i_j \, p^{i'}_i \, p^j_{j'} = p^{i'}_j \, p^j_{j'} = \delta^{i'}_{j'}.$$

C. The group properties

Tensor transformations satisfy the two so-called group properties, *symmetry* and *transitivity*. In other words, if $A^{\cdots}_{\cdots}$ transforms tensorially from one coordinate system $\{x^i\}$ to another, $\{x^{i'}\}$, then it transforms tensorially also back from $\{x^{i'}\}$ to $\{x^i\}$; and if, in addition, it transforms tensorially from $\{x^{i'}\}$ to $\{x^{i''}\}$, then it does so also directly from $\{x^i\}$ to $\{x^{i''}\}$. The 'flip' property (7.3) of the ps makes symmetry obvious:

$$A^{i'}_{j'} = A^i_j \, p^{i'}_i \, p^j_{j'} \Rightarrow A^{i'}_{j'} \, p^i_{i'} \, p^{j'}_j = A^i_j,$$

while the δ-property (7.2)(ii) ensures transitivity:

$$A^{i''}_{j''} = A^{i'}_{j'} \, p^{i''}_{i'} \, p^{j'}_{j''} = \left(A^i_j \, p^{i'}_i \, p^j_{j'} \right) p^{i''}_{i'} \, p^{j'}_{j''} = A^i_j \, p^{i''}_i \, p^j_{j''}.$$

(Higher-index tensors obviously behave in the same way.) As an important corollary, we can construct a tensor by starting with arbitrary components in one system of coordinates $\{x^i\}$ and then defining the components in all other admissible coordinate systems by tensorially transforming away from $\{x^i\}$. For if $A_{i'}$ and $A_{i''}$, say, are related to A_i tensorially, then so is A_i to $A_{i'}$ by symmetry and thus $A_{i''}$ to $A_{i'}$ by transitivity.

D. Tensor algebra

The algebra of tensors consist of four basic operations—*sum, outer product, contraction,* and *index permutation*—which all have the property of producing tensors from tensors. All can be defined by the relevant operations on the tensor components, but must then be checked for tensor character.

The *sum* $C^{i\cdots}_{k\cdots}$ of two tensors $A^{i\cdots}_{k\cdots}$ and $B^{i\cdots}_{k\cdots}$ of the same type is defined thus:

$$C^{i\cdots}_{k\cdots} = A^{i\cdots}_{k\cdots} + B^{i\cdots}_{k\cdots}.$$

Trivially it is a tensor (we again exhibit the proof for a particular case):

$$C^{i'}_{k'} = A^{i'}_{k'} + B^{i'}_{k'} = A^i_k \, p^{i'}_i \, p^k_{k'} + B^i_k \, p^{i'}_i \, p^k_{k'}$$
$$= (A^i_k + B^i_k) \, p^{i'}_i \, p^k_{k'} = C^i_k \, p^{i'}_i \, p^k_{k'}.$$

Note, however, that the sum of tensors at *different* points of V_N is not generally a tensor since in the third step above we could not generally pull out the ps. (It is this which complicates the concepts of derivative and integral in tensor analysis.) But under *linear* coordinate transformations the ps are constant, and then the sum of tensors even at different points is a tensor. Analogous remarks apply to the product of tensors to be defined next.

If $A_{...}^{...}$ and $B_{...}^{...}$ are tensors of arbitrary types, the juxtaposition of their components defines their *outer product*. Thus, for example,

$$C_{klm}^{ij} = A_k^i B_{lm}^j$$

is a tensor of the type indicated by its indices. (The simple proof is left to the reader.) As a particular case, $A_{...}^{...}$ could be a scalar. In conjunction with sum, therefore, we see that any linear combination of tensors of equal type is a tensor.

Contraction of a tensor of type (s, t) (s superscripts, t subscripts) consists in the replacement of one superscript and one subscript by a dummy index pair, and results in a tensor of type $(s - 1, t - 1)$. For example, if A_{klm}^{ij} is a tensor, then

$$B_{km}^j = A_{khm}^{hj}$$

is a tensor of the type indicated by its indices. (The proof is left to the reader.) Contraction in conjunction with outer product results in an *inner product*, for example, $C_{ikl} = A_{ij} B_{kl}^j$. A most important particular case of contraction or inner multiplication arises when no free indices remain: the result is an invariant. For example, $A_i^i, A_{ij}^{ij}, A_{ij} A^{ij}$ are invariants if the As are tensors. (A particular case: $\delta_i^i = N$.)

The last of the algebraic tensor operations is *index permutation*. For example, from a given set of tensor components A_{ijk} we can form differently ordered sets like $B_{ijk} = A_{ikj}$ or $C_{ijk} = A_{jki}$, etc., all of which constitute tensors, as is immediately clear from (7.5). Such index permutations are permissible among superscripts as well as among subscripts. As a result, 'symmetry' relations among tensor components, such as $A_{ij} = A_{ji}$ or $A_{ijk} + A_{jki} + A_{kij} = 0$, are tensor equations and thus coordinate independent.

E. Differentiation of tensors

We shall write

$$\frac{\partial}{\partial x^r} \left(A_{l...n}^{i...k} \right) = A_{l...n,r}^{i...k}.$$

[A special case of this notation has already been used in (7.7) in defining $\phi_{,i}$.] Then if $A_{l...n}^{i...k}$ is a tensor defined throughout a region, differentiation of the general tensor component transformation (7.6) yields (by use of $\partial/\partial x^{r'} = \partial/\partial x^r p_{r'}^r$):

$$A_{l'...n',r'}^{i'...k'} = A_{l...n,r}^{i...k} p_i^{i'} \cdots p_k^{k'} p_{l'}^l \cdots p_{n'}^n p_{r'}^r + P_1 + P_2 + \cdots,$$

where P_1, P_2, etc., are terms involving derivatives of the ps. [It should be noted that a product with implied summations—like the right-hand side of (7.6)—can be differentiated with complete disregard of these summations, since sum and derivative commute.] Under general coordinate transformations, therefore, $A^{i\cdots k}_{l\cdots n,r}$ is not a tensor. But under *linear* coordinate transformations (ps constant) $A^{i\cdots k}_{l\cdots n,r}$ behaves as a tensor of the type indicated by all its indices, including r, since then the Ps vanish. By a repetition of the argument, all higher-order partial derivatives,

$$A^{i\cdots k}_{l\cdots n,rs} = \frac{\partial^2}{\partial x^r \partial x^s}\left(A^{i\cdots k}_{l\cdots n}\right)$$

etc. also behave as tensors under linear transformations, each partial differentiation adding a new covariant index.

By a similar argument, the derivative of a tensor with respect to a scalar, such as $\mathrm{d}A^{\cdots}_{\cdots}/\mathrm{d}\tau$, is a tensor under *linear* coordinate transformations.

F. The metric

So far no special structure has been assumed for V^N. But many spaces in which tensors play a role are *metric* spaces; that is, they possess a rule which assigns 'distances' to pairs of neighboring points. In particular, one calls a space (*pseudo-*)*Riemannian* if there exists an invariant quadratic differential form

$$\mathbf{ds}^2 = g_{ij}\,\mathrm{d}x^i\,\mathrm{d}x^j,\tag{7.8}$$

where the gs are generally functions of position, and are subject only to the restriction $\det(g_{ij}) \neq 0$. They may, without loss of generality, be assumed to be symmetric: $g_{ij} = g_{ji}$, since the 'mixed' gs only occur in pairs, for example, $(g_{12} + g_{21})\mathrm{d}x^1\mathrm{d}x^2$. If $\mathbf{ds}^2 > 0$ when $\mathrm{d}x^i \neq 0$, the space is called *strictly* Riemannian. Euclidean N-space, which has $\mathbf{ds}^2 = (\mathrm{d}x^1)^2 + \cdots + (\mathrm{d}x^N)^2$ if the coordinates are suitably chosen, is only one example. Since we require $\mathbf{ds}^2$ to be an invariant, it follows (see Exercise 7.2) that g_{ij} must be a tensor. We call it the *metric tensor*, and (7.8) the *metric*.

In Riemannian spaces one often adopts a notation for vectors analogous to that in Euclidean spaces. Thus one writes $\mathbf{A}$ for A^i, etc., and defines the *scalar product* of two vectors as the invariant

$$\mathbf{A} \cdot \mathbf{B} = g_{ij}A^i B^j.\tag{7.9}$$

In pseudo-Riemannian spaces such as Minkowski space, the *square* of a vector,

$$\mathbf{A}^2 = \mathbf{A} \cdot \mathbf{A} = g_{ij}A^i A^j,\tag{7.10}$$

can be a positive or a negative real number. From $\mathbf{A}^2$ one defines the (non-negative) *magnitude* $|\mathbf{A}|$ or simply A, by the equation

$$A = \left|\mathbf{A}^2\right|^{1/2} \geq 0.\tag{7.11}$$

The metric ds^2 itself can be regarded as the square of the differential displacement vector $\mathbf{ds} = dx^i$, whose magnitude is denoted by ds. The reader will recall all this as being precisely what we did in the particular case of 4-vectors.

In Riemannian spaces there exists a fifth basic algebraic tensor operation, namely the *raising and lowering of indices*. For this purpose we define g^{ij} as the elements of the inverse of the *matrix* (g_{ij}). Because of the symmetry of (g_{ij}), its inverse (g^{ij}) is also symmetric. The g^{ij} are defined uniquely by the equations

$$g^{ij} g_{jk} = \delta^i_k. \tag{7.12}$$

If $g^{i'j'}$ denote the tensor transforms of g^{ij} in the $x^{i'}$ system [according to (7.4)], then, by the form-invariance of tensor component equations (since g_{ij} and δ^i_j are tensors), we have from (7.12)

$$g^{i'j'} g_{i'k'} = \delta^{i'}_{k'}.$$

But these are also the equations that uniquely define the inverse $(g^{i'j'})$ of the matrix $(g_{i'j'})$. Hence the g^{ij} defined by equations like (7.12) in all coordinate systems constitute a contravariant tensor said to be *conjugate* to g_{ij}.

Now the operations of raising and lowering indices consist in forming inner products of a given tensor with g_{ij} or g^{ij}. For example, given a contravariant vector A^i, we define its covariant components A_i by the equations

$$A_i = g_{ij} A^j. \tag{7.13}$$

Conversely, given a covariant vector B_i, we define its contravariant components B^i by the equations

$$B^i = g^{ij} B_j. \tag{7.14}$$

As can easily be verified, these operations are consistent, in that the raising of a lowered index, and vice versa, leads back to the original component. They can of course be extended to raise or lower any or all of the free indices of any given tensor: for example, if $A_{ij}{}^k$ is a tensor we can define $A^i{}_{jk}$ by the equations

$$A^i{}_{jk} = g^{ir} g_{ks} A_{rj}{}^s.$$

Note how it may sometimes be convenient, for instant recognition, to use dummies from a distant part of the index alphabet. Note also that when we anticipate raising and lowering of indices, we should write the indices in staggered form so that no superscript is directly above a subscript; for example, $C^{ij}{}_{kl}$ rather than C^{ij}_{kl}. It should also be pointed out that we think of, say, $A_{ij}{}^k$ and $A^i{}_{jk}$ as merely different descriptions of the *same* object.

Lastly, the reader should note and verify

$$g_{ij} A^i B^j = A^i B_i, \tag{7.15}$$

$$g^i{}_j = \delta^i_j, \tag{7.16}$$

the 'see-saw' rule for any dummy index pair:

$$A_i B^i = A^i B_i, \tag{7.17}$$

and the conservation of symmetries, for example:

$$A_{ij} = \pm A_{ji} \quad \Leftrightarrow \quad A^{ij} = \pm A^{ji} \quad \Leftrightarrow \quad A^i_j = A_j{}^i. \tag{7.18}$$

One use of index shifting is that it allows us to form inner products of tensors that could not otherwise be so combined (for example, $A_{ij} B^{ij}$ from A_{ij} and B_{ij}). Another use occurs in the construction of tensor equations, where all terms must balance in their covariant and contravariant indices (for example, $A_{ij} + B_{ij} = C_{ij}$ from A_{ij}, B^{ij}, and $C^i{}_j$).

G. Four-tensors

In special relativity, the underlying space V^N of the above general theory becomes Minkowski space and the 'admissible coordinates' become the standard coordinate systems

$$x^\mu = (x, y, z, ct). \tag{7.19}$$

We shall use Greek indices $\mu, \nu, \ldots$, to run from 1 to 4 and occasionally Latin indices $i, j, \ldots$, to run from 1 to 3. The relevant coordinate transformations are the general Lorentz transformations (Poincaré transformations) and quantities behaving tensorially under such transformations are called *Lorentz tensors* or *4-tensors*. We generally denote them by capital letters, such as U^μ, $E_{\mu\nu}$, etc., but the metric traditionally is $g_{\mu\nu}$. Lorentz transformations are *linear*, and so the simplifications that we alluded to above apply: tensors from *different* events can be added and multiplied together, and partial and scalar derivatives of tensors are tensors.

The relevant metric is [cf. (5.1)]

$$\mathbf{ds}^2 = g_{\mu\nu}\, dx^\mu\, dx^\nu = c^2\, dt^2 - dx^2 - dy^2 - dz^2, \tag{7.20}$$

which makes

$$g_{\mu\nu} = \text{diag}(-1, -1, -1, 1) = g^{\mu\nu} \tag{7.21}$$

in all admissible coordinate systems. The implications of these simple patterns for the raising and lowering of indices [cf. (7.13), (7.14)] are as follows:[1]

$$A_i = g_{i\mu} A^\mu = -A^i \quad (i = 1, 2, 3),$$

$$A_4 = g_{4\mu} A^\mu = A^4. \tag{7.22}$$

So, for example, if $U^\mu = \gamma(u)(\mathbf{u}, c)$—this is the index way of writing our earlier eqn (5.20)—then $U_\mu = \gamma(u)(-\mathbf{u}, c)$.

---

[1] In the case of *Cartesian tensors*, based on Euclidean space with metric $g_{ij} = \text{diag}(1, \ldots, 1) = g^{ij}$, $A_i = A^i$ and there is no difference between covariance and contravariance.

Also, whenever the $g_{\mu\nu}$ in all admissible coordinate systems are constants (as is the case here) we can raise and lower indices that precede a differentiation without ambiguity:

$$A^{\mu}{}_{,\sigma} = (A_{\nu,\sigma})g^{\nu\mu} = (A_{\nu}g^{\nu\mu})_{,\sigma}.$$

We may note from (7.21) that the 'zero-component lemma' of vector theory does not carry over without modification (cf. Exercise 7.11) to tensors; for example, $g_{12} = 0$ in all permissible coordinate systems, yet $g_{\mu\nu}$ does not vanish.

From the standard Lorentz transformation (5.11) written with 'ct' as the fourth coordinate, and from its obvious inverse, we can read off the ps that must be used in transforming all 4-tensors under standard Lorentz transformations [cf. (7.1)]:

$$p_1^{1'} = p_4^{4'} = \gamma, \qquad p_4^{1'} = p_1^{4'} = -\gamma v/c, \qquad p_2^{2'} = p_3^{3'} = 1, \qquad (7.23)$$

$$p_{1'}^{1} = p_{4'}^{4} = \gamma, \qquad p_{4'}^{1} = p_{1'}^{4} = \gamma v/c, \qquad p_{2'}^{2} = p_{3'}^{3} = 1, \qquad (7.24)$$

and all others vanish. Thus, for example, for a tensor $A^{\mu\nu}$ we have [cf. (7.4)]:

$$A^{1'2'} = A^{\mu\nu} p_{\mu}^{1'} p_{\nu}^{2'} = A^{\mu 2} P_{\mu}^{1'} = \gamma\left(A^{12} - \frac{v}{c}A^{42}\right), \qquad (7.25)$$

and so on.

According to the relativity principle, the laws of physics must have the same form in all inertial frames. This strongly suggests 4-tensor equations as the ideal expression of such laws, since they have the very property of being true in all frames if true in one. (Only for the laws of relativistic quantum theory does one need even more general 'building bricks' than 4-vectors and 4-tensors, namely spinors.)

7.3 Maxwell's equations in tensor form

Having prepared the mathematical groundwork, we are now ready to discuss electro-magnetism. Any such discussion must inevitably begin with a choice of units. The currently favored SI system (Système International) is extremely inconvenient in relativity, where it masks the inherent symmetry between the electric and the magnetic fields. Accordingly we choose the older *Gaussian* or *cgs* (centimeter-gram-second) system of units in which the charge q is defined so that the Coulomb force is $q_1 q_2/r^2$ and Lorentz's force law reads as in eqn (7.28) below. In accordance with our convention of denoting 3-vectors by lower-case letters, we write **e** and **b** for the electric and magnetic fields more usually denoted by **E** and **B**.

Maxwell's great achievement was to complete the various laws of the electro-magnetic field that had been discovered previously and to consolidate them into a self-consistent system of differential equations now known as *Maxwell's equations*:

$$\operatorname{div}\mathbf{e} = 4\pi\rho, \qquad \operatorname{curl}\mathbf{b} = \frac{1}{c}\frac{\partial\mathbf{e}}{\partial t} + \frac{4\pi\mathbf{j}}{c}, \tag{7.26}$$

$$\operatorname{div}\mathbf{b} = 0, \qquad \operatorname{curl}\mathbf{e} = -\frac{1}{c}\frac{\partial\mathbf{b}}{\partial t}. \tag{7.27}$$

Note that the first entry in each line is a scalar equation while the second is a 3-vector equation, giving us eight equations in all. Equations (7.26) connect the field to its sources—the charge density ρ and the current density $\mathbf{j}$. Equations (7.27), on the other hand, represent restrictions on the field, which, as we shall see below, amount to the necessary and sufficient condition for the existence of the usual potentials. Note that Maxwell's equations are *linear* differential equations, which has the consequence that two solutions $(\mathbf{e}, \mathbf{b}, \rho, \mathbf{j})$ can be superposed to form a third.

None of these equations says anything about the action of the field on the motion of a charged particle. That is the role of Lorentz's force law:

$$\mathbf{f} = q\left(\mathbf{e} + \frac{\mathbf{u} \times \mathbf{b}}{c}\right), \tag{7.28}$$

in which q is the charge and $\mathbf{u}$ the velocity of the particle. Unlike relativistic mass, *q is an invariant*.

There is no way to fit the five eqns (7.26)–(7.28) into the scheme of Newtonian relativity, where $\mathbf{f}' = \mathbf{f}$. What Maxwell had in fact discovered, without knowing it, was a set of equations that fit *perfectly* into the scheme of special relativity. Thus, special relativity found here a theory whose laws needed no amendment.[2] Nevertheless, recognizing the relativity of Maxwell's theory made a considerable difference to our understanding of it, and also to the techniques of problem solving in it. Moreover, its basic formal simplicity became apparent now for the first time. Within the Galilean framework, Maxwell's theory was a rather unnatural and complicated construct. Within relativity, on the other hand, it is one of the two or three simplest conceivable theories of a field of force.

To see how Maxwell's equations can be written in 4-tensorial form, thus demonstrating their Lorentz-invariance, let us begin with the Lorentz force. We have already seen [after (6.47)] that a force which acts on a particle independently of its velocity is not a Lorentz-invariant concept. Thus a relativistic force law *must* involve that velocity. From a 4-dimensional standpoint, the simplest dependence of a 4-force F^μ on the particle's 4-velocity U^μ is a linear one: $F^\mu = A^\mu{}_\nu U^\nu$, where the $A^\mu{}_\nu$ are tensorial coefficients. [Indeed, the 3-dimensional Lorentz force of eqn (7.28) is linear in the 3-velocity $\mathbf{u}$.] Let us also suppose that the force is proportional to the charge q of the particle, and that q is invariant from frame to frame. Then, lowering the index μ

---

[2] This is true, at least, of Maxwell's theory in vacuum. On the other hand, Minkowski's extension of that theory to the interior of 'ponderable' media (1908) was a purely relativistic development.

and introducing a factor $1/c$ for later convenience, we can 'guess' the tensor equation

$$F_\mu = \frac{q}{c} E_{\mu\nu} U^\nu, \tag{7.29}$$

thereby introducing the *electromagnetic field tensor* $E_{\mu\nu}$. We would surely want the force F_μ to be rest-mass preserving, which, according to (6.43), requires $F_\mu U^\mu = 0$. So we need

$$E_{\mu\nu} U^\mu U^\nu = 0 \tag{7.30}$$

for all U^μ, and hence the antisymmetry of the field tensor:

$$E_{\mu\nu} = -E_{\nu\mu}. \tag{7.31}$$

Now, antisymmetric 4-tensors have the pleasant property (cf. Exercise 7.10) that their six independent non-zero components split into two sets of three which transform as 3-vectors under rotations. (Compare this with the fact that the first three components of 4-vectors similarly transform as 3-vectors under rotations.) Let us therefore define the electric and magnetic field 3-vectors **e** and **b**, respectively, in every inertial frame, by setting

$$E_{\mu\nu} = \begin{pmatrix} 0 & -b_3 & b_2 & -e_1 \\ b_3 & 0 & -b_1 & -e_2 \\ -b_2 & b_1 & 0 & -e_3 \\ e_1 & e_2 & e_3 & 0 \end{pmatrix}, \quad E^{\mu\nu} = \begin{pmatrix} 0 & -b_3 & b_2 & e_1 \\ b_3 & 0 & -b_1 & e_2 \\ -b_2 & b_1 & 0 & e_3 \\ -e_1 & -e_2 & -e_3 & 0 \end{pmatrix}. \tag{7.32}$$

(We exhibit the contravariant form for future reference.)

With these definitions, we can verify that the tensor equation (7.29) is completely equivalent to the Lorentz force law (7.28). For, writing out the four components of the RHS of (7.29),

$$(q/c)\gamma(u) E_{\mu\nu} (\mathbf{u}, c)^\nu,$$

where the last factor denotes the νth component of its parenthesis, we get

$$(q/c)\gamma(u)(-b_3 u_2 + b_2 u_3 - ce_1, b_3 u_1 - b_1 u_3 - ce_2,$$
$$- b_2 u_1 + b_1 u_2 - ce_3, \mathbf{e} \cdot \mathbf{u}). \tag{7.33}$$

On the other hand, from (6.44)(iii) and our remark after (7.22), we have, first for *any* rest-mass preserving 3-force **f** acting on a particle with velocity **u**, and then for the Lorentz force (7.28) in particular, the corresponding (covariant) 4-force

$$F_\mu = \gamma(u)\left(-\mathbf{f}, \frac{\mathbf{f} \cdot \mathbf{u}}{c} \right) = q\gamma(u)\left(-\mathbf{e} - \frac{\mathbf{u} \times \mathbf{b}}{c}, \frac{\mathbf{e} \cdot \mathbf{u}}{c} \right). \tag{7.34}$$

But (7.33) is precisely the RHS of (7.34), and so our assertion is proved.

We may note, incidentally, that any rest-mass preserving 3-force which is *velocity-independent* in some particular inertial frame S_0 *must* be a Lorentz-type force (though not every Lorentz force is ever velocity-independent). For the corresponding 4-force then satisfies a tensor equation of type (7.29), with $E_{\mu\nu}$ defined by an array like (7.32)(i) with zero bs in S_0 and in other frames by the tensor transformation law away from S_0. In other frames $E_{\mu\nu}$ will then have non-zero bs [cf. (7.56) below] and thus give rise to a full Lorentz-type force law (7.28).

Now for Maxwell's equations. Since they are differential equations, let us consider a *continuous* distribution of sources, and, at first, one that has a *unique* 3-velocity **u** at each event. Let us define the *proper charge density* ρ_0 of this continuum as the charge density ρ measured in the local (comoving) rest-frame. Then in the lab frame, where the sources move with velocity **u**, we shall have, because of length contraction,

$$\rho = \rho_0 \gamma(u). \tag{7.35}$$

We then define the 3-*current density* as

$$\mathbf{j} = \rho\mathbf{u}, \tag{7.36}$$

and recall that the conservation of charge is expressed by the following *equation of continuity*:

$$\frac{\partial\rho}{\partial t} + \operatorname{div}\mathbf{j} = 0. \tag{7.37}$$

Since div **j** measures the outflux of charge from a (small) unit volume in unit time, this equation simply states that to the precise extent that charge leaves a small region, the total charge inside that region must decrease.

We next define the *four-current density* J^μ by the first of the following equations (provisionally—hence the brackets),

$$J^\mu = [\rho_0 U^\mu = \rho_0 \gamma(u)(\mathbf{u}, c)] = (\mathbf{j}, c\rho), \tag{7.38}$$

and note that it allows us to express the equation of continuity (7.37) in the following tensorial form [cf. (7.7)]:

$$J^\mu{}_{,\mu} = 0. \tag{7.39}$$

But while the above definition of J^μ is evidently tensorial, it is not universally applicable. In real life (for example, in a current-carrying metal wire, where the ions stand still while the electrons drift) the local charge velocities are *not* all the same. We assume that we can then divide the charges into classes which *do* have unique velocities and charge densities, and we define the *effective* ρ, **j** and **J** as the sums of the ρs, **j**s and **J**s of all the classes. These effective quantities will still satisfy (7.37), the extremities of (7.38), and (7.39), since each class satisfies these equations separately. But there will now be *no* effective 4-velocity U^μ of the charge distribution in terms of which we could write $J^\mu_{\text{eff}} = \rho_{0\,\text{eff}} U^\mu_{\text{eff}}$. For whereas the sum of any number of future-pointing timelike vectors is timelike (cf. Section 5.6), the same is not true of a *mixture* of future- and past-pointing timelike vectors. And that, in general, is just what the partial **J**s are, since the partial charge densities ρ can be positive or negative. The effective **J** can thus be timelike, spacelike, or even null, and hence not necessarily a multiple of a 4-velocity.

With these preliminaries out of the way, we are ready to 'guess' the 4-tensor forms of Maxwell's equations:

$$E^{\mu\nu}{}_{,\mu} = \frac{4\pi}{c} J^\nu \tag{7.40}$$

and

$$E_{\mu\nu,\sigma} + E_{\nu\sigma,\mu} + E_{\sigma\mu,\nu} = 0. \tag{7.41}$$

By reference to the definitions (7.32) one easily sees that the first three of the four equations (7.40) ($\nu = 1, 2, 3$) are indeed equivalent to Maxwell's equation (7.26)(ii), while the fourth ($\nu = 4$) is equivalent to (7.26)(i). Giving values 1, 2, 3, respectively, to the indices μ, ν, σ in (7.41) results in Maxwell's equation (7.27)(i), while the values 2, 3, 4; 3, 4, 1; and 4, 1, 2 give the three components of (7.27)(ii). All other sets of values either yield one of the equations already obtained or $0 = 0$.

It is very satisfactory to note that the field equation (7.40) immediately implies the equation of continuity (7.39), since $E^{\mu\nu}{}_{,\mu\nu}$ is symmetric in its subscripts but antisymmetric in its superscripts, and must therefore vanish ($E^{12}{}_{,12} = -E^{21}{}_{,21}$ etc.). This (in 3-vector language) rather than any empirical evidence, was Maxwell's reason for adding the 'displacement current' $\partial \mathbf{e}/\partial t$ to his equations, even though it puzzled some of his contemporaries. Without it, charge conservation would be violated.

Today we can give an alternative justification for the displacement current (and indeed for most of the Maxwell equations) from relativity: from the postulation that Lorentz's force law (7.28) holds in all inertial frames (which essentially just defines the **e**, **b** field in each frame), it follows that the coefficients $E_{\mu\nu}$ of its 4-dimensional formulation (7.29), (7.31) constitute a 4-tensor (cf. Exercise 7.2). Then the mere validity of Maxwell's *first* equation, div $\mathbf{e} = 4\pi\rho$ (essentially Coulomb's law), in all inertial frames, necessarily implies *all* of his second equation, (7.26)(ii), by the zero-component lemma of Section 5.6. For, as we have seen, that first equation is equivalent to the vanishing of the fourth component of the 4-vector consisting of the LHS minus the RHS of eqn (7.40), while (7.26)(ii) is equivalent to the vanishing of the remaining components. In the same way [cf. (7.59) below] the validity of the entire second (tensor-) field equation (7.41) (guaranteeing the existence of a potential) is a consequence of the validity of just div $\mathbf{b} = 0$ in all inertial frames; that is, of the absence of magnetic monopoles.

7.4 The four-potential

One of the remarkable properties of the Maxwell field is that it allows itself to be expressed in terms of a 'potential', which considerably simplifies the mathematics. In tensor language, Maxwell's potential is a covariant 4-vector Φ_μ, whose derivatives determine the field according to the equation

$$E_{\mu\nu} = \Phi_{\nu,\mu} - \Phi_{\mu,\nu}. \tag{7.42}$$

An immediate consequence of this equation is the field equation (7.41). But in the theory of differential equations the converse is also well known: eqn (7.41) is not only a necessary but also a sufficient condition for the existence of such a potential. (Compare this to the more familiar condition $g_{i,j} - g_{j,i} = 0$, or curl $\mathbf{g} = 0$, for the existence of a *scalar* potential ϕ such that $g_i = -\phi_{,i}$, or $\mathbf{g} = -\mathrm{grad}\,\phi$—a result which actually holds in all dimensions.)

Although the potential Φ_μ turns out to be *not* uniquely determined by the field tensor $E_{\mu\nu}$, picking *any* potential Φ_μ in a frame S and its tensor transforms in all other frames clearly guarantees eqn (7.42) in all frames. We may therefore take Φ_μ to be a tensor.

In terms of the 4-potential, the first tensor field equation, (7.40), which is now all that remains to be satisfied, can be re-expressed as

$$g^{\mu\sigma}(\Phi_{\nu,\sigma\mu} - \Phi_{\sigma,\nu\mu}) = \frac{4\pi}{c} J_\nu. \tag{7.43}$$

And this would be particularly simple if the second term on the LHS, $-\Phi^\mu{}_{,\nu\mu}$, were to vanish, because then these four equations for Φ_ν decouple. But this can be arranged, since the potential is not unique. Consider a second potential, $\tilde{\Phi}_\mu$, satisfying (7.42). The difference, $\Psi_\mu = \Phi_\mu - \tilde{\Phi}_\mu$, must satisfy

$$\Psi_{\mu,\nu} - \Psi_{\nu,\mu} = 0,$$

which implies that Ψ_μ is the gradient of some scalar Ψ, and so

$$\Phi_\mu = \tilde{\Phi}_\mu + \Psi_{,\mu}. \tag{7.44}$$

From an arbitrary potential $\tilde{\Phi}_\mu$ we now can, by a proper choice of Ψ, construct a new potential Φ_μ with the property

$$\Phi^\mu{}_{,\mu} = 0. \tag{7.45}$$

It is merely necessary for the scalar Ψ to satisfy

$$g^{\mu\nu}\tilde{\Phi}_{\mu,\nu} + g^{\mu\nu}\Psi_{,\mu\nu} = 0, \quad \text{that is} \quad \Box\Psi = -g^{\mu\nu}\tilde{\Phi}_{\mu,\nu}, \tag{7.46}$$

where $\Box$ denotes the D'Alembertian operator:

$$\Box \equiv {}_{,\mu\nu}g^{\mu\nu} \equiv \frac{1}{c^2}\frac{\partial^2}{\partial t^2} - \frac{\partial^2}{\partial x^2} - \frac{\partial^2}{\partial y^2} - \frac{\partial^2}{\partial z^2}. \tag{7.47}$$

In the theory of differential equations this equation is known to be in general solvable for Ψ, relative to any inertial frame S. In fact, it can be shown that

$$\Psi(\mathrm{P}) = \frac{1}{4\pi}\int \frac{[F]\,\mathrm{d}V}{r} \Rightarrow \Box\Psi = F, \tag{7.48}$$

where $F = F(x, y, z, ct)$ is any integrable function (it must be 'sufficiently small' at infinity), $[F]$ denotes the value of F 'retarded' by the light-travel time to the origin P from the position $\mathbf{r}$ of dV, and the volume integral extends over the entire 3-space in S. We can therefore assume, without loss of generality, that the potential satisfies the *'Lorenz (not Lorentz!) gauge condition'* (7.45), and then the field equations (7.43) decouple and simplify to

$$\Box \Phi_\mu = \frac{4\pi}{c} J_\mu, \tag{7.49}$$

in conjunction with (7.45)

Of course, eqn (7.49) itself can be solved by the same integral formula (7.48) as solved eqn (7.46):

$$\Phi_\mu(\mathrm{P}) = \frac{1}{c} \int \frac{[J_\mu]\,dV}{r}. \tag{7.50}$$

This is a very convenient explicit solution, since (i) it automatically satisfies the Lorenz gauge condition by virtue of eqn (7.39) (cf. Exercise 7.8), (ii) the integral can be shown to be tensorial[3] and (iii) it can be shown to be the *unique* solution in the absence of 'incoming radiation'.

Note that in charge-free regions ($J_\mu = 0$), eqn (7.49) reduces to the wave equation with speed c, showing that disturbances of the potential in vacuum are propagated at the speed of light. The potential, however, is often regarded as an 'unphysical' auxiliary. But we find at once from (7.49), (7.42), and the commutativity of partial derivatives, that when $J_\mu = 0$ the field $E_{\mu\nu}$ *itself* satisfies the wave equation

$$\Box E_{\mu\nu} = 0. \tag{7.51}$$

Hence disturbances of the *field* propagate in vacuum at the speed of light. This result, first discovered by Maxwell, was, of course, the basis for his hypothesis that light consisted of electromagnetic waves.

We finally observe that by setting

$$\Phi_\mu = (-\mathbf{w}, \phi), \tag{7.52}$$

thereby defining the 3-vector potential $\mathbf{w}$ and the 'scalar' potential ϕ (not a scalar invariant!) in each inertial frame, we can re-express our earlier eqn (7.42) in the more familiar 3-dimensional form:

$$\mathbf{b} = \mathrm{curl}\,\mathbf{w}, \qquad \mathbf{e} = -\mathrm{grad}\,\phi - \frac{1}{c}\frac{\partial \mathbf{w}}{\partial t}, \tag{7.53}$$

and similarly for eqns (7.45) and (7.49):

$$\frac{\partial \phi}{\partial t} + c\,\mathrm{div}\,\mathbf{w} = 0 \quad \text{(Lorenz gauge condition)} \tag{7.54}$$

$$\Box \phi = 4\pi\rho \qquad \Box \mathbf{w} = \frac{4\pi}{c}\mathbf{j} \quad \text{(field equations)}. \tag{7.55}$$

---

[3] Cf. W. Rindler, *Special Relativity*, Oliver & Boyd, 1966, Sec. 41.

7.5   Transformation of **e** and **b**. The dual field

In Newton's theory, if I move a point-mass through the lab, its gravitational field will be just 'a moving isotropic field', in the sense that at any frozen moment it is exactly the usual inverse-square field that would emanate from the point-mass if it were at rest. Beginning students often expect the same to hold in electromagnetism—for example, that the field of a moving bar-magnet would be just 'a moving dipole field'. This is not the case. A moving bar-magnet produces a *deformed* moving dipole field *plus* an electric field, and a moving point-charge produces a deformed moving Coulomb field plus a magnetic field.

It is the tensor property of $E^{\mu\nu}$ that determines how **e** and **b** transform from one inertial frame to another—how, for example, a pure dipole **b**-field in S′ is seen in S. To find the general transformation, let us, for example, apply (7.25) to the component $E^{1'2'}$ of (7.32)(ii), which yields

$$-b_{3'} = \gamma(-b_3 + ve_2/c).$$

In the same way we obtain all the other entries in the following list:

$$e_{1'} = e_1, \qquad e_{2'} = \gamma(e_2 - vb_3/c), \qquad e_{3'} = \gamma(e_3 + vb_2/c),$$
$$b_{1'} = b_1, \qquad b_{2'} = \gamma(b_2 + ve_3/c), \qquad b_{3'} = \gamma(b_3 - ve_2/c). \qquad (7.56)$$

The inverse transformations, as usual, are obtained by a v-reversal.

Thus on transforming from one inertial frame to another, the **e** and **b** fields get intermingled. A field which is either purely electric or purely magnetic in one frame will have both electric and magnetic components in the general frame. This 'explains' the transverse deflection of a point-charge moving through a purely magnetic field: in its rest-frame, the charge feels an electric field! For example, a field having only a b_3 component in S has a transverse electric component $e_{2'} = -(\gamma v/c)b_3$ in S′, and that is what is felt by a charge riding in S′ through S.

The transformation (7.56) is remarkable in its symmetry (apart from signs) between **e** and **b**. And Maxwell's equations themselves share this pseudo-symmetry in the absence of sources; their asymmetry arises only from the presumed non-existence of 'magnetic charge'. The formal interchange

$$\mathbf{b} \mapsto \mathbf{e}, \qquad \mathbf{e} \mapsto -\mathbf{b} \qquad (7.57)$$

leaves the set of transformation equations (7.56) invariant: writing them as $(\mathbf{e}', \mathbf{b}') = T(\mathbf{e}, \mathbf{b})$, we find $(-\mathbf{b}', \mathbf{e}') = T(-\mathbf{b}, \mathbf{e})$. Since these transformations characterize an antisymmetric tensor, the array of components formed from $E_{\mu\nu}$ by the interchange (7.57) also constitutes a tensor. It is called the *dual* of $E_{\mu\nu}$ and we denote it by $\overset{*}{E}_{\mu\nu}$

or $B_{\mu\nu}$:

$$B_{\mu\nu} = \begin{pmatrix} 0 & -e_3 & e_2 & b_1 \\ e_3 & 0 & -e_1 & b_2 \\ -e_2 & e_1 & 0 & b_3 \\ -b_1 & -b_2 & -b_3 & 0 \end{pmatrix}, \qquad B^{\mu\nu} = \begin{pmatrix} 0 & -e_3 & e_2 & -b_1 \\ e_3 & 0 & -e_1 & -b_2 \\ -e_2 & e_1 & 0 & -b_3 \\ b_1 & b_2 & b_3 & 0 \end{pmatrix}.$$

(7.58)

One more dualization brings us back to minus where we started: $\overset{*}{E}_{\mu\nu} = B_{\mu\nu}$ and $\overset{*}{B}_{\mu\nu} = -E_{\mu\nu}$.

Dualizing the pair of Maxwell equations (7.26) *without the source terms* produces (7.27)(ii) and the negative of (7.27)(i). Reference to (7.40) then shows that we can write (7.41) as $B^{\mu\nu}{}_{,\mu} = 0$. Maxwell's equations can therefore be written in the form

$$E^{\mu\nu}{}_{,\mu} = \frac{4\pi}{c} J^{\nu}, \qquad B^{\mu\nu}{}_{,\mu} = 0.$$

(7.59)

The $e - b$ pseudo-symmetry (7.57) of the theory also suggests consideration of a Lorentz-invariant force law analogous to (7.30),

$$\tilde{F}_{\mu} = -\frac{\tilde{q}}{c} B_{\mu\nu} U^{\nu},$$

(7.60)

which could apply to hypothetical magnetic point-charges $\tilde{q}$. In 3-vector form it reads, as dualization and sign-reversal of the RHS of eqn (7.28) shows at once,

$$\tilde{\mathbf{f}} = \tilde{q}\left(\mathbf{b} - \frac{\mathbf{u} \times \mathbf{e}}{c}\right).$$

(7.61)

If we accept that the north-pole of a relatively long and thin bar-magnet of length a and magnetic moment μ experiences a force $\tilde{q}\mathbf{b}$ in a pure **b**-field, where $\tilde{q} = \mu/a$, then eqn (7.61)—being valid in one inertial frame—would have to be valid in all inertial frames, and thus also in the presence of electric fields. This formula, for example, gives us directly the torque experienced by a bar-magnet moving through an electric field (cf. Exercises 1.2 and 7.12).

Just as associated with each 4-vector A^{μ} there is a scalar invariant, namely its 'square' $A_{\mu}A^{\mu}$, so also associated with each antisymmetric tensor $E_{\mu\nu}$ there are *two* and only two independent scalar invariants, which we shall call X and Y, and whose expressions in terms of **e** and **b** can be read off from (7.32) and (7.58):

$$X = \tfrac{1}{2} E_{\mu\nu} E^{\mu\nu} = \mathbf{b}^2 - \mathbf{e}^2 = -\tfrac{1}{2} B_{\mu\nu} B^{\mu\nu},$$

(7.62)

$$Y = \tfrac{1}{4} E_{\mu\nu} B^{\mu\nu} = \mathbf{e} \cdot \mathbf{b}.$$

(7.63)

Physically they tell us, for example, that if the magnitudes of the electric and magnetic fields are equal ($e = b$) at some event in one frame, they are equal at that event in all frames; and if the fields are orthogonal in one frame ($\mathbf{e} \cdot \mathbf{b} = 0$), they are orthogonal in all frames. When $X = Y = 0$ the field is said to be *null*, and **e** is perpendicular to **b** and $e = b$ in all frames. If a field is purely magnetic at some event in one frame $(X > 0)$, it cannot be purely electric in another, and vice versa. If the angle between **e** and **b** is acute in one frame $(Y > 0)$, it cannot be obtuse in another.

Can *any* field at some event be either purely electric or purely magnetic in *some* frame? Evidently not unless $Y = 0$ and $X \neq 0$. But then, indeed, there is a whole family of inertial frames, all in standard configuration with each other, and all having the *same* pure **e** or **b** field (according as $X < 0$ or $X > 0$) in the x-direction. If $X = Y = 0$ the field is null and allows no further simplification. In the general case, when $Y \neq 0$, we can at least achieve $\mathbf{e} \propto \mathbf{b}$ in a whole family of inertial frames in standard configuration, with both the **e** and **b** fields in the x-direction (cf. Exercises 7.14 and 7.15). These specializations can often simplify a calculation.

7.6   The field of a uniformly moving point charge

As a good example of the power that special relativity brought to electromagnetic theory, we shall calculate the field of a uniformly moving charge q by that typically relativistic method of looking at the situation in a frame where everything is obvious, and then transforming to the general frame. In the present case, all is obvious in the rest-frame S$'$ of the charge, where there is a simple Coulomb field of form

$$\mathbf{e}' = (q/r'^3)(x', y', z'), \quad \mathbf{b}' = 0, \quad (r'^2 = x'^2 + y'^2 + z'^2), \tag{7.64}$$

if, as we shall assume, the charge is at the origin. Now we transform to the usual second frame S in which the charge moves with velocity $\mathbf{v} = (v, 0, 0)$. We shall calculate the field everywhere in S at the instant $t = 0$ when the charge passes through the origin. Using the second line of (7.56) with $\mathbf{b}' = 0$, the inverse equations of the first line, and the Lorentz transformations of x', y', z' with $t = 0$, we find, successively,

$$\mathbf{b} = (v/c)(0, -e_3, e_2), \quad \mathbf{e} = (e_1', \gamma e_2', \gamma e_3') = (q\gamma/r'^3)(x, y, z)$$
$$r'^2 = \gamma^2 x^2 + y^2 + z^2 = \gamma^2 r^2 - (\gamma^2 - 1)(y^2 + z^2)$$
$$= \gamma^2 r^2 [1 - (v^2/c^2)\sin^2 \theta], \tag{7.65}$$

where θ is the angle between the vector $\mathbf{r} = (x, y, z)$ and the x axis. Thus,

$$\mathbf{b} = \frac{1}{c}\mathbf{v} \times \mathbf{e}, \quad \mathbf{e} = \frac{q\mathbf{r}}{\gamma^2 r^3 [1 - (v^2/c^2)\sin^2 \theta]^{3/2}}, \tag{7.66}$$

and this completes our derivation of the instantaneous field of a moving charge. It is interesting to note that the electric field at any given point in S is directed away from the point where the charge is at that instant, though (because of the finite speed of propagation of all effects) it cannot be *due* to the position of the charge at that instant and would be the same even if the charge had been deflected shortly before getting

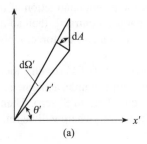
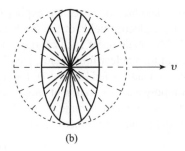

(a)   (b)

Fig. 7.1

there! Both the electric and the magnetic field strengths still fall off as $1/r^2$ in any fixed direction away from the charge. But there is now an angular dependence: the fields are strongest in a plane at right angles to $\mathbf{v}$ (where $e = \gamma q/r^2$) and weakest fore and aft (where $e = q/\gamma^2 r^2$); and this is even more true of $\mathbf{b}$ than of $\mathbf{e}$ since the former has an extra θ-dependence proportional to $\sin\theta$. The $\mathbf{b}$ field lines are circles around the line of motion.

Let us recall the field-line representation of the electric field, wherein the field strength is given by the number of lines per unit area. This is made possible by Gauss's theorem (outflux of $\mathbf{e} = 4\pi \times$ enclosed charge—the integral version of $\operatorname{div}\mathbf{e} = 4\pi\rho$) as applied to a bundle of field lines. That same theorem also tells us that the number of field lines emanating from a charge is independent of its motion; that motion, therefore, merely deforms the line pattern. Interestingly, in the case of a uniformly moving point charge, the isotropic Coulomb field-line picture gets exactly Lorentz-contracted in the direction of motion, as illustrated in Fig. 7.1(b). To see this, let us transform the Coulomb field-line picture in S' like a rigid body. In S' the solid angle of a thin pencil of lines of x-cross-sectional area dA at (x', y', z'), is given by $d\Omega' = dA\cos\theta'/r'^2 = dAx'/r'^3$ (see Fig. 7.1a). In S the corresponding solid angle is given by $d\Omega = dAx/r^3$, whence, by reference to (7.65),

$$\frac{d\Omega'}{d\Omega} = \frac{x'r^3}{xr'^3} = \frac{\gamma r^3}{r'^3} = \frac{1}{\gamma^2[1-(v^2/c^2)\sin^2\theta]^{3/2}}. \tag{7.67}$$

Comparing this with (7.66) we find for the electric field strength in S

$$e = \frac{q\,d\Omega'}{r^2\,d\Omega} = \frac{n}{d\Sigma},$$

where $n = q\,d\Omega'$ is the 'number of lines of force' in the pencil in S' and $d\Sigma = r^2\,d\Omega$ is the normal cross-section of the pencil in S. Since there are as many lines in $d\Omega$ as in $d\Omega'$, we see that the density of these lines represents the electric field strength in S as well as in S', and this establishes our assertion.

Of course, this result is purely a consequence of the laws of electrodynamics and can be obtained without the explicit use of SR [for example, using (7.50)]. Lorentz so obtained it, and thereon based an 'explanation' of the length contraction of material bodies: if the electric fields of the fundamental charges 'contract', then so must all matter, if it is made up of such charges. (Lorentz's argument, perforce, ignored the existence of the nuclear force fields, etc.)

We have already noted [cf. (7.57)] the pseudo-symmetry of Maxwell's theory in $\mathbf{e}$ and $\mathbf{b}$. Suppose a configuration of *magnetic* charges (for example, a magnetic dipole) is at rest in some inertial frame S', making the electric field in S' zero. Then in any other inertial frame S the electric and magnetic fields everywhere will be related by

$$\mathbf{e} = -\frac{1}{c}\mathbf{v} \times \mathbf{b},$$

$\mathbf{v}$ being the velocity of the sources relative to S; the proof is quite analogous to that for (7.66)(i) above, and now a consequence of $\mathbf{e}' = 0$.

7.7   The field of an infinite straight current

As another example of relativistic reasoning in electromagnetic theory, we shall derive the field of an infinite straight current—which will also throw some light on the phenomenon of length contraction. We begin by calculating the field of an infinite *static* line of charge. By symmetry, the electric field $\mathbf{e}$ must be radial, and by applying Gauss's outflux theorem to a cylinder having the line charge as axis, we find that its strength is given by

$$e = \frac{2\lambda_0}{r}, \tag{7.68}$$

where λ_0 is the line density of charge and r the radial distance. Now suppose that such an infinite line charge with proper line density λ_0 moves with velocity v relative to an inertial frame S. Because of length contraction, its line density λ in S is $\gamma\lambda_0$, and there it constitutes a current $i = \gamma\lambda_0 v$. Let us identify the location of the static line charge with the x'-axis of the usual second frame S'. By (7.68), the only non-vanishing component of the electromagnetic field at the typical point $P(0, r, 0)$ in S' is $e_{2'} = 2\lambda_0/r$. Transforming this field to S by use of the inverses of (7.56), we find as the only non-vanishing components

$$e_2 = \frac{2\gamma\lambda_0}{r} = \frac{2\lambda}{r}, \qquad b_3 = \frac{2\gamma\lambda_0 v}{cr} = \frac{2i}{cr}. \tag{7.69}$$

Note that the strength of the magnetic field is only a fraction v/c of that of the electric field, and another factor of order u/c reduces its effect, by comparison, on a charge moving with velocity u [see (7.28)]. Moreover, in a laboratory current, the electron drift velocity is only of the order of a millimeter per second. As C. W. Sherwin has said, it is hard to believe that this magnetic force, which has to suffer a denominator c^2, is the 'work force' of electricity, responsible for the operations of motors and generators. And again, considering that this force arises from transforming a purely electric field to another frame having very small velocity relative to the first, A. P. French has remarked: who says that relativity is important only for velocities comparable to that of light? The reason is that an ordinary current moves a very big charge: there are something like 10^{23} free electrons per cubic centimeter of wire. Their *electric* force, if it were not neutralized, would be enormous—of the order of two million tons of weight on an equal cubic centimeter at a distance of 10 km.

But that force *is* neutralized in a 'real' current flowing in a wire. Such a current corresponds to *two* superimposed linear charge distributions, one at rest and one in motion. The positive metal ions are at rest while the free electrons move, say, with velocity $-v$. Before the current is turned on, we can think of the ions schematically as a row of chairs on which the free electrons sit. When the current flows, the electrons play musical chairs, always moving to the next chair in unison since there can be no build-up of charge. But this means that the electrons are now as far apart *in motion* as are the stationary ions. Hence the respective line densities of ions and electrons are equal and opposite *in the lab frame*, say $\pm\lambda$, and the current is given by $i = \lambda v$. As can be seen from (7.69), the electric fields will cancel exactly, while the magnetic field is given as before by $2i/cr$.

Consider now a test charge moving with velocity **u** parallel to the wire. It experiences a force in the direction $\mathbf{u} \times \mathbf{b}$; that is, radially towards or away from the wire. In its rest-frame, where it can be affected only by **e** fields, it sees two moving lines of positive and negative charge, respectively. Without length contraction, it would always see these two lines as having equal and opposite charge densities, and so never feel a force. But, in fact, *because* of length contraction, these two charge densities differ in absolute value in all frames but the rest-frame of the wire (cf. Exercise 7.20). And it is this difference which provides the net **e** field that causes the charge to accelerate in its rest-frame. This is about the closest we get to a direct manifestation of length contraction—a contraction difference arising from a speed difference of a few millimeters per second!

7.8   The energy tensor of the electromagnetic field

In this final section we give a brief account of Minkowski's energy tensor, and show how the electromagnetic field itself possesses energy, momentum, and stress—just like a material continuum. In this way the well-validated conservation laws of mechanics can be extended to interactions between charged matter and electromagnetic fields. For example, if we simultaneously release two oppositely charged particles from rest and they accelerate towards each other, where does the kinetic energy come from? Or, relative to another inertial frame where one of these charges begins to move before the other, where does the momentum come from? One *can* use 'potential' energy and even momentum as bookkeeping devices, but there are good reasons (both formal and physical) for considering the field itself able to exchange energy and momentum with matter. According to Einstein's general relativity, energy (that is, mass), momentum, and stress all curve spacetime (measurably, in principle) at their location, and so that location is no longer a matter of convention.

Consider, then, a charged 'fluid' in the presence of an electromagnetic field but subject to no other external forces. We define a (*Lorentz-*)*4-force density vector* $\tilde{K}^\mu$ by a procedure analogous to that which we used in the definition of J^μ [cf. after (7.38)], namely dividing all moving charges into classes with unique ρ_0 and U^μ (and corresponding **u**). For each class, $\tilde{K}^\mu$ is the Lorentz force (7.29) per unit *proper* volume,[4] which thus contains a charge ρ_0; and the *effective* $\tilde{K}^\mu$ exerted on the whole fluid at any event by the field is defined as the sum of these partial $\tilde{K}^\mu$s, and hence is itself a 4-vector:

$$\tilde{K}^\mu = \sum \frac{\rho_0}{c} E^\mu{}_\nu U^\nu = \frac{1}{c} E^\mu{}_\nu J^\nu; \qquad (7.70)$$

the last equation follows from (7.38) and subsequent text. On the other hand, if we write $\tilde{\mathbf{k}}$ for the partial 3-forces per unit proper volume, we have, from (6.44),

$$\tilde{K}^\mu = \sum \gamma(u)(\tilde{\mathbf{k}}, c^{-1}\tilde{\mathbf{k}} \cdot \mathbf{u}) = (\mathbf{k}, c^{-1}\, \partial W/\partial t), \qquad (7.71)$$

where **k** is the *total* 3-force per unit *lab* volume, and $\partial W/\partial t$ is the rate of work done by the field on the fluid in a unit lab volume; for the effect of each γ-factor in the summation in (7.71) is to convert from the various unit proper volumes to the unit volume fixed in the lab.

[From here to the end of (7.77) we now are in for some serious calculating!] By use of Maxwell's equation (7.40) we can next eliminate all reference to the sources from the RHS of eqn (7.70), and write that equation in the form

$$\tilde{K}_\mu = \frac{1}{4\pi} E_{\mu\nu} E^{\sigma\nu}{}_{,\sigma} = \frac{1}{4\pi}\left[(E_{\mu\nu} E^{\sigma\nu})_{,\sigma} - E_{\mu\nu,\sigma} E^{\sigma\nu}\right]. \qquad (7.72)$$

The second term in the bracket can also be transformed into a derivative thus:

$$E_{\mu\nu,\sigma} E^{\sigma\nu} = \tfrac{1}{2}\left(E_{\mu\nu,\sigma} - E_{\mu\sigma,\nu}\right)E^{\sigma\nu} = \tfrac{1}{2}E_{\sigma\nu,\mu} E^{\sigma\nu} = \tfrac{1}{4}\left(E_{\sigma\nu} E^{\sigma\nu}\right)_{,\mu}, \qquad (7.73)$$

where in the first step we used the antisymmetry of $E^{\sigma\nu}$, in the second step the Maxwell equation (7.41), and finally the see-saw rule. So, combining (7.71), (7.72), and (7.73), we have

$$\tilde{K}^\mu = (\mathbf{k}, c^{-1}\, \partial W/\partial t) = -M^{\mu\nu}{}_{,\nu}, \qquad (7.74)$$

where

$$M^{\mu\nu} := \frac{1}{4\pi}\left(E^\mu{}_\lambda E^{\lambda\nu} + \frac{1}{4} g^{\mu\nu} E_{\lambda\rho} E^{\lambda\rho}\right). \qquad (7.75)$$

Eqn (7.74) is the Lorentz force law adapted to a charged continuum instead of a particle, with all reference to the charge or velocity of the continuum eliminated. When that charge is zero, the RHS must reduce to zero.

---

[4] What is meant by a 'quantity per unit volume' is, of course, the limit of that quantity for a finite volume V divided by V, as $V \to 0$. On the other hand, heuristically, we can regard it as the actual quantity over an actual unit volume, provided we choose our units sufficiently small; and we are always at liberty to do so. A corresponding remark applies to a 'quantity per unit time'.

The tensor $M^{\mu\nu}$ which has surfaced here, and which is easily shown to be symmetric and trace-free,

$$M^{\mu\nu} = M^{\nu\mu}, \qquad M^{\mu}{}_{\mu} = 0, \tag{7.76}$$

is the fundamentally important (Minkowski-)*energy tensor* of the electromagnetic field. For its components we find, directly from the definition and from (7.32):

$$M^{44} = \frac{1}{8\pi}(e^2 + b^2) =: \sigma = \text{energy density}$$

$$cM^{4i} = \frac{c}{4\pi}(\mathbf{e} \times \mathbf{b})_i =: s_i = \text{energy-current density (Poynting vector)}$$

$$M^{ij} = -\frac{1}{4\pi}[e_i e_j + b_i b_j + \tfrac{1}{2}g_{ij}(e^2 + b^2)] =: p_{ij} = \text{Maxwell stress tensor,} \tag{7.77}$$

where, however, the last entries in each line (the physical interpretations) have yet to be justified.

To that end, let us look at the separate components of eqn (7.74). First, setting $\mu = 4$ yields (after a sign-change)

$$-\frac{\partial W}{\partial t} = \frac{\partial \sigma}{\partial t} + \text{div } \mathbf{s}, \tag{7.78}$$

which leads us to identify σ as the energy density and $\mathbf{s}$ as the energy-current density *of the field*. For if $\partial W/\partial t$ is the rate of work done by the field on the fluid, $-\partial W/\partial t$ can be regarded as the rate of work done by the fluid on the field. And this should equal the increase of field energy, $\partial \sigma/\partial t$, in the unit volume, plus the outflux of field energy, div $\mathbf{s}$, from that volume, in unit time.

Next, let us set $\mu = i$ in eqn (7.74). This yields (again after a sign-change)

$$-k_i = \frac{\partial c^{-2} s_i}{\partial t} + \frac{\partial p_{ij}}{\partial x^j}. \tag{7.79}$$

Since $\mathbf{k}$ is the force of the field on the fluid, $-\mathbf{k}$ can be regarded as the force of the fluid on the field, and this should equal the rate at which field momentum is generated inside a unit volume. The first term on the RHS of eqn (7.79) should therefore represent the increase of field momentum inside a unit volume, and the divergence-like second term the outflux of field momentum from that volume. Accordingly we recognize $c^{-2}s_i$ as the *momentum density* of the field and p_{ij} as the *momentum-current density*; that is, the flux of i-momentum through a unit area normal to the j-direction per unit time.

We note how c^{-2} times the energy current s_i of eqn (7.78) serves as momentum density in eqn (7.79). This is a striking manifestation of Einstein's mass–energy equivalence. For an ordinary fluid we would have: energy current = energy density × velocity = c^2 mass density × velocity = c^2× momentum density. Even though in general there is *no* way to associate a unique velocity with a given electromagnetic field (cf. Exercise 7.15), or, equivalently, to assign a rest-frame to it, it still has momentum and energy current and the two are related 'as usual'.[5]

Recall that force equals rate of change of momentum. If a machine-gun fires bullets into a wooden block, that block experiences a force equal to the momentum absorbed in unit time; that is, equal to the momentum current. Maxwell accordingly regarded p_{ij} as the i-component of the total force which the field (!) on the negative side of a unit area normal to the j-direction exerts on the field on the positive side. Thus p_{ij} is called the *Maxwell stress tensor* of the field.

One can take this idea quite seriously. For example, consider a pure **e**-field parallel to the x-axis at some point of interest. Then from (7.77), $p_{11} = -(1/8)\pi e^2$. But p_{11} is pressure in the x-direction. This leads to the idea that there is *tension* (negative pressure) along the electric field lines. Such tension 'explains' the attraction between unlike charges. Similarly, $p_{22} = +(1/8)\pi e^2$. So there is *pressure* at right angles to the field lines, tending to separate them. This 'explains' the repulsion between like charges. (The reader will recall the well-known field-line patterns—materialized by iron filings—between both equal or opposite 'magnetic charges'; and the pattern is the same for electronic charges.)

7.9   From the mechanics of the field to the mechanics of material continua

It was Maxwell, Poynting, Heaviside, and J. J. Thomson who, elaborating on earlier ideas of Faraday and W. Thomson, found the mathematical expressions for the 'mechanical' characteristics of the electromagnetic field, namely its conserved energy and momentum, and its stress. Then, after the advent of special relativity, Minkowski in 1908 discovered that these mechanical field quantities combine to form a single 4-tensor and one, moreover, whose divergence links the field to any charged fluid with which it might interact [through eqn (7.74)]. That tensor showed how inextricably energy, momentum, and stress are intertwined—how each enters into the expression of any one of them in a transformation to a different inertial frame.

It is perhaps curious that the relativistic mechanics of the electromagnetic field was understood before that of down-to-earth material continua. But this stems from the accidental early close association of relativity with electromagnetism, out of whose turmoil it had been born. The realization that Minkowski's energy tensor of the field *must* carry over to ordinary matter was left to von Laue and came within a year.

From what we have seen in the preceding section it is clear that the mass density ρ, the momentum density **g** and the momentum-current density p_{ij} of a material continuum (for example, a liquid, a solid, or a gas), having necessarily the same

---

[5] This velocity-indeterminacy of an immaterial energy current is by no means peculiar to the electromagnetic field. Consider a locomotive pushing a freight car. Energy is evidently being transferred to the car across the buffers at a well-determined rate. But whether it is a big energy concentration flowing slowly or a small one flowing fast, it is impossible to decide.

transformation properties as their electromagnetic counterparts, must also combine into a 4-tensor $T^{\mu\nu}$ as in (7.77):

$$T^{\mu\nu} = \begin{pmatrix} p_{ij} & c\mathbf{g} \\ c\mathbf{g} & c^2\rho \end{pmatrix}. \tag{7.80}$$

If $\tilde{K}^\mu$, in analogy to (7.70), is the total (rest-mass preserving) external 4-force exerted on unit proper volume of fluid (and hence a vector), then we expect the analog of eqn (7.74) to hold:

$$\tilde{K}^\mu = (\mathbf{k}, c^{-1}\, \partial W/\partial t) = T^{\mu\nu}{}_{,\nu} \tag{7.81}$$

(recall that $-\tilde{K}^\mu$, not $+\tilde{K}^\mu$, was the force *on* the field, now replaced by the fluid).

The component versions of this tensor equation now read [cf. (7.78) and (7.79)]:

$$\frac{\partial W}{\partial t} = \frac{\partial c^2\rho}{\partial t} + \text{div }c^2\mathbf{g}, \tag{7.82}$$

$$k_i = \frac{\partial g_i}{\partial t} + \frac{\partial p_{ij}}{\partial x^j}. \tag{7.83}$$

And the meanings of these latter equations are clear: eqn (7.82) is the energy balance equation, which says (when the last term is transposed to the LHS) that the energy increase in a given volume equals the energy influx from the outside plus the work done by the external force on the continuum. [Cf. after (7.79).] Equation (7.83), on the other hand, is the momentum balance equation which says (again after the last term is transposed) that the rate of increase of momentum in a given volume is equal to the applied force plus any influx of momentum from outside the volume.

These are exactly the laws that we would wish to impose on the continuum in conformity with our earlier laws for systems of particles. Consequently we accept eqns (7.80)–(7.83) as the basis for relativistic continuum mechanics.

Analogously to the electromagnetic case, c^2 times the $\mathbf{g}$ from the last *column* of $T^{\mu\nu}$ serves as energy current in eqn (7.82), while the $\mathbf{g}$ from the last *row* serves as momentum density in eqn (7.83). It is therefore required by Einstein's mass–energy equivalence (as we have seen in the preceding section) that the fourth row and fourth column of $T^{\mu\nu}$ be equal in *all* inertial frames. But that can only happen if the entire tensor is symmetric (cf. Exercise 7.11), and thus also the p_{ij}:

$$T^{\mu\nu} = T^{\nu\mu}, \qquad p_{ij} = p_{ji}. \tag{7.84}$$

Note, however, that $T^\mu{}_\mu$ will in general not vanish; *that* is a peculiarity of the electromagnetic field.

Also in contrast to the electromagnetic field, the material continuum has a unique *rest-frame* at each event; that is, a unique inertial frame in which the 3-momentum vanishes: $\mathbf{g} = 0$. This is something we have noted for entire particle systems before. So if we could simply take the continuum in the infinitesimal neighborhood of each event as a particle system, the result would follow. As it is, we take it as an axiom.

Thus we can associate a unique velocity **u** with the continuum at each point, namely that of the rest-frame.

By going to this rest-frame and rotating the spatial axes to diagonalize the symmetric 3-tensor p_{ij} (cf. Exercise 7.1), we can always diagonalize $T^{\mu\nu}$:

$$T^{\mu\nu} = diag(p_1, p_2, p_3, c^2\rho_0), \tag{7.85}$$

where p_1, p_2, p_3 are the "principal pressures", and $c^2\rho_0$ is the energy density in the rest-frame. For a *perfect fluid* the principal pressures are all equal, say $p_i = p$. Then $T_{ij} = -pg_{ij}$ and thus T_{ij} is rotation-invariant in the rest frame. But then the pressure p is altogether invariant, as follows from the special f-invariance discussed after (6.47). Finally we find for $T^{\mu\nu}$ in the general frame, in which the perfect fluid moves with 4-velocity U^{μ},

$$T^{\nu\mu} = (\rho_0 + p/c^2)U^{\mu}U^{\nu} - pg^{\mu\nu}; \tag{7.86}$$

for this is a tensor equation, and it reduces to (7.85) in the rest-frame, where $U^{\mu} = (0, 0, 0, c)$.

It comes as a surprise to most beginners in relativistic continuum mechanics that, as we transform away from the rest-frame, we do not find the expected relation $\rho = \gamma^2(u)\rho_0$ between the mass density ρ_0 in the rest-frame and that, ρ, in the general frame (one γ coming from mass increase and one from length contraction). This is already clear from (7.86). Nor do we find the naïvely expected relation $\mathbf{g} = \rho\mathbf{u}$. We must, however, relegate these intriguing questions to the Exercises (see Exercises 7.26–7.28).

When the continuum under consideration carries a charge, and the only external force acting on it is an electromagnetic field, then the $\tilde{K}^{\mu}$ of eqn (7.81) is given by (7.70), which, as we have seen, is equivalent to (7.74). Equation (7.81) can then be written in the form

$$(T^{\mu\nu} + M^{\mu\nu})_{,\nu} = 0. \tag{7.87}$$

If there are non-electromagnetic external forces as well, those will appear on the RHS of (7.87). This equation shows (for $\mu = 4$ and $\mu = i$, respectively) that for the combination field-plus-continuum both energy and momentum satisfy balance equations in every volume element of the reference frame. $T^{\mu\nu} + M^{\mu\nu}$ can be regarded as the *total* energy tensor of all the energy carriers.

This is as far as we can take relativistic continuum mechanics here. But it will suffice for our purpose, which was to give the reader an understanding of the basic structure of the theory and how it connects to the rest of relativity. This theory has gained *practical* importance in recent decades, for it is no longer true to say that all continua of physical interest are non-relativistic. For example, near the horizon of a stellar black hole, or near the surface of a neutron star, or near the center of an atomic bomb, gases move under conditions so extreme that relativistic effects *can* become important. And on the other hand, the Minkowski-Laue energy tensor played a vital *theoretical* role in the birth of general relativity: it is the only quantity fit to serve as the source of the gravitational field; and its vanishing divergence, eqn (7.87), to a large extent determined the shape of Einstein's field equations. Lastly we observe that relativistic continuum mechanics is more basic and more general than relativistic particle mechanics, and the specialization from the former to the latter is much more direct than the opposite procedure. Thus from a logical point of view, the axioms of relativistic continuum mechanics provide the soundest basis for *all* of relativistic mechanics, but it is beyond our present scope to demonstrate this in detail.[6]

Exercises 7

7.1. If $T^{\mu\nu}$ is a 4-tensor, prove that under a spatial rotation T^{44} remains unchanged, T^{4i} and T^{i4} transform as 3-vectors, and T^{ij} transforms as a 3-tensor.

7.2. (i) Suppose for a set of coefficients $E_{\mu\nu}$, defined in every inertial coordinate system, the product $E_{\mu\nu}U^{\nu}$ is a covariant 4-vector for all *4-velocities* U^{μ}. Prove that the set $E_{\mu\nu}$ constitutes a tensor. [*Hint*: Prove $(E_{\mu'\nu'}p_{\nu}^{\nu'} - E_{\mu\nu}p_{\mu'}^{\mu}) U^{\nu} = 0$, write $()= \phi_{\mu'\nu}$, and pick $U^{\nu} \propto (0,0,0,1), (\frac{1}{2},0,0,1), (0,\frac{1}{2},0,1)$ and $(0,0,\frac{1}{2},1)$ in turn.]

(ii) Suppose for a set of coefficients c_{ij}, defined in every admissible coordinate system, the product $c_{ij}A^{ij}$ is invariant for *any* tensor A^{ij}. Deduce that the set c_{ij} itself constitutes a tensor.

(iii) Suppose for a set of *symmetric* coefficients $g_{ij} (= g_{ji})$ the product $g_{ij}B^{i}B^{j}$ is invariant for *any* vector B^{i}. Prove that the coefficients g_{ij} constitute a tensor. [*Hint*: Proceed as above and choose a B^{i} with only one non-zero component and then another with only two non-zero components.] The results (i), (ii), (iii) are varieties of the so-called *quotient rule*: the 'quotient' of two tensors is (under certain circumstances) itself a tensor.

7.3. If the coefficients a_{ij} are constant and symmetric, prove (and remember!) that

$$(a_{ij}A^{i}A^{j})_{,k} = 2a_{ij}A^{i}A^{j}_{,k}.$$

[*Hint*: Leibniz rule and interchange of dummy-index pairs.]

7.4. For any 2-index tensor $A_{\mu\nu}$ we define the tensors $A_{(\mu\nu)} := \frac{1}{2}(A_{\mu\nu} + A_{\nu\mu})$, $A_{[\mu\nu]} := \frac{1}{2}(A_{\mu\nu} - A_{\nu\mu})$, and call these the *symmetric* and *antisymmetric parts* of $A_{\mu\nu}$, respectively. Prove $A_{\mu\nu} = A_{(\mu\nu)} + A_{[\mu\nu]}$. If $B^{\mu\nu} = B^{\nu\mu}$ and $C^{\mu\nu} = -C^{\nu\mu}$, prove $A_{\mu\nu}B^{\mu\nu} = A_{(\mu\nu)}B^{\mu\nu}$ and $A_{\mu\nu}C^{\mu\nu} = A_{[\mu\nu]}C^{\mu\nu}$ for any $A_{\mu\nu}$.

7.5. Suppose a field of 4-force F_{μ} were derivable from a scalar potential Φ according to the equation $F_{\mu} = \Phi_{,\mu}$. Prove that the rest-mass of particles subjected to this force would have to vary as follows: $m_0 = \Phi/c^2 + \text{const}$. [*Hint*: (6.43).]

7.6. (i) A particle of rest-mass m_0 and charge q is injected at velocity **u** into a constant magnetic field **b** at right angles to the field lines. Use the Lorentz force law to establish that the particle will trace out a circle of radius $cm_0u\gamma(u)/qb$ with period

[6] See, for example, W. Rindler, *Introduction to Special Relativity*, Chapter 7.

$2\pi c m_0 \gamma(u)/qb$. (It was the γ-factor in the period that necessitated the development of synchrotrons from cyclotrons, at whose energies the γ was still negligible.)

(ii) If the particle is injected into the field with the same velocity but at an angle $\theta < \pi/2$ to the field lines, prove that the path is a helix, of smaller radius, but that the period for one complete cycle is the same as before.

7.7. Although it is natural to pair the four Maxwell equations as in (7.26) and (7.27), each pair corresponding to one of the 4-tensor equations (7.40), (7.41), yet it is also instructive to pair div $\mathbf{e} = 4\pi\rho$ with div $\mathbf{b} = 0$. These are the two equations that contain no time derivatives, and which can, in fact, be regarded as 'constraints' *on the initial conditions*. Verify that they are 'propagated' by the other two equations (the 'evolution equations') plus the equation of continuity (7.37). In other words, if initial conditions (fields and sources) are prescribed on a surface $t = t_0$, satisfying the constraints, then if we use the evolution equations to calculate the future, this will automatically continue to satisfy the constraints. [*Hint*: Consider, for a start, $(\partial/\partial t)(\text{div } \mathbf{e} - 4\pi\rho)$ and use (7.26)(ii) and (7.37).]

7.8. Prove that the retarded potential of eqn (7.50) automatically satisfies the Lorenz gauge condition (7.45). [*Hint*: Consider the past light cone with vertex at $P = (x_0^\nu)$ over which the integration is performed, and then give it and every volume element of it a displacement dx^ν in spacetime (which preserves each r and dV) so that $\Phi^\mu(x_0^\nu + dx^\nu) = c^{-1} \int J^\mu(x^\nu + dx^\nu) \, dV/r$ over the displaced cone; then expand Φ^μ and J^μ.]

7.9. Obtain the *Liénard-Wiechert potentials*

$$\phi = \left[\frac{q}{r(1 + u_r/c)}\right] \qquad \mathbf{w} = \left[\frac{q\mathbf{u}}{r(1 + u_r/c)}\right],$$

for a moving point-charge q, where the brackets indicate that the enclosed expressions are to be evaluated at the retarded event at the charge; r is the distance of that event from the observer and u_r the radial velocity of the charge away from the observer. [*Hint*: assume that the charge is moving uniformly and 'spot' that the Coulomb potential in its rest-frame satisfies the tensor equation

$$\Phi_\mu = qU_\mu/R^\nu U_\nu,$$

where U_μ is the 4-velocity of the charge and R^ν the (null) connecting vector from the retarded event at q to the observation event.] Noting that the integral formula (7.50) does not involve the acceleration of the charge, you may conclude that the above formulae hold even if the charge accelerates.

7.10. From one of the results of Exercise 7.1 and from the fact that the dual of an antisymmetric 4-tensor is itself a 4-tensor, deduce that for any antisymmetric 4-tensor with components written as in eqn (7.32), the triplets (e_1, e_2, e_3) *and* (b_1, b_2, b_3) transform as 3-vectors under spatial rotations. As a corollary, deduce that for any antisymmetric 3-tensor, with components written as in the leading matrices of eqn (7.32), the triplet (b_1, b_2, b_3) *is* a 3-vector. As an example, consider $c_{ij} = \text{curl } \mathbf{u} = u_{i,j} - u_{j,i}$.

7.11. (i) Utilizing results from the preceding exercise, prove the *zero-component lemma* for any antisymmetric 4-tensor: if any one of its off-diagonal components is zero in all inertial coordinate systems, then the entire tensor is zero.

(ii) Prove the following zero-component lemma for a *symmetric* 4-tensor $A_{\mu\nu} = A_{\nu\mu}$: if any one of its *diagonal* components is zero in all inertial systems, then the entire tensor is zero. [*Hint*: suppose $A_{44} \equiv 0$. Then the quantity $A_{\mu\nu}T^{\mu}T^{\nu}$ vanishes for all timelike 4-vectors T—why? Having shown that, put $T = (\epsilon, 0, 0, 1)$; then put $T = (\epsilon, \epsilon, 0, 1)$.]

(iii) Prove that if *one* off-diagonal component of a 4-tensor $B_{\mu\nu}$ is symmetric in all inertial frames (for example, $B_{12} \equiv B_{21}$), then the entire tensor is symmetric. [*Hint*: consider $B_{\mu\nu} = B_{(\mu\nu)} + B_{[\mu\nu]}$.]

7.12. Returning to the situation of Exercise 1.2, consider a small bar magnet of moment μ which lies on the x-axis and moves along that axis with velocity v. A stationary point-charge q is located at $(0, -r, 0)$ on the negative y-axis. Find the torque on the magnet in its rest-frame in two ways: (i) by using eqn (7.61), and (ii) by using eqn (7.66). [*Answer*: $q\mu\gamma v/cr^{2}$.]

7.13. Returning to the situation of Exercise 1.3, use the transformation equations (7.56) to find the force felt by a point charge q, at rest in an inertial frame, due to a small bar magnet of dipole moment μ pointing straight at it and moving transversely at velocity v. [*Answer*: $2q\mu\gamma v/cr^{3}$. *Hint*: the magnet's **b**-field in its rest-frame and on its axis is $2\mu/r^{3}$.]

7.14. If at a certain event an electromagnetic field satisfies the relations $\mathbf{e} \cdot \mathbf{b} = 0, e \neq b$, prove that there exists a frame in which $\mathbf{e} = 0$ or $\mathbf{b} = 0$. Then prove that infinitely many such frames exist, all in standard configuration with each other. [*Hint*: one of the desired frames moves in the direction $\mathbf{e} \times \mathbf{b}$; choose coordinates accordingly.]

7.15. If $\mathbf{e} \cdot \mathbf{b} \neq 0$, prove that there are infinitely many frames with common relative direction of motion, and only those, in which $\mathbf{e}$ is parallel to $\mathbf{b}$. [*Hint*: one of the desired frames moves in the direction $\mathbf{e} \times \mathbf{b}$, its velocity being given by the smaller root of the quadratic $\beta^{2} - Z\beta + 1 = 0$, where $\beta = v/c$ and $Z = (e^{2} + b^{2})/|\mathbf{e} \times \mathbf{b}|$. For the reality of β it is necessary to show that $Z > 2$.] Note that in all the above frames the Poynting (energy current) vector $c(\mathbf{e} \times \mathbf{b})/4\pi$ vanishes; so there can be no unique rest-frame for the electromagnetic energy.

7.16. There are good reasons for saying that an electromagnetic field is 'radiative' if it satisfies $\mathbf{e} \cdot \mathbf{b} = 0$ and $e = b$; that is, if it is null. Show that the field of a very fast moving point charge is essentially radiative in a plane which contains that charge and is orthogonal to its motion.

7.17. In a frame S, two identical point charges q move abreast along lines parallel to the x-axis, a distance r apart and with velocity v. Determine the force in S that each exerts on the other, and do this in two ways: (i) by use of the Lorentz force in conjunction with the field (7.66); and (ii) by transforming the Coulomb force from the rest-frame to the lab frame S. Note that this force is smaller than in the rest-frame, while each mass is greater. Here we see the dynamical reasons for the 'relativistic focusing' effect whose existence we recognized as inevitable by purely kinematic considerations in Section 3.5. Show that the dynamics leads to exactly the expected time dilation of an 'electron clock'.

7.18. Instead of the equal charges moving abreast as in the preceding exercise, consider now two *oppositely* charged particles moving with the same constant velocity but *not* abreast. Using any method, determine the forces acting on these charges in the lab frame, and show that they do *not* act along the line joining them (for example, along a rod that separates them) but instead constitute a torque $kr^2 \cos\theta \sin\theta v^2/c^2$ tending to turn that rod into orthogonality with the line of motion, where θ is the inclination of the rod and k is the factor multiplying $\mathbf{r}$ in (7.66). (Trouton and Noble, in a famous experiment in 1903, looked for this torque on charges at rest in the laboratory, which they presumed to be flying through the ether. The fact that the reaction forces of the rod on the charges might also not be in line with the rod was unknown then—it ultimately stems from $E = mc^2$, cf. Exercise 7.30 below—and the null result they obtained seemed puzzling. But it contributed to the later acceptance of relativity.)

7.19. Two parallel straight wires, at rest in the lab and a distance r apart, carry currents i_1 and i_2, respectively, but as usual are electrically neutral in the lab. Prove that the force which each exerts on the other is $2i_1 i_2/c^2 r$ per unit length, and that the force is attractive if the currents are in the same direction, and repulsive otherwise. [*Hint*: the Lorentz force.]

7.20. In the situation described in the final paragraph of Section 7.7, prove that in its rest-frame the test charge sees a net line density $-\lambda\gamma(u)uv/c^2$, $-v$ being the drift velocity of the electrons and λ the proper line (charge) density of the ions. [*Hint*: eqn (3.10).] Prove that the force $\tilde{\mathbf{f}}$ which it feels towards the wire in its rest-frame because of that net line charge, corresponds exactly to the force $\mathbf{f} = q\mathbf{u} \times \mathbf{b}/c$ on it in the rest-frame of the wire, when account is taken of the force transformation (6.46).

7.21. Verify from (7.77) that the energy tensor $M^{\mu\nu}$ is invariant under the dualization (7.57), so that in (7.75) we can replace $E_{\mu\nu}$ by $B_{\mu\nu}$. Referring to (7.62), derive the following alternative form for it:

$$M^{\nu\mu} = \frac{1}{8\pi}(E^{\mu}{}_{\lambda}E^{\lambda\nu} + B^{\mu}{}_{\lambda}B^{\lambda\nu}). \tag{7.88}$$

7.22. Give reasons why in a disordered (that is, random) distribution of pure radiation (a 'photon gas') the electromagnetic field components will satisfy the following relations on the (time) average:

(i) $e_1^2 = e_2^2 = e_3^2$, $b_1^2 = b_2^2 = b_3^2$,

(ii) $e_1 e_2 = e_2 e_3 = e_3 e_1 = 0$, $b_1 b_2 = b_2 b_3 = b_3 b_1 = 0$,

(iii) $e_2 b_3 - e_3 b_2 = e_3 b_1 - e_1 b_3 = e_1 b_2 - e_2 b_1 = 0$.

Deduce that the only non-zero components of the average energy tensor can then be written as $M^{44} = \sigma_0$, $M^{11} = M^{22} = M^{33} = p$, with $3p = \sigma_0$. Such a photon gas (like the famous $3°$K background radiation of the universe) *does* have a unique rest-frame; that is, the above relations hold in but one preferred frame, say S_0. Prove that in the general frame one then has

$$M^{\mu\nu} = p\left(\frac{4}{c^2}U^\mu U^\nu - g^{\mu\nu}\right),$$

where U^μ is the 4-velocity of S_0 and p transforms as an invariant.

7.23. (i) Use the energy tensor of the field to prove that the 3-force on each of two oppositely charged identical parallel plates at rest and facing each other in vacuum is $e^2 A/8\pi$, where A is the area of the plates and e is the strength of the uniform field between them, if edge effects are neglected.

(ii) If these plates are simultaneously released from rest, prove that (in the absence of gravity) the rate at which they gain kinetic energy is equal to the rate at which the field loses energy.

7.24. For the components σ and $\mathbf{s}$ of the energy tensor $M^{\mu\nu}$ establish the identity

$$\sigma^2 - \frac{s^2}{c^2} = \left(\frac{1}{8\pi}\right)^2 [(b^2 - e^2)^2 + 4(\mathbf{e} \cdot \mathbf{b})^2] =: \left(\frac{1}{8\pi}\right)^2 I^2,$$

the invariant I being defined by this equation [cf. (7.62), (7.63)]. Note that this is reminiscent of the invariance of the square of the 4-momentum $(\mathbf{p}, mc)$ of a particle, even though $(\mathbf{s}, \sigma c)$ does not represent a 4-vector. (But see the following exercise.)

7.25. Prove that the energy tensor $M^{\mu\nu}$ satisfies the *Rainich identities*

$$M^\mu{}_\sigma M^\sigma{}_\nu = \left(\frac{I}{8\pi}\right)^2 \delta^\mu_\nu, \qquad M_{\mu\nu}M^{\mu\nu} = \left(\frac{I}{4\pi}\right)^2,$$

where I is the invariant defined in the preceding exercise. [*Hint*: if $I \neq 0$, go to a frame where e_1 and b_1 are the only non-zero field components; for the case $I = 0$, appeal to continuity.]

7.26. Let a given point of a material continuum be momentarily at rest in an inertial frame S_0 and write ρ_0, t_0^{ij} for ρ and p_{ij} in S_0. (This t_0^{ij} is the elastic stress 3-tensor of classical mechanics.) Regarding S_0 as the S′ of the usual pair S, S′, and writing u for the usual v (u now being the velocity of the continuum in S), transform the relevant components of $T^{\mu\nu}$ from S_0 to S to establish the formulae

$$\rho = \gamma^2(u)(\rho_0 + u^2 t_0^{11}/c^4), \qquad g_1 = u\gamma^2(u)(\rho_0 + t_0^{11}/c^2),$$

$$g_2 = u\gamma(u)t_0^{12}/c^2, \qquad g_3 = u\gamma(u)t_0^{13}/c^2.$$

7.27. Give a physical explanation of the unexpected ρ-transformation of the preceding exercise as follows: Consider a small comoving cubical element of the continuum with its edges of proper length dl parallel to the axes of S_0, its rest-frame. Then imagine all the fluid around that cube instantaneously (and miraculously!) removed in S_0. In S_0 the energy of the cube after the miracle is still $c^2 \rho_0 \, dl^3$. But in S the 'left' face was exposed $\gamma(u)u \, dl/c^2$ seconds before the 'right' face. (Why?) During that time the material of the cube did work on the material to the right of it at the rate of $ut_0^{11} \, dl^2$. (Why?) After the removal is complete in S, the cube can be regarded as a compound particle and hence its mass in S will be $\gamma(u)\rho_0 \, dl^3$. So before the removal it must have been $[\gamma(u)\rho_0 + \gamma(u)u^2 t_0^{11}/c^4] \, dl^3$. (Why?) Complete the argument, bearing in mind the length contraction of the cube.

7.28. Give a physical explanation of the counter-intuitive **g**-transformation of Exercise 7.26 above, based on the fact that in relativity even a non-material energy current has momentum density, and that this must be added to the material momentum density $\rho\mathbf{u}$. Bear in mind that the non-material energy current in the i-direction is that which crosses a unit area normal to the i-direction and *fixed in the continuum*, and is therefore given by the scalar product of the force on that area with the velocity of that area.

7.29. A capacitor at rest in a frame S' consists of two identical parallel plates separated by a central non-conducting rod parallel to the x'-axis. When the plates are oppositely charged, there is a uniform **e**-field between them, if we ignore edge effects. By (7.77), the electromagnetic energy residing in the capacitor is given by $E = e^2 lA/8\pi$, e being the field strength, A the area of the plates, and l the length of the rod. Regarded from the usual second frame S, the electromagnetic energy is *reduced* by a γ-factor. (Why?) But clearly the total energy of this closed system should *increase* by a γ-factor. Resolve this apparent paradox[7] by including the energy of the rod and using the ρ-transformation of Exercise 7.26 above. [*Hint*: the force between the plates is $e^2 A/8\pi$, cf. Exercise 7.23 above.]

7.30. Give an approximate ('first-order') explanation (in the lab frame) of the Trouton–Noble experiment (cf. Exercise 7.18 above) along the following lines: as the rod moves forward and is subject to the squeeze of the two charges (which you can approximate by the Coulomb attraction), the force on the back end does positive work on the rod while that on the front end does negative work on it. So there is a constant immaterial energy flow along the rod from back to front and hence ($E = mc^2$!) a constant immaterial momentum **p** along the rod as well. If the rod has length r and is inclined at an angle θ, show that $p \approx qv \cos\theta/c^2 r$. As the rod moves, the *moment* of this momentum (that is, the angular momentum **L**) about a fixed point in the lab increases steadily: $d\mathbf{L}/dt = \mathbf{u} \times \mathbf{p}$. Verify that this is balanced (at least to first-order in v^2/c^2) by the torque of the charges, so that no torque is left over to rotate the rod!

[7] See W. Rindler and J. Denur, *Am. J. Phys.* **56**, 795 (1988).

Part II
General Relativity

8

Curved spaces and the basic ideas of general relativity

8.1   Curved surfaces

One of the most revolutionary features of general relativity is the essential use it makes of curved space (actually, of curved spacetime). Though everyone knows intuitively what a curved *surface* is, or rather, what it looks like, people are often puzzled how this idea can be generalized to three or even higher dimensions. This is mainly because one cannot visualize a 4-space in which the 3-space can look bent. But no such surrounding or 'embedding' space is needed to understand curvature. On the contrary, an embedding space introduces elements that are quite irrelevant to our purpose.

So let us first try to understand that part of the geometry of ordinary surfaces which is *intrinsic*; that is, independent of anything outside the surface. Intrinsic properties of a surface are those that depend only on the measure relations *in* the surface. They are those that could be determined by an intelligent race of 2-dimensional beings, entirely confined to the surface in their mobility and in their capacity to see and to measure. Intrinsically, for example, a flat sheet of paper and one bent almost into a cylinder or almost into a cone, are equivalent [see Fig. 8.1(a)]. If we closed up the cylinder or the cone, these surfaces would still be 'locally' equivalent but not 'globally'. In the same way [see Fig. 8.1(b)] a helicoid (spiral staircase) is equivalent to an almost closed catenoid (a surface generated by rotating the shape of a freely hanging chain); and so on. One way to visualize intrinsic properties, therefore, is to think of them as those that are preserved when the surface is bent without stretching or tearing. But this view has its limitations, since some surfaces, for example, spheres or closed convex surfaces in general, are 'rigid' and cannot be deformed smoothly.

The most important intrinsic feature of a surface is the totality of its *geodesics*. These are the analogs of the straight lines in the plane. They are lines of minimal length between any two of their points, provided these points are not too far apart. For example, on a sphere the geodesics are the great circles—but they are minimal only for points less than half a circumference apart. A taut string, confined to the surface, will lie along a geodesic. On a surface with a hump, such a string can loop the hump (as in Fig. 8.2) and then it clearly is not minimal from one side of the hump to the other. Figure 8.2 also illustrates the defining property of a geodesic g: if two points A and B on g are not too far apart, then all nearby lines joining A and B on the surface, like l and l', have greater length than the portion of g between A and B.

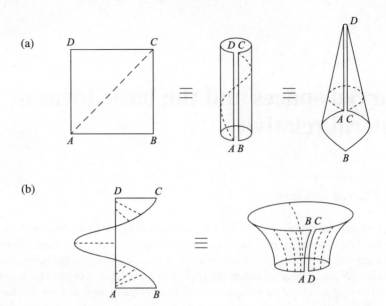

Fig. 8.1

Since their definition depends only on distance measurements *in* the surface, geodesics are clearly intrinsic; for example, they remain geodesics even when the surface is bent. (Observe this with the various dashed lines in Fig. 8.1.) Now imagine a geodesic *g* on some surface that is made of flexible material. Out of this surface cut a narrow strip which has *g* as its center-line. When that strip is placed on a flat surface, it will be perfectly straight—since geodesics are intrinsic. This points to another characteristic of geodesics: they are the straightest possible lines on a curved surface. They are the paths you describe if you were to walk on the surface and always 'followed your nose'.

From this last property the following result seems 'obvious': from each point on a smooth surface there issues a *unique* geodesic in a given direction. Another important result can also be proved: for each point P on a smooth surface there exists

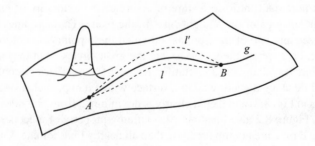

Fig. 8.2

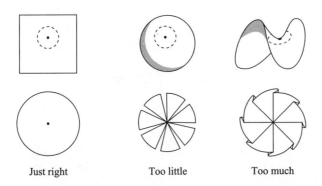

| Just right | Too little | Too much |

Fig. 8.3

a neighborhood N(P) (which can be quite large) such that every point in N(P) can be connected to P by a unique geodesic. [On a sphere, for example, N(P) is the whole sphere minus one point: the antipode of P.]

Next, let us see how the 2-dimensional beings would discover the curvature of their world. As prototypes of three different kinds of surface regions, consider a plane, a sphere, and a saddle. On each of these draw a small geodesic circle of radius r; that is, a locus of points which can be joined by geodesics of length r to some center (see Fig. 8.3). In practice this could be done by using a taut string like a tether. Then we (or the flat people) can measure both the circumference C and the area A of these circles. In the plane we get the usual 'Euclidean' values $C = 2\pi r$ and $A = \pi r^2$. On the sphere we get *smaller* values for C and A, and on the saddle we get *larger* values. This becomes evident when, for example, we cut out these circles and try to flatten them onto a plane: the spherical cap must tear (it has too little area), while the saddle cap will make folds (it has too much area).

For a quantitative result, consider Fig. 8.4, where we have drawn two geodesics subtending a small angle θ at the north pole P of a sphere of radius a. By definition, we shall assign a curvature $K = 1/a^2$ to such a sphere. At distance r along the geodesics from P, let their perpendicular separation be η. Then, using elementary geometry and the Taylor series for the sine, we have

$$\eta = \theta\left(a \sin \frac{r}{a}\right) = \theta\left(r - \frac{r^3}{6a^2} + \cdots\right) = \theta\left(r - \frac{1}{6}Kr^3 + \cdots\right), \qquad (8.1)$$

and consequently,

$$C = 2\pi\left(r - \tfrac{1}{6}Kr^3 + \cdots\right), \qquad A = \pi\left(r^2 - \tfrac{1}{12}Kr^4 + \cdots\right), \qquad (8.2)$$

where we have used $A = \int C \, dr$ to get A from C. These expansions yield the following two alternative formulae for K:

$$K = \frac{3}{\pi} \lim_{r \to 0} \frac{2\pi r - C}{r^3} = \frac{12}{\pi} \lim_{r \to 0} \frac{\pi r^2 - A}{r^4}. \qquad (8.3)$$

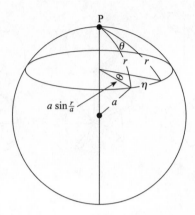

Fig. 8.4

Surprisingly, it can be proved that formula (8.1)(iii), to the order shown, holds for *any* smooth surface! In other words, the spread of neighboring geodesics from any given point P is direction-independent up to $O(r^3)$, and it is always of the form (8.1)(iii) for *some* number K that depends only on P. This number is called the (*Gaussian*) *curvature* of the surface at P. It is obviously intrinsic. Formulae (8.2) and (8.3) thus also apply generally to small geodesic circles on *all* smooth surfaces. They would enable the flat people to determine the curvature of their world at various places. Note that in convex regions, where C and A are smaller than their Euclidean values, K is positive, while in saddle-like regions it is negative.

If we differentiate (8.1)(iii) twice with respect to r we find, to lowest order,

$$\ddot{\eta} = -K\eta \quad (\cdot \equiv d/dr). \tag{8.4}$$

This is an important formula. It allows us to use the second rate of spread of neighboring geodesics issuing from a point as a measure of the curvature at that point. 'Sublinear' spread corresponds to positive curvature, 'superlinear' spread to negative curvature (cf. Fig. 8.5). In fact, Formula (8.4) applies to neighboring geodesics

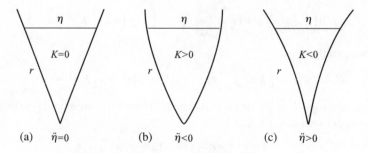

Fig. 8.5

even if they do not intersect at the point of interest; for example, all along neighboring meridian circles on a sphere (cf. Exercise 8.2). We can test this at once with eqn (8.1)(i).

As a last important result in the differential geometry of surfaces we mention the following: If the Gaussian curvature of a smooth surface vanishes everywhere, the surface is necessarily equivalent to the Euclidean plane, and if the Gaussian curvature is $1/a^2$ everywhere, the surface is equivalent to a sphere of radius a—except for possible topological modifications. In the first case, for example, the surface could also be a cone, a cylinder, a flat torus (resulting from the identification of pairs of opposite edges of a rectangle), and so on; and in the second case it could also be, for example, the 'elliptic' sphere which results from the identification of diametrically opposite points of a sphere.

8.2   Curved spaces of higher dimensions

Many of the ideas of the intrinsic differential geometry of curved surfaces can be extended to spaces of higher dimensions, such as, for example, the 3-space of our experience. In particular, geodesics in all dimensions are defined, exactly as in the 2-dimensional case, as minimal-length lines or, equivalently, as 'straightest' lines. For a sufficiently 'well-behaved' space, the two basic surface theorems then also hold: (i) there is a unique geodesic issuing from a given point in a given direction, and (ii) in a sufficiently small neighborhood of a given point P each other point can be connected to P by a unique geodesic. To define the curvature of a 3-space like ours one might then (in direct analogy to the procedure used for surfaces) construct geodesic *spheres* (instead of circles) of radius r, and compare their surface area or volume with the Euclidean values. It is logically quite conceivable that by very accurate measurements of this kind we would find *our* space to deviate slightly from flatness. The great Gauss himself made various experiments to try to find a curvature, but with the available surveying instruments (then or now) none can be detected directly.

In any case, the direct generalization of formulae (8.3) turns out to be too crude. More than one number at each point is needed to characterize fully the curvature properties of spaces of higher dimensions. Consider all the geodesics issuing from a point P in the directions of a linear 'pencil' $\lambda\mathbf{p} + \mu\mathbf{q}$ determined by *two* directions $\mathbf{p}$ and $\mathbf{q}$ at P. Such geodesics are said to generate a *geodesic plane* through P. *Its* curvature K at P is said to be the space curvature $K(\mathbf{p}, \mathbf{q})$ at P for the orientation $(\mathbf{p}, \mathbf{q})$. (In three dimensions K is completely known if it is known for 6 orientations, in four dimensions if it is known for 20.) A geodesic plane is the curved-space analog of a plane through a point, except that in general is satisfies its defining property only with respect to that one point.

An important theorem in this connection is the following: If any smooth subspace U of a larger space V contains a geodesic g of the larger space (cf. Fig. 8.6) then g is a geodesic relative to the subspace also. This follows at once from the minimal property of g. For if g is the shortest connection between A and B in V, there can be no shorter

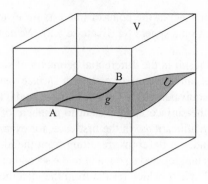

Fig. 8.6

connection in U. One consequence of this is that the geodesics of a larger space V which define a *geodesic plane* U through some point P of V must be geodesics also relative to U. So in order to determine the curvature of V at P in the orientation $(\mathbf{p}, \mathbf{q})$ we can circumvent the construction of the full geodesic plane. We need merely pick *any* pair of neighboring geodesics from the pencil $\lambda\mathbf{p} + \mu\mathbf{q}$ and apply formula (8.4) to get K. Since the pair are also U-geodesics, this K will be the required curvature of the geodesic plane at P.

If the curvature K at P is independent of the orientation, we say P is an *isotropic point*. Only then is all the curvature information contained in knowing the surface area S *or* the volume V of a small geodesic sphere in terms of its radius r (or of a 'hypersphere' in higher dimensions). In three dimensions S and V at an isotropic point are easily seen to be given by the formulae

$$S = 4\pi\left(r^2 - \tfrac{1}{3}Kr^4 + \cdots\right), \qquad V = \tfrac{4}{3}\pi\left(r^3 - \tfrac{1}{5}Kr^5 + \cdots\right). \qquad (8.5)$$

This first follows from (8.1): the ratio of S to the Euclidean value $4\pi r^2$ must equal the square of the ratio of η to *its* Euclidean value θr. The second then follows from the relation $V = \int S\,dr$. If *all* points of a space are isotropic, it can be shown that the curvature at all of them must be the same (*Schur's theorem*), and then the space is said to be *of constant curvature*.

As an interesting example, let us consider a 3-dimensional space of constant positive curvature $1/a^2$—a 'hypersphere' S^3, the 3-dimensional analog of a 2-sphere. For any two neighboring and intersecting geodesics in S^3 we must have [cf. (8.4)] $\ddot{\eta} = -(1/a^2)\eta$. By the argument of Exercise 8.2, the same relation must hold even for non-intersecting neighboring geodesics. So if we draw a geodesic plane Π at an arbitrary point P, the η of two neighboring geodesics of Π must satisfy $\ddot{\eta} = -(1/a^2)\eta$ *all along*. The curvature of Π must therefore be $1/a^2$ everywhere, and so it is a 2-sphere. Strange but true: each geodesic plane in S^3 is a 2-sphere whose 'inside' and 'outside' are identical halves of the whole space! (Perhaps an analogy will help to make this more acceptable: the 'inside' and 'outside' of each great circle on a 2-sphere are also equal halves of the full space.) Now if each geodesic plane through P is a 2-sphere of radius a, we know from (8.1)(i) *exactly* how any two neighboring geodesics issuing from P will spread. And this, in turn, allows us to find the surface area S of a 'geodesic sphere' around P of arbitrary radius r; that is, of the locus of points whose geodesic distance from P is r. We argue as we did for eqn (8.5): the ratio of S to the Euclidean value $4\pi r^2$ must equal the square of the ratio of η [which now equals $\theta a \sin(r/a)$] to *its* Euclidean value θr. This leads to the first of the following equations,

$$S = 4\pi a^2 \sin^2 \frac{r}{a}, \qquad V = 2\pi a^2\left(r - \frac{a}{2}\sin\frac{2r}{a}\right), \qquad (8.6)$$

while the second again follows from $V = \int S\,dr$. What these equations tell us is this: if we lived in a 3-sphere S^3 of curvature $1/a^2$ and drew concentric geodesic 2-spheres around ourselves, their surface area would at first increase with increasing radius r (but not as fast as in the Euclidean case), reaching a maximum $4\pi a^2$, with included volume $\pi^2 a^3$, at $r = \frac{1}{2}\pi a$. This is our 'equatorial' sphere. After that, more distant spheres have lesser and lesser surface area, until finally, at $r = \pi a$, the 'sphere' has zero area: it is a single point, our antipode. The total volume of our 3-sphere is $2\pi^2 a^3$. The picture from any other point is exactly the same. (Again, the analogy with the 2-sphere may help: If we are at the north pole and draw larger and larger geodesic circles around ourselves, their circumferences increase to a maximum at the equator, and after that a greater radius leads to a lesser circumference until, at the south pole, the circle becomes a point.) One last curiosity in S^3: Suppose I blow up a balloon of unlimited elasticity. Its surface will increase until it reaches $4\pi a^2$, at which point it has become a geodesic plane through *me*. By symmetry, it must even *look* plane to *me*. If I continue my blowing, the balloon will curve around behind me and end up enclosing me tightly! [Cf. Fig. 8.7(a).] An analogy is provided by a 2-dimensional being 'blowing up circles' on the surface of a 2-sphere, as shown in Fig. 8.7(b).

All this is not as fantastic as it may seem. The very first cosmological model put forward by Einstein in 1917 envisaged our 3-space to be precisely of this kind, a hypersphere with radius $a \approx 10^{10}$ light-years! And even today it is still considered

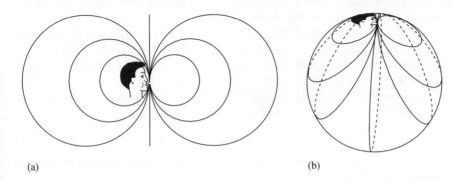

(a)   (b)

Fig. 8.7

quite possible that our universe constitutes such a hypersphere, albeit a presently expanding one. The other two modern possibilities are a flat (Euclidean) universe E^3 or one with constant *negative* curvature, H^3.

With very little extra effort we can, in fact, now derive the main geometric features of the latter also. Our arguments for S^3 apply (*mutatis mutandis*) equally to H^3. Any two neighboring geodesics of H^3 obey $\ddot{\eta} = +(1/a^2)\eta$, and so must neighboring geodesics of each geodesic plane, along their entire length. Integrating this equation yields

$$\eta = \theta a \sinh \frac{r}{a} \tag{8.7}$$

instead of our previous $\eta = \theta a \sin(r/a)$. Each geodesic plane now is a 2-space H^2 of constant negative curvature $-(1/a^2)$. (We are not as familiar with such 2-spaces, because—unlike 2-spheres—they have no simply visualizable representation in E^3. We *can* visualize them as floppy cloths throwing more and more folds as we go out from any point.)

With (8.7) instead of (8.1)(i) we find, analogously to (8.6),

$$S = 4\pi a^2 \sinh^2 \frac{r}{a}, \qquad V = 2\pi a^2 \left(\frac{a}{2} \sinh \frac{2r}{a} - r \right) \tag{8.8}$$

for the surface area and included volume of geodesic spheres of radius r in H^3. Now there is no limit to the size of these spheres, and the full space is infinite.

8.3   Riemannian spaces

In the last two sections we have rather liberally quoted, without proof, theorems from a branch of mathematics called 'Riemannian geometry', on the implicit assumption that the curved spaces under discussion were, in fact, *Riemannian*. This assumption we must now examine. On a curved surface we cannot set up Cartesian coordinates in the same way as in the plane (with the 'coordinate lines' forming a lattice of strict squares)—for if we could, we would *have* a plane, intrinsically. Certain surfaces by their symmetries suggest a 'natural' coordinatization, like the plane, or the sphere (see Fig. 8.8) on which one usually chooses co-latitude (x) and longitude (y) to specify points. On a general surface one can 'paint' two arbitrary families of coordinate lines and label them $x = \ldots, -2, -1, 0, 1, 2, \ldots$ and $y = \ldots, -2, -1, 0, 1, 2, \ldots$, respectively, and one can further subdivide these as finely as one wishes. The resulting coordinate lattice may or may not be orthogonal.

If we imagine the surface embedded in a Euclidean 3-space with coordinates (X, Y, Z) [see Fig. 8.8(c)] then it will satisfy 'parametric' equations of the form

$$X = X(x, y), \qquad Y = Y(x, y), \qquad Z = Z(x, y), \tag{8.9}$$

which we assume to be differentiable as often as required. For example, a sphere of radius a centered on the origin satisfies (cf. Fig. 8.4):

$$X = a \sin x \, \cos y, \qquad Y = a \sin x \, \sin y, \qquad Z = a \cos x.$$

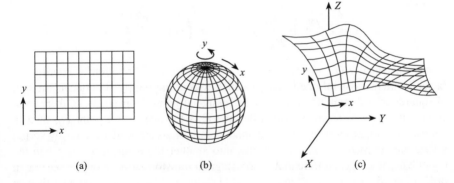

Fig. 8.8

Since the distance between neighboring points in the Euclidean space is given by

$$d\sigma^2 = dX^2 + dY^2 + dZ^2,$$

distances in the surface are given by

$$d\sigma^2 = (X_1\,dx + X_2\,dy)^2 + (Y_1\,dx + Y_2\,dy)^2 + (Z_1\,dx + Z_2\,dy)^2, \qquad (8.10)$$

where the subscripts 1 and 2 denote partial differentiation with respect to x and y, respectively. Evidently (8.10) is of the form

$$d\sigma^2 = E\,dx^2 + 2F\,dx\,dy + G\,dy^2, \qquad (8.11)$$

where E, F, G are certain functions of x and y. In the case of the sphere, for example, this method yields

$$d\sigma^2 = a^2\,dx^2 + a^2\sin^2 x\,dy^2 \qquad (8.12)$$

which can also be understood directly by elementary geometry. Whenever the squared differential distance $d\sigma^2$ is given by a homogeneous quadratic differential form in the surface coordinates, as in (8.11), we say that $d\sigma^2$ is a *Riemannian metric*, and that the corresponding surface is *Riemannian*. It is, of course, not a foregone conclusion that all metrics must be of this form: one could *define*, for example, a non-Riemannian metric $d\sigma^2 = (dx^4 + dy^4)^{1/2}$ for some abstract 2-dimensional space, and investigate the resulting geometry. Such more general metrics give rise to 'Finsler' geometry.

What distinguishes a Riemannian metric among all others is that it is *locally Euclidean*: At any given point P_0 the values of E, F, G in (8.11) are simply numbers, say E_0, F_0, G_0; thus, 'completing the square', we have, *at P_0*,

$$d\sigma^2 = \left(E_0^{1/2}\,dx + \frac{F_0}{E_0^{1/2}}\,dy\right)^2 + \left(G_0 - \frac{F_0^2}{E_0}\right)dy^2 = d\tilde{x}^2 + d\tilde{y}^2, \qquad (8.13)$$

where

$$\tilde{x} = E_0^{1/2} x + \frac{F_0}{E_0^{1/2}} y, \qquad \tilde{y} = \left(G_0 - \frac{F_0^2}{E_0} \right)^{1/2} y.$$

Hence, provided $EG > F^2$, there exists a (linear) transformation of coordinates (actually there exist infinitely many) which makes the metric 'Euclidean' (a sum of squares of differentials) at any *one* preassigned point. Conversely, *if* there exist coordinates $\tilde{x}, \tilde{y}$ in terms of which the metric is Euclidean at a point P_0, then in general coordinates it must be Riemannian at P_0; for, as the reader can easily verify, a Riemannian metric always transforms into another Riemannian metric when the coordinates undergo a differentiable non-singular transformation. Now we see that in order to predict the form of the metric (8.11) we could have dispensed with the use of the embedding space. We could simply have postulated that the surface is locally Euclidean; that is, that for any given point we can choose the coordinate system so that $d\sigma^2 = dx^2 + dy^2$ *at that point*.

All *intrinsic* properties of a surface spring from its specific metric, (8.11). For it is the metric that determines all distance relations within the surface. It is a kind of blueprint from which the surface could actually be constructed: Suppose we draw a Cartesian grid on a piece of paper, say $x, y = 0, \pm 1, \pm 2, \ldots$, using arbitrarily small units. Let us triangulate this grid by drawing one diagonal (say the one with positive slope) in each coordinate square. Then we write along the sides of all the little triangles whatever length the metric ascribes to them. This is our 'Mercator map' of the surface! By cutting out corresponding triangles that actually have the indicated dimensions, and fitting them together according to the map, we can then construct the surface.

The idea of a Riemannian metric directly generalizes to spaces of higher dimensions. Such spaces, too, can be coordinatized with arbitrary ('Gaussian') coordinates, just like a surface. In three dimensions, for example, instead of having two families of coordinate *lines*, we have three families of coordinate *surfaces*, which we can label $x, y, z = 0, \pm 1, \pm 2, \ldots$, and which provide addresses (x, y, z) for all points of the space. If there then exists a metric analogous to (8.12) which, at any given point, can be transformed to a sum of squares (a Euclidean metric), we say the space is Riemannian, or locally Euclidean.

We have already, in Chapter 7, met similar metrics and we now return to the notation introduced there [cf. eqn (7.8)] for the general, N-dimensional, case:

$$\mathbf{ds}^2 = g_{ij} \, dx^i \, dx^j \quad (i, j = 1, \ldots, N), \tag{8.14}$$

where the g_{ij} are symmetric tensor components. We also require $\det(g_{ij}) \neq 0$ to ensure that the metric is truly N-dimensional. [For example, $dx^2 + 4\,dx\,dy + 4\,dy^2 = (dx + 2dy)^2 =: dz^2$ is *not* truly 2-dimensional.] It is always possible, by a generalization of the procedure used in (8.13), to reduce any such metric at any one point to a sum of *positive and negative* squares. And although this can be done in infinitely many ways, the number of positive squares and the number of negative squares one ends up with is unique (Sylvester's theorem). For a metric to be locally *Euclidean* (all positive squares), its coefficients must satisfy certain positivity conditions; in two dimensions, as we have seen, the condition is $EG > F^2$.

Metrics of the form (8.14) are called *properly Riemannian* or just *Riemannian* if they are locally Euclidean, and *pseudo-Riemannian* otherwise. The spaces whose intrinsic geometry we discussed in the last section were tacitly assumed to be proper Riemannian spaces. Since their metric is locally Euclidean, so is their geometry. For example, on any Riemannian surface, the circumference of a small geodesic circle is $2\pi r$ to *lowest order* and thus the complete plane angle around any point is 2π. The sum of the angles in a small geodesic triangle is π to lowest order. In any Riemannian 3-space, the surface of a small geodesic sphere is $4\pi r^2$ and its volume $\frac{4}{3}\pi r^3$, to lowest order, and so forth.

In general relativity, however, we also need the geometry of curved *pseudo-Riemannian* spaces, and in particular of those that reduce locally to Minkowski space,

$$g_{\mu\nu}\, dx^\mu\, dx^\nu \longmapsto -(d\tilde{x}^1)^2 - (d\tilde{x}^2)^2 - (d\tilde{x}^3)^2 + (d\tilde{x}^4)^2; \qquad (8.15)$$

that is, whose sign distribution (or *signature*) is $(---+)$. Instead of having locally the structure of 4-dimensional Euclidean space, these spaces have, to lowest order, the structure of Minkowski space, with its light cones. Nevertheless, it is remarkable how much of the *geometry of curvature* pseudo-Riemannian and properly Riemannian spaces have in common.

To start with, all Riemannian spaces have geodesics. In pseudo-Riemannian spaces these are defined as lines of *stationary* length,

$$\delta \int ds = 0, \qquad (8.16)$$

which can be minimal, or maximal, but generally are neither (cf. Exercise 8.11). Eqn (8.16) applies to *all* Riemannian spaces. The alternative definition of geodesics as 'straightest' lines (whose exact formulation we shall see latter) also applies universally. So do the two basic theorems: a point and a direction define a unique geodesic, and all points in a sufficiently small neighborhood of a point P are connectible to P by a unique geodesic. (A 'small' neighborhood in pseudo-Riemannian spaces is defined by the smallness of the coordinate differences.) In pseudo-Riemannian spaces it can be shown that the sign of ds^2 is constant along any geodesic, but it can be positive, negative, or zero. And in all Euclidean or pseudo-Euclidean spaces [possible global metric: $\sum \pm(dx^i)^2$] the geodesics are 'straight lines' corresponding to linear equations in the Euclidean or pseudo-Euclidean coordinates.

While the directional curvature $K(\mathbf{p}, \mathbf{q})$ of pseudo-Riemannian spaces is still defined as that of the geodesic plane containing the directions $\mathbf{p}$ and $\mathbf{q}$, the curvature of the geodesic plane itself can no longer be defined (when it is not properly Riemannian) in terms of limits of small geodesic circles, because there *are* no small geodesic circles. Instead, we must fall back on the 'geodesic deviation' formula (8.4), which still applies, albeit in somewhat modified form: near every point P of a pseudo-Riemannian 2-space the metric can be reduced to that of the Minkowski x, ct plane, and so the directions at P fall into four quadrants, in two of which they are 'positive' ($\mathbf{ds}^2 > 0$) while in the other two they are 'negative' ($\mathbf{ds}^2 < 0$); and these quadrants are separated by the four null directions ($\mathbf{ds}^2 = 0$). In this general case formula (8.4) can be shown to take the form.

$$\ddot{\eta} = -\epsilon K \eta. \tag{8.17}$$

where the *indicator* ϵ takes the value 1 for two neighboring positive geodesics and -1 for two neighboring negative geodesics. The *sign* of K here loses much of its absolute significance since the overall sign of $\mathbf{ds}^2$ for pseudo-Riemannian spaces is largely conventional. Unless the planar direction is null, there will always be two opposite quadrants whose geodesics spread sublinearly as in Fig. 8.5(b) and two whose geodesics spread superlinearly as in Fig. 8.5(c). Of course, if two of these quadrants have a preferred physical significance, as in relativity, the sign of K can re-assume significance. We can and do take timelike geodesics as the positive ones, so that K will be positive if these spread sublinearly. (Not all authors follow this convention.) When the geodesic plane has an induced metric locally reducible to $-(\mathrm{d}x^2 + \mathrm{d}y^2)$, we must take $\epsilon = -1$ in all directions.

As in the case of subspaces of proper Riemannian spaces (cf. Fig. 8.6), geodesics of a pseudo-Riemannian space that lie in a subspace are geodesics in the subspace also. (And for essentially the same reason: the lengths of neighboring curves can differ at most in second-order.) So here, too, we can measure $K(\mathbf{p}, \mathbf{q})$ directly by the spread of *any* pair of neighboring geodesics of the space itself whose initial directions lie in the plane of $\mathbf{p}$ and $\mathbf{q}$.

Pseudo-Riemannian spaces can have (curvature-)isotropic points just like Riemannian spaces, and Schur's theorem applies to them as well [cf. after eqn (8.5)]. For each dimension, signature, and value of K, there is a unique space of constant curvature, except for possible topological modifications. The three locally Minkowskian spaces of constant curvature play an important role in relativity. They are: (i) Minkowski space itself ($K = 0$); (ii) *de Sitter space* ($K < 0$), whose timelike geodesics spread super-linearly; and (iii) *anti-de Sitter space* ($K > 0$), whose timelike geodesics spread sublinearly, and, in fact, 'focus' like those of a sphere. In de Sitter space it is the spatial geodesics that focus. De Sitter space is, in fact, spatially spherical and open only in the time direction, while anti-de Sitter space is temporally circular and open in all spatial directions. (These features appear clearly in the representations of these two spaces as hyperboloids of revolution—see Figs 14.1 and 14.4 in Chapter 14 below.)

We have talked earlier of the metric being the blueprint for a surface (up to bending), and the same holds for Riemannian and pseudo-Riemannian spaces of all dimensions. Spaces with identical metrics can be brought into coincidence (that is, laid on top of each other), like the corresponding surfaces in Fig. 8.1. But there is an element of arbitrariness in a Riemannian metric $g_{ij} \, \mathrm{d}x^i \, \mathrm{d}x^j$, namely the coordinates. These can be laid on the space in a largely arbitrary manner, and different choices of coordinates will lead to formally different metrics, like $\mathrm{d}x^2 + \mathrm{d}y^2$ and

$dr^2 + r^2 d\theta^2$ for the Euclidean plane, in Cartesian and polar coordinates, respectively. But as we have seen, the coefficients g_{ij} form a symmetric tensor. And it is in this so-called *metric tensor* that the intrinsic structure of a Riemannian space is fully encoded.

8.4 A plan for general relativity

We have already given a preview of general relativity in Chapter 1 (cf. Sections 1.13–1.16), although at that stage it was necessarily rather vague. The reader may nevertheless wish to look at it again from our present vantage point. We are now in a much stronger position to sketch out the theory that must result (almost inevitably) from Einstein's equivalence principle. It is not yet full GR. That only results when Einstein's field equations are added. The field equations, however, involve the curvature tensor, and so our quantitative discussion of them must wait until Chapter 10, where we shall learn more about tensors in Riemannian spaces. Until then, though we shall refer to the developing theory as GR, the reader should bear in mind that the same groundwork is shared by other 'rival' theories accepting the EP but having different field equations (for example, the theories of Nordström and of Brans-Dicke, which, however, have by now been eliminated on empirical grounds).

The vital prerequisites for Einstein's invention of GR were: (i) his EP; (ii) Newton's theory as guide and touchstone and, in particular, (iii) Galileo's Principle of shared orbits in a gravitational field; (iv) Minkowski's concept of 4-dimensional spacetime; and (v) the basic facts of Riemannian geometry.

Following Minkowski's procedure in SR, we can regard the set of all (actual or potential) events in the world as a 4-dimensional continuum (spacetime) which can be coordinatized with four arbitrary (Gaussian) coordinates x^μ ($\mu = 1, 2, 3, 4$). According to the EP, we can find near each event $\mathcal{P}$ a small freely falling box in which SR holds. This box—being a local inertial frame (LIF)—provides us with a local inertial coordinate system $\{x, y, z, ct\}$ which can be used to assign a unique squared displacement from $\mathcal{P}$ to any neighboring event according to the formula

$$\mathbf{ds}^2 = c^2 dt^2 - dx^2 - dy^2 - dz^2. \tag{8.18}$$

So there is a uniquely determinable metric structure on the global spacetime. And since the formula for the metric is locally Minkowskian, it must be globally pseudo-Riemannian with signature $(---+)$. Such spacetimes are called *Lorentzian*.

Now at each event, relative to the LIF, free particles move uniformly; that is, 4-dimensionally 'straight'. But curves that are locally straight are globally geodesic. So free particles can be expected to follow geodesics in the Lorentzian spacetime. This would generalize what we already know happens in the complete absence of gravity (that is, in the Minkowski space of SR), where, in accordance with Galileo's law of inertia, free particles have straight and therefore geodesic worldlines. Moreover, the geodesic law of motion conforms to (and in fact 'explains') Galileo's Principle, which asserts that gravitational orbits are independent of the particle that does the orbiting and are, in fact, determined by an initial velocity. For a geodesic is particle-independent and fully determined by an initial (4-dimensional) *direction*, say dx : dy : dz : $c\,dt$ in the LIF, or equivalently, by $(dx/dt, dy/dt, dz/dt)$, namely the initial *velocity*. Galileo's Principle is now seen as a mere extension of his law of inertia to curved spacetime.

GR can also predict something consistently that Newton's theory can not: the paths of light in vacuum under the influence of gravity. In Newton's theory, though one can *approximately* treat light as particles that travel at speed c, this is *not* a consistent procedure, since, by the constancy of potential plus kinetic energy, the speed of a particle in a gravitational field *must* vary as it traverses different potential levels. In GR, on the other hand, light in vacuum travels straight in the LIF, and therefore should travel geodesically in spacetime; however, while free particles have *timelike* geodesic worldlines, the worldlines of photons must be *null*.

It is logically convenient to summarize these findings (and two others) in the form of *axioms*:

1. The spacetime of events is Lorentzian; that is, pseudo-Riemannian with Minkowskian signature.
2. Free test-particles have timelike geodesic worldlines.
3. Light in vacuum follows null geodesics.
4. The arc along *any* timelike worldline corresponds to c times the proper time of an ideal point-clock that traces it out. (This will be discussed in the penultimate paragraph of this section.)
5. Einstein's field equations will relate the metric with the energy tensor of the sources.

Evidently, for the theory to be useful, it should predict the orbits in the field of a given mass distribution. Since the orbits (the geodesics) are now determined by the metric, we might expect the metric in turn to be *determined* by the sources— that is, by the gravitating matter. But there is an inherent logical obstacle to this: unless we already know the metric, we do not know the spacetime at the sources, and without that we cannot precisely *describe* extended sources. So it would seem that we are doomed to use some sort of iterative mathematical process to get from the sources to the spacetime. In practice, there are often ways to avoid this. Still, the yet-to-be-discussed field equations can do no more than *relate* the geometry and the sources.

At this stage it will be well to elaborate a number of details. First, we have used Einstein's freely-falling-box argument to justify the Lorentzian structure of spacetime; but inside matter we cannot have a freely falling box, though we can certainly imagine a freely falling coordinate system. But that is the beauty of replacing a partially *derived* result by an axiom: rather than dream up more complicated thought-experiments, like drilling little holes in the matter, etc., we simply posit the

Lorentzian structure of spacetime everywhere, and find that it leads to a self-consistent theory.

Next, we recall that in Newton's theory a 'free' particle is one subject to *no* forces, including gravity. In GR, a 'free particle' always means a freely falling one; that is, one subject to gravity *only*. In fact, in GR gravity is no longer a force, but rather part of the geometry. In Whittaker's phrase, instead of being a player, gravity has become part of the stage.

In GR, the geometry (that is, the metric $g_{\mu\nu}$), plays the part of a field. There is no action at a distance to conflict with the SR speed limit. A test particle does not 'feel' the sources directly, but rather the field which they have built up in its neighborhood. In analogy with the electromagnetic field, the metric field can transmit disturbances in the geometry (gravitational waves) at the speed of light.

If GR is correct, the spacetime we live in is curved essentially everywhere. So what happens to the neat flat-space special-relativistic laws of physics we have studied so far? The effort was not wasted. In many contexts the curvature of spacetime (which GR allows us to estimate) is so small that SR can still be used locally with great accuracy. In other cases we shall find a simple and universal procedure (suggested by the EP) for incorporating the effects of curvature into the differential formulation of the various SR laws. (It devolves on the so-called 'covariant' derivative.) So physics does *not* have to be totally revamped once more!

The last topic in need of some elaboration is the physical significance of the metric. The causal 'grain' of SR spacetime, namely the existence of little null cones at each event, and of three different kinds of displacement (timelike, spacelike, and null), is impressed on the GR spacetime through the LIFs. The little cones will no longer be 'parallel' to each other everywhere, but still, a particle worldline will be *within* the cone at each of its points (as in Fig. 5.2), and a photon will travel *along* the cones (but only in vacuum). The null-geodesic generators of the full cones can be quite tangled, and two of them issuing from one event may well meet again at another ('gravitational lensing'). As for timelike lines, they are always potential worldlines of *ideal point-clocks*—freely moving if the line is geodesic, being pushed or accelerated if the line is curved. The arc $\int ds$ along any such line is directly measured by c times the reading of the corresponding clock; for we know that this is true in SR [cf. eqn (5.6)], and the equivalence principle allows us to apply SR to infinitesimal portions of the GR path. Thus null displacements **ds** correspond to light, and timelike displacements to clock readings, but how are we to visualize a spacelike displacement **ds**? We know from SR that it corresponds to events at opposite ends of a little rod of rest-length ds, provided they are simultaneous in its rest-frame. So we can take two identical rods of length ds and slide them *slowly* over each other: the meetings of their corresponding ends will provide us with two events having the desired separation. Alternatively, there are various light-signaling methods for determining *all* types of differential interval (see, for example, Exercise 5.2). As for determining the 10 metric coefficients $g_{\mu\nu}$ at any given event $\mathcal{P}$ relative to a given Gaussian coordinate system, it will suffice to measure 10 displacements away from $\mathcal{P}$, provided they are not all null; for evidently the null intervals alone can only determine the $g_{\mu\nu}$ up to a common factor.

The reader may now be impatient to see whether indeed spacetimes exist that can be identified with known gravitational situations, and whether the geodesics in these spacetimes approximate to the Newtonian orbits. For we must not forget that Newton's theory of gravitation agrees almost perfectly with the observed phenomena throughout an enormous range of classical applications. *Any* alternative theory must yield the same predictions to within the errors of classical observations. To test this, we shall develop in the next chapter a topic that is of great importance in GR in its own right, namely the theory of static or merely stationary matter distributions. Once we have that, an encouraging parallelism of GR to Newton's theory can be very quickly established.

Exercises 8

8.1. Say which of the following are intrinsic to a 2-dimensional surface:

(i) The angle at which two curves intersect.

(ii) The property of two curves being tangent to each other at a given point.

(iii) The area contained within a closed curve.

(iv) The *normal curvature* K_n of the surface in a given direction; that is, the curvature of the section obtained by cutting the surface with a plane containing the normal at the point of interest.

(v) The *geodesic curvature* κ_g of a curve on the surface: cut out a thin strip of surface with the given curve as center-line, then lay the strip on a plane and measure its curvature; that is κ_g. Alternatively it is the curvature of the projection of the curve onto the tangent plane at the point of interest.

8.2. Generalize formula (8.4) to two neighboring geodesics g_1 and g_2 on a surface at a point where they do not intersect. [*Hint*: lay a third geodesic g at small inclination across g_1 and g_2; somewhere between the resulting intersection points let η (the normal separation between g_1 and g_2) be divided into η_1 and η_2 by g. Then apply (8.4) to η_1 and η_2.]

8.3. A paraboloid of revolution is generated by revolving the parabola $z = ax^2$ about the z-axis. Find its Gaussian curvature at the general point as a function of x. [*Hint*: consider two 'obvious' neighboring geodesics corresponding to two axial sections a small angle θ apart. *Answer*: $4a^2(1 + 4a^2x^2)^{-2}$.]

8.4. The curvature of a plane curve (or, indeed, any curve) in Euclidean space is defined as the rate of turning of the tangent (in radians) with respect to distance along the curve; it can be shown to equal $\lim (2z/r^2)$ as $r \to 0$, where r is the distance along the tangent at the point of interest and z is the perpendicular distance from the tangent to the curve. Now, referred to the tangent plane at one of its points P as the x, y plane, the equation of any regular surface near P can be written as a Taylor series in the form $z = Ax^2 + Bxy + Cy^2 +$ terms of higher order (Why?). Use this fact, and polar coordinates, to prove that the maximum and minimum of the normal curvature of a surface [cf. Exercise 8.1(iv)] occur in orthogonal directions. [*Hint*: K_n is of the form $\alpha + \beta \cos(2\theta + \gamma)$.] Gauss proved the extraordinary theorem that the product of the two (evidently non-intrinsic) extrema of the normal curvature is equal to the (intrinsic) curvature K. Thus if the spine and the ribs of the horse that fits the saddle of Fig. 8.3 locally approximate circles of radii a and b, the curvature K at the center of the saddle is $-1/ab$. Verify Gauss's result at the vertex of the paraboloid of the preceding exercise.

8.5. In the plane, the total angle Δ through which the tangent of a closed (and not self-intersecting) curve turns in one circuit is evidently always 2π. On a curved surface, the corresponding Δ (increments of angle being measured in the successive local tangent planes) generally differs from 2π. According to a beautiful theorem (Gauss–Bonnet), $2\pi - \Delta$ equals $\int K \, dS$, the integral of the Gaussian curvature over the enclosed area, provided this area is simply connected, and provided the turning is properly counted as positive or negative with reference to the 'outward' normal defined over the entire surface; at corners, the angle must lie between $\pm\pi$, and the circuit must be described in the positive sense relative to the normal on the enclosed area. Consider a sphere of radius a and on it a 'geodesic triangle' formed by three great-circular arcs making a right angle at each vertex. Test the theorem for *both* the areas that can be considered enclosed by this triangle. [Note that there is no contribution to Δ along a geodesic, since geodesics are locally straight.] Use the Gauss–Bonnet theorem to show that, remarkably, $\oint K \, dS = 4\pi$ for *any* surface topologically equivalent to a sphere, and $\oint K \, dS = 0$ for any surface topologically equivalent to a torus.

8.6. Consider a sphere a radius a and on it a 'geodesic circle' of radius r, as in Fig. 8.4. To find the total angle Δ through which the tangent of this circle turns in one circuit (increments of angle always being measured in the local tangent plane), construct the cone tangent to the sphere along the given circle and then 'unroll' this cone and measure the angle at its vertex. Thus prove $\Delta = 2\pi \cos(r/a)$. Then verify that this accords with the Gauss–Bonnet theorem of the preceding exercise—as applied to *both* possible 'insides' of the circle.

8.7. It is often thought that the sphere is the only 2-dimensional surface of *constant* curvature that is both finite and unbounded. Nevertheless a plane surface, too, can be finite and unbounded. One need merely draw rectangle in the plane, discard the outside, and 'identify' opposite points on opposite edges. The area of the resulting surface is evidently finite, yet it has no boundary. Each point is an internal point, since each point can be surrounded by a circle lying wholly in the surface: a circle around a point on an edge appears in two halves, yet is connected because of the identification; around the vertices (all identified) such a circle appears in four parts. Unlike the plane, however, this surface has the topology of a torus, and thus lacks global isotropy. (Consider, for example, the lengths of geodesics drawn in various directions from a given point.) Still, it is *locally* isotropic in its planeness.

A similar though rather more complicated construction can be made on a surface of constant negative curvature. Issuing from one of its points, we draw eight geodesics of equal length r, each making an angle of $45°$ with the next. Then we draw the eight geodesics which join their endpoints. What can you say about the angles at the vertices of the resulting 'geodesic octagon' in terms of its total area? Deduce that r can be chosen to make these angles $45°$, and chose r so. Labeling the vertices successively A, B, C, D, E, F, G, H, identify the following directed geodesics: $AB = DC, BC = ED, EF = HG, FG = AH$. Draw a picture and verify that (i) each point on an edge other than a vertex is an internal point, (ii) the eight vertices are all identified and constitute an internal point, (iii) at all these points the curvature is the same as 'inside'. Can the same trick be worked with a four- or six-sided polygon?

Analogous constructions exist, obviously, for flat 3-space and also, much less obviously, for 3-spaces of constant negative curvature.

8.8. Consider the 3-dimensional hypersphere of curvature $1/a^2$. Prove that every geodesic triangle in it must lie on a 2-sphere of curvature $1/a^2$. [*Hint*: geodesic planes.]

8.9. By reference to eqns (8.1)(i) and (8.7), prove that the metric of the 3-sphere with curvature $1/a^2$ can be written in the form

$$ds^2 = dr^2 + a^2 \sin^2 \left(\frac{r}{a}\right)(d\theta^2 + \sin^2 \theta \, d\phi^2),$$

while that of the hyperbolic 3-sphere H^3 with curvature $-1/a^2$ can be written similarly but with $\sinh^2(r/a)$ in place of $\sin^2(r/a)$. [*Hint*: Let the geodesics leaving the origin be labeled by the usual angles θ and ϕ of polar coordinates and let r be distance along these geodesics. For the small angle between neighboring geodesics, cf. the metric (8.12) of the 2-sphere.]

8.10. For a 2-space with metric $ds^2 = dr^2 + f^2(r) \, d\theta^2$ prove that $K = -f''/f$. [*Hint*: Construct a flat map of this surface with r and θ as polar coordinates; alternatively, provided $|f'| < 1$, recognize this metric as that of a surface of revolution, with θ as angle about the axis and r as distance along the meridian curves.]

8.11. To illustrate that geodesics in pseudo-Riemannian spaces generally have neither minimal nor maximal length, consider the x-axis of Minkowski space M^4. Having linear equation $x = \sigma, y = z = t = 0$ (σ being a parameter), it *is* a geodesic. Consider *neighboring* curves to it, say between $x = 0$ and $x = 2$, consisting of two straight portions: one from $(0, 0, 0, 0)$ to (a, b, c, d), and one from (a, b, c, d) to $(2, 0, 0, 0)$. Show that for suitably chosen *small* a, b, c, d, these neighbors can have greater *or* lesser length. But also prove that *timelike* geodesics in M^4 are maximal. [*Hint*: The twin paradox.]

9

Static and stationary spacetimes

9.1 The coordinate lattice

Loosely speaking, a *stationary* gravitational field is one that does not change in time and a *static* one is where the sources do not move. Examples are the following: the fields in a uniformly accelerating rocket or on a uniformly rotating disk in Minkowski space (these are a little unusual in having no sources); the fields of an arbitrary massive body at rest, or of an axially symmetric body rotating uniformly; the field of a gravitating fluid flowing uniformly around an arbitrarily shaped closed tube that is fixed in space.

In all such cases we assume that there exists at least one rigid space-filling (imagined) *lattice* that could be constructed out of massless rigid rods and that allows us to identify 'points fixed in the field' with lattice points. There may be more than one such lattice for a given source distribution; for example, for the field both outside and inside the rotating earth we could choose a lattice fixed to the earth or one fixed in space. Inside moving matter, as in the last example, the lattice must reduce to just a coordinate lattice—no rods!

We also need the concept of light signals from any fixed point to any other. If the corresponding null geodesic passes through matter, we simply replace the light by ideal neutrinos assumed to traverse even matter at the speed of light.

We now define the stationarity of a lattice by the following *light-circuit postulate*: if light is sent around any lattice polygon ABC ... A, then a standard clock at rest at A always measures the same transit time. If, in addition, that time is independent of the sense in which the polygons are traversed, we call the lattice *static*. The spacetime itself is called stationary (or static) if it contains at least one lattice that is stationary (or static).

As an example of a stationary but not static lattice, consider one rigidly attached to the rotating earth. It is easy to see that a sufficiently large lattice triangle in the equatorial plane will be traversed by light more slowly in the sense of the rotation than in the opposite sense, simply because relative to the underlying quasi-inertial background the first circuit is the longer.

There are stationary fields in which two or more different light paths are possible from point A to point B (quite apart from the fact that the spatial path of light from A to B need not coincide with that from B to A). The light-circuit postulate must then hold for each possible polygonal path.

9.2   Synchronization of clocks

The most important property of stationary spacetimes is that they admit a *preferred time*. (Four-dimensionally speaking, it is a time coordinate t such that all sections $t = $ const through the curved spacetime are identical.) As in SR, the preferred time is the result of a sensible synchronization prescription. Also as in SR, we imagine standard clocks attached to all the lattice points. But now these clocks are furnished with a lever that allows their standard rates to be altered by an arbitrary *constant* factor. One of these clocks, say that at A, is chosen to be the master clock, and is allowed to tick at its standard rate. All the other clocks are then 'rate-synchronized' with the A clock as follows. (*Full* synchronization will involve both the rate *and* the zero-point.) The master clock sends out two control signals separated by an arbitrary (but for convenience preferably small) time Δt. Thereupon, *all* other clocks adjust their rates so as to receive these signals also a time Δt apart. (If there are alternative light paths, pick an arbitrary one between any two points in each direction for the synchronization process.)

That some such rate-adjustment is in general necessary is clear from the 'gravitational-time-dilation' effect predicted by the equivalence principle (cf. Section 1.16). In fact, we already 'know' from eqn (1.11) that the B clock rate will have to be modified by a factor $e^{-\Phi_B/c^2}$, if we arbitrarily declare the potential at A to be zero. That these one-time lever adjustments will, in fact, rate-synchronize all the clocks mutually and forever and along all possible paths (in the sense that *each permanently sees every other clock tick at the same rate as itself*) is intuitively to be expected but needs to be proved.

The proof hinges on the light-circuit postulate and goes as follows: Imagine A's two control signals to an arbitrary point B reflected from B to a third point C, then back to A and once more to B (as in Fig. 9.1). Applying our postulate to the lattice

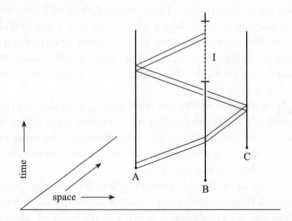

Fig. 9.1

triangles ABCA and BCAB in turn, we see that the signals get back to A *and* B (the B clock runs at a multiple of standard time) still separated by a coordinate time Δt. Because of the arbitrariness of C, B can thus be shown to be synchronous to A throughout a continuous interval I of arbitrary length. But then, by contemplating zig-zag pairs of signals bouncing back and forth between A and B for all eternity, we see, by reference to I and by repeatedly applying the postulate to the '2-gons' ABA and BAB, that B with the proper lever setting always was and always will be synchronous to A. Of course, the same signals show that A, too, is permanently synchronous to B. Now C, like B, will also be permanently synchronous to A and conversely. So the signal pair of Fig. 9.1 passes C at time separation Δt since it so arrives at A. But then the whole argument can be repeated, treating B as the master, and concluding with B and C being permanently synchronized.

Since we deliberately picked an *arbitrary* path between two points when there was a choice, we now conclude that our synchronization process ends up with each clock seeing each other clock (no matter along which path) always tick in synchrony with itself. So the same number of ticks will separate *any* two light signals along the same path at both ends. The important conclusion is that, with such rate-synchronized clocks, *the duration of any light signal from point* P *to point* Q *is always the same.*

Having rate-synchronized the lattice clocks, we are still left with the problem of synchronizing their *zero-settings*. In the general stationary case there is no uniquely preferred way of doing this, although we obviously must insist on a *continuous* time coordinate. (The 4-dimensional picture we alluded to earlier may clarify this: given a stationary spacetime, we can foliate it into identical continuous space slices in infinitely many ways—just as we can slice a long loaf of bread straight across, or at a slant, or even, if the knife is bent, with a curve.) One possible way to define a continuous time coordinate is that which Einstein used in SR. It consists in so setting the already rate-synchronized clocks that when any one of them exchanges signals with the master clock, those signals take the same coordinate time in either direction. If this is not already the case, for example, if light from A to B takes (always) a time λ and from B to A a time μ, we simply advance the setting of the B clock by $\frac{1}{2}(\mu - \lambda)$; then light in *both* directions takes a coordinate time $\frac{1}{2}(\lambda + \mu)$. In general, however, this procedure will yield different synchronizations for different choices of master clock. For if all clock pairs could be made 'Einstein synchronous', light would take the same time in both directions round all polygons and the spacetime would be static. On the other hand, that is precisely what Einstein synchronization with *one* master clock A achieves in any static spacetime. For if light in *both* directions takes a time λ between A and B, and a time μ between A and C, and a time $\lambda + \mu + \nu$ around ABCA, then it must take a time ν between B and C in both directions. (Four-dimensionally speaking, there is now a preferred foliation: orthogonally across the lattice-point worldlines. (Cf. Section 9.5 below.)

As an example, consider the clocks on a rotating disk in Minkowski space. The most convenient synchronization is that which makes them permanently agree with the clocks at rest in the underlying Minkowski space as they pass them. All clocks are then Einstein-synchronous with the clock at the center; and, more generally, any pair

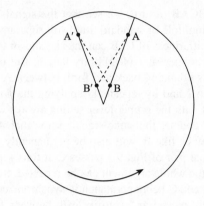

Fig. 9.2

of clocks on the same radius is Einstein-synchronous, but no other pairs. Figure 9.2 illustrates this: during a certain time-interval the radius AB moves to A′B′; in that same time light from A travels to B′ and light from B to A′, along straight lines of equal length in the underlying space.

On the rotating earth, the 'temps atomique international' (TAI) is similarly taken over, in effect, from the comoving quasi-inertial frame. Here all the clocks are Einstein-synchronous with those at the poles, and pairs of clocks on the same meridian are also Einstein-synchronous, as a slight re-interpretation of Fig. 9.2 makes clear.

There is a small gap in our synchronization argument. Some perfectly good stationary spacetimes (for example, that outside a black hole) contain point pairs that cannot exchange light (or even neutrinos) directly. The rate-synchronization must then be done patchwise, starting with a maximal patch around a first master clock A, then another patch mastered by a clock B in the first patch (which, of course, does not tick at standard rate), and so on. The light circuit postulate then guarantees synchrony between any two clocks that can exchange signals. For static spacetimes an analogous procedure for the zero-settings then yields complete Einstein-synchrony. In the general case some other continuous zero-setting must be defined.

9.3   First standard form of the metric

Let us now concentrate our attention on the coordinate lattice. We assume that the rates and zero-settings of all the lattice clocks have been synchronized in accordance with the procedure set forth in the last section. This allows us to assign a time coordinate to all events. Let the lattice itself be coordinatized by arbitrary Gaussian coordinates x^i ($i = 1, 2, 3$), which will furnish the spatial coordinates of events. We define a function of position $\Phi(x^i)$ by writing $e^{-\Phi/c^2}$ for the (time dilation) factor by which the standard clock rate at x^i has to be altered. Then for successive events at x^i we have (since c times proper time measures interval along timelike lines)

$$\mathbf{ds}^2 = e^{2\Phi/c^2} c^2 \, dt^2, \quad (x^i = \text{const}). \tag{9.1}$$

The metric of the spacetime will be, as always, a quadratic form in the coordinate differentials, with only the first coefficient determined by (9.1) so far:

$$\begin{aligned}
\mathbf{ds}^2 = {} & e^{2\Phi/c^2} c^2 \, dt^2 + A \, dx^1 \, dt + B \, dx^2 \, dt + C \, dx^3 \, dt \\
& + D(dx^1)^2 + E(dx^2)^2 + F(dx^3)^2 \\
& + G \, dx^1 \, dx^2 + H \, dx^2 \, dx^3 + I \, dx^1 \, dx^3.
\end{aligned} \tag{9.2}$$

We shall now show that stationarity (*and* the proper choice of coordinate time) imply the time-independence of all the metric coefficients, and that staticity further implies the vanishing of the time–space cross-terms: $A = B = C = 0$.

To this end, consider first a light signal ($\mathbf{ds}^2 = 0$) between neighboring lattice points that differ only in their x^1-coordinate. This signal satisfies the following equation (*with c now set equal to unity*):

$$0 = e^{2\Phi} \, dt^2 + A \, dx^1 \, dt + D(dx^1)^2, \tag{9.3}$$

which can be regarded [after division by $(dx^1)^2$] as a quadratic for dt/dx^1. Its two solutions, corresponding to the two senses of travel ($dx^1 \gtrless 0$), must be time-independent (same transit times for repeated signals), so their sum $-Ae^{-2\Phi}$ and their product $-De^{-2\Phi}$ must be time-independent also, which requires A and D to be time-independent. Similarly B and E and also C and F must be time-independent.

Next consider a light signal between neighboring points that differ in both x^1 and x^2 but not x^3, and which therefore satisfies

$$0 = e^{2\Phi} \, dt^2 + A \, dx^1 \, dt + B \, dx^2 \, dt + D(dx^1)^2 + E(dx^2)^2 + G \, dx^1 \, dx^2.$$

But the same equation must hold at a later time with the same values of dt, dx^1, and dx^2. So if G' denotes the later value of G, we have, by subtraction, $(G - G') \, dx^1 \, dx^2 = 0$, whence G is time-independent. By symmetry, of course, the same must be true of H and I, and so our first assertion is established: all metric coefficients are time-independent.

That staticity implies the vanishing of the time–space cross-terms follows from the complete Einstein synchrony: the two solutions dt/dx^1 of (9.3) are now equal and opposite and so their sum $-Ae^{-2\Phi}$ is zero. Hence A, and by symmetry also B and C must vanish. It is perhaps noteworthy that in order to derive the form of the entire metric we needed to use synchrony only for *neighboring* clocks.

We also note that a 'bad' choice of time-coordinate even in a stationary spacetime destroys the time-independence of the metric. The reader can see this quickly by checking, for example, the effect of a simple transformation like $t = \exp(x/x_0)t'$ on the usual Minkowski metric (8.18).

We postpone until Section 9.5 the proof that, conversely, every metric of the form (9.2) with time-independent coefficients represents a stationary spacetime: It represents a lattice coordinatized by the x^i and by a coordinate time t such that (i) all lattice clocks run at constant multiples of their standard rate, (ii) every light signal (and indeed every free-falling particle path) from any point P to any point Q is indefinitely repeatable over the same track and with the same duration, and (iii) in the special case $A = B = C = 0$ these signals can traverse the same track also backwards in the same time as forwards.

If for the moment we accept this, then the frequency-shift formula (1.11), namely the first of the following equations:

$$\frac{\nu_B}{\nu_A} = \frac{\exp(-\Phi_B/c^2)}{\exp(-\Phi_A/c^2)} = \sqrt{\frac{g_{tt}(A)}{g_{tt}(B)}}, \tag{9.4}$$

proved only approximately and only for static fields in Chapter 1 (with Φ as the Newtonian potential) is now seen as an exact and immediate corollary of the stationary metric (9.2.), where $\exp(2\Phi/c^2) =: g_{tt}$ is the coefficient of $c^2 \, dt^2$. For consider two consecutive wave fronts passing first A and then B. The coordinate time Δt between their passing is the same at A and B. But the frequencies recognized at A and B will be inversely proportional to the times of passing as measured by *standard* clocks. So $\nu_B/\nu_A = \exp(\Phi_A/c^2)\Delta t / \exp(\Phi_B/c^2)\Delta t$, whence (9.4).

9.4   Newtonian support for the geodesic law of motion

We shall now show how Einstein's proposed geodesic law of motion for free particles in a gravitational field finds its first *quantitative* support by leading to approximately the same orbits as Newton's theory in 'classical' situations; that is, for slow test particles in weak static fields. This is, of course, not entirely surprising, since both theories share versions of the equivalence principle, since the equivalence principle essentially forces the geodesic law in the relativistic framework, and since that framework reduces to Newton's in the classical limit. But it is encouraging, to say the least!

In the last section we found that the metric of every *static* field can be brought to the canonical form (here again *with c*)

$$\mathbf{ds}^2 = e^{2\Phi/c^2}c^2 \, dt^2 - dl^2, \tag{9.5}$$

where dl^2 is a time-independent 3-metric. We have already recognized [in the second paragraph of Section 9.2, and again in connection with (9.4)] that the Φ appearing in the time-dilation factor $\exp(-\Phi/c^2)$, and consequently in the metric (9.5), must coincide, at least approximately, with the Newtonian potential of the field. In dl we now recognize the radar distance between neighboring lattice points as determined by *standard* clocks at rest in the lattice. But radar distance along an infinitesimal rod coincides with ruler distance, even if the rod accelerates through its LIF (cf. Exercise 3.9); for the distance measured is proportional to dt while the distance moved (the error) is proportional to $(\mathrm{d}t)^2$. So dl^2 in (9.5) is just the spatial metric of the lattice. However, without the field equations we cannot predict its detailed form. So our procedure here will be to approximate the lattice with Euclidean 3-space. Obviously the lattice of such fields as that of the sun in our neighborhood cannot be very curved—or we would know it! In fact, a relativistically 'weak' field is defined precisely as one whose curvature is small; that is, one whose spacetime deviates little from Minkowski space M^4. But then, by the special-relativistic symmetry between ct on the one hand and x, y, z on the other, one would expect the deviations of *all* the metric coefficients $g_{\mu\nu}$ from their SR-values diag$(-1, -1, -1, 1)$ to be of the same order of smallness. Now in classical situations the coefficient of $c^2 \, \mathrm{d}t^2$ in (9.5), $\mathrm{e}^{2\Phi/c^2} \approx 1 + 2\Phi/c^2$, differs very little from unity. For example, throughout the exterior field of the sun the Newtonian potential satisfies $|2\Phi/c^2| < 5 \times 10^{-6}$. Suppose d$l^2$ deviates by as much from flatness. If we neglect this deviation in half of our metric, do we thereby introduce an error of 50 per cent? It depends: for light paths, sometimes yes. But for 'slow' orbits ($v \ll c$) the coefficient of $c^2 \, \mathrm{d}t^2$ contributes vastly more than the spatial coefficients. If we approximate the former with unity, we essentially throw out gravity. If we approximate dl^2 with flat space, we introduce only a small error. This can be seen as follows. A timelike geodesic is found as a curve of extremal length (actually: of maximal length) among a bundle of neighboring worldlines connecting two events in spacetime. But a slow-motion worldline in slightly deformed M^4 is almost parallel to the time axis. A deformation of the time dimension therefore has a first-order effect on the lengths of those lines, whereas a deformation of the spatial dimensions has only a second-order effect.

We can also see this quantitatively: Consider a static metric $\mathbf{ds}^2 = Ac^2 \, \mathrm{d}t^2 - B \, \mathrm{d}l^2$ (dl^2 flat). Inasmuch as they differ from unity, A and B measure deviations from M^4. The worldline of a particle moving with coordinate speed $v = \mathrm{d}l/\mathrm{d}t$ has $\mathbf{ds}^2 = \mathrm{d}t^2(Ac^2 - Bv^2)$. Thus the space and time deviations are weighted in the ratio $v^2 : c^2$. For all the sun's planets, for example, $v < 50$ km/s, so that $v^2 : c^2 < 3 \times 10^{-8}$, which illustrates the smallness of the spatial contribution. But for high-speed particles, and especially for light, this contribution can and does become significant, even in weak fields.

In light of all this, we now take the metric of a weak static field to be

$$\mathbf{ds}^2 = (1 + 2\Phi/c^2)c^2 \, \mathrm{d}t^2 - \mathrm{d}l^2, \tag{9.6}$$

where Φ is the Newtonian potential and dl^2 is flat. For a particle-worldline between two events $\mathcal{P}_1$ and $\mathcal{P}_2$ at times t_1 and t_2 we then have

$$\int_{\mathcal{P}_1}^{\mathcal{P}_2} \mathrm{d}s = \int_{t_1}^{t_2} \frac{\mathrm{d}s}{\mathrm{d}t} \, \mathrm{d}t = c \int_{t_1}^{t_2} \left(1 + \frac{2\Phi}{c^2} - \frac{v^2}{c^2} \right)^{1/2} \mathrm{d}t,$$

where $v = dl/dt$ is the coordinate velocity of the particle. Applying the binomial approximation to the last integrand, we thus find

$$\int_{\mathscr{P}_1}^{\mathscr{P}_2} ds = c \int_{t_1}^{t_2} \left(1 + \frac{\Phi}{c^2} - \frac{1}{2}\frac{v^2}{c^2}\right) dt$$

$$= c(t_2 - t_1) - \frac{1}{c}\int_{t_1}^{t_2} \left(\frac{1}{2}v^2 - \Phi\right) dt. \tag{9.7}$$

The condition that $\int ds$ be maximal is therefore equivalent to the last integral in (9.7) being minimal. But that is exactly Hamilton's Principle for the motion of a particle in a gravitational potential Φ! This shows that the 'slow-motion' geodesics of the spacetime (9.6) indeed coincide with the Newtonian orbits in a potential Φ, to first approximation. And it identifies once more the relativistic Φ (from clock synchronization) with Newton's potential.

It must have been a memorable moment when, by some such calculations as the above, Einstein first realized (probably in March 1912) that GR could 'work'. What remained was the invention of satisfactory field equations that (in some suitable limit) also reduced to Newtonian theory.

The present result illustrates well the 'man-made' character of physical theories. It is really remarkable how the same empirically known orbits can be 'explained' by two such utterly different models as Newton's universal gravitation and Einstein's curved spacetime. Nature exhibits neither potentials nor Lorentzian metrics. Yet both these human inventions lend themselves to a description of a large class of observed phenomena.

And again, GR is one of the classic examples where a pure mathematician's flight of fancy (Riemann's n-dimensional geometry of 1854) later becomes the physicists' bread and butter—a process that has often been repeated. Mathematics is the theoretical physicists' hardware store where they can buy the materials for their models!

One consequence which we can read off at once from eqn (9.5) is what has come to be called the *Shapiro time delay*. A light-ray satisfies $\mathbf{ds}^2 = 0$ and thus (with $c = 1$) $e^\Phi dt = \pm dl$, the two signs corresponding to the two possible directions of travel. Consequently a radar signal reflected by a distant object will return to its emission point after a coordinate time

$$\Delta t = 2 \int e^{-\Phi} \, dl$$

has elapsed there, the integration being performed over the path of the signal. If that passes near a concentrated mass and thus through regions of large negative Φ, the time delay can be significant. It was Shapiro who in the sixties first proposed measuring this effect for radar signals from earth to Venus, as these signals become more and more disturbed by the approaching sun. (See Section 11.2H below.)

The reader might find the concept of a 'world-movie' (somewhat similar to the 'world-map' of SR) useful in connection with static and stationary spacetimes. Imagine a movie of the 3-dimensional curved lattice, successive frames showing it at successive coordinate time intervals $t = 1, 2, 3, \ldots$. One sees particles and photons trace out identical paths again and again, always in the same time. But the most characteristic feature is the pronounced slowing down of photons (light) near heavy masses and, as we shall see later, their total standstill at horizon-edges of stationarity. Their behavior is the exact opposite of what ordinary particles do in Newtonian theory: the deeper the potential well, the faster they must go to keep the total energy constant. In the relativistic world-movie, ordinary particles behave classically for slow motions and weak potentials; but near infinite wells (horizons) they too slow down and eventually come to a complete standstill.

9.5 Symmetries and the geometric characterization of static and stationary spacetimes

Recall our *active* interpretation (in Section 2.9) of the standard Lorentz transformation as a motion of Minkowski space upon itself. Such metric-preserving motions of metric spaces upon themselves are called *isometries*, and they describe symmetries of the spaces when they exist. As other examples consider two spheres, or two planes, or two spiral staircases (half-helicoids) sliding over each other, one representing the original, the other the moved space. If the motion is continuous, like the rotation of a sphere about a fixed axis, the various points of the space trace out the so-called *group orbits* (or *trajectories*).

Under isometric mappings, all intrinsic features of a metric space are preserved, since the entire space has simply been 'moved elsewhere'. Most importantly, geodesics map into geodesics and lengths and angles remain invariant.

One particularly interesting isometry can occur under reflection in a surface or hypersurface H of one dimension less than the full space V. Let G be the congruence of geodesics orthogonal to H (that is, orthogonal to every direction *in* H) and define a mapping of points $P \mapsto P'$ as follows: P and P' shall lie on the same member of G, on opposite sides of H, and equidistant from it. If this mapping is an isometry, we say that H is a *symmetry surface* of V. It clearly divides V into two identical halves.

Suppose we have reflection isometry about H, but relative to some other congruence of curves intersecting H, not necessarily geodesic, and relative to some other parameter, not necessarily length. This *must* be equivalent to geodesic reflection isometry! For since the mapping preserves geodesics and angles, the geodesics orthogonal to H map into themselves; and since distances are preserved, equidistant points on these geodesics correspond to each other. That proves it. The members of that other congruence, incidentally, must also cut H orthogonally. For consider a small triangle ABC, where A and B lie in H while C lies on the mapping curve through A. If C' is the image of C, the triangles ABC and ABC$'$ are congruent. But since CAC$'$ is locally straight, AB must be orthogonal to it.

All this is relevant to stationary and static spacetimes, which possess some important symmetries along these lines. As is clear from (9.2) (with all the coefficients time-independent), all *time-translations*

$$(x^i, t) \longmapsto (x^i, t + a) \tag{9.8}$$

leave a stationary metric invariant. These constitute a one-parameter isometry group (with parameter a) whose orbits $t = $ var are the (timelike) worldlines of the lattice points. Additionally, *static* spacetimes possess *time-reflection* symmetry about every time-slice $t = a$: if $A = B = C = 0$ in (9.2), the transformation

$$(x^i, t) \longmapsto (x^i, 2a - t) \tag{9.9}$$

leaves the metric invariant and thus maps the spacetime isometrically upon itself. [Note that (9.9) implies $(x^i, a - b) \longmapsto (x^i, a + b)$.] By the remarks of the preceding paragraph, every time slice is then a symmetry surface, and *orthogonal* to the orbits $t = $ var.

Conversely, suppose we know that a given spacetime possesses a continuous one-parameter isometry group with timelike orbits. Then it is stationary. For proof, take the group orbits as the worldlines $x^i = $ const of 'lattice points' and the parameter as time coordinate t. Then, since a time-translated geodesic is still a geodesic over the same lattice track and with the same duration (the t-values at both ends increase equally), all light signals between two given lattice points are indefinitely repeatable and our original light-circuit postulate is satisfied, which was our criterion for stationarity. If, additionally, there exists a hypersurface H orthogonal to all the group orbits, we can re-define the zero of time by setting $t = 0$ on H. At $t = 0$ the $t = $ var lines are then orthogonal to the $t = $ const surface and this implies [as we shall see in (10.4)] that $A = B = C = 0$ in the stationary metric (9.20). But that shows $t = 0$ to be a symmetry surface relative to t (the metric being invariant under $t \mapsto -t$). So for each set of events (x^i, t) corresponding to a geodesic g from event $\mathscr{P}$ to event $\mathscr{Q}$ there is a mirror-image set $(x^i, -t)$ also forming a geodesic, g', passing over the same spatial track, from event $\mathscr{P}'$ to vent $\mathscr{Q}'$ (cf. Fig. 9.3). However, as a worldline, g' must be traversed from $\mathscr{Q}'$ to $\mathscr{P}'$; that is, into the future. So for every light signal between lattice points P and Q there is another of equal duration over the same track from Q to P. This leads to the strengthened light-circuit postulate being satisfied, which was our criterion for staticity.

So if we temporarily denote the light-circuit postulate by L and its strenthened version by LS; the metric (9.2) with time-independent coefficients by M and if A, B, C are missing, by MS; and translational symmetry with timelike orbits by T and this jointly with hypersurface-orthogonality by TS—then what we have by now established is

$$L \Rightarrow M \Rightarrow T \Rightarrow L$$

$$LS \Rightarrow MS \Rightarrow TS \Rightarrow LS.$$

Consequently for both stationary and static spacetimes all of the three relevant characterizations are equivalent.

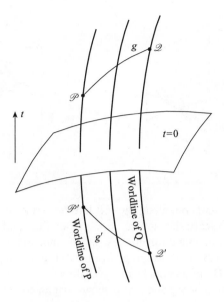

Fig. 9.3

Our arguments (two paragraphs back) have also established what we asserted earlier: *every* geodesic motion (of light *or* particles) is indefinitely repeatable in stationary
spacetimes, and even in reverse in static spacetimes.

Whilst on the subject of symmetry surfaces, it will be well to develop a few general
results which will stand us in good stead later on. An always surprising example
of symmetry surfaces is provided by the geodesic planes of 3-spheres. They are,
as we noted earlier [cf. before (8.6)], 2-spheres whose inside and outside are identical halves of the full space. We can check this from the metric of the 3-sphere
(cf. Exercise 8.9):

$$ds^2 = dr^2 + a^2 \sin^2\left(\frac{r}{a}\right)(d\theta^2 + \sin^2\theta \, d\phi^2), \qquad (9.10)$$

where θ and ϕ are the 'usual' angles at the origin. Clearly $\phi \to -\phi$ is an isometry,
so $\phi = 0$ is a symmetry surface; and it is intuitively clear that it is a geodesic plane
at the origin. Its induced metric, from (9.10), is

$$ds^2 = dr^2 + a^2 \sin^2\left(\frac{r}{a}\right)d\theta^2, \qquad (9.11)$$

which, as Fig. 8.4 shows, represents a 2-sphere of radius a. This can also be obtained
from the metric (8.12) by $x \mapsto r/a, y \mapsto \theta$.

One chief property of symmetry surfaces is that they are *totally geodesic*. By this
we mean that *each of their geodesics is also a geodesic of the full space*. For consider
a geodesic g of a symmetry surface H with initial direction **dg**. Let g' be the geodesic

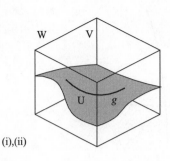 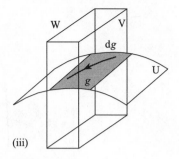

Fig. 9.4

of the full space that starts out along **dg**; if it left the surface on one side, its mirror image would have to leave the surface on the other side, and there would be two geodesics with the same initial direction, which is impossible. So g' must lie in H. But then (cf. Section 8.3, third paragraph from end) it is also a geodesic of H. Since only *one* geodesic of H can start out in any given direction, g is g', and thus a geodesic of the full space. A similar argument also shows that an equivalent characterization of a totally geodesic subspace is that *any geodesic of the full space that is tangent to it, fully lies in it.*

Totally geodesic hypersurfaces need not be symmetry surfaces: our proof would have gone through even if the symmetry extended only a little on either side of H. Also totally geodesic subspaces of an N-dimensional space can have any dimension less than N, whereas symmetry surfaces have dimension $N - 1$. And, of course, not all Riemannian spaces contain either symmetry surfaces or totally geodesic subspaces. But when they do, it can considerably shorten the calculation of geodesics.

The following three lemmas are useful in this respect: (i) *If* U *is a t.g.* (totally geodesic) *subspace of* V, *which itself is a t.g. subspace of* W, *then* U *is a t.g. subspace of* W; (ii) *if* U *is a t.g. subspace of* W, *then it is also a t.g. subspace of any subspace* V *of* W *that contains it*; (iii) *the intersection* U∩V *of any two t.g. subspaces* U *and* V *of* W *is itself t.g. Proofs* (cf. Fig. 9.4): (i) Any geodesic of U is a geodesic of V; but any geodesic of V is a geodesic of W; QED. (ii) Every geodesic of U is a geodesic of W; but every geodesic of W that lies in V is a geodesic of V (cf. Section 8.3); QED. (iii) Let **dg** be an arbitrary initial direction in U∩V: it must lie in U and in V; draw the W-geodesic in the direction **dg**: it lies totally in U and in V and thus in U∩V; it is therefore a geodesic of U∩V; so the U∩V-geodesic in the direction **dg** is a W-geodesic; QED.

By way of illustration: we have already noted that $\phi = 0$ is a symmetry surface of the 3-sphere (9.10). So is $\theta = \pi/2$, since $\theta \to \pi - \theta$ is an isometry. Hence, by lemma (iii), the radius $\phi = 0$, $\theta = \pi/2$ is t.g. By the spherical symmetry, *any* radius θ, $\phi = $ const can be isometrically rotated into that one, so *all* such radii are t.g. And a t.g. 1-space (a curve) must itself be a geodesic (since a geodesic of the big space that starts in it, stays in it.) So (to no one's surprise) all such radii are geodesics, and the surface $\phi = 0$ is indeed a geodesic plane through the origin. Einstein's static universe of 1917 has the 3-sphere as its lattice, carrying a homogeneous matter distribution (which is held in equilibrium against gravity by the so-called Λ-term in the field equations.) By the overall symmetry, all the lattice clocks can run at their standard rates, and so the metric is

$$\mathbf{ds}^2 = c^2 \, dt^2 - \left\{ dr^2 + a^2 \sin^2 \left(\frac{r}{a} \right) (d\theta^2 + \sin^2 \theta \, d\phi^2) \right\}. \tag{9.12}$$

By the same reasoning as before, the radii $\theta, \phi = $ const are now t.g. 2-spaces (in r, t). So particles and light signals that originally move radially stay on fixed radii and are geodesics of the metric $c^2 \, dt^2 - dr^2$. But this is 2-dimensional Minkowski space and its geodesics are given by $r = vt + $ const. So free particles (and photons) can go round and round great circles at constant speeds for ever. In particular, the lattice points themselves trace out timelike geodesics and are thus potential locations of free particles (galaxies!).

9.6 Canonical metric and relativistic potentials

Whereas in the case of the static metric (9.5) the 3-metric of the lattice can be read off directly as (minus) its spatial part, the same is *not* true in the general case (9.2). Why? Because there the sections $t = $ const cut *obliquely* across the worldlines $t = $ var of the lattice points; and that means (switch to SR geometry locally!) that the lattice is measured from a *moving* frame, and thus with length contraction. The way to find the lattice-metric (and more) is via what we shall call the *canonical* form of the metric, which results when we 'complete the square' in (9.2) to absorb the time–space cross-terms:

$$\mathbf{ds}^2 = e^{2\Phi/c^2} \left(c \, dt - \frac{1}{c^2} w_i \, dx^i \right)^2 - k_{ij} \, dx^i \, dx^j, \tag{9.13}$$

with $w_1 = -(c/2) A e^{-2\Phi/c^2}$, etc. Of course, Φ is the clock-rate function as it has been all along, and w_i and k_{ij} are newly arising time-independent coefficients. We shall presently recognize in w_i an analog of the Maxwellian vector potential $\mathbf{w}$ and in k_{ij} the metric of the lattice.

Our synchronization process of Section 9.2 leaves little leeway for the time coordinate t: we can change *all* clock rates by a constant factor $k > 0$, and we can change the zero points by a continuous function of position:

$$t \mapsto t' = k[t + f(x^i)]. \tag{9.14}$$

It is easy to see (cf. Exercise 9.7) that this is, in fact, the most general time transformation that leaves the metric (9.2) and hence also the metric (9.13) form-invariant; of course, the lattice coordinates can be transformed arbitrarily: $x^i \mapsto x^{i'} = x^{i'}(x^i)$. Under each of these transformations the two summands of (9.13) remain separate, and, in particular, under (9.14) we find:

$$\Phi' = \Phi - c^2 \log k, \qquad w_i' = k(w_i + c^3 f_{,i}), \qquad k_{ij}' = k_{ij}, \tag{9.15}$$

where once again we use the derivative notation introduced in (7.7). We shall refer to (9.14) and its concomitant (9.15) as *gauge-transformations* and note that physically meaningful quantities must be 'gauge-invariant'. Examples of gauge-invariant quantities are:

$$\Phi_{,i}, \qquad e^{\Phi}(w_{i,j} - w_{j,i}), \qquad k_{ij}, \qquad k^{ij}. \tag{9.16}$$

At any given lattice point we can always achieve $\Phi = w_i = 0$ by a suitable gauge-transformation and thus reduce (9.13) to $ds^2 = c^2 dt^2 - k_{ij} dx^i dx^j$, which establishes k_{ij} as the metric of the lattice [cf. after (9.5)].

A stationary metric (9.13) under certain circumstances (bad choice of clock settings) can be transformed into a *static* metric with the same lattice. This is the case whenever we can transform w_i away; that is, whenever there exists an f such that (with $c = 1$) $w_i = -f_{,i}$ ($\mathbf{w} = -\text{grad } f$), which happens (in simply connected spaces) whenever $w_{i,j} - w_{j,i}$ ('curl $\mathbf{w}$') vanishes. In fact, we shall see presently that the middle quantity in (9.16) describes the local *rotation rate* of the lattice: static lattices are characterized by non-rotation. (This is directly related to the 'hypersurface-orthogonality' of the lattice worldlines.)

To elucidate the physical significance of w_i, let us repeat our earlier weak-field slow-orbit calculation that led from (9.6) to (9.7), but this time *with* w_i. We approximate the metric (9.13) by

$$ds^2 = \left(1 + \frac{2\Phi}{c^2}\right)\left(1 - \frac{2\mathbf{w} \cdot \mathbf{v}}{c^3}\right)c^2 dt^2 - dl^2$$

$$\approx \left(1 + \frac{2\Phi}{c^2} - \frac{2\mathbf{w} \cdot \mathbf{v}}{c^3}\right)c^2 dt^2 - dl^2,$$

writing $\mathbf{w} \cdot \mathbf{v}$ for $w_i dx^i/dt$ and once again dl^2 for the metric of the lattice. This is equivalent to replacing the Φ of (9.6) by $\Phi - (\mathbf{w} \cdot \mathbf{v})/c$ and thus leads to the following generalization of (9.7):

$$\int_{\mathcal{P}_1}^{\mathcal{P}_2} ds = c(t_2 - t_1) - \frac{1}{c} \int_{t_1}^{t_2} \left(\frac{1}{2}v^2 - \Phi + \frac{\mathbf{w} \cdot \mathbf{v}}{c}\right) dt. \tag{9.17}$$

The geodesic requirement on the orbit, namely that $\int ds$ be maximal, is once again equivalent to the integral on the RHS being minimal. But an identical variational principle is known to hold in an electromagnetic field with scalar potential Φ and vector potential $\mathbf{w}$ for the non-relativistic (that is, slow) motion of a particle whose mass and charge are numerically equal ($m = q$).[1] In gravitation, the mass and 'charge' of a particle are *always* equal ($m_I = m_G$). So in a stationary field a free particle moves *like* a particle of equal charge and mass subject to a Lorentz-type force law

$$\mathbf{f} = -\text{grad } \Phi + \frac{1}{c}\mathbf{v} \times \text{curl } \mathbf{w} \tag{9.18}$$

---

[1] See, for example, J. D. Jackson, *Classical Electrodynamics*, 2nd edn, John Wiley, New York, 1975, eqn (12.9).

per unit mass. For it is this law, in conjunction with Newton's $\mathbf{f} = m\mathbf{a}$, that is equivalent to the above variational principle when $\partial \mathbf{w}/\partial t = 0$. [Cf. Jackson, *loc. cit.*, eqn (6.31).] Only now Φ and $\mathbf{w}$ are the *relativistic potentials* of the *gravitational* field, defined by the canonical metric (9.13). (The term 'gravitomagnetic potential' is also used for $\mathbf{w}$.) The slow general-relativistic motion of particles in weak stationary fields is thus seen to have pronounced Maxwellian features, which will be seen to carry over even to the field equations. (One corollary: moving matter gravitates differently from static matter.)

Now we know from classical mechanics that if a point P of a rigid reference frame L travels at acceleration $\mathbf{a}$ through an inertial frame while L rotates about P at angular velocity $\boldsymbol{\Omega}$, then a free particle of unit mass at P moving relative to L at velocity $\mathbf{v}$ experiences relative to L an 'inertial' force

$$\mathbf{f} = -\mathbf{a} + 2\mathbf{v} \times \boldsymbol{\Omega}, \tag{9.19}$$

where the last term is the well-known Coriolis force. But according to Einstein, inertial and gravitational forces are indistinguishable. Comparison of (9.19) with (9.18) thus leads to the conclusion that the *lattice* at any of its points P moves relative to the local inertial frame (LIF) with acceleration

$$\mathbf{a} = \text{grad } \Phi \tag{9.20}$$

and angular velocity

$$\boldsymbol{\Omega} = \frac{1}{2c}\text{curl } \mathbf{w}. \tag{9.21}$$

Conversely (according to the present approximative calculation) the acceleration of the LIF and its rotation rate relative to the lattice are $-\mathbf{a}$ and $-\boldsymbol{\Omega}$ respectively. The first, of course, is what would be recognized as the gravitational field relative to the lattice, and the second is the rotation rate of a 'gyrocompass' suspended at a lattice point. The *exact* calculation leads to the following (gauge-invariant!) expressions:

$$|\mathbf{a}| = \text{gravitational field} = (k^{ij}\Phi_{,i}\Phi_{,j})^{1/2} \tag{9.22}$$

$$|\boldsymbol{\Omega}| = \text{(proper-time) rotation rate of gyrocompass}$$

$$= \frac{1}{2\sqrt{2}c}e^{\Phi/c^2}\left[k^{ik}k^{jl}(w_{i,j} - w_{j,i})(w_{k,l} - w_{l,k})\right]^{1/2}, \tag{9.23}$$

where the magnitude of the gravitational field is defined as that of the proper acceleration of a lattice point. [For the proofs, see eqn (10.50) and Exercise 10.11.]

We have seen Φ in its role as clock-rate function, and, almost equivalently with that, as frequency-shift function. And both earlier and now again we have recognized its role as scalar potential. This latter, however, is to be understood as determining the *whole* field relative to the lattice—or what in Newtonian language would be the sum of the gravitational *and* the inertial force on a unit mass. In a rotating frame, for example, that includes the centrifugal force and corresponds to what on earth is called the 'geopotential'. Einstein, in his famous 1905 paper in which he already predicted time dilation by a γ-factor, suggested that this might be tested by comparing clock rates at the equator with those at the pole. His equivalence principle and with it the recognition of *gravitational* time dilation was still two years away. As a result, we know today (also from measurements) that there is *no* rate-discrepancy at all between clocks at the pole and clocks at the equator: the surface of the earth is practically an equipotential surface $\Phi = $ const! This can be seen by imagining the earth as having cooled from a rigidly rotating liquid ball; if its surface were *not* equipotential, the total field would have a component *in* the surface causing the liquid to move sideways, thus violating the assumption of equilibrium (rigid motion).

Lastly, let us take one more look at eqn (9.17) for an *arbitrary* path. It shows that the proper-time increment $c^{-1} \int ds$ measured by a slowly transported standard clock (like one of those flown by Hafele and Keating around the world—cf. penultimate paragraph of Section 3.5) differs from the coordinate-time increment $(t_2 - t_1)$ by: (i) the usual special-relativistic time dilation $- \int \frac{1}{2}(v^2/c^2)\,dt$ [cf. eqn (3.3)]; (ii) the general-relativistic 'altitude' correction $- \int (\Phi/c^2)\,dt$; and (iii) in non-static fields also a 'source-motion' correction $- \int (\mathbf{w} \cdot \mathbf{v}/c^3)\,dt$.

9.7   The uniformly rotating lattice in Minkowski space

A transparent example for some of our results on stationary fields is provided by the uniformly rotating lattice. We begin by writing the metric of Minkowski space in cylindrical coordinates r, ϕ', z (*now again with $c = 1$*):

$$\mathbf{ds}^2 = dt^2 - dr^2 - r^2\,d\phi'^2 - dz^2 \tag{9.24}$$

and introduce a new angular coordinate ϕ measured from a fiduciary half-plane that rotates about the z-axis with constant angular velocity ω; [see Fig. (9.5)]:

$$\phi = \phi' - \omega t, \qquad d\phi' = d\phi + \omega\,dt. \tag{9.25}$$

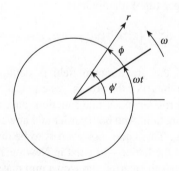

Fig. 9.5

As we mentioned in Section 9.2 (third paragraph from end), the appropriate coordinate time for a rotating disk (we now have a stack of them) is that of the underlying Minkowski space. Accordingly the only change we make to (9.24) is (9.25). A straightforward calculation then yields the following canonical form of the metric for the field associated with the rotating lattice:

$$ds^2 = (1 - r^2\omega^2)\left[dt - \frac{r^2\omega}{1 - r^2\omega^2}\, d\phi\right]^2 - dr^2 - \frac{r^2}{1 - r^2\omega^2}\, d\phi^2 - dz^2. \quad (9.26)$$

Lattice points at $r = 1/\omega$ satisfy $ds^2 = 0$; that is, they move at the speed of light. Beyond that, neither the lattice nor the metric (9.26)—though valid—is of interest.

We recognize the expression $1 - r^2\omega^2$ that occurs repeatedly in (9.26) as $(1 - v^2) = \gamma^{-2}$, γ being the Lorentz-factor of the lattice points at radius r. This explains its presence at the beginning: lattice clocks must be speeded up by a factor γ to keep in step with the underlying clocks (SR-time dilation); its presence underneath $r^2\, d\phi^2$ shows the ruler length of a lattice circle $r, z = \text{const}$ to be $\oint \gamma r\, d\phi = 2\pi r\gamma$ (SR-length contraction of comoving rulers!).

The metric of the lattice is the negative of the last three terms in (9.26) and represents a *curved* 3-space (cf. Exercise 9.11):

$$k_{ij} = \text{diag}\,(1, r^2(1 - r^2\omega^2)^{-1}, 1),$$
$$k^{ij} = \text{diag}\,(1, r^{-2}(1 - r^2\omega^2), 1). \quad (9.27)$$

And the relativistic scalar potential Φ is given by $\exp 2\Phi = (1 - r^2 w^2)$; that is,

$$\Phi = \tfrac{1}{2}\log(1 - r^2\omega^2). \quad (9.28)$$

Its gradient (the centrifugal force) is clearly in the r-direction and, since r is then ruler distance, we simply have

$$|\text{grad } \Phi| = \left|\frac{\partial \Phi}{\partial r}\right| = \frac{r\omega^2}{(1 - r^2\omega^2)}. \quad (9.29)$$

(This is also what the exact formula (9.22) yields.) So the Newtonian centrifugal force $r\omega^2$ now has a relativistic correction factor γ^2 which makes it approach infinity at $r = 1/\omega$, the limit of stationarity.

But perhaps the most interesting result to be extracted from the metric (9.26) is the so-called *Thomas precession* of special relativity. Imagine a gyrocompass suspended at one of the lattice points. Relative to the underlying space, that gyrocompass is being taken along a circular path and will not precisely return to its original orientation after a complete revolution (cf. Exercise 3.13). To calculate the effect, we first read off the relativistic vector potential components from (9.26):

$$\mathbf{w} = (0, r^2\omega(1 - r^2\omega^2)^{-1}, 0), \quad (9.30)$$

whose only non-zero derivative is

$$w_{2,1} = \frac{2r\omega}{(1 - r^2\omega^2)^2}. \tag{9.31}$$

Substituting this into (9.23) and writing v for $r\omega$, we find

$$\Omega = \frac{v}{r}(1 - v^2)^{-1} \tag{9.32}$$

exactly. [While the summation in (9.23) ranges over *all* values of i, j, k, l, only $1, 2, 1, 2$ and $2, 1, 2, 1$ give non-zero contributions.] So after one revolution the orientation of the gyrocompass *relative to the lattice* changes by an angle

$$\alpha' = \Omega\Delta\tau = \Omega\gamma^{-1}\Delta t = \Omega\gamma^{-1}2\pi\omega^{-1} = (1 - v^2)^{-1/2}2\pi \tag{9.33}$$

in the opposite sense to ω, as can be seen from (9.21) and (9.31). If this were 2π, there would be *no* precession of the gyrocompass in the underlying space. But, in fact, there *is* a precession, in the retro sense, given by

$$\alpha = \alpha' - 2\pi = 2\pi((1 - v^2)^{-1/2} - 1) \approx \pi v^2, \quad (\pi v^2/c^2) \tag{9.34}$$

per revolution. (Compare this with the result of Exercise 3.13.)

Exercises 9

9.1. Observers at two fixed points A and B in a stationary gravitational field determine the radar distance between them by use of *standard* clocks. If L_A and L_B are these determinations made at A and B respectively, prove $L_A/L_B = \exp(\Phi_A/c^2)/\exp(\Phi_B/c^2)$.

9.2. Two twins decide to 'buy youth' in two different ways. The first buys enough energy to quickly accelerate himself and his vehicle to a certain high speed, at which he then cruises through space for a long time before spending as much energy again decelerating and returning to earth. The second twin buys enough energy to lower herself and her vehicle quickly to the surface of a dense planet, where she will live for a long time before spending as much energy again to return to earth. Both return simultaneously and during their absence both have aged at only one nth the rate of their former contemporaries. But as long as the gravitational field is weak enough to be Newtonian, their expenses are essentially in the ratio $(n - 1) : \log n$ in favor of the dense planet. Prove this. (Cf. also Section 12.2 below.)

9.3. Consider a light path l from a fixed point A to a fixed point B in a stationary gravitational field with scalar potential Φ. The light was emitted by a source of proper frequency ν_0 passing A at velocity u (measured with standard clocks and rulers at rest) in a direction making an angle α with l. By using an auxiliary observer at A' near A on l and our earlier formula (4.3), find the frequency ν at which the source is seen at B.

9.4. A satellite is in circular orbit of radius r around the earth (radius R), satisfying Kepler's law $\omega^2 = GM/r^3$, where M is the mass of the earth and ω the angular velocity in the orbit. Neglecting the rotation of the earth, prove that the frequency ν observed on earth when the satellite appears directly overhead, of a source of proper frequency ν_0 on the satellite, is given by

$$\frac{\nu}{\nu_0} \approx 1 + \frac{GM}{Rc^2} - \frac{3GM}{2rc^2}.$$

Note that this exceeds unity only if $r > 3R/2$.

9.5. Two neighboring particles are released from rest on the same vertical and fall freely to earth. Use Newton's law $\ddot{r} = -GM/r^2$ to prove that the spatial separation η of these particles satisfies $\ddot{\eta} = (2GM/r^3)\eta$ (tidal acceleration!). Their worldlines are geodesics in the spacetime surrounding the earth. Draw a diagram of these (slow) geodesics in a quasi-Minkowskian spacetime to establish that ct corresponds to their arc length and η to the η of eqn (8.4). Hence deduce that $-2GM/c^2r^3$ is (at least approximately) the Gaussian curvature of earth's spacetime in an orientation determined by dt and dr. (Note the correlation of curvature with tidal force.) What is the curvature in an orientation determined by dt and a direction orthogonal to a radius?

9.6. In the Eistein universe (9.12), a rocketship moves radially, from rest, at constant proper acceleration α. How much time elapses at the starting point before the rocketship returns? [*Answer*: $2\pi(a/c)\sqrt{1 + c^2/\pi a\alpha}$.]

9.7. Prove our assertion of Section 9.6 that the most general time transformation $t = F(t', x^i)$ that leaves the stationary metric (9.2) form-invariant is that corresponding to eqn (9.14). [*Hint*: all four partial derivatives of F must be time-independent.]

9.8. Apply a gauge transformation to convert the stationary metric

$$\mathbf{ds}^2 = x^2\,dt^2 - 6x^4y\,dx\,dt - 2x^5\,dy\,dt + 6x^7y\,dx\,dy - dz^2$$

into a static metric. (The coordinate x is dimensionless, y and z are lengths, and $c = 1$.)

9.9. In the static spacetime whose canonical metric is of the form

$$\mathbf{ds}^2 = \left(1 + \frac{x^2}{a^2}\right)dt^2 - dl^2,$$

where dl^2 is an arbitrary 3-metric in x, y, z, find a subset of lattice points whose worldlines are geodesics.

9.10. In the stationary spacetime with canonical metric

$$\mathbf{ds}^2 = (dt + ay\,dx)^2 - (dx^2 + dy^2 + dz^2)$$

find the rotation of the lattice, and also prove that the lattice worldlines are geodesics. (This simple spacetime bears some resemblance to the famous *Gödel universe*.)

9.11. For the rotating lattice of Section 9.7 find the frequency-shift in the light sent from any point at radius r_0 to another at radius r_1. Compare your result with our earlier formula (4.6).

9.12. (i) By considering the lengths of circles $r, z = $ const in the lattice corresponding to the metric (9.26), prove that the Gaussian curvature of the lattice on the z-axis for the orientation orthogonal to that axis is $-3\omega^2/c^2$. (ii) Prove that the radii $r = $ var are geodesics of the lattice. [*Hint*: symmetry surfaces.] Hence find the value of its Gaussian curvature for the orientation $z = $ const as a function of r. [*Answer*: $-(3\omega^2/c^2)(1 - r^2\omega^2/c^2)^{-2}$.]

10

Geodesics, curvature tensor and vacuum field equations

10.1 Tensors for general relativity

Whereas historically SR could progress quite a long way (until 1908, to be exact) without recourse to tensors, no such elementary access to GR is possible. The very foundations of the theory, the field equations, require the language of tensors even for their enunciation.

In Chapter 7 we introduced tensors already with a view to using them *both* in SR and in GR. We regarded them as existing in an arbitrary background space V^N, coordinatized by arbitrary coordinates $\{x^i\}$, and possessing an arbitrary pseudo-Riemannian metric g_{ij}. Only after eqn (7.19) did we specialize by taking V^N to be Minkowski space, $\{x^i\}$ its *standard* coordinates (x, y, z, ct) and consequently $g_{ij} = \text{diag}(-1, -1, -1, 1)$.

Now that we have recognized *curved* spacetime as playing the key role in GR, it will be natural to take *that* as the background space for our tensors. The coordinates are no longer required to have special physical meaning (except when its suits us— for example, in the standard form of the metric for stationary spacetimes.) Arbitrary (Gaussian) coordinate systems are permitted, as long as they cover the spacetime smoothly (patchwise if necessary). Different such systems are related to each other by smooth and invertible transformations $x^{i'} = x^{i'}(x^i)$, which, of course, also form a group as did the Lorentz transformations of SR. When we transform coordinates, tensor components undergo their typical tensor transformations, but now these usually vary from point to point unlike the universal Lorentz transformations of SR.

Once again we revert to Greek indices μ, ν, $\dots$ for the range 1–4 and usually reserve Latin indices $i, j, \dots$ for the range 1–3 (or for tensors in arbitrary dimensions.) In curved spacetime we can no longer picture vectors as displacements, but physicists have little trouble picturing them as scalar multiples of *differential* displacements dx^μ. We recall the definitions (7.9)–(7.11) of scalar product, square, and magnitude of vectors:

$$\mathbf{A} \cdot \mathbf{B} = g_{\mu\nu} A^\mu B^\nu, \qquad \mathbf{A}^2 = g_{\mu\nu} A^\mu A^\nu, \qquad A = |\mathbf{A}| =: |\mathbf{A}^2|^{1/2} \geq 0. \quad (10.1)$$

These invariants can be evaluated in any coordinate system, but in the locally Minkowskian system at each point they reduce to their familiar SR forms *and* meanings. We can define the angle θ between two non-null vectors $\mathbf{A}$ and $\mathbf{B}$

by $\cos\theta = \mathbf{A} \cdot \mathbf{B}/AB$, and in properly Riemannian spaces this determines a real θ.

In particular, two *directions* $\mathbf{dx}$ and $\boldsymbol{\delta x}$ are *orthogonal* if

$$g_{\mu\nu}\, \mathrm{d}x^\mu\, \delta x^\nu = 0. \tag{10.2}$$

We speak of *coordinate hypersurfaces* $x^\nu = $ const and *coordinate lines* $x^\mu = $ var. The condition for the x^μ and x^ν lines to be orthogonal at a given point is seen to be

$$g_{\mu\nu} = 0 \quad \text{(specific } \mu \text{ and } \nu), \tag{10.3}$$

since the μth and νth components are then the only non-zero components of $\mathbf{dx}$ and $\boldsymbol{\delta x}$, respectively. We say the x^μ line is orthogonal to the x^μ hypersurface if it is orthogonal to all directions *in* the hypersurface, and for that we need

$$g_{\mu\nu} = 0 \quad \text{(specific } \mu, \text{ all } \nu \neq \mu). \tag{10.4}$$

For example, in static spacetimes with metric (9.5) the worldlines $t = $ var are orthogonal to the time slices $t = $ const, since $g_{4i} = 0$; not so in stationary spacetimes with metric (9.2), where the g_{4i} (A, B, and C) do *not* all vanish. When *all* the mixed $g_{\mu\nu}$ vanish we speak of *orthogonal coordinates* or of a *diagonal metric*. When additionally all the $g_{\mu\mu}$ are ± 1 at some point, we call the coordinates *orthonormal* or pseudo-Euclidean at that point. Since orthogonal coordinates can considerably simplify the mathematics, it is useful to know that *in 2- and 3-dimensional spaces orthogonal coordinates always exist*. In higher dimensions this is unfortunately no longer true. But at least in static spacetimes, with metric (9.5), fully orthogonal coordinates can always be presupposed since $\mathrm{d}l^2$ is diagonizable.

Only in one important respect do the 4-tensors of SR not generalize painlessly to GR. As we saw in Section 7.2, the partial differentiation of tensors is a tensorial operation *only* as long as the permitted coordinate transformations are linear—as they are for the Cartesian tensors of classical physics and the 4-tensors of SR. But in GR non-linear coordinate transformations are forced on us. And yet, neither physics nor geometry can progress far without differentiation. So a more general tensorial operation had to be found: it is called *covariant differentiation*. And since one approach to this topic is via geodesics, that is where we shall turn our attention next.

10.2   Geodesics

We have talked a lot about geodesics in the preceding two chapters but we still have no systematic method for actually *finding* the geodesics in a general curved space. This problem will now be addressed. And for the next few sections we shall again do the analysis quite generally for an N-dimensional (pseudo-)Riemannian space V^N with metric

$$\mathbf{ds}^2 = g_{ij}\, \mathrm{d}x^i\, \mathrm{d}x^j \quad (i, j = 1, \ldots, N). \tag{10.5}$$

Let us consider the geodesic between two given points A and B, and a bundle of neighboring curves also connecting A to B. Our problem is to find the curve satisfying the variational principle

$$0 = \delta \int ds = \delta \int \left| g_{ij}\, dx^i\, dx^j \right|^{1/2} = \delta \int_{u_1}^{u_2} \left| g_{ij} \frac{dx^i}{du} \frac{dx^j}{du} \right|^{1/2} du, \qquad (10.6)$$

where in the last integral u is an arbitrary parameter which continuously parametrizes the entire bundle of comparison curves (much like ct in Fig. 10.1) so as to have fixed values u_1 and u_2 at the fixed end-points of the curves. With $\dot{x}^i = dx^i/du$, eqn (10.6) becomes

$$\delta \int \left| g_{ij} \dot{x}^i \dot{x}^j \right|^{1/2} du =: \delta \int L(x^i, \dot{x}^i)\, du = 0. \qquad (10.7)$$

The reader is perhaps familiar from classical mechanics with this kind of 'variational' problem. Its solution $x^i = x^i(u)$ is found by integrating the well-known Euler–Lagrange differential equations:

$$\frac{d}{du}\left(\frac{\partial L}{\partial \dot{x}^i}\right) - \frac{\partial L}{\partial x^i} = 0. \qquad (10.8)$$

At this stage we may conveniently take u to be the arc s *along the solution curve*, provided that curve is not null. This makes $L = 1$ along the solution curve and allows us to replace the awkward Lagrangian L defined by eqn (10.7) (the square root of a metric is never pleasant) by essentially its square,

$$\mathcal{L} := g_{ij} \dot{x}^i \dot{x}^j = \pm L^2. \qquad (10.9)$$

For consider the variational principle

$$\delta \int \mathcal{L}\, ds = 0, \qquad (10.10)$$

whose solution is determined by the N equations

$$\mathbb{L}_i := \frac{d}{ds}\left(\frac{\partial \mathcal{L}}{\partial \dot{x}^i}\right) - \frac{\partial \mathcal{L}}{\partial x^i} = 0 = \frac{d}{ds}\left(2L \frac{\partial L}{\partial \dot{x}^i}\right) - 2L\frac{\partial L}{\partial x^i}, \qquad (10.11)$$

where, for future reference, we have introduced the notation $\mathbb{L}_i$ for the LHS of the ith equation. Since the solution satisfies $L = \text{const}$, we see that the Euler–Lagrange equations for $\mathcal{L}$ are equivalent to those for L. They are, in fact, the standard equations used in the practical determination of geodesics.

Mainly for theoretical purposes, we now further examine the structure of the set of eqns (10.11). With (10.9) substituted it becomes, successively (with a few index tricks),

$$\mathbb{L}_i \equiv \frac{d}{ds}(2g_{ij}\dot{x}^j) - g_{jk,i}\dot{x}^j \dot{x}^k = 0$$

$$2g_{ij,k}\dot{x}^j \dot{x}^k + 2g_{ij}\ddot{x}^j - g_{jk,i}\dot{x}^j \dot{x}^k = 0 \qquad (10.12)$$

$$(g_{ij,k} + g_{ik,j} - g_{jk,i})\dot{x}^j \dot{x}^k + 2g_{ij}\ddot{x}^j = 0.$$

So if we now define the so-called *Christoffel symbols of the first kind* by

$$\Gamma_{ijk} := \tfrac{1}{2}(g_{ij,k} + g_{ik,j} - g_{jk,i}), \tag{10.13}$$

and those *of the second kind* (also called *connection coefficients*) by raising the index i,

$$\Gamma^i_{jk} := g^{hi}\Gamma_{hjk}, \tag{10.14}$$

we can re-express eqns (10.12) (after raising i and recalling $g^i{}_j = \delta^i_j$) as

$$\tfrac{1}{2}\mathbb{L}^i \equiv \ddot{x}^i + \Gamma^i_{jk}\dot{x}^j\dot{x}^k = 0. \tag{10.15}$$

This is the alternative standard form of the set of differential equations for geodesics. It shows that geodesics are fully determined by an initial point O and an initial direction $\dot{x}^i_0$. [At least, if we assume analyticity: for eqn (10.15) yields $\ddot{x}^i_0$, (10.15) differentiated yields $\dddot{x}^i_0$, etc.]

The Christoffel symbols play an important role in differential geometry and also in GR. But, as we shall see later, *they are not tensors!* We note their symmetry:

$$\Gamma_{ijk} = \Gamma_{ikj}, \qquad \Gamma^i_{jk} = \Gamma^i_{kj}, \tag{10.16}$$

and the 'inverse' of eqn (10.13):

$$g_{ij,k} = \Gamma_{ijk} + \Gamma_{jik}. \tag{10.17}$$

[For proof, just substitute (10.13) into the RHS.] This latter equation, together with (10.13), shows that the vanishing of all the Γs at one point is equivalent to the vanishing of the derivatives of all the gs.

For the actual calculation of the Γs of a given metric (an often unavoidable task in GR) one can sometimes avoid (10.13) and simply compare the coefficients of $\dot{x}^j\dot{x}^k$ in eqns (10.15) with those in the written-out versions of eqns (10.11). This works best in the case of orthogonal coordinates when eqns (10.11) and (10.15) differ only by a factor $2g_{ii}$. But then one can also use the formulae of the Appendix of this book. (See, for example, Exercise 10.3.)

The reader may worry that eqn (10.15) 'does not know' that the parameter is supposed to be the arc. But, essentially, it knows: one can show [cf. after (10.42)] that *any solution $x^i(u)$ of (10.15) with $\dot{} = d/du$ necessarily satisfies*

$$g_{ij}\dot{x}^i\dot{x}^j = \text{const}; \tag{10.18}$$

that is, $L = \text{const}$, which was the basic assumption of our derivation.

We can re-write eqn (10.18) in the form $\mathbf{ds}^2 = (\text{const})\, du^2$ and deduce, first, that the sign of $\mathbf{ds}^2$ is necessarily constant along a geodesic, and second, that every solution parameter is 'affinely' related to the arc:

$$u = as + b. \tag{10.19}$$

Here we have assumed that there *is* an arc; that is, that our geodesic is not null. But we do not need this assumption to verify directly that the parameters of all possible solutions $x^i(u)$ of (10.15) are related affinely to each other:

$$\tilde{u} = au + b. \tag{10.20}$$

[For proof, set $\tilde{u} = f(u)$ and show that (10.15) is form-invariant if and only if $f'' = 0$.] These parameters are consequently called *affine parameters* along the geodesic.

We can now define a *null geodesic* as the limit of a family of non-null geodesics whose initial direction is gradually turned into a null direction. The following example will make this clear (cf. Fig. 10.1). Consider, in SR, the family of geodesic worldlines $x = vt$, $y = z = 0$. For each we have $s = ct/\gamma$, γ being the usual Lorentz factor. Unlike s, the affine parameter

$$u = ct \ (= \gamma s, \text{ if } s \text{ exists})$$

satisfactorily parametrizes the entire family *including* the null member. The LHSs of eqns (10.11) and (10.15) continuously change while remaining form-invariant under any such limiting procedure. These equations, with the extra requirement $\mathbf{ds}^2 = 0$, thus determine all null geodesics. [We nay note that the original variational principle (10.7), with its solution (10.8), is useless for null geodesics: for example, $\partial L/\partial x^k = \frac{1}{2}|g_{ij}\dot{x}^i\dot{x}^j|^{-1/2}g_{ij,k}$ is infinite when $L = 0$.]

An affine parameter along a null geodesic is proportional to the arc along a neighboring non-null geodesic. (In spacetime, when we speak of 'neighboring', we mean 'separated by small *coordinate* differences', provided the coordinates are continuous.) It is the nearest thing we can find to 'distance' along a null geodesic. While we cannot assign an invariant length to any segment of a null geodesic, *ratios* of lengths *can* be invariantly defined as the ratios of the corresponding increments of an affine parameter.

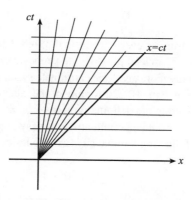

Fig. 10.1

10.3   Geodesic coordinates

Geodesics provide a convenient (though not the only) approach to an important lemma: *in a neighborhood of any given point* P *we can always construct a coordinate system such that all the Christoffel symbols vanish at* P. Any coordinate system with this property is called a *geodesic coordinate system* at P and P is said to be its *pole*. We immediately note two things: if the lemma is true, then (i) the Γs cannot be tensors, since no tensor except the zero tensor can ever have all its components zero; and (ii) in these coordinates all the $g_{ij,k}$ vanish at P.

For proof of the lemma, we first transform to orthonormal coordinates y^i at P. [cf. after (10.4)]. This can be achieved by 'completing the square' as in (8.13). Around P we then draw the *coordinate* hypersphere $\sum(y^i)^2 = \epsilon^2$ for some convenient ϵ (the analog of a circle $x^2 + c^2t^2 = \epsilon^2$ in Fig. 10.1.) We label this $u = \epsilon$, and P, $u = 0$, thereby defining an affine parameter u on every geodesic issuing from P, including possible null geodesics. (Of course, this is only one of many possibilities of 'spreading' an affine parameter continuously over all the geodesics.) There is a general theorem which states that in a sufficiently small neighborhood of P, say within our ϵ-sphere, every point Q can be connected to P by a unique geodesic. In this neighborhood we can define new coordinates x^i as follows: let $y^i(u)$ represent the unique geodesic connecting P to Q, and write

$$\left(\frac{dy^i}{du}\right)_{u=0} =: a^i \tag{10.21}$$

for its tangent vector at P. Then if Q corresponds to the parameter value u, define its x^i-coordinate thus:

$$x^i = a^i u. \tag{10.22}$$

As u varies, x ranges over the entire geodesic. Since u is an affine parameter, eqn (10.22) must satisfy the differential equation (10.15), whence

$$\Gamma^i{}_{jk} a^j a^k = 0$$

all along PQ. But this is true for every geodesic through P. So we can conclude that $\Gamma^i{}_{jk} = 0$ *at* P, and that the x^i indeed provide *one* geodesic system with pole at P.

Of course, there are many more geodesic coordinate systems at P than the presently discussed systems based on eqn (10.22), for any of the latter can be totally deformed away from P without affecting the defining property $\Gamma^i{}_{jk} = 0$ *at* P.

Having one geodesic coordinate system at P, it is easy to generate infinitely many others. For we have the following lemma: any two geodesic systems $\{x^i\}$ and $\{x^{i'}\}$ at P are related 'locally linearly':

$$(p^{i'}{}_{jk})_P = 0 \tag{10.23}$$

[in the notation (7.1)], and, conversely, any system so related to a geodesic system is itself geodesic. [Relation (10.23) is clearly an equivalence relation between coordinate systems.] For proof we have, with $\dot{} \equiv d/du$,

$$\dot{x}^{i'} = p^{i'}_j \dot{x}^j$$

$$\ddot{x}^{i'} = p^{i'}_{jk} \dot{x}^j \dot{x}^k + p^{i'}_i \ddot{x}^i. \qquad (10.24)$$

Now any geodesic $x^i(u)$ through P referred to an affine parameter must satisfy eqn (10.15), and thus $\ddot{x}^i = 0$ at P if $\{x^i\}$ is a geodesic system. For two such systems we then deduce from (10.24) that $(p^{i'}_{jk})_P$. Conversely, if $(p^{i'}_{jk})_P = 0$ and $\{x^i\}$ is geodesic at P, every geodesic in $\{x^{i'}\}$ also satisfies $\ddot{x}^{i'} = 0$ at P and so, by (10.15), $(\Gamma^{i'}_{j'k'})_P = 0$.

Next suppose we transform from a system $\{x^{i'}\}$ that is geodesic at P to an arbitrary system $\{x^i\}$. Then all geodesics through P, at P satisfy $\ddot{x}^{i'} = 0$ and thus, 'flipping' $p^{i'}_i$ in eqn (10.24),

$$\ddot{x}^i + p^{i'}_{jk} p^i_{i'} \dot{x}^j \dot{x}^k = 0.$$

Comparison with (10.15) then shows that *at* P:

$$\Gamma^i_{jk} = p^{i'}_{jk} p^i_{i'}. \qquad (10.25)$$

However, if we differentiate the relation $p^i_{i'} p^{i'}_j = \delta^i_j$ [cf. (7.2)] with respect to x^k, we find

$$p^i_{i'} p^{i'}_{jk} + p^i_{i'k'} p^{k'}_k p^{i'}_j = 0,$$

which, after a cosmetic dummy replacement, yields the following alternative to the RHS of (10.25):

$$\Gamma^i_{jk} = -p^i_{j'k'} p^{j'}_j p^{k'}_k. \qquad (10.26)$$

Both versions will be needed in the next section.

Any coordinate system in terms of which the geodesics through a given point P have equations like (10.22) are called Riemannian. If additionally they are orthonormal (that is, pseudo-Euclidean) at P, they are called *normal coordinates* with pole at P. Any two normal systems at P are related to each other globally by generalized rotations about P; that is, transformations that preserve the (pseudo-)Euclidean metric *at* P. (In spacetime these are the LTs.) For $u \propto s$, so $ds/du = s/u$ and therefore

$$(g_{ij})_P\, x^i(Q)\, x^j(Q) = (g_{ij})_P \frac{dx^i}{du} \frac{dx^j}{du} u^2 = s^2 = (g_{i'j'})_P\, x^{i'}(Q)\, x^{j'}(Q),$$

where $(g_{ij})_P$ and $(g_{i'j'})_P$ both represent the (pseudo-)Euclidean metric at P. That proves it. For example, $\sum (x^i)^2 = \sum (x^{i'})^2$ if V^N is properly Riemannian. These normal coordinates are thus the analogs of (pseudo-)Euclidean coordinates in flat space. Note also the similarity of (10.22) to the familiar relations $x = r\cos\theta$, $y = r\sin\theta$ and their analogs in higher dimensions.

In GR, *any system* $\{x, y, z, ct\}$ *that is both geodesic and orthonormal* at an event $\mathcal{P}$ is the mathematical embodiment of Einstein's 'freely falling cabin'; that is, of a local inertial frame (LIF) at $\mathcal{P}$. Its 'fixed points' $x, y, z = $ const, $ct = s$, satisfy $\ddot{x}^{\mu} = 0$, which are also the geodesic equations at $\mathcal{P}$ and approximately near $\mathcal{P}$. So the system is freely falling. The distances between neighboring fixed points near $\mathcal{P}$, $|g_{ij}\Delta x^{i}\Delta x^{j}|^{1/2}$ $(i, j = 1, 2, 3)$, remain momentarily constant because $g_{\mu\nu,\sigma} = 0$ *at* $\mathcal{P}$. So the system is also rigid and non-rotating. (Recall that a freely falling *rotating* box of loose bricks will come apart.) That makes it a LIF.

In general, if a coordinate system $\{x^{i}\}$ is both geodesic and orthonormal *at a point* P, then *near* P, by essentially the above arguments, each of the mutually orthogonal sets of coordinate lines $x^{i} = $ var is seen to correspond to parallel geodesics, and the entire coordinate net has locally the character of a (pseudo-) Euclidean coordinate net. In GR this provides the nearest thing we can get to a locally Minkowskian frame. We can do no better: the *second* derivatives of the metric, $g_{ij,kl}$, encode the curvature, and cannot generally be transformed away.

10.4   Covariant and absolute differentiation

In Section 7.2E we noted that the operation of partial differentiation is *not* a tensorial operation unless the coordinates transform linearly. The underlying reason for this is that differentiation involves taking a difference of tensors at different points, and such a difference is not, in general, a tensor (since the *p*s differ—cf. Section 7.2D). However, this is *not* a problem among the various geodesic coordinate systems with the same pole P: as we have seen in (10.23), for them $(p^{i'}_{jk})_{\mathrm{P}}$ is zero and so the first-order *p*s are essentially the same at and near P. Our plan for a tensorial and thus geometrically meaningful derivative operation is now the following: To take the derivative of a given tensor field in given coordinates at P, we first transform the field to *any* geodesic coordinate system at P, take its derivative in *that* system, and then tensorially transform the derivative back to the original system. In practice, fortunately, we shall find a formula that does all this automatically.

To simplify the manipulation, consider just a 2-index tensor field F^{i}_{j}. Under a change of coordinates it obeys the usual tensor law [cf. (7.6)]

$$F^{i'}_{j'} = F^{i}_{j}p^{i'}_{i}p^{j}_{j'}.$$

Partially differentiating both sides with respect to $x^{k'}$ yields

$$F^{i'}_{j',k'} = F^{i}_{j,k}p^{k}_{k'}p^{i'}_{i}p^{j}_{j'} + F^{i}_{j}p^{i'}_{ik}p^{k}_{k'}p^{j}_{j'} + F^{i}_{j}p^{i'}_{i}p^{j}_{j'k'}. \qquad (10.27)$$

Since the 3-index ps in the last two terms vanish between geodesic systems, $F^i_{j,k}$ is seen to behave tensorially between such systems. We now define the *covariant derivative* $F^i_{j;k}$ of F^i_j, in an arbitrary coordinate system $\{x^i\}$, as the tensor transform of the partial derivative in a geodesic system $\{x^{i'}\}$:

$$F^i_{j;k} = F^{i'}_{j',k'} p^i_{i'} p^{j'}_j p^{k'}_k.$$

Since the partial derivatives are related tensorially among geodesic systems, $F^i_{j;k}$ is related tensorially to *all* of them, by transitivity (cf. Section 7.2C). The same is true of $F^{i''}_{j'';k''}$, similarly defined in another arbitrary system $\{x^{i''}\}$. So, again by transitivity and symmetry, $F^i_{j;k}$ and $F^{i''}_{j'';k''}$ are tensorially related, which shows that our definition indeed creates a tensor.

In eqn (10.27), let us now assume the system $\{x^{i'}\}$ to be geodesic and the system $\{x^i\}$ to be arbitrary; let us write a for the dummy j in the last term, and similarly for the dummy i in the term before; if we then 'flip' $p^{i'}_i$, $p^j_{j'}$, and $p^k_{k'}$ [cf. (7.3)], we get

$$F^{i'}_{j',k'} p^i_{i'} p^{j'}_j p^{k'}_k = F^i_{j,k} + F^a_j p^i_{ak} p^i_{i'} + F^i_a p^a_{j'k'} p^{j'}_j p^{k'}_k.$$

The LHS is the required covariant derivative $F^i_{j;k}$. On the RHS we can eliminate all reference to the auxiliary geodesic coordinate system by applying our carefully prepared formulae (10.25) and (10.26). Thus we finally obtain (with some relief!)

$$F^i_{j;k} = F^i_{j,k} + F^a_j \Gamma^i_{ak} - F^i_a \Gamma^a_{jk}. \tag{10.28}$$

This formula is typical. For the general tensor field $F^{...}_{...}$ it again begins with the partial derivative, followed by *positive* Γ-terms resulting from the replacement, one after the other, of all the contravariant indices of $F^{...}_{...}$ by a dummy which is linked to a Γ that also takes over the replaced free index; similarly there is a *negative* Γ-term for each covariant index. It is an easy pattern to remember. In particular, for a scalar $\phi(x^i)$ it gives

$$\phi_{;i} = \phi_{,i}, \tag{10.29}$$

which is not surprising, since $\phi_{,i}$ is a tensor.

A pleasant property of the covariant derivative is that it satisfies the ordinary rules for differentiating sums and (inner and outer) products:

$$(S^{...}_{...} + T^{...}_{...})_{;k} = S^{...}_{...;k} + T^{...}_{...;k}, \tag{10.30}$$

$$(S^{...}_{...} T^{...}_{...})_{;k} = S^{...}_{...;k} T^{...}_{...} + S^{...}_{...} T^{...}_{...;k}. \tag{10.31}$$

For, *at the pole of geodesic coordinates, where covariant and partial differentiation are the same*, the above equations are trivially true; but, being tensor equations, they must then be true in all coordinates.

Equally pleasant is the 'covariant constancy' of the fundamental tensors g_{ij}, g^{ij}, and δ^i_j:

$$g_{ij;k} = 0, \qquad g^{ij}{}_{;k} = 0, \tag{10.32}$$

$$\delta^i_{j;k} = 0. \tag{10.33}$$

The first and third of these equations again follow immediately from their validity at the pole of geodesic coordinates. The second results on covariantly differentiating the identity $g_{ij}g^{jk} = \delta^k_i$ (and then multiplying by g^{ih}). As a consequence, covariant differentiation commutes with index shifting and contraction:

$$g_{hi}F^i_{j;k} = (g_{hi}F^i_j)_{;k} = F_{hj;k}, \tag{10.34}$$

$$\delta^j_i F^i_{j;k} = (\delta^j_i F^i_j)_{;k} = F^i_{i;k}, \tag{10.35}$$

which makes the expressions on the RHSs unambiguous, when regarded as arising from F^i_j.

We next define a concept closely related to covariant differentiation, namely *absolute differentiation*. Consider a curve C parametrically given by $x^i(u)$, and on it a tensor field; for example, $F^i_j(u)$. Then the derivative of F^i_j along C, $\mathrm{d}F^i_j/\mathrm{d}u$, is *not* a tensor (except among the geodesic coordinates at each point of C) as can again be seen by differentiating the tensor transformation law $F^{i'}_{j'} = F^i_j p^{i'}_i p^j_{j'}$, this time with respect to u. So, analogously to the covariant derivative, we now define an *absolute derivative* $\mathrm{D}F^i_j/\mathrm{d}u$ by stipulating that it be a tensor and reduce to the ordinary derivative $\mathrm{d}F^i_j/\mathrm{d}u$ at the pole of geodesic coordinates. Following a calculation analogous to that preceding eqn (10.28), we could now find the corresponding formula for $\mathrm{D}F^i_j/\mathrm{d}u$. However, we can *use* eqn (10.28) directly by applying a little trick: if F^i_j is not already defined off C, let us extend its definition arbitrarily but smoothly into a neighborhood of C. [For example, we can define $F^i_j = (F^i_j)_C$—in some arbitrary coordinate system—throughout the hypersurface generated by the geodesics orthogonal to C at a given point.] Then $\mathrm{D}F^i_j/\mathrm{d}u$ will be given by

$$\frac{\mathrm{D}F^i_j}{\mathrm{d}u} = F^i_{j;k}\dot{x}^k. \tag{10.36}$$

For the RHS *is* a tensor (from the chain rule, $\dot{x}^{i'} = p^{i'}_i \dot{x}^i$, so $\dot{x}^i$ *is* a vector.) and it *does* reduce to $\mathrm{d}F^i_j/\mathrm{d}u$ at the pole of geodesic coordinates. Substituting into (10.36) from (10.28), we then find

$$\frac{\mathrm{D}F^i_j}{\mathrm{d}u} = \frac{\mathrm{d}F^i_j}{\mathrm{d}u} + F^a_j \Gamma^i_{ak}\dot{x}^k - F^i_a \Gamma^a_{jk}\dot{x}^k. \tag{10.37}$$

In this final formula, no extension of F^i_j away from C is needed. The generalizations of both (10.36) and (10.37) to arbitrary tensor fields $F^{...}_{...}$ are obvious. In particular, and importantly, for any scalar ϕ we have $\mathrm{D}\phi/\mathrm{d}u = \mathrm{d}\phi/\mathrm{d}u$.

That absolute differentiation follows the ordinary rules for differentiating sums and products is again clear in geodesic coordinates, where $D \equiv d$; and that g_{ij}, g^{ij}, and δ^i_j are 'absolutely constant' along any curve ($Dg_{ij}/du = 0$, etc.) follows from (10.36), (10.32), and (10.33).

One particularly important vector field along a curve C is that of its tangent vector $\mathbf{T}$. If C is not null, we can take the arc as parameter and define the *unit tangent vector* as

$$T^i = \frac{dx^i}{ds}. \tag{10.38}$$

Its unicity is obvious:

$$\mathbf{T}^2 = g_{ij} \frac{dx^i}{ds} \frac{dx^j}{ds} = \frac{ds^2}{ds^2} = \pm 1. \tag{10.39}$$

The absolute derivative of $\mathbf{T}$ is called the *principal normal* $\mathbf{N}$ of C; from (10.38) and (10.37) we find

$$N^i = \frac{DT^i}{ds} = \frac{d^2 x^i}{ds^2} + \Gamma^i_{jk} \frac{dx^j}{ds} \frac{dx^k}{ds} = \frac{1}{2} \mathbb{L}^i, \tag{10.40}$$

where the last equation follows from (10.15) and implies the possibility of an alternative calculation of N^i. The magnitude of $\mathbf{N}$ is defined to be the *curvature* κ of C (cf. Exercise 10.9):

$$\kappa = |g_{ij} N^i N^j|^{1/2}. \tag{10.41}$$

That $\mathbf{N}$ is indeed *a* normal can be seen by absolutely differentiating (10.39):

$$0 = \frac{D}{ds}(g_{ij} T^i T^j) = 2g_{ij} N^i T^j = 2\mathbf{N} \cdot \mathbf{T} \tag{10.42}$$

From the last equation in (10.40) we see that a geodesic has zero curvature. It is this property that allows us to characterize geodesics as 'straightest' lines.

If we define $T^i(u)$ and $N^i(u)$ as in (10.38) and (10.40), but for an *arbitrary* parameter u instead of the arc s, eqn (10.42) without the initial '$0 =$' and with u for s, still applies. So if $\mathbb{L}_i(u) = 0$, and consequently $N^i(u) = 0$, we shall have, from (10.42), $g_{ij} \dot{x}^i \dot{x}^j = \text{const}$, since the absolute derivative of a scalar coincides with its ordinary derivative. This establishes our earlier assertion [cf. the paragraph containing (10.18)] that any solution of the geodesic equation $\mathbb{L}^i(u) = 0$ necessarily satisfies (10.18). Recall that this implies the constancy of the *sign* of ds^2 all along every geodesic; in spacetime, therefore, geodesics fall into three categories: timelike, spacelike, and null.

In Euclidean 3-space, if along a straight line we define a unit tangent vector $\mathbf{t} = d\mathbf{r}/ds$, then $\mathbf{t}$ is constant, $d\mathbf{t}/ds = 0$; and this holds even if we define $\mathbf{t} = d\mathbf{r}/du$ for some affine parameter u. But in terms of an *arbitrary* parameter w the tangent vector $\mathbf{t} = d\mathbf{r}/dw$ would vary in magnitude and so we would have $d\mathbf{t}/dw = \phi \mathbf{t}$ for some function ϕ along the line; and conversely, such an equation still implies a straight line. In just the same way, in curved space, the equation $\mathbf{N}(w) \propto \mathbf{T}(w)$, or

$$\frac{D}{dw}\mathbf{T}(w) = \phi(w)\,\mathbf{T}(w), \quad \text{or} \quad \frac{d^2x^i}{dw^2} + \Gamma^i_{jk}\frac{dx^j}{dw}\frac{dx^k}{dw} = \phi\frac{dx^i}{dw} \tag{10.43}$$

still implies a geodesic. (Evidently the w here is a non-affine parameter, otherwise the RHS would be absent.) For proof, suppose $\mathbf{T}(w) = dx^i/dw$, $\mathbf{T}(u) = dx^i/du$, for two different parameters u and w, so that

$$\mathbf{T}(w) = \mathbf{T}(u)\frac{du}{dw} =: \psi(w)\,\mathbf{T}(u). \tag{10.44}$$

Substituting this into (10.43)(i) and performing the differentiation, we have

$$\psi^2\frac{D}{du}\mathbf{T}(u) + \frac{d\psi}{dw}\mathbf{T}(u) = \phi(w)\,\psi(w)\,\mathbf{T}(u).$$

We can now choose $\psi(w)$ so as to make the second and third terms in this equation equal:

$$\log\psi = \int \phi(w)\,dw,$$

and this, by (10.44), yields $u = \int \psi(w)\,dw$ as an affine parameter. Conversely, if we differentiate (10.44) absolutely with respect to w and assume w to be an *affine* parameter $[\mathbf{N}(w) = 0]$, we find $\mathbf{N}(u) \propto \mathbf{T}(u)$ in terms of a general parameter u.

An important conclusion that can be drawn from (10.40) and the orthogonality (10.42) (even if s is replaced by an affine parameter u) is the following: $\mathbb{L}_i(u)\,T^i(u) = 0$. This tells us that there is a linear relation connecting the N differential equations $\mathbb{L}_i(u) = 0$. So if $N-1$ of them are satisfied, the Nth will be satisfied automatically— *unless* $T^N \equiv dx^N/du = 0$; that is, unless $x^N = $ const along the curve. This often allows us to discard whichever is the 'worst' of the differential equations and to replace it by the universally valid relation $\mathscr{L} = $ const $(= \pm 1$ if $u = s)$, which contains only *first* derivatives.

Let us now look at the tangents and normals of worldlines $x^\mu(\tau)$ of particles in curved spacetime ($\tau = s/c$ being proper time). Naturally we define the *generalized 4-velocity* $\mathbf{U}$, *4-acceleration* $\mathbf{A}$, and the *proper acceleration* α of such particles as follows:

$$U^\mu := \frac{dx^\mu}{d\tau} = cT^\mu \tag{10.45}$$

$$A^\mu := \frac{DU^\mu}{d\tau} = c^2 N^\mu = \frac{1}{2}c^2\mathbb{L}^\mu \tag{10.46}$$

$$\alpha := |g_{\mu\nu}A^\mu A^\nu|^{1/2} = c^2\kappa. \tag{10.47}$$

For these are tensorial definitions and they reduce in LIF coordinates (cf. penultimate paragraph of Section 10.3)—that is, when seen in the freely falling cabin—to exactly the familiar SR definitions. Apart from factors c, U^μ is thus seen to be the tangent vector, A^μ the normal vector, and α the curvature of the worldline! Just as we would expect, geodesic worldlines are characterized by $A^\mu = 0$ or, equivalently, by $\alpha = 0$: geodesics correspond to non-accelerated particles.

As a simple example, let us find the 4-acceleration A^μ of a lattice point in the stationary field (9.13). It will provide us with a measure of the gravitational field itself. From (10.46) and (10.40), now with $\dot{} = d/ds$ and temporarily with $c = 1$, we have

$$A_\mu = g_{\mu\nu}\ddot{x}^\nu + \Gamma_{\mu\nu\rho}\dot{x}^\nu\dot{x}^\rho.$$

But for a lattice point ($x^i = \text{const}, i = 1, 2, 3$) in (9.13) we also have

$$\dot{x}^i = 0, \qquad \dot{x}^4 = e^{-\Phi} = \text{const}, \qquad \ddot{x}^\mu = 0,$$

so that [cf. (10.13)]

$$A_\mu = \Gamma_{\mu 44}e^{-2\Phi} = -\tfrac{1}{2}g_{44,\mu}e^{-2\Phi} = (-\Phi_{,i}, 0). \tag{10.48}$$

As long as we preserve the canonical form (9.13), the quantity

$$\mathbf{a} = \Phi_{,i} \tag{10.49}$$

is a 3-vector relative to the spatial metric k_{ij}. Reference to (5.23) and (7.22) then identifies it as the 3-acceleration of the lattice point relative to the rest-LIF. It is the force on our feet when we stand in the lattice. Its negative is the gravitational field. Its magnitude α, the proper acceleration, can be evaluated *either* relative to the metric k_{ij}, or as that of A_μ relative to the metric (9.13). Either approach yields

$$\alpha = (\Phi_{,i}\Phi_{,j}k^{ij})^{1/2}. \tag{10.50}$$

To see the latter, observe that $\Phi_{,i}\Phi_{,j}k^{ij}$ is gauge invariant [cf. (9.16)] and equals $-A_\mu A_\nu g^{\mu\nu}$ when the gauge is $\Phi = w_i = 0$. Note also from dimensional considerations that eqns (10.50), (10.49), and the extremities of eqn (10.48) are valid even if $c \neq 1$.

One further geometric aspect of absolute differentiation needs to be addressed briefly. It concerns the *parallel transport* of a vector V^i along a curve $x^i(u)$, which is defined by

$$\frac{DV^i}{du} = 0. \tag{10.51}$$

In flat space, referred to (pseudo-)Euclidean coordinates, this reduces to the constancy of the components. On a 2-surface, if we cut out a strip bounded by the curve and flatten it out in a plane, a parallel vector field along the curve becomes parallel in the plane. Note, from (10.40), written with u for s, that geodesics (even null geodesics) transport their own tangents parallely. The scalar product of two parallely transported vectors remains constant:

$$\frac{D}{du}(g_{ij}V^iW^j) = g_{ij}\frac{DV^i}{du}W^j + g_{ij}V^i\frac{DW^j}{du} = 0,$$

and so does the magnitude of any one such vector (put $V^i = W^i$). Hence the angle between two parallely transported vectors remains constant, and a vector transported parallely along a geodesic subtends a constant angle with its tangent. Parallel transport of a vector from point A to point B generally depends on the path chosen, and parallel transport around a closed loop generally returns the vector in a different orientation.

As an application, consider a null geodesic $g : x^\mu = x^\mu(u)$ representing a ray of light or the path of a photon, u being an affine parameter. At each of its points the wave vector $\mathbf{L}$ [cf. (5.31)] must be a multiple of the tangent vector $\mathbf{T} = \mathrm{d}x^\mu/\mathrm{d}u$: $\mathbf{L} = \theta(u)\mathbf{T}$. Suppose the light was emitted by a source of proper frequency ν_0 and 4-velocity $\mathbf{V}_0$ at point $\mathcal{P}_0$ of g, so that $\nu_0 = \mathbf{L}_0 \cdot \mathbf{V}_0 = \theta_0 \mathbf{T}_0 \cdot \mathbf{V}_0$. (For proof, evaluate the product in the rest-frame of the source.) Now let $\tilde{\mathbf{V}}$ be the parallel vector field along g determined by $\mathbf{V}_0$. Since consecutive vectors of $\tilde{\mathbf{V}}$ correspond to parallel worldlines in the LIFs along g, all observers crossing g with velocities $\tilde{\mathbf{V}}$ see the source still at frequency $\nu_0 : \theta(u)\mathbf{T} \cdot \tilde{\mathbf{V}} = \nu_0$. But $\mathbf{T} \cdot \tilde{\mathbf{V}}$ is constant along g, whence $\theta = $ const. So the frequency ν seen by an arbitrary observer crossing g with velocity $\mathbf{V}$ is given by

$$\frac{\nu}{\nu_0} = \frac{\mathbf{L} \cdot \mathbf{V}}{\mathbf{L} \cdot \tilde{\mathbf{V}}} = \frac{\mathbf{T} \cdot \mathbf{V}}{\mathbf{T} \cdot \tilde{\mathbf{V}}} = \frac{\mathbf{T} \cdot \mathbf{V}}{\mathbf{T}_0 \cdot \mathbf{V}_0}. \tag{10.52}$$

This formula holds in *arbitrary* spacetimes; in stationary spacetimes we already have a simpler method (cf. Exercise 9.3). The reader may have felt uneasy at our use (now that we are in curved spacetime) of the *special*-relativistic wave vector $\mathbf{L}$ and the *special*-relativistic result $\nu = \mathbf{L} \cdot \mathbf{V}$. But bear in mind that a tensor is defined by its components in any *one* coordinate system, which can always be taken to be the LIF; and there our SR results are assumed to hold.

There is one variation of parallel transport that plays an important role in GR. How, for example, are the axes of a gyrocompass transported along some arbitrary worldline it is forced to follow (as when a gyrocompass is fixed to a lattice point in a stationary field)? Such 'transport without rotation' is called *Fermi–Walker (FW-)transport*. Given any timelike worldline C with tangent $\mathbf{T}$ ($\mathbf{T}^2 = 1$) and normal $\mathbf{N}$, we require the following properties of FW-transport:

 (i) $\mathbf{T}$ itself is FW-transported
 (ii) $\mathbf{V} \cdot \mathbf{T} = 0$ and $\mathbf{V}$ FW-transported $\Rightarrow \overset{*}{\mathbf{V}} \propto \mathbf{T}$ ($* = D/\mathrm{d}s$)
(iii) Scalar products are preserved.

Condition (i) ensures that in an FW-transported Minkowskian reference frame the particle (gyro) is always momentarily at rest; (ii) requires any FW-transported vector in the 3-space orthogonal to C to vary only in the direction of $\mathbf{T}$; that is, not to rotate *about* $\mathbf{T}$; and (iii) of course, also implies that magnitudes are preserved. Now we can check immediately that the following differential equation, which *defines* the FW-transport of a vector $\mathbf{V}$, fulfills all three conditions:

$$\overset{*}{\mathbf{V}} = (\mathbf{V} \cdot \mathbf{T})\mathbf{N} - (\mathbf{V} \cdot \mathbf{N})\mathbf{T}. \tag{10.53}$$

[For (iii), consider $(\mathbf{V} \cdot \mathbf{W})^* = \overset{*}{\mathbf{V}} \cdot \mathbf{W} + \mathbf{V} \cdot \overset{*}{\mathbf{W}}$]. When C is a geodesic, $\mathbf{N} = 0$ and the RHS of (10.53) vanishes; FW-transport then reduces to parallel transport.

10.5 The Riemann curvature tensor

Let us introduce the following notation for the repeated covariant derivative of a tensor:

$$(T^{\cdots}_{\cdots;i})_{;j} = T^{\cdots}_{\cdots;ij},$$

and similarly for higher derivatives. (This parallels the notation we have already introduced in Section 7.2E for repeated *partial* derivatives: $T^{\cdots}_{\cdots,ij}$.) From elementary calculus we are familiar with the commutativity of partial derivatives: $T^{\cdots}_{\cdots,ij} = T^{\cdots}_{\cdots,ji}$. But the corresponding statement for covariant derivatives is generally false. Only in *flat* space is it true that

$$T^{\cdots}_{\cdots;ij} = T^{\cdots}_{\cdots;ji} \quad \text{(flat space)}. \tag{10.54}$$

For then we can always choose (pseudo-)Euclidean coordinates ($g_{ij} = \pm\delta^i_j$), in which the Γs vanish globally, and in which even repeated covariant differentiation therefore reduces to repeated partial differentiation; in these coordinates, (10.54) is true, and being tensorial, it must then be true in all coordinates. In the general case we cannot prove (10.54) by going to the pole of geodesic coordinates: for while the Γs vanish there, the same is not true of their derivatives. And it is these which cause the inequality.

Let us do the calculation for the simplest case, a vector V^h. We have

$$V^h_{;j} = \left[V^h_{,j} + V^a \Gamma^h_{aj}\right] =: \begin{bmatrix} h \\ j \end{bmatrix}, \text{ say}$$

$$V^h_{;jk} = \begin{bmatrix} h \\ j \end{bmatrix}_{,k} + \begin{bmatrix} b \\ j \end{bmatrix} \Gamma^h_{bk} - \begin{bmatrix} h \\ b \end{bmatrix} \Gamma^b_{jk}.$$

If we write this out in full, reverse j, k and subtract, we find

$$V^h_{;jk} - V^h_{;kj} = -V^a R^h_{ajk}, \tag{10.55}$$

where

$$R^h_{ijk} = \Gamma^h_{ik,j} - \Gamma^h_{ij,k} + \Gamma^h_{aj}\Gamma^a_{ik} - \Gamma^h_{ak}\Gamma^a_{ij}. \tag{10.56}$$

Since V^a in (10.55) is an arbitrary vector, it follows from the quotient rule (cf. Exercise 7.2) that R^h_{ijk} must be a tensor. It is, in fact, one of the most important tensors in Riemannian geometry, the so-called *Riemann curvature tensor*. Its relation to the curvature at a given point will become apparent a little later. In flat space it clearly vanishes. And conversely, its global vanishing can be shown to imply flat space.

Equation (10.55) can be generalized to arbitrary tensors:

$$T^{ij\cdots}_{k\cdots;lm} - T^{ij\cdots}_{k\cdots;ml} = -T^{aj\cdots}_{k\cdots} R^i{}_{alm} - T^{ia\cdots}_{k\cdots} R^j{}_{alm} - \cdots$$

$$+ T^{ij\cdots}_{a\cdots} R^a{}_{klm} + \cdots \qquad (10.57)$$

Note how this pattern is reminiscent of the covariant derivative formula: Here we have a *negative* R-term for each contravariant index and a positive R-term for each covariant index of the original tensor. If we lower h in eqn (10.55), see-saw a, and anticipate the symmetry (10.63) of the Riemann tensor, we see (10.57) validated for V_h. (Recall that index shifting commutes with covariant differentiation.) For scalars, (10.57) yields

$$\Phi_{;ij} = \Phi_{;ji}, \qquad (10.58)$$

which can also be seen directly, since $\Phi_{;ij} = \Phi_{,i;j} = \Phi_{,ij}$ at the pole of geodesic coordinates.

The fully covariant version of the curvature tensor,

$$R_{hijk} = g_{ha} R^a{}_{ijk}, \qquad (10.59)$$

will exhibit most clearly its many symmetries. (These are to be expected, since surely $4^4 = 256$ independent components would be a bit much for describing the curvature at a point in 4-space, say!) A straightforward calculation converts the RHS of (10.59) into the following alternative forms:

$$R_{hijk} = \Gamma_{hik,j} - \Gamma_{hij,k} + \Gamma^a_{ij}\Gamma_{ahk} - \Gamma^a_{ik}\Gamma_{ahj}, \qquad (10.60)$$

$$R_{hijk} = \tfrac{1}{2}(g_{hk,ij} + g_{ij,hk} - g_{hj,ik} - g_{ik,hj}) + \Gamma^a_{ij}\Gamma_{ahk} - \Gamma^a_{ik}\Gamma_{ahj}. \qquad (10.61)$$

At the pole of geodesic coordinates all the undifferentiated Γ's vanish, which makes it easy to read off the following symmetries:

$$R_{hijk} = -R_{hikj}, \qquad (10.62)$$

$$R_{hijk} = -R_{ihjk}, \qquad (10.63)$$

$$R_{hijk} = R_{jkhi}, \qquad (10.64)$$

$$R_{hijk} + R_{hjki} + R_{hkij} = 0. \qquad (10.65)$$

[The first and last of these follow most easily from (10.60), the second and third, from (10.61).] As a consequence of these symmetries, it can be shown that the number of independent components of R_{hijk} in N dimensions is reduced to $N^2(N^2 - 1)/12$; that makes 20 for $N = 4$, 6 for $N = 3$, and only one for $N = 2$. (In the last case all non-zero components equal $\pm R_{1212}$.)

Again, at the pole of geodesic coordinates, where the Γ's vanish and the covariant derivative equals the partial, we have from (10.56),

$$R^h{}_{ijk;l} = \Gamma^h_{ik,jl} - \Gamma^h_{ij,kl}. \qquad (10.66)$$

This provides a simple way to establish an additional *differential* symmetry of the curvature tensor, independent of the previous algebraic symmetries, namely the *Bianchi identity*:

$$R^h{}_{ijk;l} + R^h{}_{ikl;j} + R^h{}_{ilj;k} = 0. \tag{10.67}$$

By successive contraction of the Riemann tensor we obtain two further curvature quantities of great importance, especially in GR. First, the *Ricci tensor*:

$$R_{ij} := R^h{}_{ijh} = R_{jh}{}^h{}_i = R_j{}^h{}_{hi} = R^h{}_{jih} = R_{ji}, \tag{10.68}$$

and second, the *curvature scalar*:

$$R := R^i{}_i = R^{ij}{}_{ji}. \tag{10.69}$$

To establish the symmetry of the Ricci tensor, we have used the interchange symmetry (10.64), the see-saw rule, and the skew-symmetries (10.62) and (10.63) simultaneously. Of the other two possible contractions of the Riemann tensor, one vanishes: $R^h{}_{hjk} = 0$, because of (10.63), and the other, $R^h{}_{ihj} = -R^h{}_{ijh}$, is the negative of the Ricci tensor.

This may be a good place to warn the reader of an unfortunate sign confusion in the literature. About fifty per cent of authors define $R^h{}_{ijk}$ as $\Gamma^h{}_{ij,k} - \cdots$ instead of our $\Gamma^h{}_{ik,j} - \cdots$, and another fifty per cent define R_{ij} as $R^h{}_{ihj}$. *Caveat lector!*

If in the Bianchi identity (10.67) we contract h with k, then raise i and contract it with j, also writing $R_{,l}$ for $R_{;l}$ (since R is a scalar), we get the following 'twice-contracted' Bianchi identity:

$$R_{,l} - 2R^j{}_{l;j} = 0. \tag{10.70}$$

For the construction of his full field equations, Einstein needed to find a curvature-related symmetric two-index tensor of 'zero divergence'. It was eqn (10.70) that led the way to what is now known as the *Einstein tensor*:

$$G_{ij} := R_{ij} - \tfrac{1}{2}g_{ij}R, \qquad G^i{}_{j;i} = 0. \tag{10.71}$$

The reader will find at the end of this book an Appendix from which to read off, without undue labor, the Christoffel symbols, as well as the components of the Riemann, Ricci, and Einstein tensors, plus the curvature scalar, for diagonal metrics of dimension 2, 3, and 4.

So far, our discussion of the curvature tensor has been mainly formal. We shall now exhibit, without proof, three standard formulae that show more directly its connection with the curvature of the underlying space. The first of these (and the easiest to establish—see Exercise 10.15) gives the change ΔV that a vector V suffers when it is parallely transported around a small parallelogram spanned by the displacements dx and δx, and going along dx first:

$$\Delta V^h = -R^h{}_{ijk}V^i \, dx^j \, \delta x^k. \tag{10.72}$$

Note that, since ΔV^h is the difference between two vectors at *one* point, it is itself a vector. [The connection of (10.72) with curvature is discussed in Exercise 10.16.]

The second important formula[1] gives the curvature $K(\mathbf{p}, \mathbf{q})$ of a space V^N for the orientation $\mathbf{p}$, $\mathbf{q}$ (cf. Section 8.2):

$$K(\mathbf{p}, \mathbf{q}) = \frac{R_{hijk} p^h q^i p^j q^k}{(g_{hj} g_{ik} - g_{hk} g_{ij}) p^h q^i p^j q^k}. \tag{10.73}$$

At an isotropic point, where the curvature is the same for all orientations, we must then have

$$R_{hijk} = K(g_{hj} g_{ik} - g_{hk} g_{ij}). \tag{10.74}$$

(While the sufficiency of this condition is evident, its necessity is a little more tedious to establish—cf. Exercise 10.17.) In particular, for a 2-surface, where every point is isotropic (there is only one orientation!), this yields

$$K = \frac{R_{1212}}{(g_{11} g_{22} - g_{12}^2)} = \frac{R_{1212}}{\|g_{ij}\|}. \tag{10.75}$$

By raising h and contracting it with k in (10.74), we find, at a general isotropic point,

$$R_{ij} = -(N - 1) K g_{ij}, \tag{10.76}$$

and doing this once more, with i and j,

$$R = -N(N - 1) K. \tag{10.77}$$

Suppose every point of some V^N is an isotropic point; we can then covariantly differentiate (10.76) and (10.77) (treating K a priori as variable) and substitute into (10.70). That yields

$$(N - 1)(N - 2) K_{,l} = 0, \tag{10.78}$$

and consequently *Schur's theorem*: if $N > 2$ and every point of V^N is isotropic, K must be constant.

We note from (10.76) that every space of constant curvature and every 2-space is an *Einstein space*, namely a space satisfying the proportionality

$$R_{ij} = \phi g_{ij} \tag{10.79}$$

for some scalar ϕ. (Contraction shows $\phi = R/N$.) An argument identical to that which leads to Schur's theorem shows that for every Einstein space ϕ must be constant unless $N = 2$, in which case $\phi = -K$, by (10.76). It can also be shown (cf. Exercise 10.19) that every Einstein space of dimension 3 (but no other) is necessarily of constant curvature.

---

[1] See, for example, J. L. Synge and A. Schild, *Tensor Calculus*, University of Toronto Press, Toronto, 1949, p. 95.

The third in this set of important curvature formulae is that for *geodesic deviation* (cf. Synge and Schild, *loc. cit.*, p. 93). It concerns two neighboring geodesics g_1 and g_2 separated by an infinitesimal connection vector field η^h along and orthogonal to g_1. If $x^i(s)$ is the coordinate representation of g_1 in terms of its arc s, the formula in question is

$$\frac{D}{ds}\left(\frac{D\eta^h}{ds}\right) = (R^h{}_{ijk}\dot{x}^i\dot{x}^j)\eta^k, \qquad (10.80)$$

overdots denoting d/ds. Its analogy to our earlier formula (8.17) (and thus its connection with curvature) is evident, but note that here we differentiate the connecting *vector* and not just its length. If we specialize (10.80) to 4-dimensional spacetime, and consider g_1 and g_2 as the worldlines of two neighboring free particles in a gravitational field, the formula gives their *relative* acceleration vector. The relative acceleration is a manifestation of the so-called *tidal field*, which is what remains of the gravitational field in a freely falling box at the first particle. [The earth (apart from its rotation and its own gravity) is essentially such a freely falling box in the combined gravitational field of the sun and the moon, and it is the tidal force between the center and the surface that pulls up the tides.] In GR there is no tensorial measure of the gravitational field itself, since that can always be transformed away by going to the LIF. The quantity that *does* have a tensorial measure is the tidal field, and, as seen from (10.80), the Riemann tensor is its measure. If we throw out a handful of test particles and measure their relative accelerations, we could, in principle, use (10.80) to determine $R^\mu{}_{\nu\rho\sigma}$ at any event in spacetime.

10.6   Einstein's vacuum field equations

Our final task in this chapter is to set up the vacuum field equations of GR, for which our study of the curvature tensor has prepared us.

The starting point *must* be Newton's inverse-square gravitational theory. Its tremendous success in classical situations makes it imperative that any new gravitational theory reduce to Newton's 'in the classical limit'. Newton's inverse-square law can be expressed in differential form as *Poisson's equation*,

$$-\text{div}\,\mathbf{g} \equiv \text{div}\,\text{grad}\,\Phi \equiv \sum \Phi_{,ii} = 4\pi G\rho, \qquad (10.81)$$

and this is what the GR field equations must spring from.

Accordingly, let us consider a *weak* gravitational field, with *slowly* (that is, non-relativistically) moving sources, and therefore slowly changing field components. From the relativistic point of view, this shall mean that spacetime is globally quasi-Minkowskian, with coordinates $\{x^i, ct\}$ such that

$$g_{\mu\nu} \approx \text{diag}(-1, -1, -1, 1), \qquad g_{\mu\nu,4} \approx 0. \qquad (10.82)$$

We then expect the Newtonian equation of motion,

$$\frac{d^2 x^i}{dt^2} = -\Phi_{,i} \tag{10.83}$$

and the Einsteinian equation of motion [cf. (10.15)],

$$\frac{d^2 x^\mu}{d\tau^2} + \Gamma^\mu_{\rho\sigma} \frac{dx^\rho}{d\tau} \frac{dx^\sigma}{d\tau} = 0 \tag{10.84}$$

to be essentially equivalent for *slowly* moving test particles. For such particles we have $t \approx \tau$ (proper time), and consequently, with $u \ll c$,

$$\frac{dx^\mu}{d\tau} \approx (\mathbf{u}, c), \qquad \frac{d^2 x^i}{d\tau^2} \approx \frac{d^2 x^i}{dt^2}, \qquad \frac{d^2 t}{d\tau^2} \approx 0. \tag{10.85}$$

With that, the equivalence of (10.83) and (10.84) is assured if

$$c^2 \Gamma^i_{44} = \Phi_{,i} \quad \text{and} \quad \Gamma^4_{44} = 0, \tag{10.86}$$

since all other Γs have negligible effect on the orbits, their contribution being diminished by small velocity components u_i. (Recall—from Section 9.4—how the spatial geometry was unimportant in determining the *slow* orbits in static spacetimes.) And (10.86)(ii) is automatic, since we assume $g_{\mu\nu,4} \approx 0$.

Now Φ satisfies Poisson's equation (10.81), and so we have, from (10.86)(i),

$$\sum \Phi_{,ii} = c^2 \Gamma^i_{44,i} = 4\pi G\rho. \tag{10.87}$$

The last equation in this line represents the Newtonian field equation in 'relativistic' though not yet tensorial form. However, referring to (10.56), and using (10.87)(i), we find

$$R_{44} = R^\mu{}_{44\mu} = R^i{}_{44i} = -\Gamma^i_{44,i} = -c^{-2} \sum \phi_{,ii}, \tag{10.88}$$

since in a *weak* field we ignore products of Γs (we assume the $g_{\mu\nu,\rho}$ to be small and their products to be negligible), and in a *slowly* varying field we ignore $g_{\mu\nu,4}$. So if there is to be correspondence with Newtonian theory, (10.88) must be satisfied in lowest approximation. At first, we shall be interested only in *vacuum* fields, such as the field around the sun, for which $\rho = 0$. The relevant Newtonian field equation, $\sum \phi_{,ii} = 0$, by (10.88), then implies $R_{44} = 0$. This *suggests*

$$R_{\mu\nu} = 0 \tag{10.89}$$

as a candidate for the vacuum field equation of GR, which, of course, must be tensorial. And that, indeed, was Einstein's proposal (1915). It has been strikingly vindicated: not only does GR, completed by this field equation[2] (and its generalization to the non-vacuum case, see Section 14.2), reproduce within experimental errors all those

---

[2] Sometimes we think and speak of (10.89) as a single tensor *equation*, sometimes as ten component *equations*.

Newtonian results that agree so well with observation, but where GR differs observably from Newton's theory (as in the precession of the orbit of Mercury or in the bending of starlight around the sun) it is GR that is found to be correct. A later 'crucial' effect that also validated GR was the Shapiro light retardation in the field of the sun (see Section 11.2H). And more recently there have been indirect indications for the existence of black holes and gravitational radiation—further support of the field equation (10.89). The well-established gravitational Doppler shift, occasionally also referred to as a crucial effect, is, in fact, not a test of the field equation but merely of the equivalence principle—at least in lowest order, which is all that can be observed at present.

But how is it that instead of the *one* field equation (10.81) of Newtonian theory there should be *ten* in GR? (The Ricci tensor, because of its symmetry, has 10 independent components.) The reason is that the field equations must determine the whole metric; that is, the $g_{\mu\nu}$. And there are just ten of these. Once we lay an arbitrary coordinate system over our spacetime, the ten functions $g_{\mu\nu}$ should be uniquely determined. In fact, these ten tensor components $g_{\mu\nu}$ are the analogs of the Newtonian scalar potential Φ (for which there is just one field equation) and of the 4-vector potential components Φ_μ of Maxwell's theory (for which there are four field equations). This is why Einstein chose the symbol g (for gravitational potentials) to denote the metric. We have already seen how in the static metric (9.5) g_{44} is directly related to Newton's Φ, and how in the stationary metric (9.13) the $g_{\mu 4}$ are analogous to Maxwell's Φ_μ.

The analogy of the three theories can be seen quite clearly both in the field equations and in the equations of motion. Consider first the (vacuum) field equations:

$$\text{Newton}: \quad g^{ii}\Phi_{,ii} = 0 \qquad [\text{cf. (10.81)}]$$

$$\text{Maxwell}: \quad g^{\mu\nu}\Phi_{\rho,\mu\nu} = 0 \qquad [\text{cf. (7.49)}]$$

$$\text{Einstein}: \quad g^{\mu\nu}g_{\rho\sigma,\mu\nu} + \cdots = 0 \qquad [\text{cf. (10.61)}].$$

And next, the equations of motion (of which only Newton's are velocity-independent):

$$\text{Newton}: \quad \frac{d^2x^h}{dt^2} = -g^{hi}\Phi_{,i}$$

$$\text{Maxwell}: \quad \frac{d^2x^\mu}{d\tau^2} = -\frac{q}{cm_0}g^{\mu\alpha}(\Phi_{\alpha,\beta} - \Phi_{\beta,\alpha})\frac{dx^\beta}{d\tau}$$

$$[\text{cf. (7.30), (7.42)}]$$

$$\text{Einstein}: \quad \frac{d^2x^\mu}{d\tau^2} = -\frac{1}{2}g^{\mu\alpha}(g_{\alpha\beta,\gamma} + g_{\alpha\gamma,\beta} - g_{\beta\gamma,\alpha})\frac{dx^\beta}{d\tau}\frac{dx^\gamma}{d\tau}$$

$$[\text{cf. (10.15)}]$$

GR is thus a field theory with a tensor potential. That is why its presumed quantum, the *graviton*, would have spin 2. One enormous difference from the other two theories is that GR is *non-linear*: though its field equations are linear in the *second* derivatives of the $g_{\mu\nu}$, as is seen from eqn (10.61), they are non-linear in the first derivatives. This not only greatly complicates their mathematical solution, but it also robs us of the possibility of 'adding' solutions. One reason for the non-linearity is mathematical: there are *no* tensor fields that are linear in the $g_{\mu\nu}$ and their partial derivatives. But actually, non-linearity is needed by the physics: it allows gravity itself to gravitate—as demanded by $E = mc^2$. We have already noted (in Section 6.3) how the negative binding energy of atomic nuclei diminishes the total mass of the constituent nucleons, thus causing the well-known 'mass defect'. In much the same way, we would expect the gravitational forces that hold a massive body together to diminish the mass of its constituents. The field of two massive balls which stick together by mutual gravity should be less than twice the field of one ball. Yet the full field equations do not treat the gravitational field itself as a source: it is the non-linearity that miraculously takes account of it. (Cf. end of Subsection 11.2B.)

From the theoretical point of view, therefore, the field equations $R_{\mu\nu} = 0$ seem just right. Certainly there exist none that are simpler and still consistent with the fundamental ideas of GR. And it should be remembered that field equations are a matter of choice and not of proof. They belong among the axioms of a theory. The next logical step is to see whether they correctly predict verifiable results. In order to show this, we begin in the next chapter to develop some of the consequences of the theory.

Exercises 10

10.1. Consider the approximate static metric (9.6) for the case of a spherically symmetric body of mass m:

$$ds^2 = (1 + 2\Phi)\, dt^2 - dr^2 - r^2(d\theta^2 + \sin^2\theta\, d\phi^2), \qquad \Phi = -Gm/r \quad (c = 1).$$

Apply eqns (10.11) to orbits in the symmetry surface $\theta = \pi/2$. Verify that they imply (at least approximately) the conservation of angular momentum and of energy, the inverse-square law for radial fall, and, for circular orbits, Kepler's third law $(d\phi/dt)^2 = Gm/r^3$. [*Hint:* $\mathcal{L} = (1 + 2\Phi)\dot{t}^2 - \dot{r}^2 - r^2\dot{\phi}^2 = 1$.]

10.2. In the spacetime with metric $ds^2 = dt^2 - dl^2$, where dl^2 is an arbitrary and possibly time-dependent 3-metric, prove that the coordinate lines $t = $ var are geodesics. If dl^2 is time-independent, prove that every geodesic of ds^2 follows a geodesic spatial track in the metric dl^2. [*Hint:* (10.11) and text after (10.18).]

10.3. Use the method suggested in the text [after (10.17)] to calculate all the Γ^i_{jk} of the metric of the 3-sphere given in Exercise 8.9; for simplicity put $a = 1$. [*Hint:* for example, $(\partial\mathcal{L}/\partial\dot{r})^{\bullet} - \partial\mathcal{L}/\partial r = 2\ddot{r} - 2\sin r\cos r(\dot{\theta}^2 + \sin^2\theta\dot{\phi}^2)$; so $\Gamma^1_{11} = 0$, $\Gamma^1_{22} = -\sin r\cos r$, $\Gamma^1_{33} = -\sin r\cos r\sin^2\theta$ and $\Gamma^1_{ij} = 0$ $(i \neq j)$; etc.]

10.4. Two metrics $ds^2 = g_{ij}\, dx^i\, dx^j$ and $\widetilde{ds}^2 = \tilde{g}_{ij}\, dx^i\, dx^j$ on the same coordinate net are said to be *conformally related* if $\tilde{g}_{ij} = \psi g_{ij}$ for some positive function ψ of the coordinates. Prove that if the curve $x^i = x^i(u)$ is a *null* geodesic of the metric ds^2, then it is also a null geodesic of the metric $\widetilde{ds}^2$; that is, conformal spaces share their null geodesics. [*Hint*: prove $\tilde{\Gamma}^i_{jk}\dot{x}^j\dot{x}^k = \Gamma^i_{jk}\dot{x}^j\dot{x}^k + (\psi^{-1}\psi_{,j}\dot{x}^j)\dot{x}^i$ and use the result of the paragraph containing eqn (10.43).]

10.5. With $\dot{} \equiv d/dw$, prove that eqn (10.43) can alternatively be written in the form $\mathbb{L}_i \propto \partial \mathscr{L}/\partial \dot{x}^i$.

10.6. Prove that for any vector A_i and for any antisymmetric tensor T_{ij},

$$A_{i;j} - A_{j;i} = A_{i,j} - A_{j,i}$$

$$T_{ij;k} + T_{jk;i} + T_{ki;j} = T_{ij,k} + T_{jk,i} + T_{ki,j}.$$

10.7. Given the rule for differentiating a determinant $a = \|a_{ij}\|$: $a_{,k} = a\sum_{ij} A_{ji}a_{ij,k}$, where A_{ij} is the inverse matrix of a_{ij}, prove the often useful result

$$\Gamma^i_{ji} = \frac{1}{2}\frac{1}{g}\frac{\partial g}{\partial x^j} = (\log\sqrt{|g|})_{,j}, \quad g = \|g_{ij}\|.$$

[*Hint*: the first expression is invariant under $g \mapsto -g$.]

10.8. The familiar operations $\nabla\cdot$ and ∇^2 on Euclidean 3-vectors and scalars, respectively, can be generalized to arbitrary coordinates and arbitrary spaces: $\nabla\cdot\mathbf{A} := A^i_{\;;i}$ and $\nabla^2\Phi := g^{ij}\Phi_{;ij}$. Prove

$$\nabla\cdot\mathbf{A} = A^i_{\;,i} + \tfrac{1}{2}g^{-1}g_{,i}A^i = |g|^{-1/2}(|g|^{1/2}A^i)_{,i},$$

$$\nabla^2\Phi = |g|^{-1/2}(|g|^{1/2}g^{ij}\Phi_{,i})_{,j},$$

and use the second of these formulae to express $\nabla^2\Phi$ in cylindrical polar coordinates ρ, ϕ, z in Euclidean 3-space. Finally, A^{ij} is antisymmetric, prove

$$A^{ij}_{\;;j} = |g|^{-1/2}(|g|^{1/2}A^{ij})_{,j} \quad \text{and} \quad A^{ij}_{\;;ij} = 0.$$

10.9. Fill in the details of the following proof of a formula for the curvature κ of a curve C in a space V^N. Let g be its tangent geodesic at the point P. Choose Riemannian coordinates so that

$$C : x^i = (\dot{x}^i)_P s + \tfrac{1}{2}(\ddot{x}^i)_P s^2 + \cdots$$

$$g : \tilde{x}^i = (\dot{x}^i)_P s,$$

s being arc along g *and* C. Then near P the orthogonal distance η between g and C is given by

$$\eta^2 = g_{ij}(x^i - \tilde{x}^i)(x^j - \tilde{x}^j) = \tfrac{1}{4}(g_{ij}\ddot{x}^i_P\ddot{x}^j_P)s^4 = \tfrac{1}{4}(g_{ij}N^i_P N^j_P)s^4 = \tfrac{1}{4}\kappa^2 s^4,$$

so that $\kappa = \lim_{s\to 0} 2\eta/s^2$, just as in the plane, where g is simply the tangent.

10.10. In any stationary spacetime (9.13), prove directly that the initial acceleration $d^2x^i/d\tau^2$, relative to the lattice, of a test particle released from rest, equals $-\Phi_{,i}$ in orthonormal coordinates. [*Hint*: in eqn (10.15) put $(\dot{x}^i)_0 = 0$, $(\dot{x}^4)_0 = e^{-\Phi}$, and cf. (10.48).]

10.11. In the stationary spacetime (9.13), consider the infinitesimal Fermi–Walker transport of a vector V^μ along the worldline of a lattice point, $x^i = $ const. Choosing the gauge $\Phi = 0$, $\mathbf{w} = 0$, $k_{ij} = \text{diag}(1, 1, 1)$ at the event of interest, $c = 1$ and $V^\mu = (1, 0, 0, 0)$, prove $dV^\mu/ds = (\frac{1}{2}(w_{1,i} - w_{i,1}), 0)$ and verify that this is consistent with (9.21). [*Hint*: a position vector $\mathbf{r}$ being rotated about the origin with angular velocity $\mathbf{\Omega}$ changes by $d\mathbf{r} = \mathbf{\Omega} \times \mathbf{r}\, dt$.] By repeating this verification with $V^\mu = (0, 1, 0, 0)$ and $V^\mu = (0, 0, 1, 0)$—which essentially follows from symmetry—we can in this way *establish* (9.21) in the present gauge. But then (9.23) necessarily follows from gauge invariance.

10.12. (i) Compute R_{1212} for the metric $y^2\, dx^2 + x^2\, dy^2$ (do *not* use the Appendix for this except as a check) and so verify that it represents the Euclidean plane. (ii) Do the same for the metric $y\, dx^2 + x\, dy^2$ and deduce that it represents a *curved* surface.

10.13. By reference to the definitions (10.60), (10.68), verify that when a metric (of arbitrary dimensions) is the sum of two mutually independent submetrics, the Riemann and Ricci tensor components of the submetrics together constitute all non-zero components of these tensors in the full metric. Also show that the geodesics of the full metric are of the form $x^i = x^i(u)$, $x^\alpha = x^\alpha(u)$, where the $x^i(u)$ are geodesics of the submetric involving the x^i, and $x^\alpha(u)$ are geodesics of the submetric involving the x^α, and u is an affine parameter for both submetrics *and* the full metric.

10.14. In Riemannian coordinates the equation of a geodesic through the pole is $x^i = a^i s$, where $a^i = (\dot{x}^i)_0$. Hence $(\ddot{x}^i)_0 = 0$, $(\dddot{x}^i)_0 = 0$, etc. But $\ddot{x}^i = -\Gamma^i_{jk}\dot{x}^j\dot{x}^k$, so $(\dddot{x}^i)_0 = -(\Gamma^i_{jk,l})_0 a^j a^k a^l$—why? Deduce that $P(\Gamma^i_{jk,l})_0 = 0$ and $P(\Gamma_{ijk,l})_0 = 0$, where P denotes the sum over all permutations of j, k, l. Use this identity in proving that

$$g_{ij} = (g_{ij})_0 + \tfrac{1}{3}(R_{hijk})_0 x^h x^k + O(x^3).$$

This formula allows us, for example, to estimate the useful extent of a LIF. [*Hint*: convert $\frac{1}{2}(g_{ij,hk})_0 x^h x^k$ into the curvature term.[3]

10.15. Consider a small parallelogram PQRS spanned by two displacements $\mathbf{dx}$ and $\mathbf{\delta x}$ away from a pole P of geodesic coordinates. Transport a vector V^h parallelly from P to the opposite vertex R along the two alternative paths and prove that

$$V_R^h(\mathbf{dx}\ \text{then}\ \mathbf{\delta x}) - V_R^h(\mathbf{\delta x}\ \text{then}\ \mathbf{dx}) = -R^h{}_{ijk}V^i\, dx^j\, \delta x^k,$$

and that, consequently, for a round-trip from R, with $-\mathbf{dx}$ first, the increment in V^h is given by (10.72).

10.16. On a 2-surface, the 'total angle Δ through which a closed curve turns' (cf. Exercise 8.5) is measured against the standard of a parallel vector field along the curve. So if, after a complete circuit, for example the parallely transported initial

---

[3] Cf. L.P. Eisenhart, *Riemannian Geometry*, Princeton University Press, Princeton 1960, pp. 252–3.

tangent vector returns rotated through an angle θ in the sense in which the curve is described, the curve has turned by that much less than 2π : $\Delta = 2\pi - \theta$. By the Gauss–Bonnet theorem of Exercise 8.5, therefore, $\theta = 2\pi - \Delta = \int K \, dS$. Verify this last formula, for an infinitesimal rectangle, from (10.72). [*Hint*: use orthonormal coordinates and let $\mathbf{V} = \mathbf{dx}$.]

10.17. Establish formula (10.74) along the following lines: Let T_{hijk} be LHS–RHS of (10.74), so that (10.73) reads $T_{hijk}p^h q^i p^j q^k = 0$. Verify that T_{hijk} has the same symmetries (10.62)—(10.65) as R_{hijk}. Now give $\mathbf{p}$ and $\mathbf{q}$ the specific values $p^1 = 1, q^2 = 1$, all other components zero, and deduce $T_{1212} = 0$; so evidently *all* components with two equal index pairs vanish. Next put $p^1 = p^2 = 1, q^3 = 1$ and deduce $T_{1323} = 0$; so evidently all components with one equal index pair vanish. Lastly, put $p^1 = p^2 = 1, q^3 = q^4 = 1$, and deduce $T_{1324} + T_{1423} = 0$; interchange 2 and 3: $T_{1234} + T_{1432} = 0$; add these two equations and deduce the antisymmetry $T_{1234} = -T_{1324}$; now show that $T_{1234} + T_{1342} + T_{1423} = 3T_{1234} = 0$; so evidently all components with four unequal indices vanish.

10.18. Derive the metric of the paraboloid that results when the parabola $z^2 = 4a(x - a), y = 0$, is revolved about the z-axis in Euclidean 3-space, using $r = (x^2 + y^2)^{1/2}$ and the azimuthal angle ϕ as coordinates. [*Answer*: $dl^2 = (1 - a/r)^{-1} dr^2 + r^2 \, d\phi^2$.] Prove that the Gaussian curvature of this surface is given by $K = -\frac{1}{2}a/r^3$. [*Hint*: find R_{1212} using the Appendix.]

10.19. For every properly Riemannian 3-space prove that at a point where the coordinates are orthonormal the following relations hold: $R_{3123} = R_{12}, R_{1221} = \frac{1}{2}(R_{11} + R_{22} - R_{33})$ and similarly for the other non-zero components of R_{hijk}. Hence prove that every properly Riemannian 3-dimensional Einstein space is of constant curvature. By writing ϵ_i for the sign of g_{ii}, generalize the argument and prove the result also for pseudo-Riemannian 3-spaces.

10.20. Prove that every vacuum spacetime ($R_{\mu\nu} = 0$) whose metric has the form $A(x) \, dt^2 - dx^2 - dy^2 - dz^2$, where $A(x)$ is an arbitrary positive function of x, is necessarily flat. [*Hint*: Use the Appendix to show that R_{1414} is essentially the only possible non-zero component of $R_{\mu\nu\rho\sigma}$ and then apply the field equations.] Show also that $A(x) = x^2$ is essentially the only solution. The metric represents a flat lattice with all the field lines parallel to the x-direction; and the result shows that the only static vacuum field of this nature is that of the uniformly accelerating rocket—for which indeed we saw that the 'gravitational field' was proportional to $1/x$, as it is here (cf. end of Section 3.7).

10.21. For the approximative metric of a weak static field $\mathbf{ds}^2 = (1 + 2\Phi/c^2)c^2 \, dt^2 - \sum dx^2$ [cf. (9.6)], verify that $R_{44} \approx -\sum \Phi_{,ii}/c^2$, in conformity with (10.88). [*Hint*: use the Appendix.]

11

The Schwarzschild metric

11.1 Derivation of the metric

It was left to Schwarzschild to find, early in 1916, the first and still the most important exact solution of the Einstein vacuum field equations. It represents the field outside a spherically symmetric mass in otherwise empty space. (Later it was also recognized as the solution representing both the outside and the inside of a non-rotating black hole.) Schwarzschild's solution allows the exact calculation of several of the 'post-Newtonian' effects of GR, including the precession of planetary orbits, the bending of light around the sun, the exact gravitational frequency shift, the Shapiro time delay of light passing near the sun, and the precession of orbiting gyroscopes. Curiously, Einstein himself had apparently not attempted to find this relatively straightforward and elegant solution (perhaps doubting its existence) and had contented himself with approximative calculations to derive the advance of the perihelion and the bending of light.

In re-deriving Schwarzschild's metric, say for the spacetime outside a stable non-rotating star, which is clearly static, we can fall back on our findings of Chapter 9. With a suitably adapted time coordinate t, the sought-for solution will be a special case of the static metric (9.5). As for the lattice, spherical symmetry implies that it will be a radial distortion of Euclidean 3-space E^3. The metric of E^3 in polar coordinates is

$$dl^2 = dr^2 + r^2(d\theta^2 + \sin^2\theta \, d\phi^2), \tag{11.1}$$

where r is distance from the origin, θ the inclination *from* the vertical, and ϕ the ('azimuthal') angle *around* the vertical. The angular part is the metric on the 2-sphere of radius r, with the quantity in parenthesis measuring the square of the angular distance between neighboring points. In a *curved* spherically symmetric space, the area of successive parallel spheres will not necessarily increase as the square of the distance. It will be convenient, nevertheless, to have the area of the sphere at radial coordinate r still be $4\pi r^2$, thus defining the coordinate r as '*area distance*', an intrinsic quantity. The radial distance increments will then generally be of the form $F(r) \, dr$ [or, for calculational convenience, $e^{B(r)} \, dr$] rather than just dr. Thus, *if we set $c = 1$*, spherical symmetry allows us to specialize (9.5) to the form

$$\mathbf{ds}^2 = e^{A(r)} \, dt^2 - e^{B(r)} \, dr^2 - r^2 \, (d\theta^2 + \sin^2\theta \, d\phi^2), \tag{11.2}$$

since the potential $\Phi = \frac{1}{2}A$ must be independent of θ and ϕ. This metric contains only two unknowns, the function $A(r)$ and $B(r)$, which must now be determined by the field equations.

Because diagonal metrics like (11.2) occur frequently, we have listed in the Appendix once and for all the curvature tensor components for such metrics. From that list we now find for the metric (11.2):

$$R_{rr} = \tfrac{1}{2}A'' - \tfrac{1}{4}A'B' + \tfrac{1}{4}A'^2 - B'/r \tag{11.3}$$

$$R_{tt} = -e^{A-B}\left(\tfrac{1}{2}A'' - \tfrac{1}{4}A'B' + \tfrac{1}{4}A'^2 + A'/r\right) \tag{11.4}$$

$$R_{\theta\theta} = e^{-B}\left[1 + \tfrac{1}{2}r(A' - B')\right] - 1 \tag{11.5}$$

$$R_{\phi\phi} = R_{\theta\theta}\,\sin^2\theta \tag{11.6}$$

$$R_{\mu\nu} = 0 \quad \text{when } \mu \neq \nu, \tag{11.7}$$

where primes denote d/dr and where we have introduced an often useful notation: if, for example, the indices 1, 2, 3, 4 refer to the coordinates r, θ, ϕ, t in this order, then $R_{rt} = R_{14}$, $R_{\phi\phi} = R_{33}$, etc. The advantage of this notation is that it is independent of which numbers are assigned to which coordinates. It is the analog of writing $\mathbf{a} = (a_x, a_y, a_z)$.

The vacuum field equations require $R_{\mu\nu} = 0$ for all indices. Thus (11.3) and (11.4) yield

$$A' = -B' \tag{11.8}$$

whence $A = -B + k$, k being a constant. Reference to (11.2) shows that a simple change in time scale, $t \mapsto e^{-k/2}t$, will absorb this k, and then

$$A = -B. \tag{11.9}$$

With that, (11.5) yields
$$e^A(1 + rA') = 1, \tag{11.10}$$

or, setting $e^A = \alpha$,
$$\alpha + r\alpha' = (r\alpha)' = 1. \tag{11.11}$$

This equation can at once be integrated, giving

$$\alpha = 1 - \frac{2m}{r}, \tag{11.12}$$

where at this stage $2m$ is merely a constant of integration of the dimension of a length. But, as the notation suggests, m will presently turn out to be the mass of the central body in 'relativistic units' (which make $c = G = 1$). Substituting our findings into (11.2), we have now obtained the *Schwarzschild metric*

$$\mathbf{ds}^2 = \left(1 - \frac{2m}{r}\right)dt^2 - \left(1 - \frac{2m}{r}\right)^{-1}dr^2 - r^2(d\theta^2 + \sin^2\theta\,d\phi^2). \tag{11.13}$$

However, since we have used the field equations (11.3) and (11.4) only in combination, we must yet verify that (11.13) satisfies these equations separately. Happily it does, except, of course, at $r = 0$ and $r = 2m$, where the metric becomes meaningless. (These loci turn out to have physical importance and will be discussed later.) Note that far out ($r \to \infty$) the metric becomes Minkowskian. This 'asymptotic flatness' is a *result*; that is, it was not included as an *assumption*. It implies that asymptotically t and r have their special-relativistic significance.

11.2   Properties of the metric

A. The field strength and the meaning of m

Comparison of (11.13) with (9.5) shows that the relativistic potential Φ of the Schwarzschild metric (temporarily *with c*) is given by

$$\Phi = \frac{c^2}{2}\ln\left(1 - \frac{2m}{r}\right) = c^2\left(-\frac{m}{r} + \cdots\right). \tag{11.14}$$

From this we can calculate the field strength g. The field itself is radial, of course, and from (9.20) we find

$$g = |\text{grad } \Phi| = \frac{d\Phi}{dl} = \frac{d\Phi}{dr}\frac{dr}{dl} = \frac{mc^2}{r^2}\left(1 - \frac{2m}{r}\right)^{-1/2}, \tag{11.15}$$

where l is radial ruler distance and $dr/dl = (1 - 2m/r)^{1/2}$ by (11.3). The same exact result can be obtained from (9.22).

Note how the field becomes infinite at the so-called *Schwarzschild radius* $r = 2m$ and Newtonian for large r. Indeed, if we set

$$m = \frac{GM}{c^2}, \tag{11.16}$$

we seen that, for large r, $g \sim GM/r^2$. Consequently we identify M with the mass of the central body, so that both theories agree asymptotically for the weak far field. In relativistic units ($c = G = 1$) we simply have $m = M$.

B. Birkhoff's theorem

In 1923 Birkhoff discovered a very important extension of the validity of Schwarzschild's solution. He showed that *even without the assumption of staticity, Schwarzschild's solution is the unique vacuum solution with spherical symmetry*. The proof is not even all that difficult, but we shall omit it here; basically one adapts Schwarzschild's proof to the case where A and B in (11.2) are allowed to be functions of r and t.

Birkhoff's theorem has significant implications. Suppose, for example, the central spherical body were to start pulsating or exploding or imploding in a spherically symmetric (radial) way: the external field would show no trace of a response.

(Just as in Newton's theory!) In particular, no spherically symmetric gravitational radiation (being a progressive disturbance of the vacuum field) is possible. Again, the field inside a spherical vacuum cavity (if its center is regular) must be flat, even if the surrounding spherically symmetric matter were to move radially. For the spacetime is regular (that is, it has finite curvature) at $r = 0$ only if $m = 0$ in which case the Schwarzschild metric becomes the Minkowski metric. Somewhat more generally, consider a vacuum zone between concentric spheres $\sum_1$ and $\sum_2$ with spherically symmetric but possibly radially moving matter both inside $\sum_1$ and outside $\sum_2$. In between, the Schwarzschild metric applies. Specifically, consider a massive star surrounded, beyond a concentric sphere, by a radially expanding isotropic universe. Its field is that of Schwarzschild and its planetary (test-)orbits 'feel' nothing of the expanding universe.

Lastly, consider a large cold ball of matter. Suppose this ball is allowed to shrink under its own gravity, and suppose that much of its original potential energy is thereby used to heat it up. Because of $E = mc^2$, the heated material has greater mass. Birkhoff's theorem shows that the field equations (at least in this case) allow for the 'gravity' of the gravitational field.

C. The Schwarzschild radius

The reader has no doubt observed a most un-Newtonian feature of the Schwarzschild spacetime, namely the coordinate singularity at the Schwarzschild radius $r = 2m$. A first indication of what goes on there is the fact that the g-field (11.15) becomes infinite. An infinite g-field in a static metric means that for a particle to remain at rest in the lattice it must have infinite proper acceleration, that is, it must be a photon! So it would appear that the locus $r = 2m$ is potentially an outwardly directed spherical light-front that doesn't get anywhere! It will turn out to be the 'horizon' of a 'black hole' whenever the vacuum continues through it. But for the moment it will not concern us. For the moment, we shall be interested in the gravitational field outside a star like our sun or a planet like our earth (idealized as non-rotating). And then the locus $r = 2m$ turns out to be irrelevant. For the Schwarzschild vacuum solution terminates at the surface of the central body, which in general is far beyond the critical value $r = 2m$. Inside the body an entirely different metric takes over, depending somewhat on its equation of state, but in any case *regular* throughout. The Schwarzschild radius $2m$, or equivalently $2GM/c^2$, for the sun turns out to be 2.9 km, whereas the exterior field terminates at $r_\odot \approx 7 \times 10^5$ km. So the sun would have to shrink by a linear factor of the order of 10^5 for its Schwarzschild radius to become relevant! For the earth, the Schwarzschild radius is 0.88 cm, and for a proton it is 2.4×10^{-52} cm.

11.3 The geometry of the Schwarzschild lattice

In order to apply the Schwarzschild metric to physical problems it is important to understand its geometry. One way to visualize any curved spherically symmetric

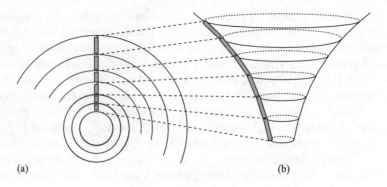

Fig. 11.1

3-space like that of the Schwarzschild lattice, whose metric is given by

$$dl^2 = \frac{dr^2}{(1 - 2m/r)} + r^2(d\theta^2 + \sin^2\theta \, d\phi^2), \qquad (11.17)$$

is to pretend that it is really flat (that is, E^3), but that rulers in it behave strangely. In the case of (11.17), we pretend that by some action of the central mass little *radial* rulers shrink by a factor $(1 - 2m/r)^{1/2}$—as all rulers might shrink if, for example, it were colder near the center. Accordingly we can draw any plane through the origin of this E^3, say the plane $\theta = \pi/2$, and when we measure it out with little rulers its geometry is curved. Moreover, by the spherical symmetry, *all* such central planes are identically curved. Figure 11.1(a) shows a typical one. The little rulers shown along one radial direction are intrinsically all of the same length. This section therefore has the same ruler geometry as the surface of revolution (actually a paraboloid) shown in Fig. 11.1(b), where the rulers are undisturbed and the circles have the same circumference as in (a). [(a) can also be regarded as a view of (b) 'from the top'.] But note how much easier it is to visualize a family of plane central sections with equally distorted rulers, than a family of paraboloids that can be rotated into each other!

The central sections here discussed, of which the loci $\theta = \pi/2$ or $\phi = $ const are typical ones, must a priori be symmetry surfaces (cf. Section 9.5) of both the full Schwarzschild spacetime (11.13) and of its 3-dimensional lattice. Formally this is most easily established for the 'plane' $\phi = 0$ since $\phi \mapsto -\phi$ leaves both the full and the spatial metric invariant; and by a coordinate rotation any other central section can be isometrically transformed into this one.

The way to determine the exact surface of revolution isometric to the central section $\theta = \pi/2$ [cf. Fig.11.1(b)] is to compare the metric of that section, $dl^2 = (1 - 2m/r)^{-1} dr^2 + r^2 \, d\phi^2$, with the general metric of a surface of revolution, $dl^2 = (z'^2 + 1) \, dr^2 + r^2 \, d\phi^2$, where $z = z(r)(r^2 = x^2 + y^2)$ is the curve that, when rotated

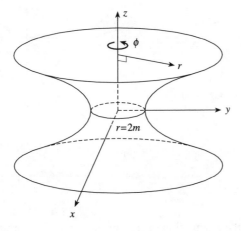

Fig. 11.2

about the z-axis, generates the surface. In this way one finds the generating curve

$$z^2 = 8m(r - 2m), \qquad (11.18)$$

and so the surface is a paraboloid, usually referred to as *Flamm's paraboloid* (cf. Fig.11.2). Of course, only half of this paraboloid (the upper or the lower) is relevant for our present purposes, and of *that* only the part corresponding to r-values greater than the radius of the central body. The curvature of the paraboloid can be computed to be $-m/r^3$ at coordinate r (cf. Exercise 10.15). At the surface of the earth this is $-2 \times 10^{-27}\,\mathrm{cm}^{-2}$ and at the surface of the sun it is $-4 \times 10^{-28}\,\mathrm{cm}^{-2}$. Of course, it is in the nature of paraboloids to flatten out at large distances from the axis and become effectively plane.

11.4 Contributions of the spatial curvature to post-Newtonian effects

(The less geometrically inclined reader can omit this section without loss of continuity.) Minute though it is, the spatial curvature of the Schwarzschild metric contributes significantly to four 'post-Newtonian' effects of GR, namely the bending of light around the sun, the Shapiro light echo delay for sun-grazing radar signals to Venus, the advance of the perihelia of the planets, and the de Sitter geodetic precession. If one calculates the geodesics of (11.13) with simply dr^2 in place of $dr^2/(1 - 2m/r)$, one gets only two-thirds of the advance of the perihelia, one-half of the bending of light and of the echo delay, and only one-third of the geodetic precession. We shall here show directly how the spatial geometry makes its contributions.

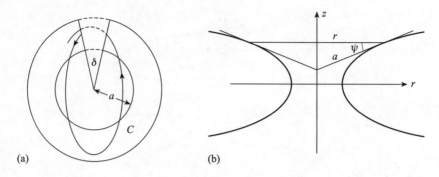

Fig. 11.3

Suppose that on the assumption of flat-space geometry an orbit is nearly circular, with mean radius a, and possibly with some perihelion advance, like the curve C in Fig. 11.3(a). To first approximation, the 'plane' of the orbit is really the tangent cone to Flamm's paraboloid at radius $r = a \cos \psi$ [see Fig. 11.3(b)], where the small angle ψ is given by

$$\psi \approx \frac{dz}{dr} = \frac{4m}{z}, \tag{11.19}$$

as we calculate easily from (11.18). To make the plane of the flat-space calculation into this cone, we must cut out of it a wedge of angle δ such that

$$a(2\pi - \delta) = 2\pi r = 2\pi a \cos \psi$$
$$\approx 2\pi a\left(1 - \tfrac{1}{2}\psi^2\right). \tag{11.20}$$

Clearly, δ will be the contribution of the spatial geometry to the perihelion advance. Solving (11.20) and substituting from (11.19) and (11.18) (with $r \gg m$), we get

$$\delta \approx \frac{2\pi m}{r}, \tag{11.21}$$

which is one-third of the full perihelion advance, as we shall see in Section 11.3.

The contribution of the space geometry to the bending of light can be understood in the same kind of way. If a long, thin, rectangular strip of paper, with a straight line drawn down its middle (corresponding to a straight light path in flat space), is glued without wrinkles to the upper half of Flamm's paraboloid, and then viewed from the z-axis at large z, the center-line will appear bent: this is precisely the contribution of the space geometry to the bending of light. If the center-line is already slightly bent relative to the strip, as implied by the EP for a light ray, then it will appear even more bent when applied to the paraboloid. The EP, as we have seen in Section 1.16, predicts the exact curvature of light in the static space *locally*, which corresponds to little tangent-plane elements on Flamm's paraboloid. We have needed the field equations to tell us how these tangent-plane elements are fitted together, namely into a portion of a paraboloid.

As for the signal running time, the space geometry contributes to its lengthening simply by lengthening the track. For consider a string stretched across the top of the paraboloid of Fig. 11.2, approaching the z-axis no closer than $2m$. The space geometry forces the string onto the paraboloid, thereby lengthening it.

The angle δ evaluated in (11.21) and illustrated in Fig. 11.3 is also fairly obviously the contribution of the space geometry to the advance of the axis of a test gyroscope in circular orbit around a mass m at radius r, if the axis lies in the plane of the orbit (or of the projection of the axis onto this plane, otherwise). There is a second contribution to this advance, namely the so-called *Thomas precession*, which is a flat-space phenomenon [cf. Section 9.7 especially eqn (9.34)]. In the present case, by Kepler's third law ($\omega^2 = GM/r^3$), this amounts to $\pi m/r$ numerically. However, for a freely falling gyroscope the retrograde sense is reversed: it is the frame of the field that Thomas-precesses around the gyroscope, which itself is 'free'. The total effect, geometric and Thomas, gives the well-known *de Sitter precession* of $3\pi m/r$, in the same sense as the orbit. (Cf. Section 11.13 below.)

There is one basic fact that is illustrated incidentally by the 'paper-strip' argument we employed two paragraphs up: a light signal in general does *not* follow a spatial geodesic in static or stationary fields. For whereas the spatial geodesic corresponds to the center-line of the strip, the light path is bent relative to that line whenever the gravitational field has a component orthogonal to it; that is, the light path has *spatial curvature*, as we deduced from the equivalence principle in Section 1.16.

11.5 Coordinates and measurements

The relation of the coordinates to the time and distance measurements that can actually be performed in a spacetime is always entirely encoded in the metric. For example, the coordinates θ and ϕ in the Schwarzschild metric—and indeed in any spherically symmetric metric like (11.2)—are very simply related to observations. By an argument analogous to that given at the end of Section 9.5, the radii $\theta, \phi = $ const are totally geodesic and hence potential light paths. So if the central body were transparent, and if we could observe from its center, the θ and ϕ of any event would simply be the usual polar angles subtended at that center by the light whereby it is seen. As for the time coordinate t, we already understand its relation to clock time from the general discussion of static fields in Chapter 9: it is the time shown by standard lattice clocks that have been speeded up by a factor $(1 - 2m/r)^{-1/2}$.

What about the distance between two widely separated events? The primary definition of distance in a static (or stationary) lattice is the integral of ruler distance, $\int dl$, along the shortest geodesic joining two points. But in astronomical applications of GR, where ruler measurements are neither practicable nor relevant, other methods of determining distances must be used. These include radar distance, distance from apparent size or luminosity, and distance by parallax. Classically all these methods would be equivalent, but in curved spacetime they are not, except in the local limit.

As an illustration, we briefly compare radial coordinate distance, ruler distance and radar distance in the Schwarzschild metric. Along any radius $\theta, \phi = \text{const}$ we have, from (11.13),

$$dl = \left(1 - \frac{2m}{r}\right)^{-1/2} dr \approx \left(1 + \frac{m}{r}\right) dr$$

whence

$$\int_{r_1}^{r_2} dl \approx r_2 - r_1 + m \ln \frac{r_2}{r_1}. \tag{11.22}$$

In the sun's field the excess $m\ln(r_2/r_1)$ of ruler distance over coordinate distance from the sun's surface to earth is only about 8 km, a discrepancy of one part in 2×10^7.

Radar distance R is defined as cT, $2T$ being the proper time elapsed at the observer between emission and reception of a radio echo. A radial signal satisfies $\mathbf{ds}^2 = 0$ and thus, by (11.13) (now *with c*),

$$c\,dt = \pm\left(1 - \frac{2m}{r}\right)^{-1} dr \approx \pm\left(1 + \frac{2m}{r}\right) dr, \tag{11.23}$$

the two signs corresponding to the two possible directions of motion. Converting from coordinate to proper time at the location of the observer, say r_1, by multiplying by the time dilation factor $(1 - 2m/r_1)^{1/2} \approx 1 - m/r_1$, we have

$$R \approx \left(1 - \frac{m}{r_1}\right) \int_{r_1}^{r_2} c\,dt \approx \left(1 - \frac{m}{r_1}\right)\left(r_2 - r_1 + 2m \ln\frac{r_2}{r_1}\right). \tag{11.24}$$

Returning to the observer at the center of a transparent central body, we note that the 'distance from apparent size' of a small object, namely the square root of the ratio of its cross-sectional area to the solid angle at which it is seen, coincides with the coordinate r. For, since the entire sphere at coordinate r has area $4\pi r^2$, an nth part of that sphere has area $4\pi r^2/n$ and is seen at solid angle $4\pi/n$.

11.6　The gravitational frequency shift

According to our general discussion of static metrics in Chapter 9 and, in particular, using eqn (9.4), we can read off directly from the Schwarzschild metric (11.13) the frequency shift between any two of its lattice points A and B:

$$\frac{\nu_B}{\nu_A} = \left(\frac{1 - 2m/r_A}{1 - 2m/r_B}\right)^{1/2}. \tag{11.25}$$

Equivalently, as we have discussed in Section 1.16, this formula also gives the gravitational time dilation. It is an exact result. Its validity in lowest order,

$$\frac{\nu_B}{\nu_A} = 1 - \frac{m}{r_A} + \frac{m}{r_B} + \cdots,$$

is simply a consequence of the equivalence principle and Newton's theory, as we have seen in Section 1.16. Its experimental verification at that level, which by now is very satisfactory (cf. Section 1.16), is therefore a test of the equivalence principle but not of 'full' GR. Full GR is distinguished among all possible gravitational theories that accept the EP by its specific field equations. It is these that determine the exact coefficient of dt^2 in the Schwarzschild metric and hence (11.25). Only a second-order verification of that formula would provide specific support for GR, but that, unfortunately, is presently unattainable.

11.7   Isotropic metric and Shapiro time delay

By a simple change of the radial coordinate (cf. Exercise 11.3) we can recast the Schwarzschild metric (11.13) into the following so-called *isotropic form*:

$$\mathbf{ds}^2 = \frac{(1 - m/2\bar{r})^2}{(1 + m/2\bar{r})^2}\, dt^2 - (1 + m/2\bar{r})^4 (dx^2 + dx^2 + dz^2), \qquad (11.26)$$

where $\bar{r}^2 = x^2 + y^2 + z^2$. Suppose we now consider a light signal passing a spherical mass m at distance R, as in Fig. 11.8 below. In first approximation we can take its path as given by $y = R, z = 0, \bar{r} = \sqrt{x^2 + R^2}$. Then if, in (11.26), we neglect $O(m^2/\bar{r}^2)$ and set $\mathbf{ds}^2 = 0$ for our signal, we find

$$dt \approx \pm\left(1 + \frac{2m}{\bar{r}}\right) dx = \pm\left(1 + \frac{2m}{\sqrt{X^2 + R^2}}\right) dx.$$

Integrating this equation between 0 and X gives us the coordinate time Δt for the signal to travel from the point of closest approach to $x = X$ (or vice versa):

$$\Delta t \approx X + 2m \log \frac{X + \sqrt{X^2 + R^2}}{R} \approx X + 2m \log \frac{2X}{R},$$

where for the last approximation we assumed $R^2/X^2 \ll 1$. So if the signal actually travels from X_1 on one side to X_2 on the other side, the total coordinate time Δt_{12} is given by

$$\Delta t_{12} \approx (X_1 + X_2) + 2m \log \frac{4X_1 X_2}{R^2}.$$

The logarithmic term is the Shapiro time delay. It arises from both the spatial and temporal coefficients in the metric and can thus serve as a true test of the GR field equations, as was first proposed by Shapiro in 1964. If one measures the radar distance to Mercury or Venus or even to an artificial satellite for a period that includes a close conjunction with the sun, that distance will show a sharp temporary increase at conjunction owing to the time-delay term. For example, in the case of Mercury this can amount to some 66 extra kilometers.

First verifications to an accuracy of ~20 per cent (using Mercury and Venus) were announced by Shapiro in 1968 and to ~5 per cent in 1971. Use of transponders on the Viking spacecraft in orbit around Mars and on the ground on Mars enabled Reasenberg and Shapiro to increase the accuracy of the test to 0.1 per cent by 1979.

11.8   Particle orbits in Schwarzschild space

The variety of qualitatively different orbits in Schwarzschild space is much larger than in the case of a Newtonian central field, where all the orbits are conics. In Schwarzschild space, for example, there are *spiraling* orbits. While a detailed discussion of all possible orbits is beyond our present scope, we shall nevertheless provide an overview.

Recall that the 'plane' $\theta = \pi/2$ (and every other such central plane obtainable from this one by a rotation) is a symmetry surface. (The metric is invariant under a reflection in it: $\theta \mapsto \pi - \theta$.) Now, every possible initial direction of an orbit lies in precisely one such plane. [The viewpoint of flat space and shrunk rulers is very useful here— cf. after (11.17).] But since a symmetry surface is totally geodesic, a geodesic that starts out in such a plane must remain in it, which shows that *all orbits are plane*. Without loss of generality we can take $\theta = \pi/2$ as the plane of our orbits, and work out the geodesics of the full space as the geodesics of this totally geodesic subspace, which has the metric

$$\mathbf{ds}^2 = \alpha \, dt^2 - \alpha^{-1} \, dr^2 - r^2 \, d\phi^2, \qquad \alpha = 1 - \frac{2m}{r}. \tag{11.27}$$

The Lagrangian $\mathscr{L}$ [cf. (10.9)] is then given by

$$\mathscr{L} = \alpha \dot{t}^2 - \alpha^{-1}\dot{r}^2 - r^2\dot{\phi}^2, \tag{11.28}$$

where $\dot{} \equiv d/ds$. This Lagrangian is independent of t and ϕ, whence the corresponding Euler–Lagrange equations (10.11) have two immediate first integrals:

$$\alpha \dot{t} = k, \tag{11.29}$$

$$r^2\dot{\phi} = h. \tag{11.30}$$

The third Euler–Lagrange equation reads

$$(\alpha^{-1}2\dot{r})^{\cdot} - \left\{ \frac{2m}{r^2}\dot{t}^2 - \frac{\partial}{\partial r}(\alpha^{-1})\dot{r}^2 - 2r\dot{\phi}^2 \right\} = 0. \tag{11.31}$$

We have exhibited this last equation for *one* purpose only, namely to extract from it the circular orbits, $r = $ const. With $\dot{r} \equiv 0$, this equation yields (and the preceding two equations permit) the relation

$$\omega^2 = \frac{m}{r^3}, \tag{11.32}$$

where $\omega = \mathrm{d}\phi/\mathrm{d}t$. This is Kepler's third law! Of course, such an exact correspondence is a mere accident of our choice of t- and r-coordinates. Substitution of (11.32) into (11.27) yields

$$\mathbf{ds}^2 = \left(1 - \frac{3m}{r}\right)\mathrm{d}t^2, \tag{11.33}$$

which shows that for massive particles we must have $r > 3m$. The orbit $r = 3m$ evidently corresponds to a circular *light* path!

For *all* orbits, the eqns (11.29) and (11.30) are the relativistic analogs of the Newtonian equations of conservation of (kinetic plus potential) energy and angular momentum, respectively. In the case of eqn (11.30), the analogy is obvious, and we speak of h as the *specific* angular momentum (angular momentum per unit rest-mass). To see the connection of (11.29) with energy, suppose the orbit goes to, or comes from, infinity. At infinity $\alpha = 1$, Schwarzschild space reduces to Minkowski space, and $\dot{t} = \mathrm{d}t/\mathrm{d}s = \gamma$, the Lorentz factor. So k is then the specific *total* energy of the particle (remember we are working in units that make $c = 1$). When the orbit does *not* include infinity we can *define* k to be the total energy. (Cf. Exercises 11.10 and 11.11.)

As we have remarked after eqn (10.42), a solution of $N - 1$ of the Euler–Lagrange equations automatically satisfies the Nth, unless that solution includes $x^N = \mathrm{const}$. Since we have already dealt with $r = \mathrm{const}$, we can now forget about eqn (11.31). Instead, we use the metric itself. For the moment we are interested in timelike orbits and so we can set $\mathbf{ds}^2 = \mathrm{d}s^2$ in (11.27) and $\mathscr{L} = 1$ in (11.28). If we then multiply by α, substitute from (11.29) and (11.30) for $\dot{t}$ and $\dot{\phi}$ and rearrange terms, we can obtain

$$\dot{r}^2 = k^2 - \frac{1}{r^3}(r - 2m)(r^2 + h^2) =: k^2 - V(r), \tag{11.34}$$

where the last equation defines the 'effective potential' $V(r)$, in analogy to 1-dimensional motion. [Like the first integrals (11.29) and (11.30), the equation obtained in this way from the metric is always free of second derivatives!] Note that $V(r)$ contains the angular momentum h as a parameter. Figure 11.4(a) shows a typical graph of $V(r)$ for some specific choice of h.

For comparison, let us consider the Newtonian central field. The equations for energy and momentum conservation now read (with $G = 1$)

$$\frac{1}{2}(\dot{r}^2 + r^2\dot{\phi}^2) - \frac{m}{r} = E, \tag{11.35}$$

$$r^2\dot{\phi} = h, \tag{11.36}$$

where $E = $ specific energy and the overdot here denotes $\mathrm{d}/\mathrm{d}t$. Elimination of $\dot{\phi}$ between these two equations gives us the analog of eqn (11.34):

$$\dot{r}^2 = 2E - \frac{1}{r^2}(h^2 - 2mr) =: 2E - V_N(r), \tag{11.37}$$

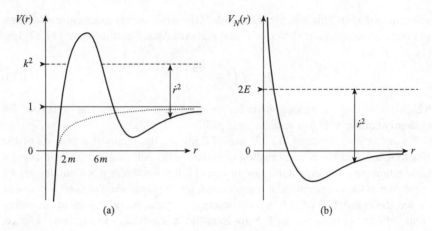

Fig. 11.4

where the last equation defines the effective Newtonian potential $V_N(r)$. (This is actually double of what one would normally call the potential.) Figure 11.4(b) shows a typical graph of this function.

Though the graphs in Fig. 11.4 do not tell the whole story, there is much to be learned from them. A help in their interpretation is the equation we get from (11.34) [and similarly from (11.37)] by differentiating with respect to s and then canceling $\dot{r}$:

$$2\ddot{r} = -\frac{dV}{dr}. \tag{11.38}$$

Now, a particular orbit is determined by a coice of h and k. The former determines the exact form of $V(r)$. Let us draw a horizontal line [dashed in Fig. 11.4(a)] at height k^2. The depth of $V(r)$ below this line is $\dot{r}^2$. So only those portions of such a dashed line that lie *above* $V(r)$ correspond to possible orbits. Consider a line like the one marked $2E$ in Fig. 11.4(b). This corresponds to a scattering orbit: as r comes in from infinity, the radial velocity component $|\dot{r}|$ increases until r passes the bottom of the potential well; then it decreases to zero as r reaches the intersection point of $V(r)$ and $2E$, at which $\ddot{r}$ is positive by (11.38), and so r has a minimum there, turns around, and goes back to infinity. Analogous orbits are possible also in the relativistic case. Here, however, with the *same* energy and angular momentum we also have a bound orbit (to the left of the potential hump) that spirals out of $r = 2m$, reaches a maximum value of r, and then spirals back into the horizon.

As this assertion implies, our discussion is, in fact, also applicable to orbits around a complete Schwarzschild white hole–black hole, where the vacuum extends through the horizon $r = 2m$. Particles can then come out of the horizon whilst it is still a white hole and fall back in again when it has meanwhile become a black hole, but not vice versa. (Chapter 12 will clarify these remarks!)

One thing that can *not* be read off easily from the potential diagram is the angular behavior (although we know $\dot{\phi} = h/r^2$.) For example, we cannot tell that the

Newtonian scatter orbit is a hyperbola (a parabola if $E = 0$) or that the oscillating orbits (corresponding to energy horizontals within the potential well) are ellipses in the Newtonian case and precessing pseudo-ellipses in the relativistic case.

The bottom point of the potential well corresponds to a stable circular orbit, since any disturbance just modulates it into a nearby ellipse (in Newton's theory) or pseudo-ellipse (in GR). But the top of the potential hump in the relativistic case corresponds to an *unstable* circular orbit; any disturbance produces something like the time-reversal of the orbits corresponding to the horizontal tangent: spiraling into the circular orbit from below or from above. For very small h ($h < 4m$) the top of the hump dips below the horizontal asymptote $V = 1$, which enables particles coming out of the horizon to spiral out to arbitrarily large distances before turning round and reversing course.

For the minimum point of $V_N(r)$ [that is, for the zero of $(d/dr)V_N(r)$], we find

$$r = h^2/m, \tag{11.39}$$

and so there *is* a minimum (horizontal tangent) whenever $h \neq 0$. Not so in the relativistic case. Here the condition for $(d/dr)V(r) = 0$ is found to be

$$r = \frac{h^2 \pm \sqrt{h^4 - 12m^2h^2}}{2m}, \tag{11.40}$$

and this has no solution if $h < 2\sqrt{3}m = 3.464\,m$. In that case the graph of $V(r)$ looks like the dotted curve in Fig. 11.4(a), and there is no 'centrifugal barrier' at any energy level: every incoming particle falls into the black hole, or hits the central body if there is one.

Note from (11.40) that the smallest r-value for a minimum (and the largest for a maximum) is $r = 6\,m$ so this is the lower bound for stable circular orbits, and the upper bound for unstable ones.

11.9   The precession of Mercury's orbit

Our next task is to examine in detail the changes that relativity has wrought in planetary orbit theory. Minute though these changes are in the case of our solar system, one such change in particular, the precession of the orbit of Mercury, holds an important place in the history of the acceptance of general relativity (cf. Section 1.14).

Again, for comparison purposes, we begin with the Newtonian orbits, where the basic equations are (11.35)–(11.37). Suppose we are now interested only in the *shape* of the orbit and not in its temporal development. Then we could eliminate t from (11.36) and (11.37) by forming $\dot{r}^2/\dot{\phi}^2 = (dr/d\phi)^2$. However, in practice it has been found convenient here to work with the reciprocal of r,

$$u := \frac{1}{r}. \tag{11.41}$$

If we convert eqn (11.37) into an equation in u, and then eliminate t between *that* equation and (11.36), we easily find

$$\left(\frac{\mathrm{d}u}{\mathrm{d}\phi}\right)^2 + u^2 = \frac{2mu}{h^2} + \frac{2E}{h^2}. \tag{11.42}$$

It turns out that the best way to integrate this differential equation is to differentiate it first, so as to get rid of the squared derivative. If we do this, and cancel a factor $2\mathrm{d}u/\mathrm{d}\phi$, we get the simple differential equation

$$\frac{\mathrm{d}^2u}{\mathrm{d}\phi^2} + u = \frac{m}{h^2}. \tag{11.43}$$

Its general solution is of the form

$$u = \frac{m}{h^2}(1 + e\cos\phi), \tag{11.44}$$

apart from the freedom $\phi \mapsto \phi + \text{const}$. Equation (11.44) represents a conic of eccentricity e with focus at the origin of r, and hence all Newtonian orbits are such conics. In the case of the planets, the orbits are ellipses ($e < 1$).

If we now apply the same successful procedure (converting to $u = 1/r$ and forming $\dot{u}^2/\dot{\phi}^2$) to the relativistic equations (11.34) and (11.30), we are led to the following differential equation:

$$\frac{\mathrm{d}^2u}{\mathrm{d}\phi^2} + u = \frac{m}{h^2} + 3mu^2. \tag{11.45}$$

The extra term on the RHS is the obvious cause for the existence of all the non-classical orbits. But in the case of the solar planets, it merely acts as a minute 'correction' term. For consider the ratio $3mu^2 : m/h^2$ of the two terms on the RHS. In full units this reads

$$\frac{3}{r^2} : \frac{c^2}{r^2v^2} \approx 3\frac{v^2}{c^2}, \tag{11.46}$$

where v is the orbital velocity, making $h \approx rv$. For Mercury, the fastest of the planets, we have $3v^2/c^2 \approx 10^{-7}$. This relative smallness of the correction term allows us to substitute for the u in it the Newtonian value (11.44). With that, eqn (11.45) reads

$$\frac{\mathrm{d}^2u}{\mathrm{d}\phi^2} + u = \frac{m}{h^2} + \frac{3m^3}{h^4}(1 + 2e\cos\phi + e^2\cos^2\phi). \tag{11.47}$$

Linear differential equations of this type are solved by adding to the general solution of the equation with only some or none of the terms on the RHS, *particular integrals* corresponding to the remaining terms, taken one at a time [as exemplified by eqns (11.43) and (11.44)]. So we need particular integrals of the following three types of equations,

$$\frac{\mathrm{d}^2u}{\mathrm{d}\phi^2} + u = A, \quad = A\cos\phi, \quad = A\cos^2\phi,$$

where each A is, in fact, a constant of order m^3/h^4. These must then be added to (11.44). As can be verified easily, such particular integrals are, respectively,

$$u = A, \quad = \tfrac{1}{2} A\phi \sin \phi, \quad = \tfrac{1}{2} A - \tfrac{1}{6} A \cos 2\phi. \tag{11.48}$$

Of these, the first simply adds a minute constant to the Newtonian solution (11.44), while the third adds a minute constant and a periodic 'wiggle', all quite unobservable. But the *second* adds something that does *not* have period 2π and which is, in fact, responsible for an ultimately observable precession. So we take as our approximative solution of (11.47) the following:

$$
\begin{aligned}
u &= \frac{m}{h^2}\left(1 + e\cos\phi + \frac{3m^2}{h^2}e\phi\sin\phi\right) \\
&\approx \frac{m}{h^2}\left[1 + e\cos\left(1 - \frac{3m^2}{h^2}\right)\phi\right],
\end{aligned}
\tag{11.49}
$$

where we have used the formula $\cos(\alpha - \beta) = \cos\alpha\cos\beta + \sin\alpha\sin\beta$, and the approximations $\cos\beta \approx 1$, $\sin\beta \approx \beta$ for a small angle β. The equation shows u (and therefore r) to be a periodic function of ϕ with period

$$\frac{2\pi}{1 - 3m^2/h^2} > 2\pi. \tag{11.50}$$

Thus the values of r, which of course trace out an approximate ellipse, do not begin to repeat until somewhat *after* the radius vector has made a complete revolution. Hence the orbit can be regarded as an ellipse that rotates ('precesses') about one of its foci (see Fig. 11.5) by an amount

$$\Delta = \frac{2\pi}{1 - 3m^2/h^2} - 2\pi \approx \frac{6\pi m^2}{h^2} \approx \frac{6\pi m}{a(1 - e^2)} \tag{11.51}$$

per revolution. Here we have used the Newtonian relation

$$\frac{2m}{h^2} = \frac{1}{r_1} + \frac{1}{r_2} = \frac{2}{a(1 - e^2)}, \tag{11.52}$$

which follows from (11.44) on setting $\phi = 0, \pi$; a is the semi-major-axis, and r_1, r_2 are the maximum and minimum values $a(1 \pm e)$ of r. This Δ is the famous Einsteinian advance of the perihelion.

From our approximative treatment it is by no means obvious that the relativistic quasi-elliptic orbits are strictly periodic, which in fact they are. But for bound orbits the *exact* solution of the GR analog to the Newtonian equation (11.42) [which we did not exhibit but nevertheless used en route to (11.45)] is an elliptic function $u(\phi)$ with real period. Alternatively, we can read off the strict periodicity from the potential diagram, Fig. 11.4(a), where the orbits in question correspond to k^2-levels within the

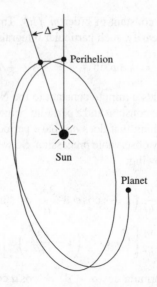

Fig. 11.5

potential well. Thus after each cycle r and $\dot{r}$ return to their 'initial' values $r_{\max}$ and zero, which, together with the constant values of h and k, uniquely determine the orbit in space and time; so each cycle repeats exactly.

If we apply the procedure that led to (11.51) to the flat-space metric (9.6), which was based solely on the equivalence principle, we get only two-thirds of the full GR precession. The missing third comes from the spatial geometry, as we have already indicated in Section 11.4 at least for nearly circular orbits.

The accuracy with which planetary perihelion shifts can be observed depends not only on the shift per orbit, but also on the frequency of revolution (which multiplies the effect) and on the ellipticity of the orbit, which sharpens the definition of the perihelion. The case of Mercury is by far the most favorable. Its actually observed precession, relative to the Newtonian local inertial frame of the solar system, is $\sim574''$ per terrestrial century. All but $\sim43''$ of this can be accounted for by the action of the other planets (for example, $277''$ from Venus, $153''$ from Jupiter, etc.) The best modern value for the irreducible residue is $42''.98 \pm 0.04$. And $42''.98$ is also the prediction of general relativity!

Of course, all the other planets precess too, and can be expected to have similar irreducible conflicts with Newtonian theory. With modern observational and computational techniques it has been possible to determine these residues for Venus, Earth, and Mars, and they are found to agree with the GR predictions ($8''.62$, $3''.84$, $1''.35$, respectively) to better than 1 per cent.

Although our analysis is not directly applicable to close binary star systems (some of which have pronounced eccentricities and periods of as little as a few hours) it can at least give us some order-of-magnitude estimates of what to expect. For example, for a system consisting of two solar masses orbiting each other one solar diameter apart (whose period is about five hours), eqn (11.51) with $m = 2M_\odot$ suggests an orbital precession of $\sim 4 \times 10^{-5}$ radians per revolution, which adds up to $\sim 4°$ per year. Indeed, the famous 'binary pulsar' discovered by Hulse and Taylor in 1974 and intensively studied ever since (it has been called an extraterrestrial laboratory for general relativity!) has an orbital period of ~ 8 h and a whopping orbital precession of $4°.22663$ per year. If one uses the appropriate analysis, the GR expression for the precession turns out to be a function of the total mass M, which in the case of the binary pulsar was unknown a priori. The incredibly accurate determination of the precession therefore was used to fix M, and time-dilation measurements on the period of the pulsar (as its speed and distance from the companion vary) then allowed the determination of the separate masses ($1.44\ M_\odot$ for the pulsar, $1.39\ M_\odot$ for the companion.) This, in turn, allowed a prediction to be made of the rate of energy loss due to gravitational radiation. And the prediction was verified precisely by the observed minute continuous increase in the orbital frequency. This constitutes the strongest support available so far for the general-relativistic theory of gravitational radiation.

11.10   Photon orbits

As we have seen in Section 10.2, photon orbits (null geodesics) satisfy the same geodesic differential equations as ordinary particles, except that we must use an affine parameter instead of the arc length, and set $\mathbf{ds}^2 = 0$. In this way we obtain from the metric (11.27) the following three equations in analogy to (11.29), (11.30), and (11.34):

$$\alpha \dot{t} = \bar{k} \tag{11.53}$$

$$r^2 \dot{\phi} = \bar{h} \tag{11.54}$$

$$\dot{r}^2 = \bar{k}^2 - \frac{\bar{h}^2}{r^2}\left(1 - \frac{2m}{r}\right) =: \bar{k}^2 - V_P(r), \tag{11.55}$$

where now the overdot denotes differentiation with respect to the affine parameter, and where the last equation defines $V_P(r)$, the effective photon potential. [In the Newtonian case, inasmuch as one can speak of photon orbits as those satisfying $v \approx c$ ($v = c$ is impossible to maintain), the potential is simply a special case of that defined in (11.37) and illustrated in Fig. 11.4(b).] As in the case of ordinary particles, the potential diagram, Fig. 11.6, holds the key to the classification of the orbits. Note that a photon starting in an outward direction ($\dot{r} > 0$) keeps going outward for ever, or until its $\bar{k}^2$-line intersects the potential curve; and similarly for incoming rays. The sign of $\dot{r}$ can change only at those intersection points, and since eqn (11.38) still holds, now for V_P, those intersection points correspond to maxima or minima of r.

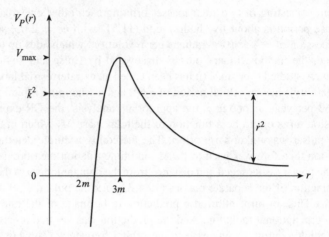

Fig. 11.6

On differentiating $V_P(r)$ we find

$$V_{\max} = \frac{\bar{h}^2}{27m^2} \tag{11.56}$$

at $r = 3m$. The circular light path corresponding to $r \equiv 3m$, which we already discovered after (11.33), is evidently unstable; a slight increase in r causes it to spiral out to infinity, a slight decrease and it will spiral into the horizon (or the central mass). $V_{\max}$ is evidently the critical value for $\bar{k}^2$: while orbits with greater $\bar{k}^2$ do not encounter the potential barrier, those with lesser $\bar{k}^2$ get reflected at that barrier.

Following Misner, Thorne, and Wheeler,[1] we shall find it of interest to classify the orbits relative to the angle ϑ which a ray 'initially' makes with the radial direction. Directly from the spatial part of the metric (11.27), as applied to an infinitesimal portion of the orbit [see Fig. 11.7(a)], we find

$$\sin^2 \vartheta = \frac{r^2\,d\phi^2}{(\alpha^{-1}\,dr^2 + r^2\,d\phi^2)} = \alpha\left(\alpha + \frac{1}{r^2}\frac{dr^2}{d\phi^2}\right)^{-1}. \tag{11.57}$$

On the other hand, from (11.54),

$$\dot{r}^2 = \frac{dr^2}{d\phi^2}\dot{\phi}^2 = \frac{dr^2}{d\phi^2}\frac{\bar{h}^2}{r^4}. \tag{11.58}$$

In conjunction with (11.55) this yields

$$\frac{1}{r^2}\frac{dr^2}{d\phi^2} = \frac{r^2}{\bar{h}^2}\dot{r}^2 = \frac{\bar{k}^2}{\bar{h}^2}r^2 - \alpha, \tag{11.59}$$

---

[1] C. W. Misner, K. S. Thorne, J. A. Wheeler, *Gravitation*, W. H. Freeman, San Francisco, 1973, p. 675.

(a)

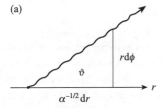

(b)

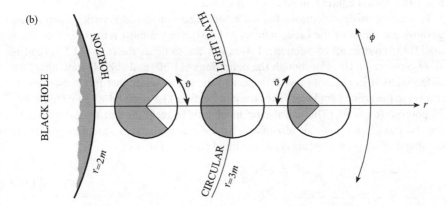

Fig. 11.7

whence, by (11.57),

$$\sin^2 \vartheta = \frac{\alpha}{r^2} \frac{\bar{h}^2}{\bar{k}^2}. \tag{11.60}$$

And this formula applies equally to the angle between a direction and the outward *or* the inward radius [since $\sin(\pi - \vartheta) = \sin \vartheta$]. We see from Fig. 11.6 that if a photon is emitted at an r-value *greater* than $3m$ in an *outward* direction ($\dot{r} > 0$), it will go to infinity. But if emitted inwards, its fate depends on the angle ϑ which the initial direction makes with the inward radius. For large $\bar{k}^2$-values (small ϑ) it falls into the horizon; for small $\bar{k}^2$-values (large ϑ) it gets scattered. If the photon is emitted at an r-value *less* than $3m$ in an *inward* direction, it will go through the horizon. But if emitted outwardly, it will go to infinity for large $\bar{k}^2$ (small ϑ); for small $\bar{k}^2$ (large ϑ) it will attain a maximum radius, turn back and spiral into the horizon. In both cases the critical $\bar{k}^2$-value is $\bar{k}^2 = V_{\max} = \bar{h}^2/27m^2$, and so, by (11.60), the critical ϑ-value is given by

$$\sin \vartheta = \frac{\alpha^{1/2}}{r} 3\sqrt{3}m. \tag{11.61}$$

Figure 11.7, drawn in the plane of the orbit, illustrates these results. Initial directions within the black sectors correspond to those that will propel the photon into the black hole, directions within the white sectors send the photon to infinity, and the critical directions lead to orbits that spiral onto the circle $r \equiv 3m$. At $r = 3m$ the critical angle is $90°$.

11.11   Deflection of light by a spherical mass

As our discussion in the preceding section showed, only those photon orbits that come within a few multiples of $r = 2m$ show strongly non-Newtonian features, such as capture by the horizon, or by the circular orbit at $r = 3m$. Still, even in such problems as the deflection of starlight by the field of the sun (see Fig. 11.8), though the relativistic scattering orbit is similar to the Newtonian, namely almost a straight line, the actual scattering angle can be doubled.

To analyze this problem, we fall back once again on the relativistic equation that governs the *shape* of the orbit, namely (11.45). But we must remember that the h in (11.45) (and in all of Section 11.8) is *not* the same as the $\bar{h}$ in (11.55) (and in all of Section 11.10). For though the definitions (11.30) and (11.54) *look* alike, the differentiation in (11.30) is with respect to the arc s, while that in (11.54) is with respect to the affine parameter. Thus the h in Section 11.8 must be set equal to infinity for photons ($ds = 0$). (This can also be understood by recalling that h is the relativistic angular momentum per unit *rest*-mass.) So the exact relativistic equation governing the shape of *all* photon orbits is the modified eqn (11.45):

$$\frac{d^2u}{d\phi^2} + u = 3mu^2. \tag{11.62}$$

For solving the present problem we shall rely once more on the smallness of the term on the RHS, this time relative to the u on the LHS. Without it, a suitable solution of (11.62) is the 'straight line' $R/r = \sin\phi$ (cf. Fig. 11.8), or

$$u = \frac{\sin\phi}{R}, \tag{11.63}$$

where R can be regarded as the radius of the sun. Substituting this first approximation to the orbit into the RHS of (11.62) gives

$$\frac{d^2u}{d\phi^2} + u = \frac{3m}{R^2}(1 - \cos^2\phi),$$

of which a particular integral is [cf. (11.48)]

$$u = \frac{3m}{2R^2}\left(1 + \frac{1}{3}\cos 2\phi\right).$$

Adding this to (11.63) yields the second approximation

$$u = \frac{\sin\phi}{R} + \frac{3m}{2R^2}\left(1 + \frac{1}{3}\cos 2\phi\right). \tag{11.64}$$

For large r the first and dominant term on the RHS shows that ϕ is very small, and so $\sin\phi \approx \phi$, $\cos 2\phi \approx 1$. Going to the limit $u \to 0$ in (11.64), we thus find $\phi \to \phi_\infty$ (see Fig. 11.8), where

$$\phi_\infty = -\frac{2m}{R}.$$

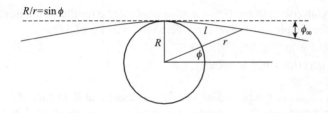

Fig. 11.8

Consequently the magnitude of the total deflection of the ray, by symmetry, is

$$\Theta = \frac{4m}{R} \quad \left(= \frac{4Gm}{c^2 R} \right) \tag{11.65}$$

in radians, the expression in parenthesis being in full units.

For a ray grazing the sun, for example, this amounts to $1''.75$. Some 10 expeditions attempting to observe stars near the sun during a total solar eclipse have been launched since Eddington's historic first in 1919, the latest to Mauritania in 1973 by a team from the University of Texas. In spite of all these efforts it has proved impossible to reduce the 20 per cent uncertainty of Eddington's original observations to much below 10 per cent—within which agreement with Einstein's prediction was indeed found.

But a major step forward came in 1969 with an entirely new method that relied on radio signals and thus did not involve waiting for, and traveling to, a solar eclipse. On its path around the sun the earth each year passes locations where the sun aligns with close configurations of distant radio-emitting quasars. As the earth passes, their relative angular separations change—in perfect accord with Einstein's bending formula. The latest (1991) results by Robertson *et al.*, using Very Long Baseline Interferometry (VLBI), verified Einstein's prediction to a previously undreamt of accuracy of 10^{-4}.

As we mentioned earlier, on the basis of Newtonian theory one gets a deflection of light that is only one-half as big as that predicted by GR. An instructive way to obtain this result is to integrate the *local* bending expression (1.15) we obtained in Chapter 1 (and which applies equally in GR and Newton's theory) over *flat* space. According to (1.15), disregarding the sign, writing l for the arc along the orbit and ψ for its inclination to the initial tangent, we have, for the configuration illustrated in Fig. 11.8,

$$\kappa = \frac{d\psi}{dl} = g \cos\theta \approx \frac{m}{r^2} \cdot \frac{R}{r} \approx \frac{mR}{(l^2 + R^2)^{3/2}}.$$

From this we find

$$\frac{1}{2}\Theta = \int d\psi = \int_0^\infty \frac{mR}{(l^2 + R^2)^{3/2}} \, dl = \left[\frac{ml}{R(l^2 + R^2)^{1/2}} \right]_0^\infty = \frac{m}{R}, \tag{11.66}$$

which bears out our assertion. Of course, for highly relativistic scattering orbits, for example for rays approaching a concentrated mass to within almost $3m$, neither the above relativistic nor the Newtonian approximative results would be expected to apply, nor the simple factor 2 relating them (cf. Exercise 11.14).

11.12   Gravitational lenses

From the analysis of the preceding section it is clear that if a beam of parallel light of sufficient cross-section is directed at a massive object, those rays that do not hit the object but pass around it get converged [as in Fig. 11.9(a)]. This is an example of a 'gravitational lens', so called because (like a glass lens) it converges the light. *But that is also the whole extent of the analogy.* Gravitational lenses do not bring the light to a unique focus, produce neither real nor virtual images, and, in contrast to convex glass lenses, their bending decreases with distance from the center. Even so, as we shall see, they can serve as important observational tools in modern cosmology.

Figure 11.9(a) shows the simplest configuration of gravitational lensing. A point-source lies directly behind a spherical gravitating body, and an observer sees it as a luminous ring (an 'Einstein ring') around the lens. The shortest distance D that an observer can be from the lens to see such a ring occurs when the source is essentially at infinity and the light just grazes the lens. Then, using (11.65), we find

$$D = \frac{R}{\Theta} = \frac{R^2}{4m},$$ (11.67)

where R (the distance of closest approach) is now the radius of the lens, and m is its mass. If the lens is a star like the sun, $\Theta = 1''.75$, $R = R_\odot$, and $D \approx 10^{-2}$ light-years. Observers farther from the lens see the same source by rays that have passed the lens farther out and which have therefore been less bent. So the ring they see has a lesser angular diameter. In fact, from (11.67), that diameter is given by $2\Theta = 4\sqrt{m/D}$.

While such perfect alignment as is needed for the occurrence of an Einstein ring is extremely unlikely with stellar lenses, approximate Einstein rings *have* been observed with whole galaxies acting as lenses and radio galaxies as sources. A more likely situation is illustrated in Fig. 11.9(b), where the observer sees *two* images of the source. ('Image' is used here in the sense of an image on the retina, as when I see a single image of a star in the ordinary way of looking at it, by catching a bundle of rays.) One image in that figure is seen by more or less direct viewing, the other by light that has detoured around the lens.

Thin cones of rays emanating at the source with the same solid angle may well arrive at the observer from different directions with *different* cross-sections. Accordingly the two images may have different brightnesses. Occasionally a lensed image is thus greatly magnified. Another type of magnification occurs when the apparent diameter of an extended source is stretched; that is, when the lensed cone of rays from the rim of the source to the eye subtends a greater solid angle than if the source were viewed directly. In fact, the two types of magnification go hand in hand, by a classical

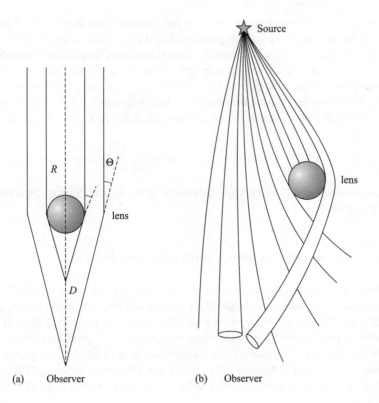

Fig. 11.9

theorem of Etherington [cf. after (17.20) below.] Some of the most distant known galaxies were observed only because foreground clusters magnified their images.

With point-like lenses (because the light paths are plane) there can only be *two* paths (except when the alignment is perfect) for the light from a source to us. But with *extended transparent* lenses, such as galaxies or clusters of galaxies, there can be more than two images, depending on the exact configuration; and at least one of them will be magnified ('magnification theorem').

The way multiple images can be identified as such is by comparing their redshifts (which must be the same), their spectra, and also their light curves, if available. Fortunately, quasars undergo irregular but rapid changes in luminosity. So if two images display the same redshift and congruent light curves (possibly displaced by a constant time interval Δt of a few years corresponding to the difference in the optical path length) they can be assumed to represent the same source. The number of confirmed cases of double or multiple quasar images grows yearly and lies in the dozens.

Lensing, when perfected (though preliminary results exist already), should allow among other things direct determinations of cosmological distances and thus of the

Hubble expansion parameter, as well as independent (non-dynamical) determinations of the masses of the lensing galaxies. We shall indicate without proof (but see Exercise 11.16), and only in principle, how this works. Neglecting the expansion of the universe, any possible non-sphericity of the lens, and the Shapiro effect, and assuming the source to be essentially at infinity, one can quite easily obtain the following two formulae relating the (small) angular distances θ_1, θ_2 of the images from the center of the lens, the mass m of the lens, our distance D from it, and the signal separation Δt:

$$\theta_1\theta_2 = \frac{4m}{D}, \qquad D = \frac{\Delta t}{|\theta_1^2 - \theta_2^2|} \qquad (11.68)$$

(in units making $c = G = 1$). Thus, by observing θ_1, θ_2, and Δt, we can in principle find D and m.

11.13   De Sitter precession via rotating coordinates

De Sitter precession (also known as *geodetic precession*) is a general-relativistic phenomenon without classical equivalent. Its mathematical existence was discovered as early as 1916 by de Sitter, who found that a gyroscope in free (and therefore 'geodetic') orbit around a massive body will precess. We shall here re-derive this result (for the special case of circular orbits) by the method of rotating coordinates that we already used in Section 9.7 to derive the Thomas precession. (This method is good for finding circular geodesics and gyroscopic precession in all axisymmetric spacetimes.[2])

As in Section 9.7, but now in the Schwarzschild rather than the Minkowski metric, we introduce a uniformly rotating coordinate system relative to which Schwarzschild spacetime is no longer static but still stationary. Writing ϕ' for the original Schwarzschild coordinate ϕ in (11.13) and introducing a new angular coordinate ϕ measured from a fiducial half-plane that rotates about the 'vertical' with constant angular velocity ω (cf. Fig. 9.3), we have

$$\phi = \phi' - \omega t, \qquad d\phi' = d\theta + \omega\, dt, \qquad (11.69)$$

and consequently—after setting $\theta = \pi/2$ for a typical orbital plane, and doing a little algebra,

$$\mathbf{ds}^2 = \left(1 - \frac{2m}{r} - r^2\omega^2\right)\left(dt - \frac{r^2\omega}{1 - (2m/r) - r^2\omega^2}\, d\phi\right)^2$$
$$- \left(1 - \frac{2m}{r}\right)^{-1} dr^2 - \frac{r^2 - 2mr}{1 - (2m/r) - r^2\omega^2}\, d\phi^2. \qquad (11.70)$$

---

[2] Cf. W. Rindler and V. Perlick, *Gen. Rel. and Grav.* **22**, 1067 (1990).

This is a stationary metric whose lattice rotates with angular velocity ω relative to the original lattice. The first parenthesis corresponds to the factor $\exp(2\phi/c^2)$ of the canonical form (9.13) (now with $c = 1$). We know from (9.22) that at points where $\Phi_{,i} = 0$ there is *no* gravitational field; so a free particle can remain at rest there, and the worldline of the lattice point is a geodesic. Here this happens where $\omega^2 = m/r^3$. So we have recovered our earlier result (11.32) without solving a differential equation. (This is how rotating coordinates can determine the circular geodesics in *all* axisymmetric metrics.)

The geodetic precession on a circular orbit can be obtained by applying formula (9.23), for the rotation rate of a gyrocompass fixed in the lattice, to the metric (11.70) at a freely orbiting lattice point. Comparing (11.70) with the canonical form (9.13), and using $\omega^2 = m/r^3$ where convenient, we find

$$e^{2\Phi} = 1 - \frac{3m}{r},$$

$$\omega_3 = r^2\omega\left(1 - \frac{2m}{r} - r^2 w^2\right)^{-1}, \qquad \omega_{3,1} = 2r\omega\left(1 - \frac{3m}{r}\right)^{-1}, \qquad (11.71)$$

$$k^{11} = 1 - \frac{2m}{r}, \qquad k^{33} = \left(1 - \frac{3m}{r}\right)\left(1 - \frac{2m}{r}\right)^{-1} r^{-2},$$

with the indices 1,2,3 referring to r, θ, ϕ, respectively. Equation (9.21) shows that $-\Omega$, the angular velocity of the gyrocompass relative to the rotating lattice, in the present case points in the negative direction relative to the orbit. (This is also clear physically, since in Newton's theory the gyrocompass would not precess at all relative to the *original* lattice.) For the magnitude of Ω we substitute from (11.71) into (9.23) and find

$$\Omega = e^{\Phi}(k^{11}k^{33}w_{3,1}^2)^{1/2} = \omega. \qquad (11.72)$$

(This coincidence of Ω with ω is only apparent, since Ω is a *proper* rotation rate while ω is a *coordinate* rate.) We can now repeat, *mutatis mutandis*, the argument that led from (9.32) to (9.34). Using $\Delta\tau = (1 - 3m/r)^{1/2}\Delta t$ which follows from (11.33), we thus find for the precession of the gyrocompass per orbital revolution, relative to the rotating lattice:

$$\alpha' = -2\pi\left(1 - \frac{3m}{r}\right)^{1/2} \qquad (11.73)$$

without approximation, and therefore, relative to the original lattice:

$$\alpha = -2\pi\left[\left(1 - \frac{3m}{r}\right)^{1/2} - 1\right] \approx \frac{3\pi m}{r} = 3\pi v^2. \qquad (11.74)$$

Because of the positive sign, this precession is in the *same* sense as that in which the orbit is described. Note from (11.73) that, as $r \to 3m$, the precession relative to the rotating lattice ceases altogether and the gyroscope turns a constant face towards the center.

De Sitter had, in fact, discovered the precession now named after him by calculating the motion of the earth–moon 'gyroscope' in its orbit around the sun. The minute theoretical precession of about $0.02''$ per year was hopelessly beyond verification at the time. Nevertheless, it is precisely this instance of de Sitter precession that has been verified recently with modern methods of laser ranging and radio interferometry: by Bertotti *et al.* in 1987 to an accuracy of $\sim$10 per cent, by Shapiro *et al.* in 1988 to $\sim$2 per cent, and by Dicke *et al.* in 1994 to $\sim$1 per cent.

As a footnote, we may mention a mild inconsistency in the term 'geodetic precession': while in GR *non-spinning point* particles unquestionably follow geodesics, spinning particles theoretically follow slightly different orbits, as has been shown by Papapetrou and others. Still, the spin would have to be quite extreme before it could affect our results measurably.

Exercises 11

11.1. Prove that the GR vacuum field equations do not permit the existence of a static *cylindrical* spacetime or portions thereof (that is, nested 2-spheres of equal radius.) Its metric would be of the form

$$\mathbf{ds}^2 = A(z)\, dt^2 - dz^2 - a^2(d\theta^2 + \sin^2\!\theta\, d\phi^2),$$

where z is ruler distance along the generators $\theta, \phi =$ const of the 3-cylinder. [It is really this result—apart from common sense—that allows us to pick the radius r of the nested spheres in (11.2) as a variable.] [*Hint*: There is an important theorem (cf. Exercise 10.13) to the effect that if a metric splits into the sum of two independent metrics of lower-dimensional spaces, the Riemann and Ricci tensors of the full space are simply those of the subspaces. For example, the $R_{z\theta}$ of the above metric automatically vanishes while its $R_{\theta\theta}$ is just the $R_{\theta\theta}$ of the negative 2-sphere metric. Use the Appendix to show $R_{\theta\theta} = -1$. Also note from the Appendix (or directly from the definitions) that the R_{ijkl} of two metrics $\mathbf{ds}^2$ and $k\,\mathbf{ds}^2$ are in the ratio $1 : k$ while the R_{ij} are the same.]

11.2. What is the radial *coordinate* velocity dr/dt of light in Schwarzschild spacetime (11.13)? Note how this decreases as $r \to 2m$. If light is sent radially towards a concentrated spherical mass and reflected by a stationary mirror back to where it came from, what can you say about the total time taken as the mirror approaches $r = 2m$? If a uniformly emitting light-source falls freely along a radius towards and through the horizon at $r = 2m$, what can you say *qualitatively* about its apparent velocity, brightness, and color as observed from the point where it was dropped?

11.3. By introducing a new radial coordinate $\bar{r}$ defined by the relation

$$r = \left(1 + \frac{m}{2\bar{r}}\right)^2 \bar{r},$$

which is strictly monotonic for $r > 2m (\bar{r} > \frac{1}{2}m)$, transform the Schwarzschild metric (11.13) into the following well-known 'isotropic' form:

$$\mathbf{ds}^2 = \frac{(1 - m/2\bar{r})^2}{(1 + m/2\bar{r})^2}\, dt^2 - (1 + m/2\bar{r})^4 [d\bar{r}^2 + \bar{r}^2(d\theta^2 + \sin^2\theta\, d\phi^2)].$$

One can then go from $\bar{r}$, θ, ϕ to x, y, z by the usual transformation from polar to Cartesian coordinates, and so obtain (11.26).

We have seen in Section 11.3 that we arrive at the correct ruler geometry if we pretend that space is flat and that transverse rulers are unaffected by the gravitational field while radial rulers shrink by a factor $(1 - 2m/r)^{1/2}$. Now we see that this was *really* just pretending and not some genuine physical effect: with the isotropic metric we arrive at the correct ruler geometry if we pretend that rulers in *any* direction shrink by a factor $(1 + m/2\bar{r})^{-2}$.

11.4. The frequency-shift formula (9.4) for stationary spacetimes applies when source and observer are at rest in the lattice. If either or both of them move, we can locally apply a previous SR result (cf. Exercise 5.21): The ratio of the frequencies which two momentarily coincident observers with 4-velocities U_1 and U_2 ascribe to a wave train with wave vector $\mathbf{L}$ that passes them is given by $\nu_1/\nu_2 = \mathbf{N} \cdot \mathbf{U}_1/\mathbf{N} \cdot \mathbf{U}_2$, $\mathbf{N}$ being any null vector parallel to $\mathbf{L}$. Use this to show that, if a source of proper frequency ν_A at lattice point A has 4-velocity $\mathbf{U}_A$ and an observer at B has 4-velocity $\mathbf{U}_B$, the observed frequency ν_B is given by

$$\frac{\nu_B}{\nu_A} = \frac{(\mathbf{N}_B \cdot \mathbf{U}_B)(\mathbf{N}_A \cdot \tilde{\mathbf{U}}_A)}{(\mathbf{N}_B \cdot \tilde{\mathbf{U}}_B)(\mathbf{N}_A \cdot \mathbf{U}_A)} \cdot \sqrt{\frac{g_{tt}(A)}{g_{tt}(B)}},$$

where $\mathbf{N}_A$ and $\mathbf{N}_B$ refer to the connecting ray at A and B, respectively, and $\tilde{\mathbf{U}}_A$, $\tilde{\mathbf{U}}_B$ are the 4-velocities of the lattice points A and B. In static spacetimes (9.5) show that if we standardize each $\mathbf{N}$ to $\hat{\mathbf{N}} := dx^\mu/dt$ for the ray in question, the formula simplifies to

$$\frac{\nu_B}{\nu_A} = \frac{\hat{\mathbf{N}}_B \cdot \mathbf{U}_B}{\hat{\mathbf{N}}_A \cdot \mathbf{U}_A} \cdot \frac{g_{tt}(A)}{g_{tt}(B)}.$$

11.5. Two free particles are in circular orbit in Schwarzschild space (11.13), one at radius r_1 and the other at radius $r_2(> r_1)$. At some instant, the inner particle sees the outer particle by light that has traveled radially. If the proper frequency of the light emitted at the outer particle was ν_2, what is the frequency ν_1 seen at the inner particle? [*Hint:* use the last formula of the preceding exercise. *Answer:* $\nu_1/\nu_2 = (1 - 3m/r_2)^{1/2}/(1 - 3m/r_1)^{1/2}$.]

11.6. Verify that the coordinate lines $r = $ var (θ, ϕ, $t = $ const) are geodesics of the Schwarzschild metric (11.13). [*Hint:* Either use the equations, or appeal to what we know about totally geodesic subspaces, cf. Section 9.5.]

11.7. Consider radial free fall in Schwarzschild spacetime (11.13) as a particular case of the eqns (11.29), (11.30), and (11.34). Find the exact escape velocity at radius r, as measured by *standard* clocks and rulers locally. [*Hint:* standard clocks in the neighborhood of a given lattice point read time $T = e^\Phi t$. *Answer:* $(2m/r)^{1/2}$, just as in Newton's theory. Note that this tends to the speed of light at the horizon.]

11.8. Using (11.35) and the *Newtonian* potential diagram Fig. 11.4(b), show that the type of orbit is fully determined by an initial velocity at any given point, independently of the initial direction: if the initial velocity equals or exceeds the escape velocity $(2m/r)^{1/2}$, the orbit goes to infinity, and otherwise it is closed. Discuss how the general relativistic situation differs from this.

11.9. At radial coordinate r in outer Schwarzschild space an otherwise freely moving particle is projected with initial proper angular velocity $d\phi/d\tau = \omega$ (in the plane $\theta = \pi/2$) at right angles to the radial direction. Find the minimum value of ω for the particle to reach infinity. What is the corresponding value of ω in Newton's theory? Do the two results agree for large r? [*Answer:* $\omega^2 = r^{-2}(\alpha^{-1} - 1)$.]

11.10. Prove that, in first approximation, the quantity $\alpha\dot{t}$ of eqn (11.29) which is strictly conserved along any geodesic in Schwarzschild spacetime (11.13), represents the sum of the internal energy, the kinetic energy, and the potential energy, per unit rest-mass, of the freely falling particle. [*Hint:* Prove $\alpha\dot{t} \approx 1 + \frac{1}{2}v^2 - m/r$.]

11.11. Prove that the relativistic energy, relative to an observer at rest in the lattice, of a particle of rest-mass m_0 moving arbitrarily through Schwarzschild space (11.13) is $m_0\alpha^{1/2}\dot{t}$. Prove also that if this energy could be totally converted into radiation and sent to infinity, its magnitude on arrival would be $m_0\alpha\dot{t}$, and thus always the same, if the particle moves freely. [*Hint:* Apply locally the SR formula $\mathbf{U}_{observer} \cdot \mathbf{P}_{particle} =$ (energy relative to observer)—cf. (6.15).]

11.12. A particle and its antiparticle, both of rest-mass m_0, orbit a spherical mass m along the same circular orbit of ruler circumference $2\pi r$, but in opposite directions. They collide and annihilate each other. How much radiative energy is liberated, as measured by an observer at rest in the lattice where the collision occurs? [*Answer:* $2m_0(1 - 3m/r)^{-1/2}(1 - 2m/r)^{1/2}$.]

11.13. Suppose that at the instant a particle in circular orbit around a spherical mass passes through a point P, another freely moving particle passes through P in a radially outward direction, at precisely the velocity necessary to ensure that it falls back to P when also the orbiting particle again passes through P after a complete orbit. Without detailed calculations, show that it is certainly possible for the two particles to take different *proper* times between their encounters. We have here a version of the 'twin paradox' where neither twin experiences proper acceleration.

11.14. Derive the exact Newtonian scattering angle Θ for a ray of light approaching the center of a spherical mass m to within a distance R, and do this twice: once on the assumption that the 'light-particles' have velocity c at the point of closest approach, and once on the perhaps more reasonable assumption that they have velocity c at infinity. Work in units making $c = G = 1$. [*Answers:* $\sin\frac{1}{2}\Theta = (R/m \mp 1)^{-1}$. *Hint:* Let the path be described by (11.44), so that $\cos\phi_\infty = -e^{-1}$. In the first case, $h = R$. In the second case use (11.37) to determine E, then (11.42) to find h^2.]

11.15. Prove that every GR orbit that contains a point where $\dot{r} = 0$, is mirror-symmetric about the radius to that point. [*Hint:* Start with the reflection invariance of (11.45).] Use this result to give an alternative argument for the strict periodicity of quasi-elliptic orbits.

11.16. Derive the lensing formulae (11.68). [*Hint*: Let Q be a point-source essentially at infinity, C a point-lens, and O the observer. Let the (small) angle between OC and CQ be ϕ. Then $\theta_1 - \phi = 4m/D\theta_1$, $\theta_2 + \phi = 4m/D\theta_2$—why? The first result now follows. Next, from any point P on the continuation of the straight line QC an Einstein ring is seen; so in a 2-dimensional diagram the 'left' and 'right' light arrives simultaneously. Suppose P and O are on the same wave front of the *right* light. Let ψ be the (small) angle between the left and right wave fronts at P, which, to sufficient accuracy, equals the corresponding angle at O, and so $\psi = \theta_1 + \theta_2$. Then $\Delta t = \text{PO} \cdot \psi \approx D\phi\psi$—why? The second result now follows.]

11.17. Find the unique radius for a free circular light path $r = \text{const}$ in the spacetime with metric (in units making $c = 1$):

$$\mathbf{ds}^2 = \exp(r^2/a^2)\, \mathrm{d}t^2 - \mathrm{d}r^2 - r^2(\mathrm{d}\theta^2 + \sin^2\!\theta\, \mathrm{d}\phi^2).$$

[*Answer*: $r = a$. *Hint*: rotating coordinates.]

11.18. Find the geodetic precession per orbital revolution of a gyroscope in free circular orbit $r = \text{const}$ in the spacetime of the preceding exercise. [*Answer*: $2\pi\{1 - (1 - r^2/a^2)\exp(r^2/2a^2)\}$ in the same sense as the orbit.]

12

Black holes and Kruskal space

12.1   Schwarzschild black holes

A. Formation of horizons

We have already on several occasions noted the special role that the locus $r = 2m$ (the *Schwarzschild horizon*) plays in the Schwarzschild metric. It is now time to study it systematically.

In full units [cf. (11.16)] the Schwarzschild radius $\tilde{r}$ for a spherical mass M of uniform density ρ and radius r is given by

$$\tilde{r} = \frac{2GM}{c^2} = \frac{8\pi}{3}\frac{G}{c^2}\rho r^3, \tag{12.1}$$

if we ignore curvature corrections. So we have

$$\frac{\tilde{r}}{r} = \frac{8\pi}{3}\frac{G}{c^2}\rho r^2, \tag{12.2}$$

which shows that, for any density ρ, however small, we *can* have $\tilde{r} > r$ if only r is large enough; that is, the horizon *can* be outside the mass. For the sun we saw (cf. end of Section 11.2C)

$$\frac{\tilde{r}_\odot}{r_\odot} \approx \frac{3}{7} \times 10^{-5}. \tag{12.3}$$

But consider a (quite unrealistic!) *spherical and non-rotating* galaxy containing about 10^{11} suns, like ours, but equally spaced throughout and initially at rest. Since $\tilde{r} \propto m$, we would have

$$\tilde{r}_{\text{Gal}} \approx 10^{11}\tilde{r}_\odot \approx \frac{3}{7} \times 10^6 r_\odot, \tag{12.4}$$

by (12.3). So the ratio of the volume inside $\tilde{r}_{\text{Gal}}$ to the volume of a sun is given by

$$\left(\frac{\tilde{r}_{\text{Gal}}}{r_\odot}\right)^3 \approx 10^{17}. \tag{12.5}$$

If this were 10^{11}, the suns would have to be packed like stacks of cannon balls to fit into the horizon sphere. But we have a factor 10^6 to spare; that is, such a stack expanded by a linear factor of 10^2 would still fit into the galactic horizon. In other words, if that galaxy were to collapse to a volume where the individual stars were still 100 stellar diameters apart, it would already be inside its horizon.

Intuition tells us that (given the assumed initial conditions) such a collapse would surely occur. (Only rotation and centrifugal force can save most galaxies from this fate.) Intuition also tells us that, as the last of the stars fall into the horizon, nothing very special would occur there. The essence of the horizon has already been foreshadowed in Section 11.2C: it is a potential outwardly directed spherical light front of *constant* radius. But that is a *global* property; locally, to a freely falling observer, the horizon is a light front like any other.

B. Regularity of the horizon

Let us look at the Schwarzschild metric (11.13) over the entire range $r > 0$. Examination of the steps leading to (11.13) shows that it satisfies the vacuum field equations for $r < 2m$ just as well as for $r > 2m$. To see that nothing untoward occurs *at* the horizon, we could work out (rather tediously, with the help of the Appendix, or quite easily with a computer) the 14 independent invariants of the Riemann curvature tensor and so verify that the curvature remains finite there. (The separate *components* of the curvature tensor do not tell the whole story, since they have no intrinsic significance.) For example, we can find

$$R_{\mu\nu\rho\sigma} R^{\mu\nu\rho\sigma} = 48\frac{m^2}{r^6}, \qquad R_{\lambda\mu}{}^{\nu\rho} R_{\nu\rho}{}^{\sigma\tau} R_{\sigma\tau}{}^{\lambda\mu} = 96\frac{m^3}{r^9}. \qquad (12.6)$$

So at least two out of the fourteen are well-behaved at $r = 2m$. But either one of these invariants already shows that we have a genuine curvature singularity at $r = 0$.

A simpler way to show that the spacetime is regular at $r = 2m$ is to find an alternative coordinate system that bridges the Schwarzschild singularity regularly. One such is the *Eddington–Finkelstein* system r, θ, ϕ, v, where r, θ, ϕ are the original Schwarzschild coordinates but v is defined by

$$t = v - 2m \log \left| \frac{r}{2m} - 1 \right| - r. \qquad (12.7)$$

We immediately find

$$dt = dv - \alpha^{-1} dr \quad (\alpha = 1 - 2m/r), \qquad (12.8)$$

which, when substituted into (11.13), gives the Eddington–Finkelstein form of the Schwarzschild metric:

$$\mathbf{ds}^2 = \alpha\, dv^2 - 2\, dv\, dr - r^2(d\theta^2 + \sin^2\theta\, d\phi^2). \qquad (12.9)$$

It is regular for *all* $r > 0$, and this shows the horizon events to be ordinary. (Note that if we put $v = R+T, r = R-T$, then at the horizon where $\alpha = 0$ the first part of the metric reads $2(dT^2 - dR^2)$.) The Schwarzschild singularity is thus recognized as a mere *coordinate singularity*: it can be transformed away.

For $r > 2m$ and for $r < 2m$ the Eddington–Finkelstein metric clearly satisfies the vacuum field equations, since these are tensorial and therefore form-invariant under a regular coordinate transformation, which (12.7) is, except at $r = 2m$. But since the $g_{\mu\nu}$ of (12.9) together with their inverses and first and second derivatives are continuous at $r = 2m$, the field equations must be satisfied there also, by continuity. So the entire spacetime described by the Schwarzschild metric is regular and vacuum down to $r = 0$, where the curvature becomes infinite.

C. Infalling particles

Another way to see the 'normalcy' of the horizon is to examine the free fall of particles through it, most easily along a radius. Going back to eqn (11.34), and setting $h = 0$ for radial fall [cf. (11.30)], we find, writing $2E$ for $k^2 - 1$,

$$\dot{r}^2 = \frac{2m}{r} + 2E. \tag{12.10}$$

This equation has exactly the Newtonian form (kinetic plus potential energy = const). Of course, here the derivative is with respect to proper time rather than Newton's time, but this form makes it a priori clear that *all* proper fall times are finite. For a particle dropped from rest at $r = r_0 > 2m$ we have $E = -m/r_0$, and, if it falls to $r = r_1$, the proper time elapsed is seen to be

$$\Delta\tau = \int_{r_1}^{r_0} \frac{dr}{\sqrt{2m/r - 2m/r_0}} = \int_{r_1}^{r_0} \frac{\sqrt{rr_0}}{\sqrt{2m(r_0 - r)}}\,dr. \tag{12.11}$$

As expected, this is finite all the way down to $r_1 = 0$. In particular, as shown by eqn (12.13) below, if the particle is released from rest at the horizon (or rather, infinitesimally outside of it), $\Delta\tau$ down to $r = 0$ is πm.

On the other hand, the external *coordinate* time t elapsed in a fall to the horizon is always infinite, which most importantly means that an external observer never *sees* the particle cross the horizon. (Coordinate time increases along a light signal.) The reason for this divergence is that the integral for Δt looks just like that in (12.11) except for an extra factor

$$\frac{k}{1 - 2m/r}$$

in the integrand, as follows from (11.29).

D. Non-staticity of inner Schwarzschild space

In our original derivation of the Schwarzschild metric we looked for a *static* spherically symmetric field. General relativity, through Schwarzschild, tells us that such fields cannot be continued to arbitrarily small radii. The horizon is the limit of staticity. No part of the spacetime inside it is static, or even stationary. To see this, note that both g_{tt} and g_{rr} in (11.13) change sign *at* $r = 2m$; so inside the horizon the dr^2-term is the only positive one. Consequently r cannot stand still for a particle- *or* photon worldline, which must satisfy $ds^2 \geq 0$. But not being able to stand still is characteristic of *time*! And indeed, inside the horizon, r is *the* time coordinate (though it also retains its geometrical significance). As such it can only go *one* way: the future time direction inside a black hole is that of *decreasing* r. Since at a lattice point of a stationary field the curvature would have to remain constant, eqns (12.6) show that r would have to remain constant. But lattice points are potential particles, so no lattice can exist.

Even more directly, the non-stationarity of any part of the inner Schwarzschild region becomes apparent from the fact that the total length of any particle-worldline inside the horizon is at most πm (whereas in stationary spacetimes all lattice points have infinite worldlines). For, integrating along a worldline inside the horizon, we have, from (11.13),

$$\Delta s = \int \left\{ \left(\frac{2m}{r} - 1\right)^{-1} dr^2 - \left(\frac{2m}{r} - 1\right) dt^2 \right.$$
$$\left. - r^2(d\theta^2 + \sin^2\theta \, d\phi^2) \right\}^{1/2}. \tag{12.12}$$

Any variation in t, θ, or ϕ decreases the value of this integral. It is therefore maximal for a worldline $t, \theta, \phi = $ const (which, by this very argument, must be a timelike geodesic, cf. also Exercise 11.6). And that maximum is given by

$$\Delta s_{\text{max}} = \int_0^{2m} \left(\frac{2m}{r} - 1\right)^{-1/2} dr = \pi m \tag{12.13}$$

(as can be checked by the substitution $r = 2m \sin^2 u$). The particle cannot oscillate between two r-values. Time cannot turn around. It begins at $r = 2m$ and ends at $r = 0$. [Note that (12.13) is a special case of (12.11) corresponding to $r_0 = 2m, r_1 = 0$.]

For example, once all the stars have crossed $\tilde{r}_{\text{Gal}}$ in our earlier thought experiment, they have at most a proper time $10^{11}\pi m_\odot$ to live before being annihilated by the infinite tidal forces [that is, curvature—cf. after (10.80)] at $r = 0$. In full units, this maximal proper time would be $10^{11} m_\odot c^{-1} \approx 1.55 \times 10^6$ s or ~ 18 days. This is, in fact, *identical* with the time required in Newtonian theory for a ball of mass $10^{11} M_\odot$ and radius $\tilde{r}_{\text{Gal}} = 2 \times 10^{11} G M_\odot c^{-2}$ to collapse from rest—perhaps not surprisingly, in view of the shared equation (12.10). The Newtonian time for the original non-rotating hypothetical galaxy to collapse (if it had a typical radius of $\sim 10^{23}$ cm) would be $\sim 10^8$ years. On the other hand, a star more or less like the sun, once inside its 6 km-horizon, would collapse in $\sim 10^{-11} \times 18$ days $\approx 10^{-5}$ s!

E. Funnel geometry

In spite of all classical analogs, what happens inside the horizon is very far from being Newtonian or Euclidean, especially as we proceed inwards and especially in the final stages. We have seen that the lines $t, \theta, \phi = $ const are possible radial particle geodesics in the vacuum inner Schwarzschild space. Consider two neighboring concentric infalling spherical shells of test particles so moving, one with $t \equiv t_0$ and the other with $t \equiv t_0 + \delta t$. Along a radius the metric reads

$$\mathbf{ds}^2 = \left(\frac{2m}{r} - 1\right)^{-1} dr^2 - \left(\frac{2m}{r} - 1\right) dt^2. \qquad (12.14)$$

Applying SR locally, we recognize the last term as ruler distance squared in the LIF. Consequently the ruler distance δl between the infalling shells, as measured in the LIF, is given by

$$\delta l = \left(\frac{2m}{r} - 1\right)^{1/2} \delta t. \qquad (12.15)$$

And this tends to infinity as the 'time' r progresses from $2m$ to zero. There is little difference between a shell so moving [which corresponds to $k = 0$ in Fig. 11.4(a)] and one released from rest just outside the horizon (which corresponds to $k = $ infinitesimal). So we can approximate the latter with the former. A useful picture of inner Schwarzschild spacetime is therefore an infinite succession of such freely falling test-spheres peeling off from just outside the horizon and disappearing down an infinitely long funnel.

The smoothed-out *inside* of our hypothetical collapsing-ball galaxy [cf. after (12.3)] is not a vacuum and so the vacuum Schwarzschild metric does not apply to it. As the ball contracts it creates the funnel between itself and the horizon. Its own geometry, however, provides a rounded cap to the funnel's ever lengthening and narrowing neck. [Cf. end of Section 12.5.] Note, incidentally, how futile it would be to try to define a gravitational field strength in a non-stationary spacetime like a black hole neck.

F. Formation of black holes

If our collapsing-ball galaxy continuously emits spherical light fronts from its surface, then, as that surface approaches the horizon, the light takes ever longer coordinate times to reach an observer at rest in the external lattice (cf. Exercise 11.2). The horizon-crossing event itself is seen only in the infinite future. Human and other activities on planets (as seen by the external observer) are gradually slowed down to a standstill at the horizon-crossing. The redshift of the stars has by then become infinite. The outward light front emitted *at* the horizon remains stuck there for ever and the infalling stars by their own reckoning obviously recede from it at the speed of light. Outward light fronts emitted later contract immediately. No more communication with the outside is possible: a *black hole* has been formed. The collapse can no longer be halted, no matter what internal forces might oppose it. For particles *and* photons in the vacuum between the fall and the horizon, r must steadily decrease. It is sometimes said that an irresistible gravitational force pulls everything, including light, inwards. That force is nothing but *time* (in the guise of r), which inexorably sweeps everything before it into the future.

It is believed that there may well be old and inactive black holes containing millions of solar masses at the core of many galaxies, including our own. The active collapse of such a *galactic black hole* and the energy generated in the process (for example, through rotation, magnetic fields, and synchrotron radiation) may account for the extreme luminosity of quasars. On the other hand, single stars can collapse to form *stellar black holes*. Millions of these may exist in our galaxy alone. It is expected that when a star's nuclear fuel is depleted and thermal pressure can no longer resist gravity, then, after various possible more or less violent events (for example, supernovae), the star will end up 'dead' in one of three final configurations: a white dwarf, a neutron star (pulsar), or a black hole. The matter in a white dwarf has density $\sim 10^5$ g/cm^3 and consists of a (Fermi-) gas of electrons intermingled with a gas of nuclei. Stiffness against gravity is provided by electron degeneracy. It can be shown by quantum-mechanical considerations that the maximum mass of a white dwarf is $\sim 1.4\, M_\odot$ (the 'Chandrasekhar limit'). More massive stars can overcome this electron pressure and be halted next by the pressure of neutron degeneracy, when all the electrons have been squeezed into the protons to form neutrons. The density of a neutron star is $\sim 10^{14}$ g/cm^3 and its radius is of the order of 10–20 km. Again, there is a quantum-mechanical limit of $\sim 3\, M_\odot$ to the mass of neutron stars. [The Schwarzschild limit is less crucial here: eqn (12.1) yields a maximum mass of $\sim 6\, M_\odot$ for balls of neutron-star density.] When even neutron pressure cannot halt the collapse, a black hole would seem to be the inevitable outcome. There are now on record several binary-star systems with one invisible component whose calculated mass exceeds these limits and which is therefore presumed to be a black hole.

We may note that, since most stars and galaxies at formation possess angular momentum, their collapse, if it occurs, could be preceded by a period of rapid rotation—to which the spherically symmetric Schwarzschild metric would apply only very approximatively. It is the *Kerr metric*, which also possesses a horizon, that then takes over. [Cf. after (15.92).]

12.2 Potential energy; A general-relativistic 'proof' of $E = mc^2$

We have already seen several roles played by the relativistic scalar potential Φ in static and stationary spacetimes—for example, in connection with clock rates and light frequencies, but also as yielding the gravitational field by its gradient. Does it additionally measure (as in Newton's theory) the energy needed to move a unit mass from point A to point B? The answer is both yes and no and has implications for black holes.

As we proved in (10.49), $-\Phi_{,i}$ is the gravitational field and so, by our special-relativistic result $\mathbf{f} \cdot \mathbf{dr} = \mathrm{d}E$ [cf. (6.45)], the energy needed in the local rest-frame to move a particle of unit rest-mass through a displacement dx^i is $\Phi_{,i}\,dx^i$. The total energy needed to move the particle from point A to point B, *if acquired en route*, is therefore $\int \Phi_{,i}\,dx^i = \int d\Phi = (\Phi_B - \Phi_A)$, just as in classical mechanics.

Suppose, however, that we (standing at B) wish to pull a particle of unit rest-mass up along a field line from a point A of lower potential, with a massless string. The string may have to be frictionlessly constrained to follow the field line across the lattice. Suppose we pull the string with force f through a ruler distance dl, thereby doing work $dE_B = f\,dl$. The particle gains energy $dE_A = g\,dl$ locally, g being the gravitational field strength at A. But if, alternatively, our work were used to generate an amount dE_A of radiative energy at A, which was then beamed to B, it would arrive at B diminished to $dE_A\exp(\Phi_A - \Phi_B)$, by the usual redshift formula (9.4) and Planck's relation $E = h\nu$. Since energy is conserved, this must equal dE_B. So we have

$$\frac{f}{g} = \frac{dE_B}{dE_A} = e^{(\Phi_A - \Phi_B)} < 1. \tag{12.16}$$

To digress for a moment: In Schwarzschild spacetime, (12.16) yields, together with (11.15) and (11.16),

$$f = \frac{e^{\Phi_A}}{e^{\Phi_B}}g = \frac{\alpha_A^{1/2}}{\alpha_B^{1/2}} \cdot \frac{GM}{r_A^2}\alpha_A^{-1/2} = \frac{GM}{r_A^2}\alpha_B^{-1/2}. \tag{12.17}$$

So the force f_∞ needed at very large r to dangle a particle of unit rest-mass at the horizon $r_A = 2GM$ is given by $f_\infty = (4GM)^{-1}$, a quantity referred to as the *surface gravity* of the black hole.

Continuing with our previous calculation, and writing $\Phi_{,i}\,dx^i$ for the local energy needed to raise the particle at the general point along the way, we have, from (12.16),

$$\int dE_B = e^{-\Phi_B}\int_A^B e^\Phi \Phi_{,i}\,dx^i = e^{-\Phi_B}\int_A^B d(e^\Phi) = e^{-\Phi_B}(e^{\Phi_B} - e^{\Phi_A}).$$

Thus the total energy expended at B in pulling the particle up from A is given by

$$E = 1 - e^{(\Phi_A - \Phi_B)} = 1 - \frac{e^{\Phi_A}}{e^{\Phi_B}}. \tag{12.18}$$

Conversely, this is also the energy per unit mass we would *extract* at the end of the string if we slowly *lowered* the particle into the potential well.

Let us take the concrete example of Schwarzschild spacetime (11.13), where

$$e^\Phi = \left(1 - \frac{2m}{r}\right)^{1/2}, \qquad \Phi = \frac{1}{2}\log\left(1 - \frac{2m}{r}\right).$$

No particle could afford to pay along the way to extract itself from near the horizon: $\Phi_B - \Phi_A$ (in particular $-\Phi_A$) becomes infinite as A approaches the horizon. On the other hand, rope-rescue is affordable: eqn (12.18) yields $E = 1$ for A *on* the horizon. In full units this reads $E = m_0 c^2$ for a particle of rest-mass m_0. The converse is even more interesting: By slowly lowering a particle of rest-mass m_0 to the horizon, we could extract from it *all* its energy $m_0 c^2$!

A fair question: have we really extracted this energy from the particle, using the field as a mere catalyst, or have we extracted it from the field? This is best answered by playing the game with a thin spherical *shell* of mass m_0 at $r = r_B$, say, and demonstrating that the field remains unchanged. After the shell is lowered slowly to the horizon (so as to arrive there without kinetic energy), an amount of energy $m_0 c^2$ (and thus a mass m_0) has been accumulated at its original location. The shell itself can then be discarded into the horizon. By Birkhoff's theorem, the field outside the reconstituted shell has not changed. But the field between the black hole and that shell has not changed either! For if it had, we could repeat the process indefinitely and so build up an arbitrarily large field *difference* across the shell at r_B, which is absurd.

As a corollary, we note that the mass of the black hole is unchanged in spite of having absorbed a mass m_0 'from rest at the horizon'. This is what is meant by saying that the number of elementary particles inside a black hole is indeterminate.

12.3   The extendibility of Schwarzschild spacetime

Let us consider what might be called a 'Schwarzschild diagram', Fig. 12.1, in analogy to the Minkowski diagram of special relativity. This is an r, t map of motions along a typical radius $\theta, \phi = $ const in Schwarzschild spacetime (11.13). The heavy line at $r = 2m$ represents the horizon. The most important feature of the diagram are the light

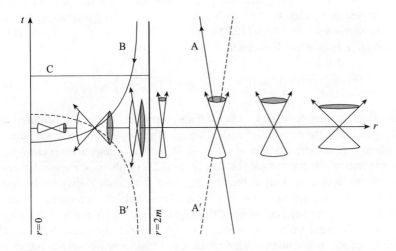

Fig. 12.1

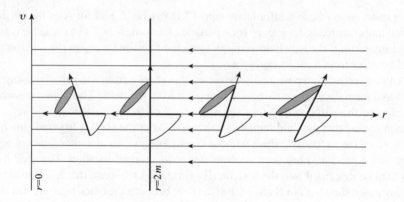

Fig. 12.2

cones at various r-values. For radial light rays ($\mathbf{ds}^2 = 0$) the metric (11.13) yields

$$\frac{dr}{dt} = \pm\left(1 - \frac{2m}{r}\right). \tag{12.19}$$

This corresponds to the *slopes relative to the t-axis* of inwardly and outwardly emitted signals. For large r, these slopes approach ± 1, as in the Minkowski diagram. As we come in to the horizon, however, the slopes approach zero: plots of light signals *near* the horizon are almost vertical on *both* sides of the horizon. But, as we cross the horizon, the 'insides' of the cones, namely the regions where $\mathbf{ds}^2 > 0$ which contain all worldlines through the vertex, change discontinuously. And so does the direction of the future, as indicated by the little arrows on the cones.

For comparison, Fig. 12.2 shows an 'Eddington–Finkelstein diagram', which nicely demonstrates the continuity of the light cone structure in coordinates that bridge the horizon continuously. From eqn (12.9) we have, for the two radial light signals through each event, in Eddington–Finkelstein coordinates,

$$dv = 0, \qquad \frac{dr}{dv} = \frac{1}{2}\left(1 - \frac{2m}{r}\right). \tag{12.20}$$

In this diagram, therefore, all $v = $ const lines correspond to inwardly emitted radial light signals. For the outwardly emitted signals, the slope tends to $\frac{1}{2}$ for large r, zero at the horizon, and tends to $-\infty$ as $r \to 0$. Both diagrams make it clear that for *all* worldlines inside the horizon, be they of particles or photons, r must decrease.

In spite of its discontinuity at the horizon, the Schwarzschild diagram has certain advantages, and we now return to it. Consider a typical infalling-particle worldline (a geodesic), say the one consisting of the segments marked A and B in the diagram, which may be regarded as connected at $t = \infty$. Since the entire Schwarzschild metric is reflection-symmetric in the coordinate t (that is one of its advantages), we know from general theory (cf. Fig. 9.3 and accompanying text) that the t-reflected segments A$'$, B$'$ are also geodesics. A$'$ represents a free particle moving away from the horizon. Where has it come from? Evidently from inside the horizon, moving along B$'$ *towards* the horizon. Yet the inner Schwarzschild space we have discussed so far does not permit such motions and its horizon is not penetrable in that direction. The only way out of this dilemma is to have a second copy of the inner region joined to the outer region at the horizon, one in which time runs in the sense of *increasing* r and which is bounded by an *outwardly* penetrable horizon. [Indeed that must be where the various particles 'coming out of the horizon' in Fig. 11.4(a) all come from.] But that is not all. In the original inner spacetime, B$'$ traversed *towards* $r = 0$ must be a possible particle worldline, and *it* evidently comes from A$'$ traversed in the sense of decreasing t. So we also need a second copy of outer Schwarzschild spacetime, one in which time runs in the sense of decreasing t. All four of these regions are joined at $r = 2m$.

This mysterious topology will become quite transparent when all four regions are seen as portions of *Kruskal space*, which removes the distortions inherent in the Schwarzschild coordinates. (The Eddington–Finkelstein coordinates are only a half-way measure). The most misleading feature of the Schwarzschild diagram is the line $r = 2m$. Its entire finite portion will be seen to correspond to a single event (inaccessible to particles from the outside), while the portions at $\pm\infty$ correspond to light signals to and from this event! We recall that $t = \text{const}$ corresponds to a radial particle geodesic in inner Schwarzschild spacetime. A typical such trajectory is marked C in Fig. 12.1. Does it suddenly stop at the horizon? No: it goes out to the horizon in one copy of inner space and comes back in the other. It is such apparent stopping in mid-space of geodesics that is referred to as the *extendibility* of the original Schwarzschild solution. A spacetime is said to be *extendible* if it is a mere portion of a larger spacetime where geodesics do *not* stop in mid-space. (Geodesics stopping at a spacetime *singularity* do not indicate extendibility.) That larger space is called the *maximal extension* of the smaller space. In the case of Schwarzschild, the maximal extension is Kruskal space.

12.4 The uniformly accelerated lattice

In preparation for the discussion of Kruskal space, we return to special relativity and the uniformly accelerated rocket of Section 3.8. Looking at Minkowski space, M^4, from the lattice of such a rocket will shed much light on the relation between Schwarzschild and Kruskal space.

Recall that the motion of the entire rocket was characterized by eqn (3.19),

$$x^2 - t^2 = X^2 \tag{12.21}$$

(now written with $c = 1$), each point of the rocket corresponding to some fixed value of the parameter $X > 0$, and having constant proper acceleration $1/X$. Consider now the following transformation from the standard coordinates x, y, z, t of M^4 to new coordinates X, Y, Z, T:

$$t = X \sinh T, \quad x = X \cosh T, \quad y = Y, \quad z = Z \quad (X > 0). \quad (12.22)$$

This implies

$$x^2 - t^2 = X^2, \qquad x/t = \coth T \quad (12.23)$$

and also the following transformation of the metric:

$$\mathbf{ds}^2 = dt^2 - dx^2 - dy^2 - dz^2 = X^2 \, dT^2 - dX^2 - dY^2 - dZ^2, \quad (12.24)$$

as can be verified at once. The new metric is also static, and also has a Euclidean lattice, in which X, Y, Z measure ruler distance. Comparing (12.23)(i) and (12.21), we see that each of its lattice points execute hyperbolic motion in the x-direction of M^4 and constitutes, in fact, a fixed point in the rocket. Our lattice *is* the rocket of Section 3.8! Quite independently of what we proved there, all the properties of this rocket now follow from our general knowledge of static spacetimes. Obviously the lattice 'moves rigidly'. Its relativistic potential Φ is given by $X^2 = \exp 2\Phi$, $\Phi = \log X$ and so the exact gravitational field strength is given by

$$g = \left| \frac{d\Phi}{dX} \right| = \frac{1}{X}, \quad (12.25)$$

which is also the proper acceleration of the corresponding lattice points.

Although the 'rocket' defined by (12.22) coincides with fully one-half $(x > 0)$ of the lattice of M^4 at $t = 0$, we shall mostly prefer to visualize just a typical part of it, something like a tall thin skyscraper on its side, accelerating along the positive x-axis of M^4. Infinitely many such skyscrapers side by side, all infinitely tall, then constitute the full rocket lattice.

Quadrant I of Fig. 12.3 is the Minkowski diagram of the rocket. It shows the world-lines (hyperbolae) of the lattice points, as well as the adapted-time cuts $T = $ const. As in all static spacetimes, these time cuts are orthogonal to the lattice worldlines.

The $\pm 45°$ lines through the origin are potential light fronts representing the limits of the rocket lattice. They correspond to $X = 0$ and thus to $g = \infty$. They will soon be recognized as *horizons*. One of these light fronts accompanies the rocket during the 'first half of eternity' $(t < 0)$ when the rocket decelerates and moves oppositely to the acceleration which always acts in the positive x-direction. At $t = 0$ this light front peels off and is replaced by another that accompanies the accelerating rocket henceforth. The ruler distance of either front, while it is 'attached' to the rocket, from any rocket point remains constant and equal to X, the point's coordinate. (Cf. end of Section 2.9.)

As the Minkowski diagram shows, this one rocket provides alternative coordinates for only one quadrant of M^4, namely that marked I in the diagram. (This 'wedge', so coordinatized, is occasionally referred to as *Rindler space*; for reasons that will presently become clear, it has also been called a 'toy black hole'.) It is easy enough to coordinatize quadrant III with a similar rocket moving in the opposite direction,

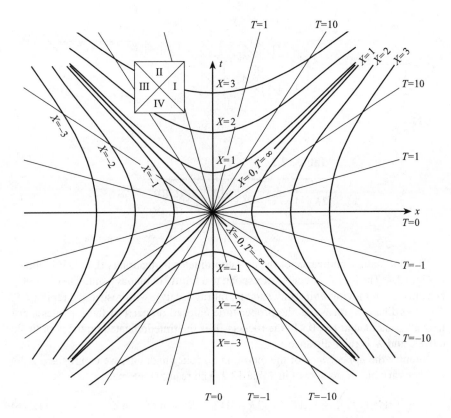

Fig. 12.3

Table 12.1

	I $(X > 0)$ and III $(X < 0)$	II $(X > 0)$ and IV $(X < 0)$
$t =$	$X \sinh T$	$X \cosh T$
$x =$	$X \cosh T$	$X \sinh T$
$x/t =$	$\coth T$	$\tanh T$
$x^2 - t^2 =$	X^2	$-X^2$
$dt^2 - dx^2 =$	$X^2 \, dT^2 - dX^2$	$-X^2 \, dT^2 + dX^2$

by choosing $X < 0$ in (12.22). Equations (12.23) and (12.24) remain valid. What about quadrants II and IV? These can be coordinatized in *formally* the same way, using the same Lorentz-invariant and mutually orthogonal pattern of hyperbolae and radii. Table 12.1 summarizes the $x, t \mapsto X, T$ transformation in all four quadrants. The 'trivial' part of the transformation, $y = Y, z = Z$, remains valid everywhere.

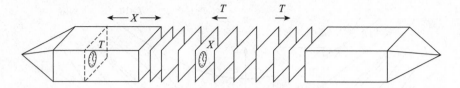

Fig. 12.4

Table 12.2

$$2R - 1 = x^2 - t^2 = \begin{cases} X^2 & \text{(I, III)} \\ -X^2 & \text{(II, IV)} \end{cases}$$

The kinematic significance of the new coordinates in quadrants II and IV is shown in Fig. 12.4. The loci $T = \text{const}$ correspond to a family of planes parallel to the rocket bottoms, each moving uniformly (that is, falling freely), with respective rapidity T. The coordinate X measures the proper time elapsed on each of these planes since they all coincided at $t = 0$. (In the rockets, T is the time indicated by the coordinate clocks, and X is ruler distance.)

To unify the *appearance* of the metric in all four quadrants, we can replace X by another variable, R, as shown in Table 12.2. The result is the metric

$$\mathbf{ds}^2 = (2R - 1)\,dT^2 - (2R - 1)^{-1}\,dR^2 - dY^2 - dZ^2, \tag{12.26}$$

valid *everywhere*.

The Schwarzschild metric has a curvature singularity. Minkowski space (no matter how we change the coordinates) has none. So in order to aid the analogy, we produce an artificial singularity: we truncate M^4 at $R = 0$. As shown in Fig. 12.5, the space-time regions corresponding to $R < 0$ are declared not to exist, and the two branches of the hyperbola $R = 0$ become the past and future *edges* of spacetime. Truncated Minkowski space is, in a sense, a universe of finite duration. Free paths all begin on the past edge and end on the future edge. Infinite existence can be had only outside the horizons (rocket bottoms), and only by ceaselessly accelerating so as not to fall in. (This produces the ever-increasing time dilation needed for unlimited existence.) Once in quadrant IV, a particle *must* hit the future edge.

The analogy we wish to draw is between our rocket in quadrant I and a skyscraper outside a Schwarzschild black hole, as shown in Fig. 12.6, where struts connect the whole family skyscrapers sideways to prevent their falling into the horizon.

For us, Minkowski space is easy to understand—even if truncated. But in its totality it is not quite so easy to understand from the vantage point of the rocket inhabitants. They live in a changeless world. Their most convenient metric is (12.24)(ii), but for our benefit they are also happy with (12.26). They can draw a diagram very much like the Schwarzschild diagram of Fig. 12.1 and they can figure out that particles

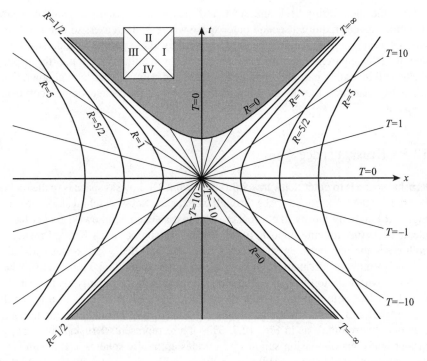

Fig. 12.5

dropped down an open elevator shaft fall into an 'inner' space $R < \frac{1}{2}$, where R is time and steadily decreases. But they also see the need for a second inner space where *R increases*, to explain where incoming particles come from. (They may even observe free particles being shot up into their open elevator shafts, reaching some maximum

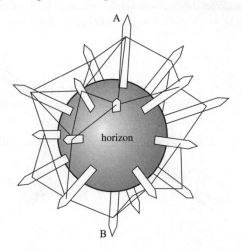

Fig. 12.6

height, and then falling back and out.) And, just as in the Schwarzschild case, the two inner spaces require a second outer space where T decreases, so that geodesics will not stop in mid-space. It is particularly hard for these people to understand how their horizon apparently permits particles to cross it in either direction, little suspecting that all incoming particles cross *one* horizon and outgoing particles another. Only when they learn how they fit into Minkowski space does the rocket people's confusion evaporate.

12.5   Kruskal space

Kruskal space is to outer Schwarzschild space what Minkowski space is to the rocket. It serves as the background space through which each Schwarzschild skyscraper (cf. Fig. 12.6) actually *moves* like a rocket. And just as each Minkowski rocket has an identical partner moving oppositely (cf. Fig. 12.4), so does each Kruskal rocket. In both cases the gap between such rocket pairs can be filled with geodesically (that is, freely) moving test matter. Of course, while the individual Minkowski rockets all move parallelly to the x-axis, the Kruskal rockets move radially, as indicated in Fig. 12.6. But note that the partner of the rocket labeled A in Fig. 12.6 is *not* the rocket labeled B, diametrically across the horizon from A, but rather A′ on the other side of a 'wormhole', as in Fig. 12.7. This figure represents Kruskal space at one instant, with one dimension suppressed (circles are really spheres!). It is similar to the cross-section of Schwarzschild space shown in Fig. 11.1, but now *with* the bottom half of the funnel, as in Fig. 11.2. As time goes on, the spool lengthens and the rockets fly through space. The horizons stay of constant diameter and separate 'at the speed of light', while the test matter in between (shown as a series of circles) moves geodesically, as in Fig. 12.4.

To facilitate the analogy with our Minkowski rocket, we shall choose the units of length, time, *and* mass in the Schwarzschild metric (11.13) so as not only to make $c = G = 1$ but also $m = \frac{1}{4}$! With that, and writing T, R for the Schwarzschild t and

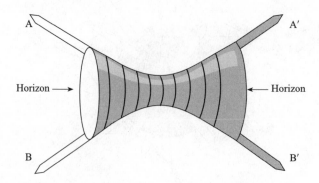

Fig. 12.7

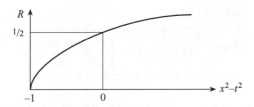

Fig. 12.8

r, (11.13) becomes

$$\mathbf{ds}^2 = \left(1 - \frac{1}{2R}\right) dT^2 - \left(1 - \frac{1}{2R}\right)^{-1} dR^2 - R^2(d\theta^2 + \sin^2\theta \, d\phi^2). \quad (12.27)$$

Note that the horizon is now at $R = \frac{1}{2}$. In the same units, the metric that defines Kruskal space is the following:

$$\mathbf{ds}^2 = \frac{1}{2Re^{2R}}(dt^2 - dx^2) - R^2(d\theta^2 + \sin^2\theta \, d\phi^2), \quad (12.28)$$

where R is a monotonic positive function of $x^2 - t^2$, implicitly defined by the equation

$$e^{2R}(2R - 1) = x^2 - t^2. \quad (12.29)$$

Unfortunately, this cannot be solved explicitly for R in terms of elementary functions. But Fig. 12.8 illustrates the relation between R and $x^2 - t^2$ schematically.

Note (from its last parenthesis) that the metric (12.28) describes a spherically symmetric spacetime (nested parallel 2-spheres), in which R measures area distance [cf. after (11.1)]. The coordinate x in (12.28) is to be regarded as an alternative radial distance measure, whose relation to R changes with time. [Our use of the letters x and t in (12.28), while suggestive for our purposes, is non-standard: most authors use u and v instead.]

Kruskal space, K^4, is, in fact, made up of two outer and two inner Schwarzschild spacetimes, all separated by horizons. Each outer region can be regarded as made up of rockets flying through space; or perhaps even better, since the radially flying rockets never separate, as stationary rockets with space streaming past them. Each inner region can be filled with freely moving test *spheres* (instead of the parallel test planes of Fig. 12.4), which are represented by circles in Fig. 12.7.

To justify these assertions, let us consider a *Kruskal diagram* in the coordinates x and t. It looks identical to the Minkowski diagram of Fig. 12.5, but has a different interpretation. That same diagram now represents a single *radial* line θ, $\phi = $ const of K^4 rather than a line parallel to the x-axis of M^4. We see from the form of the metric (12.28) that all $\pm 45°$ lines in the diagram ($dx = \pm dt$) are light paths—just as in the Minkowski diagram. The $\pm 45°$ lines through the origin, which will again turn out to

Table 12.3

$$e^{2R}(2R - 1) = x^2 - t^2 = \begin{cases} X^2 & \text{(I, III)} \\ -X^2 & \text{(II, IV)} \end{cases}$$

be horizons (rocket bottoms), again divide the diagram into four distinct quadrants. Now apply the transformation of Table 12.1 of the preceding section to the Kruskal metric (12.28)! This transforms its $(dt^2 - dx^2)$ part into one involving the auxiliary variable X, to which Kruskal's R of eqn (12.29) is now related as indicated in Table 12.3. (Cf. Table 12.2.) If we finally use this table to eliminate X in favor of R, the result is Schwarzschild's metric (12.27) *in all four quadrants*! [Compare with (12.26).] By an argument analogous to that of the last paragraph of Section 12.1B, *all* of Kruskal space except for its two singular extremities $R = 0$ satisfies Einstein's vacuum field equations.

Quadrants I and III represent two outer Schwarzschild regions $R > \frac{1}{2}$, with future senses corresponding to T increasing and decreasing, respectively. Quadrants II and IV represent two inner Schwarzschild regions $0 < R < \frac{1}{2}$ with future senses corresponding to R decreasing and increasing, respectively. Note how particles and light from IV can move to I and III, and from there to II. Particles can also move directly from IV to II through the origin; free worldlines of this nature correspond to the typical one marked C in Fig. 12.1. Region IV is called a *white hole*, since every particle or photon is expelled from it; region II is called a *black hole*, since nothing can escape from it.

The very shape of the hyperbolic worldlines $R = $ const in quadrant I suggests a rocket flying through Kruskal space. That these rockets move rigidly is clear from the rigidity of the Schwarzschild lattice, and that each point moves with constant proper acceleration is implicit in the staticity of the Schwarzschild metric.

But there is a more direct way of seeing this. Just like M^4, K^4 transforms into itself isometrically under a homogeneous Lorentz transformation in x and t! For such a transformation preserves both $dt^2 - dx^2$ *and* $t^2 - x^2$ (and thus R) in (12.28). Under such an active LT, the worldlines $R = $ const $(x^2 - t^2 = $ const$)$ transform into themselves and so their proper acceleration is everywhere the same as at the vertex, and therefore constant. Moreover, any two cuts $T = $ const through quadrant I are Lorentz transformable into each other and therefore isometric. Hence we have a rigidly moving uniformly accelerating rocket.

There are other consequences of the radial Lorentz invariance of Kruskal space: (i) Since $x = 0$ is evidently a symmetry surface of (12.28), the line $x = 0$, that is, $T = 0$ (and $\theta, \phi = $ const) in quadrants II and IV is a geodesic; but all other lines $T = $ const in these two quadrants are Lorentz-transformable into this one and hence geodesics too. They are the worldlines of the free test particles moving between the rockets. (ii) Lorentz invariance also makes it obvious that the total proper lifetimes along all these geodesics are the same. (iii) Under any isometry, geodesics map into geodesics. It follows, in particular, that all the geodesics leaving any one of the hyperbolic worldlines in quadrant I tangentially can be mapped into the one at the vertex. Consequently, for example, all free particles dropped from rest at the same point in outer Schwarzschild space take the same proper time to reach the singularity, in spite of the non-staticity of the inner region. [The same result is implicit in eqns (12.11) and (12.13).] Evidently $T \mapsto T + \text{const}$ is an isometry for the entire Schwarzschild metric (12.27), mapping geodesics into geodesics; in the outer regions this is a simple time translation, while in *all* regions it is a Lorentz transformation in x and t (cf. Exercise 12.6). (iv) As we mentioned in the preceding paragraph, any two cuts $T = \text{const}$ in region I (but, of course, continuing to region III) are isometric. We know this a priori, since, via (12.27), any such cut consists of two entire outer Schwarzschild lattices; they join together at $R = \frac{1}{2}$; that is, at the horizon. With one dimension suppressed, any cut $T = \text{const}$ through Kruskal space is thus a *full* Flamm paraboloid—top and bottom.

How did we arrive at the wormhole of Fig. 12.7 as a typical time slice of Kruskal space? In non-stationary spacetimes (like Kruskal space) the time coordinate is largely arbitrary, though one would probably still want to adapt it to whatever symmetries the spacetime possesses. Kruskal's original time coordinate t is not the best choice for understanding the geometrical evolution of this spacetime: any section $t = \text{const} > 1$ (or < -1) cuts the spacetime in two! A more convenient slicing is provided by the hyperbolae confocal with the hyperbola of the singularity $R = 0$ (see Fig. 12.9). This

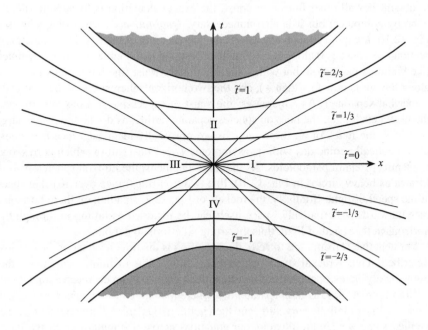

Fig. 12.9

entire family is characterized by the equation

$$\frac{t^2}{a^2} - \frac{x^2}{2-a^2} = 1,$$ (12.30)

the distance between the foci being $2\sqrt{2}$, and the parameter a varying (for the hyperbolae of interest) between one and zero. The asymptotes of all these hyperbolae intersect at the origin, and as $a \to 0$ the asymptotes open up while the vertices approach the origin. Kruskal's time t on the t-axis can serve to calibrate these hyperbolae, so that our new time coordinate $\tilde{t}$ coincides with the parameter a in eqn (12.30), provided we assign the negative value to it below the x-axis. The entire Kruskal 'universe' then comes into being at time $\tilde{t} = -1$ and ceases to exist at time $\tilde{t} = 1$. We can regard the infinite proper lifetimes outside the horizon as due to fast motion and consequent time-dilation, just as in the Minkowski rockets.

The R in the Kruskal diagram (Fig. 12.5) tells us the radius of the 2-sphere $R =$ const of full Kruskal space through the event in question. In fact, each point in the diagram can be regarded as representing such a 2-sphere. As we move along any line $\tilde{t} =$ const in the Kruskal diagram from left to right, R decreases from infinity to a minimum at $x = 0$, and then increases back to infinity, giving us the wormhole geometry. The distance between successive spheres $R =$ const corresponds to the arc along the line $\tilde{t} =$ const. Along $\tilde{t} = 0$, this is easy to calculate (from the Schwarzschild form of the metric), and that is why the equation of the Flamm paraboloid is so easy to obtain. For all other lines $\tilde{t} =$ const, the exact calculation is forbidding (unless done by computer) but it is also unnecessary. *Qualitatively* correct diagrams (see Fig. 12.10) are quite enough to give us an understanding of the Kruskal geometry. A section $\tilde{t} =$ const > 0 close to $\tilde{t} = 0$ will correspond to a double-trumpet very much like Flamm's paraboloid but with a slightly thinner waist (the minimum value of R along the cut is now less than $\frac{1}{2}$), and the two horizon spheres $R = \frac{1}{2}$ have already somewhat separated. As $\tilde{t}$ increases, the waist of successive sections gets thinner, the neck gets longer, the horizon spheres separate farther, as do the flares. For large R-values, the flares of all these sections are practically identical, since $\tilde{t} =$ const then practically coincides with its asymptotes $T =$ const, all of which correspond to identical Flamm paraboloids. As the neck lengthens, the horizon spheres $R = \frac{1}{2}$ (drawn as heavy circles in Fig. 12.10, like two iron rings) race over Kruskal space at the speed of light. Similarly, the rockets of Fig. 12.7, standing on these horizons, race outwards. Alternatively, space itself can be regarded as falling in through the horizons at the speed of light, thus thwarting all efforts to get out.

For negative $\tilde{t}$, from $-\infty$ to zero, the evolution is the time reversal of the one just described. Thus Kruskal space begins as an infinite line of infinite curvature; this immediately flares open into a long-drawn-out double trumpet, reaches maximum girth and zero neck as a Flamm paraboloid, whereupon the entire sequence is reversed, back to a line of infinite curvature, and then nothing. (Or, at least, nothing that we can predict, since we cannot integrate our equations across a singularity.) As the flares open up, the horizons bounding the white-hole region IV move inwards; at half-time

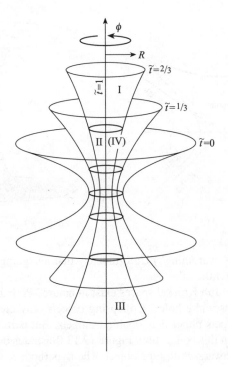

Fig. 12.10

they cross over each other, region IV disappears and the black-hole region II begins to develop.

There is an alternative (though related) slicing of Kruskal space that is also of some interest. Take the same confocal hyperbolae $\tilde{t} =$ const as in Fig. 12.9 between the horizons, but *just* outside the horizons, where they intersect a fixed hyperbola $R = \frac{1}{2} + \epsilon$, continue them with $T =$ const. A typical such slice is shown as a heavy line in Fig. 12.11(a). It corresponds to two exact Flamm-paraboloid halves ('minus ϵ') joined by the neck ('plus ϵ') that corresponds to the $\tilde{t}$-hyperbola under consideration, as shown in Fig. 12.11(b). As time progresses, the paraboloid halves remain unchanged. Only the neck between them contracts from infinity to zero, at which time the horizons are interchanged, whereupon the neck lengthens back to infinity. An observer on one of the paraboloids (outer Schwarzschild space!) is oblivious to the dynamic behavior of the neck. Nevertheless this is the only way a vacuum *can* behave to create the static conditions outside.

John Wheeler and his school at one time hoped to construct a *geometric* theory of elementary particles in which Kruskal spaces together with their 'electrically charged' generalizations, and a spacetime honeycombed with Kruskal-type worm-holes would play a basic role. ('Matter without matter', 'charge without charge',

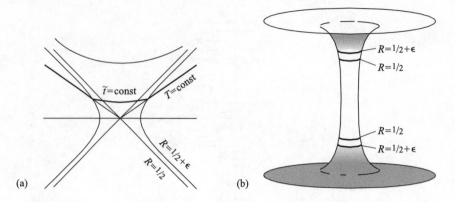

Fig. 12.11

'vacuum geometry is everything'.) Unfortunately that intriguing idea eventually ran into unsurmountable difficulties.

Other than that, do full Kruskal spaces exist in nature? Probably not. They would have to be created as white holes. Collapsing objects only create *partial* Kruskal spaces—single trumpets rather than double trumpets. But partial Kruskal diagrams are still very useful in this connection. Figure 12.12 illustrates on a Kruskal diagram the collapse of a Schwarzschild-type object. The hyperbola $R = R_s$ represents the object's original surface. At B the collapse BCD begins. The part of the diagram to the left of the surface locus ABCD is inapplicable and would have to be replaced by some representation of the interior metric. But the part to the right of that line is fully operational and could be used, for example, to study how an outside observer sees the collapse. In Fig. 12.11(b) we could imagine a somewhat curved black rubber disk covering the hole and representing the object before collapse; the trumpet below it is missing. As the collapse begins, the disk shrinks and drops down the neck, which only comes into existence as needed. Eventually the disk passes the horizon at $R = \frac{1}{2}$, whereupon continued collapse becomes inevitable. The neck now lengthens and narrows indefinitely, with the ever smaller rubber cap rounding it off at the bottom.

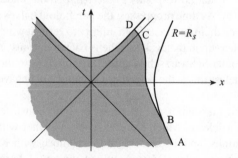

Fig. 12.12

12.6   Black Hole Thermodynamics and Related Topics

In this section we report, in a purely anecdotal way, some of the later developments in black hole theory, of which the reader should at least be aware.[1]

A kind of revolution began in theoretical general relativity in 1965 with the introduction of global topological techniques by Penrose, and it continued later with the introduction of quantum-mechanical methods by Hawking in 1974. Already in 1963, general relativity had been enriched by Kerr's discovery of a new exact stationary vacuum solution of Einstein's field equations that corresponds to the axially symmetric field around some steadily rotating source. It contains just two free parameters: one, m, representing the effective mass of the source, and another, a, representing its angular momentum per unit mass. When a is set to zero, Kerr's solution reduces to that of Schwarzschild. It gradually came to be understood, and proved, that it represents a rotating black hole, and, in fact, the unique final equilibrium configuration of *any* electrically neutral fully collapsing body, no matter how irregular, after all the wobbling and gravitational radiating has died off. Like the Schwarzschild field, Kerr's field also has a horizon and an inner curvature singularity. It soon became the basis of all further black-hole research.

What Penrose established in 1965 was the first "singularity theorem". Before that, it was thought that Schwarzschild- and even Kerr-collapse were very special and artificially symmetric cases, and that perhaps in more realistic situations the collapsing matter might somehow swirl around and avoid ending in a singularity. Penrose showed that this was not so. He invented the concept of a "trapped surface". This is a closed surface, such that all light emitted on it moves inward; which is just what happens over any surface $r = \text{const} < 2m$ in Schwarzschild space. (Technically, one requires both inward and outward directed light rays to "converge".) And then he proved that under some very reasonable assumptions (positivity of energy, etc.) a singularity must form inside any trapped surface. Not all of the matter necessarily ends up at that singularity, but at least some will. Later Penrose and Hawking, Hawking alone, and others, extended the original singularity theorems and proved, for example, that our expanding universe, even under less symmetric initial conditions than are usually assumed, must have had a singularity in the past, and will have another in the future, if it ever recollapses.

These global topological studies led Hawking in 1970 to the theorem that the area of a black hole (namely, of its horizon) can never decrease, and that, for example, when two black holes merge, the area of the resultant black hole exceeds the sum of the previous areas. This was reminiscent of the second law of thermo-dynamics, the inevitable increase of entropy. Christodoulou, then a graduate student at Princeton, had already pointed out the similarity of several of the black-hole equations to those of thermodynamics. Bekenstein, another Princeton student, now became convinced that black holes actually *must* have entropy, and that from Hawking's result one could identify this entropy with a multiple of the hole's area. He had long been worried by

---

[1] For a fuller and very readable overview, see J.D. Bekenstein, *Physics Today*, Jan. 1980.

the aspect of black holes as unlimited sinks for entropy, in violation of the second law. Here was the solution! He found a formula for this entropy, which turned out to give huge values for stellar black holes. This he did by clever thought experiments, for example, the slow lowering of tepid gas in a box to the horizon (cf. our Section 12.2), and thus extracting most of its energy but not all (since the box must not quite touch the horizon.)

But now came a paradox: having entropy, the black hole would have to have a temperature; and having a temperature, it would have to radiate. Yet radiate it cannot: nothing can escape a black hole.

Or can it? After much original opposition to Bekenstein's ideas from the "establishment" (including Hawking), Hawking himself in 1974 made the startling discovery that black holes can and do radiate—by quantum-mechanical processes. Moreover, the radiation (*Hawking radiation*) will have a blackbody spectrum. A "poor man's version" of the process is that vacuum fluctuations just outside the horizon create virtual photon pairs, which the tidal forces not only pull apart but also convert into real photons. One is swallowed, the other escapes, and has robbed the black hole of the energy needed to make it real; so the black hole shrinks ever so little. In this way the liaison between quantum mechanics and general relativity (it is not yet a full marriage!) saved general relativity from the embarrassment of allowing objects (black holes) that could defy the second law of thermodynamics.

The Bekenstein-Hawking formulas for the temperature and entropy of a black hole are as follows:

$$T_{BH} = \frac{\hbar c^3}{8\pi GkM}, \qquad S_{BH} = \frac{kc^3 A}{4G\hbar}, \tag{12.31}$$

where k is Boltzmann's constant, M is the mass of the black hole, and A its surface area. As the radiation carries away mass (at first at an incredibly slow rate for "normal" holes), the temperature rises, and the mass-loss rate speeds up. One speaks of black-hole *evaporation*. In the end (aeons down the road), the tidal forces get strong enough to pull apart even massive virtual particle pairs (such as electrons) and the black hole ends in a violent explosion. One can calculate its total lifetime to be given by

$$t_{BH} \approx 1.5 \times 10^{66} \left(\frac{M}{M_\odot}\right)^3 y. \tag{12.32}$$

From the analogy of the accelerating reference frame discussed in Section 12.4 above, it may now come as no complete surprise that even in that frame a stationary observer will measure a temperature; that is, an observer accelerating through the Minkowski vacuum sees a virtual heat bath! (Because of the absence of tidal forces, the photons never become real.) This effect was discussed by Davies in 1975 and analysed by Unruh in 1976. It is referred to as *Unruh radiation*, and its temperature is given by

$$T_U = \frac{\hbar \alpha}{2\pi ck}, \tag{12.33}$$

α being the proper acceleration of the observer. For an acceleration of 1 g, the temperature is a mere 4×10^{-20} K. But for the high accelerations occurring in particle accelerators it may become measurable; and there are, indeed, indications of its reality.[2]

Just as the existence of black holes without temperature and without entropy would disturb the fabric of thermodynamics, so the existence of "naked" singularities would disturb not only thermodynamics but all of physics. For the very essence of physics is that it allows us to predict the evolution of systems. Naked singularities, unguarded by outwardly impassable horizons—like the initial singularity of a Kruskal "white hole"—would foil all predictions, since no laws govern what comes out of them. Guided by various plausibility calculations, Penrose in 1969 proposed what he called the *cosmic censorship hypothesis*: Nature forbids nakedness! The only permitted naked singularity is the big bang itself; here we believe we know what came out. According to Penrose's hypothesis, no singularity apart from the big bang will ever be "seen" by any observer. Realistic mass distributions imploding to a singularity will always develop a horizon, that is, will form a black hole. Although some theoretical counterexamples have been constructed, the general concensus is that these are all too artificial to occur naturally. With the censorship hypothesis, general relativity and the rest of physics are in harmony. As Penrose has pointed out, the hypothesis even seems to be required in order that the modern discussion of black holes can be carried through—like Hawking's area increase theorem or the settling down into a Kerr black hole rather than something worse in generic gravitational collapse. But a rigorous proof is still outstanding, and is regarded as one of the most urgent problems in general relativity.

Exercises 12

12.1. Using the last formula of Exercise 11.4, prove that an observer falling freely from rest at infinity towards a Schwarzschild black hole of mass m receives a radially infalling photon at frequency $v = v_0(1 + \sqrt{2m/r})^{-1}$, where $r > 2m$ is the observer's radial coordinate at reception and v_0 is the photon's frequency at infinity. Give reasons why this formula must continue to hold even for $0 < r \leq 2m$. What is v at the horizon? What is v at $r = 0$? [*Hint*: (12.10), (11.29).]

12.2. In Newtonian theory, consider the gravitational collapse from rest of a homogeneous ball of dust of mass M and original radius r_0 and density ρ_0. Prove that the total time Δt of the collapse is given by

$$\Delta t = \frac{\pi}{2} r_0^{3/2} (2GM)^{-1/2} = (3\pi/32 G\rho_0)^{1/2}.$$

[*Hint*: Consider the motion of a particle of the surface; a differential equation of the form $\ddot{r} = f(r)$ ($\dot{} = d/dt$) is solved by using an integrating factor $\dot{r}$; and for the resulting integral cf. (12.13).]

---

[2] J.S. Bell and J.M. Leinaas, *Nucl. Phys.* B284, 488 (1987).

12.3. In Newtonian theory, prove that the collapse time of two point masses m, originally at rest a distance $2r_1$ apart, is given by

$$\Delta t = \pi r_1^{3/2} (2Gm)^{-1/2}.$$

12.4. A small electric generator lights an incandescent lamp at the bottom of a deep shaft on earth. If the same generator is placed at the top of the shaft and connected to the lamp by resistance-less wires, by what factor does the lamp shine more brightly? [*Hint*: cf. Section 12.2.]

12.5. By examining the metric (12.9) and Fig. 12.2, find an Eddington–Finkelstein type coordinate system r, θ, ϕ, u that regularly covers a 'white hole'; that is, an outer Schwarzschild space where the positive sense of t determines the future and an inner Schwarzschild space where the positive sense of r determines the future. (The original system covered a *black* hole.) Also draw the new system's light cone diagram analogous to Fig. 12.2.

12.6. Prove that for coordinates x, t and X, T related as in Table 12.1 (be it in Minkowski or in Kruskal space) a T-translation $T \mapsto T - V$ is equivalent to a standard Lorentz transformation in x and t with V as the hyperbolic parameter called ϕ in our earlier eqn (2.16). [*Hint*: Set $t' = X \sinh(T - V)$, etc., and recall the rule for translating trigonometric into hyperbolic formulae: $\cos \mapsto \cosh$, $\sin \mapsto \sinh$, $\sin \sin \mapsto -\sinh \sinh$; for comparison, add and subtract eqns (2.16) (i) and (ii).]

12.7. In the Minkowski-space rocket described in Section 12.4, prove that all light paths (except those in the X-direction) are *semi-circles* relative to the Euclidean lattice coordinates X, Y, Z, with typical equation $X^2 + Y^2 = \text{const}$, $X > 0$. Similarly, prove that all free-particle paths (other than those parallel to the X-direction) are *semi-ellipses*. [*Hint*: Use Table 12.1 to translate from the underlying x, y, z, t system—but preferably after applying a suitable Lorentz transformation (aberration!).]

12.8. In truncated Minkowski space *all* timelike geodesics, and, in particular, those parallel to the spatial x-axis, end on the future 'edge'. Show that, by contrast, in Kruskal space timelike radial geodesics do not necessarily end on the future singularity. (This essentially stems from the fact that under a $1/r^2$ law there is an escape velocity, while under a $1/r$ law there is none.) On the Kruskal diagram, sketch such an unending geodesic, bearing in mind that eventually it must surpass every positive R- and every positive T-value.

12.9. In the Kruskal diagram, qualitatively sketch a sequence of worldlines of free particles dropped from rest at some given lattice point P in region I; that is, issuing tangentially from one of the hyperbolae $R = \text{const}$. Recall that all such geodesics are Lorentz transformable into each other. In particular, prove: (i) If one of these worldlines were to have an inflection point, so would all. [*Hint*: An inflection point would lie between two points at which the tangents are parallel.] (ii) All these worldlines start off *between* the hyperbola corresponding to P and its tangent at the event of release. [*Hint*: Show that for the special radial geodesics of (12.28) which start with $\dot{x} = 0$ at $t = 0$, $\ddot{x}$ is initially positive.] Observe that *all* the geodesics which impinge the locus $R = 0$

at any one of its points, are divided by the geodesic $T = $ const into two classes, coming from quadrants I and III respectively. They are further subdivided into those that ultimately come from II, and those that do not. [In time reversal, consider escape velocity.]

12.10. Two particles are dropped, one after the other, from rest at the same lattice point of outer Schwarzschild space, and both fall through the horizon to the singularity $R = 0$. Use a Kruskal diagram to answer the following questions: (i) Do the particles see each other at all times? (ii) Does the second particle see the first hit the singularity? (iii) Is there a finite portion of the first particle's history that cannot be seen by the second? Compare this situation with Newtonian infall to a point mass.

12.11. Describe the geometry of inner Schwarzschild space (that is, region II in the Kruskal diagram), as a succession of cuts $R = $ const, from $R = \frac{1}{2}$ to $R = 0$. What can you say about the loci $T = $ const which all coincided at $R = \frac{1}{2}$? Can an observer, riding on such a locus, at all times see all other such observers? [*Hint*: For the metric of these cuts it is convenient to choose the original Schwarzschild coordinates, as in (12.27), writing $\tilde{T}$ for R and $\tilde{R}$ for T.]

12.12. (*Denur's paradox*) Consider a Schwarzschild black hole of radius r_0. An infalling concentric massive shell of negligible thickness at radius $r_1 > r_0$ has just created another horizon at r_1. Consider a particle between these two horizons. Before the shell sweeps over it, it lives in an outer Schwarzschild region (by Birkhoff's theorem), and can therefore remain at constant r (for example, using a jet engine). Yet it is inside the outer horizon. How can a particle 'stand still' inside that horizon? On an $r - \tau$ diagram analogous to Fig. 12.1, but with a conventional coordinate τ instead of t that remains finite when the shell crosses both horizons, sketch the history of the infalling shell as well as the complete history of the two horizon light fronts, both before and after they are crossed by the shell. Draw little future light cones at various points along all three of these worldlines and also in the regions between. [*Answer*: As the light cone structure will demonstrate, a particle *can* stand still in the diminishing zone outside the inner horizon and inside the shell. All along the worldline of the infalling shell, between the two horizons, the 'right' side of the light cone points straight up. Cf. J. Ehlers and W. Rindler, *Phys. Lett.* A**180**, 197 (1993).]

12.13. According to the Stefan-Boltzmann law, the rate of energy loss from a hot surface at temperature T is given by σT^4 per unit area, where the Stefan-Boltzmann constant σ has the value $5.670 \times 10^{-5} \, \text{erg cm}^{-2} \text{s}^{-1} \text{K}^{-4}$. Use this information to derive (12.32). [*Hint*: take the surface area of the black hole to be $4\pi r^2$ with $r = 2\,GM/c^2$.]

12.14. Taking the age of the universe to be $\sim 2 \times 10^{10} y$, what would have to be the initial mass of a primordial black hole (a "minihole")—formed from fluctuations soon after the big bang—for it to explode today? [*Answer*: $\sim 10^{-19} \, \text{M}_{\odot}$ or $\sim 10^{14}$ g.] Astromomers have looked for minihole explosions but found no evidence of their existence.

13

An exact plane gravitational wave

13.1   Introduction

In this short chapter we shall exhibit and discuss one very special gravitational wave, not so much because in itself it has any practical importance, but rather to familiarize the reader with the very idea of gravity waves and with some of the topological subtleties that may arise. In this particular case these are quite as surprising as was the topology of Kruskal space.

One might well expect that curvature disturbances in spacetime would propagate at the speed of light, and so give rise to 'gravitational radiation'. That this is, in fact, the case has been shown by many theoretical investigations, using mainly approximative methods. However, there also exist certain exact solutions of Einstein's vacuum field equations that clearly represent gravity waves, and we shall here examine one of the simplest of these, a special plane 'sandwich' wave. It is not the kind of wave that could be generated by any reasonable source distribution; rather, it would have to be created *in toto*, much like full Kruskal space. Nevertheless it exhibits some interesting properties which might be expected to apply also in more realistic radiative situations.

Note that a gravity wave in vacuum satisfies the *vacuum* field equations. Although it certainly carries energy and momentum, such gravitational energy and momentum are not (and cannot be) treated as source terms in the field equations; they are nevertheless included in the physics though the non-linearity of those equations.

13.2   The plane-wave metric

Any scalar wave profile $p = f(x)$ propagating in the x direction at speed c will, after time t, be given by the equation $p = f(x - ct)$. We shall here take $c = 1$ and also, following tradition, write $p = p(t - x)$, where $p(x) = f(-x)$. Guided by the electromagnetic analogy, we might expect gravitational waves to be transverse; that is, to distort spacetime only in directions orthogonal to the direction of propagation. In this way we are led to consider plane-wave metrics of the form

$$\mathbf{ds}^2 = dt^2 - dx^2 - p^2(u)\,dy^2 - q^2(u)\,dz^2, \tag{13.1}$$

where $p(u)$ and $q(u)$ are assumed to be positive functions depending only on the 'null coordinate'

$$u := t - x. \tag{13.2}$$

For simplicity we assume the absence of a cross-term $r^2(u)\,dy\,dz$. Such waves are homogeneous on all instantaneous 2-planes $x =$ const. By reference to the Appendix, it is straightforward to compute the components of the Riemann tensor $R_{\mu\nu\rho\sigma}$ and the Ricci tensor $R_{\mu\nu}$ for the metric (13.1). In this way we obtain the *flatness condition*

$$\ddot{p} = \ddot{q} = 0 \quad \Leftrightarrow \quad R_{\mu\nu\rho\sigma} = 0 \tag{13.3}$$

(where overdots denote differentiation with respect to u), and the *vacuum condition* (field equation)

$$\frac{\ddot{p}}{p} + \frac{\ddot{q}}{q} = 0 \quad \Leftrightarrow \quad R_{\mu\nu} = 0. \tag{13.4}$$

We now specialize the metric (13.1) to represent a *sandwich* wave, namely a region of curvature sandwiched between two parallel null planes $u = 0$ and $u = -a$, say, outside of which spacetime is flat, namely Minkowskian (cf. Fig. 13.1). In 3-dimensional language this *appears* to correspond to a single 3-dimensional flat background space through which two parallel planes orthogonal to the x-axis and bounding a region of curvature travel at the speed of light in the positive x-direction (cf. Fig. 13.2). However, as we shall see, this picture will have to be taken with a grain of salt.

According to our plan, we must choose the functions $p(u)$ and $q(u)$ so as to satisfy the vacuum condition (13.4) everywhere and the flatness condition (13.3) outside the region bounded by the planes $u = 0$ and $u = -a$. Along these planes the metric must satisfy certain smoothness conditions, which here simply amount to p, q and $\dot{p}, \dot{q}$ being continuous. The flatness condition demands that p and q be linear functions of u outside the wave zone. For reasons that will become clear presently, we choose

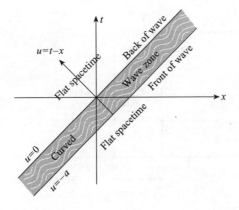

Fig. 13.1

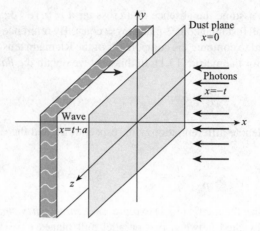

Fig. 13.2

these linear functions as follows:

$$p = q = 1 - u \quad \text{(when } u > 0) \tag{13.5}$$

and

$$p = p_0 = \text{const}, \quad q = q_0 = \text{const} \quad \text{(when } u < -a). \tag{13.6}$$

Exactly how p and q can be continued (in infinitely many ways) inside the wave zone so as to satisfy the vacuum field equation (13.4) and to join smoothly with (13.5) at $u = 0$ and to go smoothly over into *some* constant values at $u = -a$ in accordance with (13.6), is shown in Section 13.7 below, so as not to break the continuity here. But Fig. 13.3 should make it plausible that a continuation *can* be found. By (13.4), $\ddot{p}$

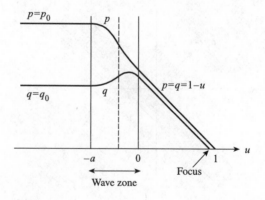

Fig. 13.3

and $\ddot{q}$ must have opposite signs, so p and q must inflect at a common u value; and the larger of the two must have a proportionally larger second derivative. Let us assume that a definite choice for these functions in the wave zone has been made.

13.3   When wave meets dust

To analyze what happens when this wave encounters a sheet of stationary test dust (that is, dust so 'light' as to make no contribution to the spacetime geometry), and later also an oncoming sheet of test photons (cf. Fig. 13.2), we need the following two lemmas:

Lemma I: *Particles satisfying the equations*

$$x, y, z = \text{const}$$

follow timelike geodesics of the metric (13.1).

Lemma II: *'Particles' satisfying the equations*

$$x = \pm t; \qquad y, z = \text{const}$$

follow null geodesics of the metric (13.1). *In both cases, t is an affine parameter.* For proof we need only check (with the help of the Appendix) that

$$\Gamma^{\mu}_{xx} = \Gamma^{\mu}_{xt} = \Gamma^{\mu}_{tt} = 0; \tag{13.7}$$

the worldlines of Lemma I imply $d^2x^{\mu}/dt^2 = 0$, $dy/dt = 0$, $dz/dt = 0$, and so do those of Lemma II; so each of these sets of worldlines satisfies the geodesic equation (10.15) with t as affine parameter. And substitution in the metric shows the former lines to be timelike and the latter to be null.

Now consider a typical plane $x = 0$ of test dust particles all of which are initially at rest in the Minkowski space with metric

$$\mathbf{ds}^2 = dt^2 - dx^2 - p_0^2\,dy^2 - q_0^2\,dz^2 \tag{13.8}$$

ahead of the wave. By virtue of the continued validity of the equations $x = 0$; $y, z = $ const for the motion of each such particle through and beyond the wave zone (according to Lemma I), and by our choice of the functions $p(u)$ and $q(u)$, all these particles will focus (meet) at one single event $\mathcal{F}$ having $(x, y, z, t) = (0, 0, 0, 1)$. For the dx, dy, dz of any two neighboring dust particles remain constant (with $dx = 0$) and the square of their spatial separation at all times after leaving the wave zone is given by $dx^2 + (1-t)^2(dy^2 + dz^2)$, which vanishes at $t = 1$. Since the particle having $x = y = z = 0$ is present at $\mathcal{F}$, so must all the others be. On the other hand, we note that the distance between any two imagined neighboring dust particles having only an x-separation, say dx, never changes.

For readers returning to the present section after covering Section 16.3 below, we may point out that the effect of the passing wave on the sheet of dust at $x = 0$ is to convert it into a 2-dimensional, flat, collapsing Milne universe with FRW metric

$$\mathbf{ds}^2 = dt^2 - (1 - t)^2(dy^2 + dz^2). \tag{13.9}$$

We shall presently develop a method for finding the exact paths of these particles in the post-wave space, and by that method also show that a plane of photons meeting the wave head-on gets similarly focused (though not at $\mathcal{F}$). The same method will also provide answers to some troubling questions: (i) What about the singularity of the metric (13.1) at $u = 1$ where, by (13.5), the functions $p(u)$ and $q(u)$ vanish? (However, since the space is flat arbitrarily close to this event, it is at least not a curvature singularity.) (ii) How is it that dust particles arbitrarily far away from the x-axis at impact can travel to arrive at $\mathcal{F}$ a time $1 + a$ later without breaking the relativistic speed limit? And (iii) why does the focus lie on the x-axis rather than off it, seeing that there is complete homogeneity in the y and z directions?

13.4   Inertial coordinates behind the wave

The method we have alluded to consists in going to a new set of coordinates X, Y, Z, T in the post-wave space, which converts the metric (13.1) there into the familiar Minkowskian form and so provides a direct physical picture of what is going on. One possible such coordinate set is given by the transformation

$$
\begin{aligned}
T &= t - \tfrac{1}{2}(1 - u)(y^2 + z^2) \\
X &= x - \tfrac{1}{2}(1 - u)(y^2 + z^2) \\
Y &= (1 - u)y \\
Z &= (1 - u)z,
\end{aligned}
\tag{13.10}
$$

which transforms (13.1) with (13.5) into the metric

$$\mathbf{ds}^2 = dT^2 - dX^2 - dY^2 - dZ^2 \tag{13.11}$$

in the post-wave space (as can be checked at the cost of a little algebra). This immediately answers the first of our 'troubling questions': there are *no* intrinsic singularities in the post-wave zone.

The front of the wave, $u = -a = t - x$, is seen as a 2-plane orthogonal to the x-axis and traveling at the speed of light in the x-direction in the inertial frame (13.8) ahead of the wave. Similarly, the back of the wave, $u = 0 = t - x = T - X$, is seen as a 2-plane traveling at the speed of light in the X-direction in the inertial frame (13.11) behind the wave.

As we shall show in Section 13.7, we can have p_0 and q_0 arbitrarily close to unity by taking the width of the wave zone small enough. To simplify our further discussion we accordingly go to the limit and treat the wave as a 'delta wave', with essentially zero width and $p_0 = q_0 = 1$. The wave is then axially symmetric around the x-axis.

Let us now fix our attention on a *typical* radial line of dust particles in the plane $x = 0$, say

$$x = 0, \qquad y = \eta, \qquad z = 0, \tag{13.12}$$

with η serving as parameter. By Lemma I, eqns (13.12) permanently characterize any one of these particles. After emerging from the wave, by (13.10), the η particle satisfies the equations of motion (with t as parameter)

$$\begin{aligned} T &= t - \tfrac{1}{2}(1 - t)\eta^2 \\ X &= -\tfrac{1}{2}(1 - t)\eta^2 \\ Y &= (1 - t)\eta \\ Z &= 0. \end{aligned} \tag{13.13}$$

In particular, it exits the wave at $u = 0$ (and $t = 0$), at the event

$$(X, Y, Z, T) = \left(-\tfrac{1}{2}\eta^2, \eta, 0, -\tfrac{1}{2}\eta^2 \right) \tag{13.14}$$

and travels forward along the line

$$Y = -\frac{2}{\eta}X, \qquad Z = 0. \tag{13.15}$$

The effect of the passing wave is therefore to 'kick' the particle *forward* and towards the X-axis. (A similar effect can occur with electromagnetic waves: at first, the passing electric field kicks a stationary charge sideways, and once it moves the magnetic field kicks it longitudinally.)

We observe, from (13.13), that all the particles of the original plane $x = 0$ will be present at the focus event $\mathscr{F}$ corresponding to the parameter value $t = 1$:

$$(X, Y, Z, T) = (0, 0, 0, 1). \tag{13.16}$$

The squared displacement for each particle from exiting the wave to $\mathscr{F}$ is therefore given by

$$\begin{aligned} \Delta s^2 &= \Delta T^2 - \Delta X^2 - \Delta Y^2 - \Delta Z^2 \\ &= \left(1 + \tfrac{1}{2}\eta^2\right)^2 - \left(\tfrac{1}{2}\eta^2\right)^2 - \eta^2 = 1. \end{aligned} \tag{13.17}$$

It is positive: none has broken the speed limit! The mechanism whereby this is possible is the most surprising result of this analysis. While the particles all enter the wave simultaneously at $t = 0$ as judged in the front inertial frame S, they leave it at *different* times as judged in the back inertial frame Š: one ring at a time, as we see from (13.14), the farther from the X-axis, the earlier. It is this that gives them the necessary head

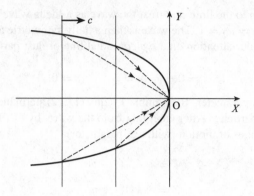

Fig. 13.4

start to make it to the focus event. By (13.14), the emergence events all lie on the paraboloid of revolution (cf. Fig. 13.4)

$$Y^2 + Z^2 = -2X. \tag{13.18}$$

As the wave plane passes over this stationary paraboloid, it releases particles at its intersection ring with the paraboloid, whereupon these particles set out on their journey to the focus. Emerging rings have the same diameter as when they entered, since, by (13.10), $Y = y$ at $u = 0$.

Let us now return to the question of why the focus lies on the X-axis and not on some other point of the plane $X = 0$, since the physics itself does not single out the X-axis. The answer is that our choice (13.10) of coordinates for the post-wave inertial frame $\tilde{S}$ favors the particle originally on the x-axis: $\tilde{S}$ is that particle's eventual rest-frame, and so all other particles travel to that one. Every one of the original particles in this way determines a post-wave inertial frame in which *it* would be at rest at emergence while the rest of the picture would be identical. (Readers who have studied Milne's universe will readily understand this.)

13.5   When wave meets light

An analysis very similar to that for a stationary plane of dust can be given for an incident plane $x = -t$ of test photons meeting the wave head-on at $x = t = 0$. Instead of (13.12), we now have

$$x = -t, \qquad Y = \eta, \qquad z = 0 \tag{13.19}$$

for a typical radial line of photons. And by our Lemma II, eqns (13.19) will permanently characterize a given photon. After emerging from the wave this photon, by

(13.10), therefore satisfies the equations of motion (again with t as parameter)

$$T = t - \tfrac{1}{2}(1 - 2t)\eta^2$$
$$X = -t - \tfrac{1}{2}(1 - 2t)\eta^2 \qquad (13.20)$$
$$Y = (1 - 2t)\eta$$
$$Z = 0.$$

In particular, it exits the wave at $u = 0$, (and $t = 0$), hence at the event

$$(X, Y, Z, T) = \left(-\tfrac{1}{2}\eta^2, \eta, 0, -\tfrac{1}{2}\eta^2 \right). \qquad (13.21)$$

Not surprisingly, we have exactly the same exit pattern for the photons as for the dust: one ring of photons exits after another, and all the exit events lie on the paraboloid (13.18), as before.

After emerging from the wave zone, the photons must, of course, travel rectilinearly, which is consistent with (13.20). Like the dust particles swallowed up by the wave at the same time as the photons, the latter also converge onto a common focus event, though a different one from that of the dust. Inspection shows that the paths (13.20) all contain the event

$$(X, Y, Z, T) = \left(-\tfrac{1}{2}, 0, 0, \tfrac{1}{2} \right), \qquad (13.22)$$

which, in fact, occurs at the same value of the affine parameter, $t = \tfrac{1}{2}$, for all of them. This is the focus event for the photons.

13.6   The Penrose topology

A plane delta wave with flat spacetime in back and in front of it might be pictured (wrongly) as a simple Minkowski space with a 'crack' filled with curvature cutting it in half, as shown in Fig. 13.5. In this picture, two events on either side of the crack,

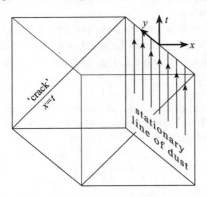

Fig. 13.5

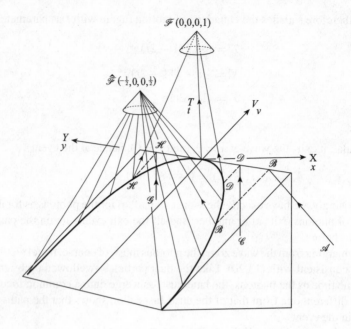

Fig. 13.6

which are close in the diagram, would be close also in a physical sense. But just by the focusing property of the wave, even if it were only focusing in one direction (as it would be without special adjustment of p and q—cf. Exercise 13.5), this picture cannot be right. For consider a single infinite line of dust particles, as in eqn (13.12), whose worldlines are marked in Fig. 13.5, entering the wave. With the simple topology of Fig. 13.5, such a line would emerge from the crack also as an infinite straight line. But then the particles could not possibly all travel to a focus in the finite future without breaking the speed limit (that is, without some worldlines being inclined to the time direction at more that 45°).

The true situation, as was first pointed out by Penrose,[1] is that the two half-Minkowski spaces on either side of the crack are joined with a more complicated topological identification of points (cf. Fig. 13.6). Let us take y, z, and

$$v = t + x \tag{13.23}$$

as coordinates for the lower crack surface, whose equation is $x = t$, or $u = 0$ (since we assume $a = 0$). And similarly take Y, Z, and

$$V = T + X \tag{13.24}$$

---

[1] R. Penrose, in *Battelle Rencontres*, ed. C. M. Dewitt and J. A. Wheeler, W. A. Benjamin, New York, 1968, p. 198; also R. Penrose, *Int. J. Theor. Phys.* **1**, 61 (1968).

as coordinates for the upper crack surface. Then, by (13.10), we have the relation

$$V = v - Y^2 - Z^2. \tag{13.25}$$

(In Figs 13.5 and 13.6 the z and Z dimensions are, of necessity, suppressed.) All straight lines $v = $ const on the lower crack surface [such as the line of particles (13.12) just entering the wave] correspond to, and must be topologically identified with, the respective parabolae $V = v - Y^2$ on the upper crack surface. It is easy to convince oneself that each such parabola lies entirely within the past light cone of the corresponding focus event, so that the focusing can be achieved legally. In fact, by (13.17), each such parabola is the intersection of the crack plane with the hyperboloid of revolution $\Delta s^2 = 1$ centered on the corresponding particle focus $\mathscr{F}$. It is also the intersection of the crack plane with the past light cone of the corresponding photon focus $\hat{\mathscr{F}}$. The diagram shows one typical photon path ($\mathscr{ABF}$) and two typical particle paths ($\mathscr{CDF}$ and $\mathscr{GHF}$) through the wave.

In sum, then, the topology of our particular delta wave consists of two half-Minkowski spacetimes connected across a null plane (the crack) with a parabolic relative displacement between identified points of the lower and upper crack surfaces. In full dimensions, the crack is 3-dimensional (the history of a 2-plane traveling at the speed of light), and the topological displacement (13.25) is a paraboloid of revolution. The details of the identification depend on the particular wave, but there must always be a displacement involved if the wave is homogeneous and focusing.

13.7 Solving the field equation

In this section we shall justify our solution (13.5) and (13.6) of the field equation (13.4) for the metric (13.1)—a task we postponed at the time to streamline the argument. For this purpose, we introduce two auxiliary functions $L(u)$ and $m(u)$ by the equations

$$p = Le^m, \qquad q = Le^{-m}. \tag{13.26}$$

It is then straightforward to verify that the field equation (13.4) corresponds to

$$\ddot{L} + L\dot{m}^2 = 0. \tag{13.27}$$

Our choice (13.5) for the region $u > 0$, namely $p = q = 1 - u$, corresponds to $L = 1 - u$ and $m = 0$, while the choice (13.6) for the region $u < -a$, namely $p = p_0$, $q = q_0$, corresponds to $L, m = $ const (cf. Fig. 13.7). The results we anticipated were that p and q could be suitably continued across the wave zone and also that $p_0, q_0 \rightarrow 1$ as $a \rightarrow 0$.

We now join the two prescribed pieces of L across the wave zone by the quartic

$$L = 1 - u + \frac{1}{a^2}u^3 + \frac{1}{2a^3}u^4, \tag{13.28}$$

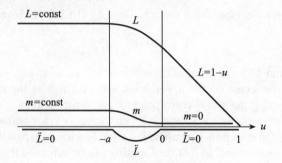

Fig. 13.7

which makes L as well as

$$\dot{L} = -1 + \frac{3}{a^2}u^2 + \frac{2}{a^3}u^3, \tag{13.29}$$

and

$$\ddot{L} = \frac{6}{a^3}(au + u^2) \tag{13.30}$$

continuous at both ends $u = 0$ and $u = -a$. [Note that, by (13.27), we need $\ddot{L}$ continuous for $\dot{m}$ to be continuous.] The function $m(u)$ can then be obtained from (13.27) by quadrature; by (13.27) and (13.30) it will satisfy $\dot{m} = 0$ at $u = 0, -a$. Thus, via (13.26), p and q are now satisfactorily connected across the wave zone. Our next interest is in the inequalities

$$1 \le L < 1 + \tfrac{3}{2}a \tag{13.31}$$

and

$$|\ddot{L}| \le \frac{3}{2a} \tag{13.32}$$

valid throughout the wave zone so that, from (13.27),

$$|\dot{m}| \le \sqrt{\tfrac{3}{2}a^{-1}}. \tag{13.33}$$

For proof of (13.31) we observe that all terms but the third on the RHS of (13.28) are non-negative in the wave zone, so that L is overestimated by their maximum, $1 + \tfrac{3}{2}a$. On the other hand, the sum of the two middle terms is non-negative, so that L is underestimated by 1. For proof of (13.32) we find the minimum of the RHS of (13.30) by differentiating with respect to u (cf. Fig. 13.7).

The inequality (13.33) limits the change of m across the wave zone to $|\Delta m| \le \sqrt{\tfrac{3}{2}a}$. With decreasing a, therefore, we have $L(-a) \to 1$ and $m(-a) \to 0$, whence $p_0, q_0 \to 1$, as we wished to show.

Exercises 13

13.1. Verify eqns (13.3) and (13.4).

13.2. Verify that eqn (13.10) leads to (13.11).

13.3. Verify eqn (13.27).

13.4. Prove that at any instant $T = $ const the entire set of already emerged particles originally at rest at $x = 0$ lies on the ellipsoid of revolution

$$2\left(X + \frac{1-T}{2}\right)^2 + Y^2 + Z^2 = \frac{1}{2}(1-T)^2,$$

which, as one would expect, shrinks to a point when $T = 1$. [*Hint*: eliminate t and η from eqns (13.13).]

13.5. Consider a plane sandwich wave of type (13.1), flat outside the wave zone which here we take as $0 < u < a$, and defined inside by $p = \cos ku, q = \cosh ku$. (i) What are the functional forms of p and q outside the wave zone? (ii) How must k be chosen so that when the wave becomes a delta wave ($a \to 0$), $p = 1 - u$, and $q = 1 + u$ behind the wave? (iii) Describe the exact effect of this delta wave on a sheet of stationary test dust at $x = 0$, and on a plane $x + t = 0$ of oncoming test photons. (iv) Describe the Penrose topology of this delta wave.

13.6. Use the machinery developed in Section 13.7 to prove that it is impossible to construct a plane sandwich wave which, instead of fully focusing, fully diverges an incoming sheet of test dust—which would typically require $L = 1 + u, m = 0$, behind the wave. [*Hint*: sketch the graph of $L(u)$ and consider eqn (13.27).]

14

The full field equations; de Sitter space

14.1   The laws of physics in curved spacetime

We have not yet seen the full version of Einstein's field equations, only the special version that applies in vacuum. For the exterior Schwarzschild problem that was quite sufficient, since symmetry did most of the work for us; and in the case of gravitational waves it was also sufficient, as long as propagation rather than generation was our interest. But if, for example, we want to calculate the spacetime around (and inside) an arbitrary blob of matter, or the gravitational waves emitted by some double star system, then we need to know how matter interacts with spacetime, or in other words, the field equations with source terms.

The analogous Newtonian field equations are the 'vacuum' Laplace equation $\sum \Phi_{,ii} = 0$ and the 'full' Poisson equation $\sum \Phi_{,ii} = 4\pi G\rho$ [cf. (10.81)]. The former, in conjunction with spherical symmetry, is quite enough to yield the inverse-square character of the field around a spherical mass M: $\Phi \propto 1/r$, $\mathbf{g} \propto \hat{\mathbf{r}}/r^2$. But the latter is needed to fix the constant ($\mathbf{g} = -GM\hat{\mathbf{r}}/r^2$) or, for example, to find the field *inside* the mass.

Einstein's full field equations must take the place of Poisson's equation. But here we run into a problem, namely how to treat the sources. For these have no choice but to live in the *curved* spacetime which they at least partly generate. In fact, we have *two* problems, one of which is deep and has no general solution. It is the problem of circularity we have already alluded to in Section 8.4: we need the sources before we can solve for the spacetime, but we need the spacetime before we can even properly *describe* the sources. An iterative or an approximative approach will often surmount this difficulty. The other problem is easier: what consistency laws must the sources satisfy in curved spacetime (for example, conservation of energy and momentum)?

To address this second problem, we first widen it: what are the exact laws of physics in curved spacetime? For example, what laws govern particle collisions or electromagnetic fields in Schwarzschild space? Happily, it turns out that the process of adapting the SR laws to curved spacetime is fairly straightforward, as long as the laws are local. The key ideas in this transition are the following:

(i) The definition of physical quantities, and the laws governing them, are in the nature of axioms; their formulation and adoption are matters of judgement rather than proof.

(ii) All we can logically require of a curved-spacetime law or definition is that it be coordinate independent, that it be as simple as possible, and that it reduce to the corresponding SR law or definition (if there is one) in the special case of Minkowski spacetime.

(iii) In Minkowski spacetime, referred to standard coordinates, absolute and covariant derivatives of tensors reduce to ordinary and partial derivatives, respectively, since the Christoffel symbols vanish.

Accordingly, the standard procedure is *first* to express the SR definition or law in fully tensorial form (that is, a form not restricted to the use of standard coordinates); this usually requires substituting absolute and covariant derivatives in place of ordinary and partial derivatives. Once this is done, we can simply accept the same tensorial definitions and laws in curved spacetime. We have already seen this procedure applied in the definitions of the generalized 4-velocity and 4-acceleration [cf. (10.45), (10.46)] of a particle in curved spacetime.

Einstein's equivalence principle *also* tells us how to do physics in curved spacetime, namely to transfer to a freely falling Einstein cabin (or LIF) and there to use SR. The process described above is essentially equivalent to this, and saves the detour to the LIF. For, as we have seen (at the end of Section 10.3), a LIF at some event $\mathcal{P}$ corresponds to an orthonormal geodesic system of coordinates with pole at $\mathcal{P}$; and, in general, if we follow the standard procedure, the curved-spacetime law will reduce to the SR law at the pole of such coordinates. However, the present discussion also shows up one of the limitations of the equivalence principle. Suppose the SR law involves *second* partial derivatives of a tensor, for example, $F^{\mu}{}_{,\rho\sigma}$. When we generalize this to $F^{\mu}{}_{;\rho\sigma}$, that will *not* reduce to $F^{\mu}{}_{,\rho\sigma}$ at the LIF pole. For at any point where there is curvature, we can make at most the Γs but not their derivatives vanish. In such cases, the equivalence principle just cannot be fully obeyed: the inevitable tidal forces, described by the derivatives of the Γs, make themselves felt even in the freely falling cabin.

There is another problem with SR laws involving higher partial derivatives of tensors: whereas, for example, $F^{\mu}{}_{,\rho\sigma} = F^{\mu}{}_{,\sigma\rho}$, we have $F^{\mu}{}_{;\rho\sigma} = F^{\mu}{}_{;\sigma\rho} - F^{\nu} R^{\mu}{}_{\nu\rho\sigma}$ [cf. (10.55)]. So the procedure may become ambiguous, and then extraneous criteria must come into play.

A variation on this theme is provided by Maxwell's equations. Generalizing them in their potential form [cf. (7.45), (7.49)]

$$\Phi^{\mu}{}_{,\mu} = 0, \qquad \Phi^{\mu}{}_{,}{}^{\nu}{}_{\nu} = \frac{4\pi}{c} J^{\mu}, \tag{14.1}$$

to curved spacetime appears to be straightforward: just replace the commas by semicolons (without ambiguity, since $\Phi^{\mu}{}_{;}{}^{\nu}{}_{\nu} = \Phi^{\mu}{}_{;\nu}{}^{\nu}$). But then charge conservation, $J^{\mu}{}_{;\mu} = 0$ [cf. (7.39)] would no longer be implicit. What has gone wrong? Generalizing the original Maxwell equation (7.41) leaves it unchanged (cf. Exercise 10.6) and so in curved spacetime, too, it implies the existence of a potential [cf. (7.42) and

Exercise 10.6]:

$$E_{\mu\nu} = \Phi_{\nu,\mu} - \Phi_{\mu,\nu} = \Phi_{\nu;\mu} - \Phi_{\mu;\nu}. \tag{14.2}$$

Substituting that into the other original Maxwell equation (7.40) and choosing a generalized Lorenz gauge, $\Phi^{\mu}{}_{;\mu} = 0$, we find

$$E^{\nu\mu}{}_{;\nu} = \Phi^{\mu}{}_{;}{}^{\nu}{}_{\nu} - \Phi^{\nu}{}_{;}{}^{\mu}{}_{\nu} = \frac{4\pi}{c} J^{\mu}. \tag{14.3}$$

But whereas in SR the Lorenz gauge makes the second Φ term vanish, here it merely allows us to replace that term by $\Phi^{\nu} R^{\mu}{}_{\nu}$. So the proper generalization of Maxwell's equations in potential form is

$$\Phi^{\mu}{}_{;\mu} = 0, \qquad \Phi^{\mu}{}_{;}{}^{\nu}{}_{\nu} + \Phi^{\nu} R^{\mu}{}_{\nu} = \frac{4\pi}{c} J^{\mu}, \tag{14.4}$$

and this *does* imply $J^{\mu}{}_{;\mu} = 0$ (cf. Exercises 14.1, 14.2).

Returning now to the specific problem that triggered our excursion into extending SR laws in general, let us recall the SR mechanics of continua discussed in Section 7.9. A continuum is the most general distribution of matter. And, as we have seen, in SR the state of a continuum is fully described by its energy tensor $T^{\mu\nu}$. This specifies the energy density, the momentum density, and the various stress components (relative to a given inertial frame) according to the scheme (7.80). No one of these parts has intrinsic meaning by itself; all enter into the transformation of each one. Hence nothing less than the full energy tensor can figure in any invariant equation—unless it be just its trace, $T = T^{\mu}{}_{\mu}$, a scalar.

If we have a continuum in *curved* spacetime, we can define its energy tensor $T^{\mu\nu}$ through its components (7.80) in any LIF. This allows us to determine the components of $T^{\mu\nu}$ in *all* coordinate systems by tensorially transforming away from the LIF. Clearly the symmetry property $T^{\mu\nu} = T^{\nu\mu}$ (which ultimately depends on $E = mc^2$) will hold generally, and will prove to be vital in the construction of the full field equations.

In the absence of an external force density $\tilde{K}^{\mu}$ [cf. (7.81)], $T^{\mu\nu}$ satisfies the four conservation equations $T^{\mu\nu}{}_{,\nu} = 0$ for energy and momentum [cf. (7.82), (7.83)] locally in each LIF. Note that gravity does not enter as an external force, since in the LIF there *is* no gravity. In general coordinates the local conservation equations will read

$$T^{\mu\nu}{}_{;\nu} = 0. \tag{14.5}$$

And this is the restriction on the source distribution that we have been looking for, which will influence the form of the GR field equations.

The one external force that may act on our continuum, if it is charged, is an electromagnetic force. As we have seen in eqn (7.75), we must then add the energy tensor $M^{\mu\nu}$ of the electromagnetic field to that of the matter in order to get the joint conservation equation $(T^{\mu\nu} + M^{\mu\nu})_{,\nu} = 0$. In curved spacetime this becomes

$$(T^{\mu\nu} + M^{\mu\nu})_{;\nu} = 0. \tag{14.6}$$

Since the electromagnetic field carries energy, and energy is equivalent to mass, we certainly expect the electromagnetic field to influence spacetime *just like matter*, through its energy tensor.

14.2 At last, the full field equations

In Section 10.6 we saw reasons why the GR analog of Laplace's equation $\sum \Phi_{,ii} = 0$ is $R_{\mu\nu} = 0$. What, then, is the analog of Poisson's equation $\sum \Phi_{,ii} = 4\pi G\rho$? In Newtonian gravity the only source is the mass density ρ; the instantaneous field is the same whether the mass moves or not, or whether it is squeezed or not. Already in Maxwell's theory the *motion* of a charge makes a difference to the field. And now it looks as though in GR even the state of stress of the matter will affect the field, since only $T^{\mu\nu}$ as a whole offers itself as the source term.

One's first try at generalizing Poisson's equation would surely be $R_{\mu\nu} = \text{const} \cdot T_{\mu\nu}$, as is permitted by the symmetry of $T_{\mu\nu}$. (For brevity, we assume the absence of an electromagnetic energy tensor $M_{\mu\nu}$; if present, it simply gets added to $T_{\mu\nu}$.) But just as Maxwell's field equations (14.4) are constructed so as to be consistent with charge conservation, $J^{\mu}{}_{;\mu} = 0$, so the GR field equations should be consistent with energy and momentum conservation, $T^{\mu\nu}{}_{;\nu} = 0$. This is where our first try fails. $R^{\mu\nu}{}_{;\nu} = 0$ is certainly not an identity. The need for it to be nevertheless satisfied would raise to an impossible 14 the number of independent conditions on the 10 'unknown' functions $g_{\mu\nu}$. On the other hand, as we have seen in (10.71), there *does* exist a modified version of $R_{\mu\nu}$, namely the *Einstein tensor* (or 'trace-reversed Ricci tensor') $G_{\mu\nu}$, which is symmetric *and* divergence-free:

$$G_{\mu\nu} = R_{\mu\nu} - \tfrac{1}{2}g_{\mu\nu}R, \qquad G^{\mu\nu}{}_{;\nu} = 0. \tag{14.7}$$

It is this which needs to go on the LHS of the field equations:

$$R_{\mu\nu} - \tfrac{1}{2}g_{\mu\nu}R = -\kappa T_{\mu\nu}, \tag{14.8}$$

where κ is a universal constant which will be determined by the classical limit. These are the full GR field equations that Einstein proposed in 1915. The conservation equation $T^{\mu\nu}{}_{;\nu} = 0$ is now an automatic consequence of the field equations (as is the analogous conservation equation in Maxwell's theory), rather than a separate restriction.

The choice of $G_{\mu\nu}$ for the LHS of the field equations may at first seem somewhat arbitrary. But it can be shown [cf. D. Lovelock, *J. Math. Phys.* **13**, 874 (1972)] that $G_{\mu\nu}$ is the *only* tensorial and divergence-free function of the $g_{\mu\nu}$ and at most their first and second partial derivatives, that, when put on the LHS of the field equations, allows Minkowski space as a solution in the absence of sources. Since, as we have recognized on several occasions (cf. Section 10.6), the $g_{\mu\nu}$ are the relativistic potentials, the analogy with Newton's and Maxwell's theory (or just the usual demand for simplicity) makes it reasonable to look for no higher than second-order differential field equations.

And it is the uniqueness of $G_{\mu\nu}$—the non-existence of an alternative that is linear in the $g_{\mu\nu}$ and their derivatives—which forces the non-linearity of the field equations.

Let us next determine the value of κ. For this and many other purposes it is convenient to cast the field equations into an alternative standard form. Raising μ and contracting it with ν in (14.8) yields

$$R = \kappa T, \tag{14.9}$$

so that the field equations (14.8) can be rewritten in the form

$$R_{\mu\nu} = -\kappa(T_{\mu\nu} - \tfrac{1}{2}g_{\mu\nu}T). \tag{14.10}$$

Thus to correct the 'naive' version $R_{\mu\nu} = -\kappa T_{\mu\nu}$, we can replace *either* side by its trace-reversal. The form (14.10) makes it immediately clear that the full equations reduce to the previously accepted vacuum equations $R_{\mu\nu} = 0$ in the absence of sources. To determine κ we can now pick *any* convenient known situation. So let us consider a weak static source distribution, having negligible stress and momentum, in a quasi-Minkowskian background. Then $T_{\mu\nu} = \text{diag}(0, 0, 0, c^2\rho)$, $T = c^2\rho$, and so (14.10) yields $R_{44} = -\tfrac{1}{2}\kappa c^2\rho$. On the other hand, for just such a situation we saw in (10.88) (from a comparison with Newton's theory) that $R_{44} = -\sum \phi_{,ii}/c^2$. It therefore follows from comparison with Poisson's equation (10.81) that

$$\kappa = \frac{8\pi G}{c^4} = 2.073 \times 10^{-48}\,\text{s}^2\,\text{cm}^{-1}\,\text{g}^{-1}, \tag{14.11}$$

and this is the accepted value of *Einstein's gravitational constant*.

According to the field equations (14.8), the only explicit sources of the gravitational field $g_{\mu\nu}$ are the material energy tensor $T_{\mu\nu}$ and, if present, the electromagnetic energy tensor $M_{\mu\nu}$. So does the gravitational field itself have no energy? Does gravity not gravitate? We have already broached this question at the end of Section 10.6. If two massive balls stick together gravitationally, we expect their joint field to be less than twice the field of one ball, by the field of the negative binding energy. Yet it would be hard to include that with the sources, since at the outset the gravitational field is the unknown. Moreover, it *cannot* have an energy tensor analogous to that of the electromagnetic field, since there is no intrinsic gravitational *field strength*: at each event gravity can be transformed away by going to the LIF. But in Section 10.6 we have also indicated the resolution of this apparent problem. The gravity of gravity is 'miraculously' taken care of by the non-linearity of the field equations—their most significant difference from those of Newton and Maxwell. Non-linearity 'spoils' the additivity of solutions. The field of two balls is not twice that of one ball. The difference, we trust, corresponds to the gravity of gravity.

The non-linearity of the field equations also permits them to *imply* the equations of motion. In electromagnetism, for example, the field equations are perfectly satisfied by the superimposed Coulomb fields of two identical little balls of charge a small distance apart and eternally at rest. What makes them move apart is the *additional*

Lorentz force law. Not so in GR. Already the *field* equations would not tolerate two Schwarzschild fields side by side: solutions cannot be added. Each particle 'feels' the presence of the other. Einstein suspected early on that the originally hypothetical geodesic law of motion might, in fact, be implicit in the field equations. And in a series of complicated papers stretching from 1927 to 1940 he and collaborators proved this to be the case.

We can illustrate this in a simple situation. A continuum modeled by non-interacting particles moving in unison at each event is technically referred to as 'dust'. (Any random motion would contribute stresses, whereas 'dust' is stressless.) Its energy tensor is given by

$$T^{\mu\nu} = \rho_0 U^\mu U^\nu, \tag{14.12}$$

ρ_0 being its proper density and U^μ its 4-velocity. Equation (14.12) implies that $T^{\mu\nu} = \text{diag}(0, 0, 0, c^2\rho_0)$ in every rest LIF. The field equations, as we have seen, imply $T^{\mu\nu}{}_{;\nu} = 0$; that is,

$$U^\mu(\rho_0 U^\nu)_{;\nu} + \rho_0 U^\mu{}_{;\nu} U^\nu = 0. \tag{14.13}$$

Multiplying this by U_μ yields

$$c^2(\rho_0 U^\nu)_{;\nu} + \rho_0 U_\mu U^\mu{}_{;\nu} U^\nu = 0. \tag{14.14}$$

But $U^\mu{}_{;\nu} U^\nu = A^\mu$, the 4-acceleration of the continuum [cf. (10.46) and (10.36)], and so the second term in (14.14) vanishes, being $\rho_0 \mathbf{U} \cdot \mathbf{A}$. Hence the first term vanishes also. When that is substituted into (14.13), we get $A^\mu = 0$. So the dust moves geodesically. Now if every particle in even a very *small* dust cloud floating in vacuum follows a geodesic, surely it seems very likely that a free test particle does the same.

We have already loosely indicated in Section 10.6 how GR and Newtonian gravitational theory converge in the limit of classical mechanics. In fact, this correspondence can be established much more carefully and turns out to be extremely close, as long as the field is weak ($\Phi \ll c^2$), the sources move slowly ($u \ll c$), the energy-density term predominates in the energy tensor, and only orbits of slow particles are considered. Consequently all the classical observations of celestial mechanics, which are in such excellent agreement with Newtonian theory, can be adduced as support for Einstein's theory too, at least at one end of the applications spectrum. It would, of course, be desirable to test the full field equations in non-Newtonian situations. But though they are being used routinely in calculating neutron star models and similarly extreme astrophysical objects, quantitative confirmation is still sparse. (The famous 'binary pulsar' has been a fruitful laboratory of GR.)

Additional confirmation may soon come in connection with 'gravitomagnetism' [cf. after (9.18)]. Whereas Newton's theory can be regarded as a 'first approximation' to GR, there is a definite sense in which a Maxwell-like gravitational theory can be regarded as a kind of second approximation. The full field equations imply, as we noted before, that moving matter gravitates differently from matter at rest. We shall see (in

Section 15.5 below) that moving matter generates a 'gravitomagnetic' field which pushes moving test masses sideways, just as a magnetic field pushes moving charges sideways. And there is hope that we can detect, for example, the gravitomagnetic field generated by the rotating earth through satellite experiments.

Perhaps a word about light propagation inside matter will be in order here. Unless the matter is extremely tenuous (as it is in cosmology), we would not expect the light to follow null geodesics—whose speed in every LIF is c. But since there is an extension of Maxwell's theory within moving media to curved spacetime, one can study wave propagation in media in curved spacetime much as one does in Minkowski space. Such an analysis would be unavoidable in order to predict the exact rays. We may also state without proof that Maxwell's curved-spacetime *vacuum* equations lead precisely to the null-geodetic propagation of electromagnetic waves in curved vacuum.

To end this section, we return to the problem of circularity inherent in the full field equations. They can *not* be regarded as straightforward prescriptions for finding the metric $g_{\mu\nu}$ corresponding to a *given* matter distribution $T_{\mu\nu}$. We can lay an arbitrary coordinate system over our spacetime, but then we cannot, in general, describe the sources without reference to a preexisting metric. Even for so simple a material as 'dust' [cf. (14.12)], how could we specify a velocity vector U^μ of magnitude c, or how pick a density ρ_0 satisfying the conservation equation $T^{\mu\nu}{}_{;\nu} = 0$, without reference to a metric? In general, therefore, we must regard the field equations simply as ten restrictions on the tensor *pair* $(g_{\mu\nu}, T_{\mu\nu})$. Occasionally a high degree of symmetry (spherical, axial, temporal, etc.) inherent in a problem, may suggest preferred coordinates in terms of which the metric may be determined up to just a few unknown functions [as we saw in the case of spherical symmetry, cf. (11.2)]. These unknowns would then enter *both* sides of the field equations, and be determined by them, thus leading to an exact solution.

Observe how in their original form (14.8) the field equations immediately yield the components $T_{\mu\nu}$ once a coordinate system and a metric have been chosen. This might appear to offer a promising way of generating any number of exact solution pairs $(g_{\mu\nu}, T_{\mu\nu})$, from which one could then pick the one most suitable for one's purposes. But the trouble with this idea is that most randomly chosen metrics yield completely unphysical energy tensors $T_{\mu\nu}$—for example, ones with regions of negative energy, or of unreasonably large momentum- or stress components. Nevertheless, an ever-growing number of physically acceptable 'exact solution' pairs $(g_{\mu\nu}, T_{\mu\nu})$ are known by now, and these provide valuable insights into the potentialities of GR.

Still, when it comes to tackling real-life GR problems, such as arise in astrophysics, one of two practical methods is usually employed. The first is an approximative method called *linearized* GR, which plays out in Minkowski space and works along the lines of a classical linear field theory. This will be the subject of our Chapter 15. The other method, which takes advantage of the emergence of powerful computers, is called *numerical relativity*. Although this has many variations, the basic idea is to choose some 'initial' hypersurface corresponding to the world at one instant $x^4 = 0$, to consistently specify the metric and energy tensor on that surface, and then to 'evolve' both to successive future surfaces. It turns out that four of the ten field equations can be used as consistency conditions on the initial surface, while the remaining six determine the evolution.

14.3 The cosmological constant

The field equations (14.8) were constructed so as to allow flat Minkowski space as a solution in the absence of all sources. If we drop this requirement, then a slightly more general symmetric and divergence-free tensor can be put on the LHS (also uniquely, as was later shown by Lovelock, *loc. cit.*):

$$R_{\mu\nu} - \tfrac{1}{2} g_{\mu\nu} R + \Lambda g_{\mu\nu} = -\kappa T_{\mu\nu}, \qquad (14.15)$$

where Λ is a universal constant. This could be positive, zero, or negative, and, like R, has the dimensions of a curvature, namely (length)$^{-2}$. It has come to be called the *cosmological constant*, because only in cosmology does it play a significant role. There, however, it may well be forced on us by the observations. Current cosmological data limit its magnitude to

$$|\Lambda| < 10^{-55} \, \text{cm}^{-2}, \qquad (14.16)$$

and we shall see that this makes it totally negligible in all non-cosmological contexts.

It was Einstein himself who, in 1917, added the 'cosmological term' to his original field equations, for the sole purpose of making a *static* universe possible. When he later came to accept the irrefutable evidence for the expansion of the universe, he scrapped the term for good in his own work, and called it 'the biggest blunder of his life'. From today's perspective, Einstein's real blunder was to have insisted on a static universe in the face of what the field equations were clearly telling him. The Λ term, on the other hand, seems to be here to stay; it belongs to the field equations much as an additive constant belongs to an indefinite integral.

Analogously to (14.9), we find from (14.15) that

$$R = \kappa T + 4\Lambda, \qquad (14.17)$$

which allows us to rewrite (14.15) in the alternative form

$$R_{\mu\nu} = \Lambda g_{\mu\nu} - \kappa \left(T_{\mu\nu} - \tfrac{1}{2} g_{\mu\nu} T \right). \qquad (14.18)$$

Clearly these equations do not permit flat spacetime in the absence of sources, when they reduce to

$$R_{\mu\nu} = \Lambda g_{\mu\nu}, \qquad (14.19)$$

which implies curvature. [The reader may recall, cf. (10.79), that spaces satisfying a relation like (14.19) are called *Einstein spaces*.] In the total absence of sources and waves we would expect spacetime to assume the maximal symmetry of a space of constant curvature K. Comparing (14.19) with (10.76), we see that this curvature would be $K = -\tfrac{1}{3}\Lambda$. A spacetime having negative constant curvature is called de Sitter space,[1] and we shall discuss it in Section 14.4 below. If we accept Einstein's modified field equations with positive Λ, as cosmological evidence seems to suggest, it would be de Sitter space rather than Minkowski space which corresponds to 'undisturbed' spacetime. Because of its miniscule curvature, the difference would be locally quite undetectable.

In eqn (10.88) we saw that, in the Newtonian limit, $R_{44} = -\sum \Phi_{,ii}/c^2$; and after (14.10) we saw that, in the same limit, the 4,4-component of the RHS of eqn (14.10) reduces to $-4\pi G\rho/c^2$. Substituting these values in the modified field equations (14.18) yields *their* Newtonian equivalent, namely the *modified Poisson equation*:

$$\sum \Phi_{,ii} + \Lambda c^2 = 4\pi G\rho. \tag{14.20}$$

Even in Newton's theory, with such a field equation, a static universe (Φ = const) becomes possible; one merely needs the precarious balance: $\Lambda = 4\pi G\rho/c^2$, as, indeed, did Einstein.

Equation (14.20) shows that Λ is equivalent to an all-pervasive negative 'space' density $\rho_\Lambda = -\Lambda c^2/4\pi G$, and thus, by Gauss's outflux theorem, to a repulsive gravitational field $\frac{1}{3}\Lambda c^2 r$ away from any center. (Note how this is reminiscent of a *tidal* force.) As an example, the Λ force away from the sun at the location of the earth, would be about 10^{-21} of the attraction of the sun itself, if we take Λ at its maximal value of 10^{-55} cm^{-2}.

14.4　Modified Schwarzschild space

In Section 11.1 we derived the Schwarzschild metric (11.13), the unique spherically symmetric vacuum solution of Einstein's field equations without Λ term. (Though we assumed staticity in the derivation, Birkhoff's theorem asserts that the Schwarzschild solution is the unique consequence of spherical symmetry.) Let us now do the corresponding analysis for the *modified* field equations, once more assuming staticity in the 'outer' region. This will lead us to a first rough bound on the value of Λ, as well as to the important de Sitter- and anti-de Sitter spacetimes.

We can again begin with the spherically symmetric metric (11.2) and its Ricci tensor (11.3)–(11.7), and again it will be convenient to work in 'relativistic units' making $c = G = 1$. But now we require $R_{\mu\nu} = \Lambda g_{\mu\nu}$ instead of $R_{\mu\nu} = 0$. Nevertheless, as before, we find $A' = -B'$ and $A = -B$. The equation $R_{\theta\theta} = \Lambda g_{\theta\theta}$ then yields

$$e^A(1 + rA') = 1 - \Lambda r^2,$$

or, setting $e^A = \alpha$,

$$\alpha + r\alpha' = (r\alpha)' = 1 - \Lambda r^2.$$

---

[1] It should be noted that for authors who choose the opposite basic signature to ours, namely $(+++-)$, de Sitter space has *positive* curvature.

Hence

$$\alpha = 1 - \frac{2m}{r} - \frac{1}{3}\Lambda r^2, \tag{14.21}$$

where $-2m$ is again a constant of integration. It is easily verified that this solution indeed satisfies *all* the equations $R_{\mu\nu} = \Lambda g_{\mu\nu}$. Thus we have found the following metric:

$$\mathbf{ds}^2 = \left(1 - \frac{2m}{r} - \frac{1}{3}\Lambda r^2\right) dt^2 - \left(1 - \frac{2m}{r} - \frac{1}{3}\Lambda r^2\right)^{-1} dr^2 - r^2(d\theta^2 + \sin^2\!\theta\, d\phi^2),$$

$$\tag{14.22}$$

representing *Schwarzschild–de Sitter space*, so-called because for 'small' r it approximates Schwarzschild space and for 'large' r, de Sitter space if Λ is positive (as we shall see). We can think of it as de Sitter space disturbed by the presence of a single spherical mass.

Once again, there is a theorem analogous to Birkhoff's: even without the assumption of staticity, (14.22) results *almost* uniquely as a consequence of spherical symmetry; but now, with Λ, there also exists a solution consisting of successive *identical* spheres; that is, a cylindrical universe [cf. W. Rindler, *Phys. Lett.* A**245**, 363 (1998), and Exercise 14.6]. This, of course, could not be confused with the field around a mass.

By comparison with the standard metric for stationary fields, (9.13), we can read off the 'relativistic potential' Φ of the metric (14.22):

$$\Phi \approx -\frac{m}{r} - \frac{1}{6}\Lambda r^2, \tag{14.23}$$

valid as long as $\Phi \ll 1$. This once more establishes m as the mass of the central body, and the effect of Λ as that of a repulsive force of magnitude $\frac{1}{3}\Lambda r$.

The effect of Λ on planetary orbits can be found by retracing the steps that led to the previous orbit equations (11.30) and (11.45). Even with Λ the former remains unchanged, but instead of the latter we now find (cf. Exercise 14.8)

$$\frac{d^2u}{d\phi^2} + u = \frac{m}{h^2} + 3mu^2 - \frac{\Lambda}{3h^2u^3}. \tag{14.24}$$

The main effect of the extra Λ term in this equation can be shown to be an *additional* advance of the perihelion by an amount

$$\Delta = \frac{\pi\Lambda h^6}{m^4} = \frac{\pi\Lambda a^3(1 - e^2)^3}{m}. \tag{14.25}$$

In the case of Mercury, for example, this would be one second of arc per century if Λ were $\sim 5 \times 10^{-42}$ cm^{-2}. Since this would have been detected, Λ *cannot* be that big.

14.5   de Sitter space

As a second application of the metric (14.22) we specialize it to the case $m = 0$, when it is called the *de Sitter metric*:

$$ds^2 = (1 - \tfrac{1}{3}\Lambda r^2)\, dt^2 - (1 - \tfrac{1}{3}\Lambda r^2)^{-1} dr^2 - r^2(d\theta^2 + \sin^2\theta\, d\phi^2). \quad (14.26)$$

This represents the unique spherically symmetric spacetime that satisfies the modified vacuum field equations and has a center ($r = 0$) where it is regular. Since we expect the undisturbed vacuum to have the maximal symmetry of a spacetime of constant curvature $-\tfrac{1}{3}\Lambda$ [cf. after (14.19)], we expect the metric (14.26) to describe such a spacetime: it is called *de Sitter space*, D^4, if $\Lambda > 0$, and *anti-de Sitter space*, $\tilde{D}^4$, if $\Lambda < 0$. These spaces are of importance in cosmology.

Note that, for positive Λ, the metric (14.26) has at least a *coordinate* singularity at $r = \sqrt{3/\Lambda}$; yet for both smaller and larger r it satisfies the field equations, as is clear from the derivation. And if our conjecture that (14.26) is a space of constant curvature is correct, then there are no *geometric* singularities. Still, whenever a physically meaningful coordinate system has a metric singularity, something of physical interest is going on. So also here. It turns out that the locus $r = \sqrt{3/\Lambda}$ has many analogies to the Schwarzschild horizon $r = 2m$: it is a static light front (the field strength becomes infinite) and bounds the static lattice, which here is *inside* of it; and also, half-way through eternity, it changes direction. Observe that, just like the Schwarzschild metric *inside* its horizon, the de Sitter metric *outside* its horizon becomes non-static: the dr^2 term is then the only positive term, so r cannot stand still for a particle worldline.

We shall now take an instructive roundabout route to establish that the metric (14.26) represents a spacetime of constant curvature. Let us begin by considering the 4-dimensional hypersurface Σ defined by the equation

$$x^2 + y^2 + z^2 + u^2 + v^2 = a^2 \quad (14.27)$$

in 5-dimensional Euclidean space E^5 with metric

$$\mathbf{ds}^2 = dx^2 + dy^2 + dz^2 + du^2 + dv^2. \quad (14.28)$$

All planar directions anywhere in Σ are equivalent. For any one such direction *at the E^5 origin* can be transformed into any other by a suitable rotation, which leaves (14.27) and (14.28) unchanged; and vector transformations are independent of location, by the linearity of the transformation. Thus every point of Σ is an isotropic point, and so, by Schur's theorem, Σ is a space of constant curvature. The cuts $u = 0$ and $v = 0$ are evidently symmetry surfaces of Σ, whence their intersection, $x^2 + y^2 + z^2 = a^2$, $\mathbf{ds}^2 = dx^2 + dy^2 + dz^2$, is totally geodesic, and consequently a geodesic plane of Σ. But it is just a sphere of curvature $1/a^2$, which is therefore the curvature of Σ. So Σ is the 4-sphere S^4 of radius a.

Now for a little trick: we apply the coordinate transformation $v \mapsto it$! Equation (14.27) becomes

$$x^2 + y^2 + z^2 + u^2 - t^2 = a^2, \quad (14.29)$$

and the metric (14.29) becomes

$$\mathbf{ds}^2 = \mathrm{d}x^2 + \mathrm{d}y^2 + \mathrm{d}z^2 + \mathrm{d}u^2 - \mathrm{d}t^2. \tag{14.30}$$

But a mere coordinate transformation (even if it is complex) cannot change the validity of a tensor equation, and so, as reference to (10.73) shows, all sectional curvatures are preserved. So the hypersurface (14.29) in the 5-dimensional Minkowski-type space (14.30) still has curvature $1/a^2$. However, if we adopt the opposite metric

$$\mathbf{ds}^2 = \mathrm{d}t^2 - \mathrm{d}x^2 - \mathrm{d}y^2 - \mathrm{d}z^2 - \mathrm{d}u^2, \tag{14.31}$$

the subspace (14.29) has *negative* curvature, $-1/a^2$. For a metric reversal

$$g_{\mu\nu} \mapsto -g_{\mu\nu} \tag{14.32}$$

has the effect of changing the sign of every covariant curvature–tensor component, as can be seen by inspecting eqn (10.60). Consequently all sectional curvatures (10.73) change sign, and this establishes our assertion. (It has relevance for Footnote 1 above.)

The next step is to prove that the space characterized by the de Sitter metric (14.26) with positive Λ can be mapped isometrically into the subspace (14.29) of 5-dimensional Minkowski space M^5 with metric (14.31). Let us replace the Euclidean coordinates y, z, u by polar coordinates r, θ, ϕ in the usual way, so that $y^2 + z^2 + u^2 = r^2$ and $\mathrm{d}y^2 + \mathrm{d}z^2 + \mathrm{d}u^2 = \mathrm{d}r^2 + r^2(\mathrm{d}\theta^2 + \sin^2\theta\,\mathrm{d}\phi^2)$. Then eqn (14.29) reads

$$x^2 + r^2 - t^2 = a^2, \tag{14.33}$$

and the M^5 metric (14.31) reads

$$\mathbf{ds}^2 = \mathrm{d}t^2 - \mathrm{d}x^2 - \mathrm{d}r^2 - r^2(\mathrm{d}\theta^2 + \sin^2\theta\,\mathrm{d}\phi^2). \tag{14.34}$$

We shall temporarily write T for t in (14.26) and also set $3/\Lambda = a^2$, so that de Sitter's metric reads:

$$\mathbf{ds}^2 = (1 - r^2/a^2)\,\mathrm{d}T^2 - (1 - r^2/a^2)^{-1}\,\mathrm{d}r^2 - r^2(\mathrm{d}\theta^2 + \sin^2\theta\,\mathrm{d}\phi^2). \tag{14.35}$$

The mapping in question leaves r, θ, ϕ unchanged. But we now map into a 5-dimensional space: $(T, r, \theta, \phi) \mapsto (t, x, r, \theta, \phi)$. The $(T, r) \mapsto (t, x)$ mapping is essentially our old rocket mapping, whose details depend on which quadrant of the (t, x) plane we map into [see Fig. 14.1(a) and compare Tables 12.1, 12.2, and 12.3 of Chapter 12]. Here we first define the auxiliary function X and note its differential:

Table 14.1

$$a^2 - r^2 = \begin{cases} X^2 \text{(I, III)} \\ -X^2 \text{(II, IV)}, \end{cases} \qquad \mathrm{d}X^2 = \left(\frac{a^2}{X^2} \mp 1\right)\mathrm{d}r^2$$

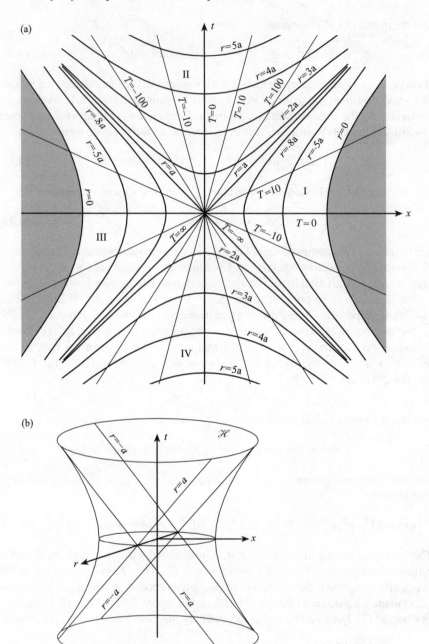

Fig. 14.1

Then we define the transformation $(T, r) \mapsto (t, x)$ by the first two lines of Table 14.2:

Table 14.2

	I($X > 0$) and III($X < 0$)	II($X > 0$) and IV($X < 0$)
$t =$	$X \sinh(T/a)$	$X \cosh(T/a)$
$x =$	$X \cosh(T/a)$	$X \sinh(T/a)$
$x/t =$	$\coth(T/a)$	$\tanh(T/a)$
$x^2 - t^2 =$	X^2	$-X^2$
$dt^2 - dx^2 =$	$(X^2/a^2)\,dT^2 - dX^2$	$-(X^2/a^2)\,dT^2 + dX^2$

The $x^2 - t^2$ entries in this table, together with Table 14.1, confirm that the map point always lies on the surface (14.33), while the $dt^2 - dx^2$ entries confirm (after a little algebra) that the mapping is isometric; that is, (14.35) goes over into (14.34).

So we have mapped the original de Sitter space (14.35) isometrically into the constant-curvature hypersurface (14.29) in M^5. This surface is a (hyper-)hyperboloid analogous to the one-sheeted hyperboloid in Fig. 5.1, being the locus of points with constant squared displacement $-a^2$ from the origin. Without regard to θ and ϕ, its equation is (14.33) and *that* locus is the hyperboloid of revolution $\mathcal{H}$ shown in Fig. 14.1(b). All points of the full surface (14.29) which have the same r-value are here condensed into a single point, so that each point of $\mathcal{H}$ stands for an entire 2-sphere having the radius of its r coordinate. Alternatively we can regard $\mathcal{H}$ as the *full* representation of a *single* radial direction $\theta, \phi = $ const, $-\infty < r < \infty$, of the original de Sitter space (14.35), the front half of $\mathcal{H}$ representing the positive radius, the back half of $\mathcal{H}$ the negative radius.

Actually, $\mathcal{H}$ is made up of *two* 'inner de Sitter spaces' ($|r| < a$) and *two* 'outer de Sitter spaces' ($|r| > a$), as described by the original metric (14.35). Quadrant I represents an inner space where T increases into the future, quadrant III one where T *decreases* into the future; quadrant II represents an outer space where r increases into the future, quadrant IV one where r *decreases* into the future.

The reader has undoubtedly noticed that the relation between the original de Sitter space (14.35) and the full hyperboloid (14.29) is closely analogous to the relation between Schwarzschild space and Kruskal space. The full hyperboloid represents the 'maximal extension' of the original de Sitter space. An examination of the 'whence and whither' of geodesics (just as in the case of Schwarzschild) makes it clear that at least two inner and two outer de Sitter spaces must be joined together to allow the continuity of geodesics, and the hyperboloid not only shows how to do this, but also that the result is maximal.

The surface of revolution $\mathcal{H}$ is quite analogous to the flat Kruskal diagram, Fig. 12.5, in which all $\pm 45°$ lines correspond to light paths. (Only the horizons are actually marked.) The $\pm 45°$ (to the vertical) straight-line generators of $\mathcal{H}$, typified by the pair of lines $r = a$ [and thus $t = \pm x$, by (14.33)], also represent light-paths, being both null and straight in M^3 (with metric $ds^2 = dt^2 - dx^2 - dr^2$). Note that, by symmetry, all geodesics (and also all worldlines of uniformly accelerated particles) through the spatial origin of de Sitter space (14.35) lie on a radius and thus on the hyperboloid $\mathcal{H}$. But since de Sitter space has constant curvature, its spatial origin is arbitrary, and thus *each* of its geodesics (and uniformly accelerated worldlines) lies on a hyperboloid like $\mathcal{H}$.

Observe that $\mathcal{H}$ can be mapped isometrically onto itself by rotations in (x, r) and standard LTs in (t, x) and (t, r), all of which leave (14.33) and (14.34) (without the angular term) invariant. This allows us to prove two interesting results: (i) all sections of $\mathcal{H}$ by planes through its center are geodesics, and (ii) all other plane sections, if timelike, are potential worldlines of particles moving with constant proper acceleration. To prove (i), we need only observe that the particular section $r = 0$, being a symmetry surface, is clearly geodesic, and that, by an LT in (t, r) and a rotation about the t-axis, this can be transformed into *any* other timelike central section. (Spacelike geodesics arise similarly from the waist circle $t = 0$.) To prove (ii), consider any of the hyperbolic sections $r = \text{const} < a$ shown in Fig. 14.1(a); by an LT in (t, x) any of its points can be brought to the vertex, leaving the hyperbola as a whole unchanged. It consequently represents a worldline with constant proper acceleration. By choosing various r values we get planes variously distant from the center, which can then be brought into coincidence with any other relevant plane by an LT in (t, r) followed by a rotation about the t-axis.

There is also an analogy to our earlier rocket space: quadrants I and III can be regarded as filled with rockets. Again the rockets 'stand' on the spherical horizon as in Fig. 12.6, but this time they point *inwards*.

Lastly we show how 'maximal de Sitter space' (that is, the entire hyperboloid $\mathcal{H}$, which should really have a name of its own, as does Kruskal space!) can be visualized (much like Kruskal space) as the time evolution of a spacelike section. Here each section $t = t_0 = \text{const}$ is simply a 3-sphere, as is clear from (14.29) and (14.31):

$$x^2 + y^2 + z^2 + u^2 = a^2 + t_0^2$$

$$\mathbf{ds}^2 = \mathrm{d}x^2 + \mathrm{d}y^2 + \mathrm{d}z^2 + \mathrm{d}u^2. \tag{14.36}$$

(The intrinsic geometry of a space is never affected by a metric reversal $\mathbf{ds}^2 \mapsto -\mathbf{ds}^2$; and for proper Riemannian spaces one naturally prefers the positive version of the metric.) That this *is* a 3-sphere follows as after (14.27), (14.28). We can represent it by one of its typical symmetry surfaces, $u = 0$, a 2-sphere in E^3 of radius $\sqrt{(a^2 + t_0^2)}$: $x^2 + y^2 + z^2 = a^2 + t_0^2$. The horizons $r = a$ correspond to the circles $y^2 + z^2 = a^2$, $x = \pm t_0$ on this 2-sphere (see Fig. 14.2). As time t progresses from $-\infty$ through zero to $+\infty$, this sphere shrinks from infinite extension to a minimum radius a at $t = 0$, whereupon it expands to infinity once again. Meanwhile the horizon-circles (light fronts), having *fixed* radius a, approach each other for 'the first half of eternity', during which time they are penetrable inwards (that is, towards the respective origin-observers O, O') by particles and light; they meet and cross each other at $t = 0$,

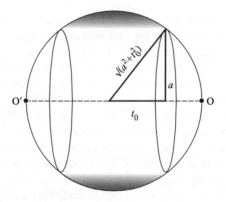

Fig. 14.2

and then they separate again for the second half of eternity when they are penetrable outwards. We can think of them as running over the 'material' of the balloon (space) at the speed of light; when the balloon shrinks, they run outward, when it expands, they run inward.

The above considerations also allow us to conclude that the spatial geometry of the static lattice of inner de Sitter space is that of a hypersphere S^3 *of radius a*, or rather a *hemi*-hypersphere, whose equator is the horizon. [More experienced readers may have recognized this directly from the form of the spatial part of the metric (14.35).] For the geometry of the lattice corresponds to that of *any* spacelike cut $T = $ const through the spacetime [cf. Fig. 14.1(a)]. But the particular cut $T = 0$ coincides with $t = 0$, and *that*, as we have seen in (14.36), is a 3-sphere of radius a, in which the two horizons just touch at the equator; so inside each horizon there is a hemi-hypersphere of radius a.

The movie-like time development of de Sitter space can be generalized to Schwarzschild–de Sitter space. Instead of a regular center, the spacetime (14.22) has a Kruskal wormhole. So the de Sitter sphere of Fig. 14.2 must now have a wormhole at the center of each horizon, which naturally joins to another de Sitter sphere, and so on ad infinitum, since this 'extension' process has no natural end (see Fig. 14.3). At

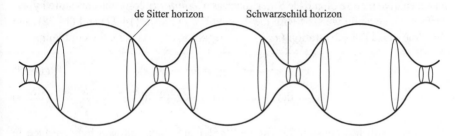

de Sitter horizon   Schwarzschild horizon

Fig. 14.3

$t = \pm\infty$, the spheres are infinitely large and the wormholes infinitely thin and long. In between they shrink to a minimum, and then expand again. On each sphere, and on each wormhole, there are two horizons of constant radius.

The metric (14.26) was found by de Sitter in 1917, the year of Λ and of Einstein's static universe. It, too, was put forward as a static universe (up to the horizon), albeit one of vanishing density. However, it was soon realized that free particles would stream away from the center, driven by the Λ force, and de Sitter space eventually became a model for an *expanding* universe. It still plays a role in modern cosmology as the limit of a whole family of non-empty models. Its horizon was at first misunderstood just like that of Schwarzschild space, but not for nearly as long. Already by 1920 Eddington clearly understood it—and missed an equal understanding of Schwarzschild's singularity by a hair. Unaccountably (like so many things that seem easy in hindsight), *that* had to wait another 13 years—for Lemaître in 1933.

14.6   Anti-de Sitter space

The cosmological significance of the spacetime (14.26) with $\Lambda < 0$ (known as *anti-de Sitter space*) is distinctly inferior but still not negligible. This spacetime has no coordinate singularity and no horizon; it is globally static; and it is maximal (that is, inextensible). And in its simplest topological form it possesses, as we shall see, an interesting feature shared with a few other GR universes: closed timelike lines, which allow one to travel into one's past.

To study it, we again temporarily write T for t in (14.26) and this time set $3/\Lambda = -a^2$:

$$\mathbf{ds}^2 = (1 + r^2/a^2)\, \mathrm{d}T^2 - (1 + r^2/a^2)^{-1}\, \mathrm{d}r^2 - r^2(\mathrm{d}\theta^2 + \sin^2\!\theta\, \mathrm{d}\phi^2). \quad (14.37)$$

It is algebraically clear that this spacetime has constant *positive* curvature $1/a^2$ if de Sitter spacetime has constant *negative* curvature $-1/a^2$. For any curvature formula previously involving a^2 must now involve $-a^2$ and be otherwise the same. For the same reason, *the global spatial lattice of anti-de Sitter space must be a 3-space of constant negative curvature $-1/a^2$*. But we shall still find it instructive to obtain an alternative picture of anti-de Sitter space as a *manifestly* constant-curvature hypersurface. Once again we start with the 4-sphere of radius a, (14.27) and (14.28), and this time make the substitution $(x, y, z, u, v) \mapsto (x, iy, iz, iu, t)$, thus obtaining

$$t^2 + x^2 - y^2 - z^2 - u^2 = a^2, \quad (14.38)$$

$$\mathbf{ds}^2 = \mathrm{d}t^2 + \mathrm{d}x^2 - \mathrm{d}y^2 - \mathrm{d}z^2 - \mathrm{d}u^2. \quad (14.39)$$

This space still has constant curvature $1/a^2$! It now appears as a hypersurface in 5-dimensional *pseudo*-Minkowskian space $\tilde{M}^5$ characterized by the metric (14.39).

As before, we can go over to polar coordinates in place of y, z, u:

$$t^2 + x^2 - r^2 = a^2, \tag{14.40}$$

$$\mathbf{ds}^2 = \mathrm{d}t^2 + \mathrm{d}x^2 - \mathrm{d}r^2 - r^2(\mathrm{d}\theta^2 + \sin^2\theta\,\mathrm{d}\phi^2). \tag{14.41}$$

Then the required mapping between anti-de Sitter space [as described by (14.37)] and this hypersurface is the following:

$$t = (a^2 + r^2)^{1/2} \cos \frac{T}{a} \tag{14.42}$$

$$x = (a^2 + r^2)^{1/2} \sin \frac{T}{a}, \tag{14.43}$$

while r, θ, ϕ go over unchanged. This mapping evidently satisfies (14.40), and a simple calculation shows that it also takes the metric (14.37) into (14.41). It is, in fact, one-to-one for the full range of all the coordinates involved except for T: all points (events) whose T coordinates differ by multiples of $2\pi a$ map into the *same* point.

Let us look at this situation geometrically. Without regard to θ and ϕ, the equation of the hypersurface (14.38) is (14.40), whose locus is the hyperboloid of revolution $\widetilde{\mathcal{H}}$ shown in Figure 14.4. All points of the full hypersurface which have the same r coordinate are condensed into a single point of $\widetilde{\mathcal{H}}$, so that each point of $\widetilde{\mathcal{H}}$ stands for an entire 2-sphere having the radius of its r coordinate (just as in the case of our earlier $\mathcal{H}$.) Thus the full space (14.38) is a 4-dimensional hyper-hyperboloid. But once again we can alternatively regard $\widetilde{\mathcal{H}}$ as the full representation of a single radius $\theta, \phi = \text{const}$,

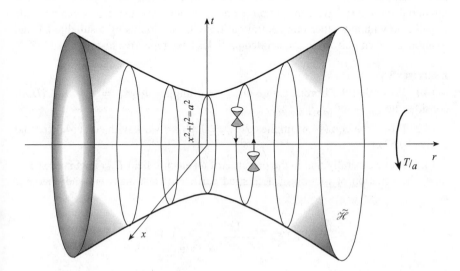

Fig. 14.4

$-\infty < r < \infty$, of the original space (14.37), the right half of $\tilde{\mathcal{H}}$ now corresponding to the positive direction of the radius, the left half to its negative direction. And once again, all geodesics and all uniformly accelerated particle worldlines lie on a hyperboloid like $\tilde{\mathcal{H}}$.

The worldlines $r = $ const of all the lattice points of the static metric (14.37) are circles in Figure 14.4, T/a measuring the angle about the r axis. After a time lapse $\Delta T = 2\pi a$ the entire universe repeats its history! Since in Fig. 14.4 there are *two* time coordinates, t and x, it is possible to orient *all* the light cones consistently (simply in the direction of increasing T), and for an entire circle to be a timelike worldline ($\mathbf{ds}^2 > 0$), with all the logical paradoxes which that involves. However, it is by no means necessary to accept the *possibility* offered by this spacetime to have closed timelike lines. All we need to do to get a physically more reasonable spacetime is to regard $\tilde{\mathcal{H}}$ as an infinite *scroll*, like a roll of paper. After one complete circuit we simply find ourselves on the next layer. Then the map from the original space (14.37) to the hyper-hyperboloid is strictly one-to-one.

As in the de Sitter case, there are obvious isometries that take the space (14.40), (14.41) into itself, namely rotations in (t, x) and standard LTs in (t, r) and (x, r). And again we can use these to establish that timelike *central* sections of $\tilde{\mathcal{H}}$ correspond to geodesic worldlines, while timelike non-central sections correspond to uniformly accelerated worldlines. The waist circle $r = 0$, being a symmetry surface of $\tilde{\mathcal{H}}$, is clearly geodesic, and also timelike; by an LT in (t, r) it can be arbitrarily tilted and then by a rotation in (t, x) it can be moved into any other timelike central section. So *all* timelike geodesics are closed curves in the non-scrolled spacetime, and all free particles repeat their histories. Once again, all the generators of the hyperboloid, typified by the line pair $x = a, t = \pm r$, are null geodesics (being both straight and null). Next, all sections $r = $ const of $\tilde{\mathcal{H}}$ (worldlines of the lattice points), by their rotational symmetry have constant proper acceleration. And these, too, can be made to coincide with any other non-central but timelike section using a suitable LT and rotation. So even uniformly accelerated particles have repetitive histories.

Exercises 14

14.1. Prove that the Lorenz gauge $\Phi^{\mu}_{;\mu} = 0$ implies $\Phi^{\nu,\mu}_{\nu} = -\Phi^{\nu} R^{\mu}_{\nu}$. [*Hint*: consider $\Phi^{\mu}_{;\rho\sigma} - \Phi^{\mu}_{;\sigma\rho}$.]

14.2. Prove that eqn (14.4) implies $J^{\mu}_{;\mu} = 0$. [*Hint*: work with eqn (14.3)(ii) and apply (10.57).]

14.3. Prove carefully that in the presence of an electric field $E_{\mu\nu}$ in curved space-time, the equation of motion of a charged particle of rest-mass m_0 and charge q is

$$\frac{d^2 x^{\mu}}{d\tau^2} + \Gamma^{\mu}_{\rho\sigma} \frac{dx^{\rho}}{d\tau} \frac{dx^{\sigma}}{d\tau} = \frac{q}{cm_0} E^{\mu}_{\rho} \frac{dx^{\rho}}{d\tau}.$$

[*Hint*: (10.46).]

14.4. For 'dust', we have seen that the first term in eqn (14.14) must vanish. Interpret this as the energy balance equation in the rest-LIF. (The geodesic motion of the dust takes care of momentum balance.)

14.5. We have already seen (in Exercise 10.20) that Einstein's vacuum field equations do not permit the existence of a constant parallel vacuum gravitational field g, whose metric would have to be of the form [cf. (10.49), (10.50)] $ds^2 = \exp(2gx)\,dt^2 - dx^2 - dy^2 - dz^2$. Use the full field equations to determine the energy tensor $T_{\mu\nu}$ corresponding to this metric, and decide whether it represents a 'reasonable' matter distribution inside of which the constant parallel field could exist. [*Hint*: use the Appendix. *Answer*: $T_{\mu\nu} = \mathrm{diag}(0, g^2, g^2, 0)$; not reasonable.]

14.6. Prove that spherically symmetric (cylindrical) *Bertotti–Kasner space*, defined by the metric

$$ds^2 = dt^2 - e^{2\sqrt{\Lambda}t}dr^2 - \frac{1}{\Lambda}(d\theta^2 + \sin^2\theta\,d\phi^2)$$

satisfies Einstein's vacuum field equations with Λ term. [*Hint*: Appendix. You may use the important result that when a metric splits into two unrelated metrics, the Riemann and Ricci tensor components of the two submetrics constitute the entire set of non-zero Riemann and Ricci tensor components of the full metric.]

14.7. Consider the 'interior' Schwarzschild metric:

$$ds^2 = \left[\tfrac{3}{2}\sqrt{1 - Ar_0^2} - \tfrac{1}{2}\sqrt{1 - Ar^2}\right]^2 dt^2 - (1 - Ar^2)^{-1}\,dr^2 - r^2(d\theta^2 + \sin^2\theta\,d\phi^2),$$

in the region $0 \le r \le r_0$, where A is a positive constant such that $A < 8/9r_0^2$. Observe that it is static. Observe that its spatial lattice is a 3-space of constant curvature A (similar to that of de Sitter space, cf. Section 14.4). Observe that, if we set $A = 2m/r_0^3$, then *at* $r = r_0$ this metric matches the 'exterior' Schwarzschild metric. By working out its Ricci tensor components [with the help of eqns (11.3)–(11.7)], verify that the interior Schwarzschild metric represents a ball of radius r_0 of *constant* density $\rho = 3A/\kappa$, and pressure p given by

$$p = \frac{3A}{\kappa}\frac{\alpha - \alpha_0}{3\alpha_0 - \alpha} \qquad (\alpha^2 = 1 - Ar^2, \alpha_0^2 = 1 - Ar_0^2),$$

which varies from zero at $r = r_0$ to a maximum at the center. For $r > r_0$ we assume vacuum and the exterior Schwarzschild metric.

14.8. Verify eqn (14.24). Then fill in the details of the following proof for eqn (14.25): As a first approximation, as before, substitute the Newtonian solution (11.44) into (14.24), ignoring the term $3mu^2$, which has already been dealt with. Expand in powers of e, assuming $e \ll 1$; that is, assuming a nearly circular orbit. Then it is the term $(\Lambda h^4/m^3)e\cos\phi$ that causes the precession. And since earlier the term $(6m^3/h^4)e\cos\phi$ in eqn (11.47) was seen to cause a precession $\Delta = 6\pi m^2/h^2$ [cf. (11.51)], the present Λ term causes the precession given by (14.25).

14.9. For the de Sitter and anti-de Sitter spacetimes (14.22) prove that the gravitational field in the static region of the metric (which in the anti-de Sitter case covers the whole spacetime) is given by

$$g = \tfrac{1}{3}\Lambda r / \left(1 - \tfrac{1}{3}\Lambda r^2\right)^{1/2},$$

and do this twice: once by analogy with eqn (11.15), and once by calculating $g_{\mu\nu} A^\mu A^\nu$ for a lattice point. Observe, in particular, what happens at the horizon of de Sitter space (interpretation?) and at infinity in anti-de Sitter space. For any two neighboring free particles a distance r apart (either of which can be regarded as the orgin), observe that the relative acceleration is $\Lambda r/3$. By reference to (8.4) show that this is consistent with constant curvature $-\Lambda/3$.

14.10. Explain how to read off from Fig. 14.4 the fact that the total lifetime ΔT for any free photon in anti-de Sitter space is πa. Verify this by calculation from the metric (14.37).

14.11. (i) By reference to eqns (14.27) and (14.28), prove that every symmetry surface of every *properly Riemannian* space of constant *positive* curvature is itself a space of the same constant curvature. [*Hint*: $v = 0$ is a typical symmetry surface.]

(ii) To investigate *properly Riemannian* spaces of constant *negative* curvature, start with eqns (14.27) and (14.28), and make the transformation $(x, y, z, u, v) \mapsto (ix, iy, iz, iu, t)$, which yields $t^2 - \sum x^2 = a^2$, $\mathbf{ds}^2 = \mathrm{d}t^2 - \sum \mathrm{d}x^2$. But this hypersurface is *spacelike*, since its position vector $\mathbf{V} = (t, x, \ldots)$ is timelike and displacements in it are orthogonal to $\mathbf{V}$: $t\,\mathrm{d}t - \sum x\,\mathrm{d}x = 0$. So the hypersurface is a *properly* Riemannian space of constant curvature $1/a^2$. Yet for all displacements in it, $\mathbf{ds}^2 < 0$. For properly Riemannian spaces we always take $\mathbf{ds}^2 > 0$, so here we must reverse the metric and take $\mathbf{ds}^2 = \sum \mathrm{d}x^2 - \mathrm{d}t^2$. Then the hypersurface is seen to have curvature $-1/a^2$. (In two dimensions it corresponds to the 2-sheeted hyperboloid of Fig. 5.1.) Now prove the analog of (i) for properly Riemannian spaces of constant negative curvature.

(iii) Prove as directly as possible (that is, without the detour to higher dimensions) that the static lattice of de Sitter (and anti-de Sitter) space, whose metric can be read off from (14.26), is a space of constant curvature $\Lambda/3$.

14.12. In a homogeneous cosmological model, the geodesic worldlines of all the galaxies are on the same footing; that is, each galaxy sees the same. So there must be isometries of the background spacetime which map the whole congruence of these 'fundamental' worldlines onto itself and any one of them into any other. Three essentially different such congruences can be laid on de Sitter space. The first consists of all the sections of the hyperboloid $\mathcal{H}$ of Fig. 14.1(b) which contain the t-axis; rotations about that axis are the corresponding isometries. This universe contracts from infinity to a minimum (at the waist of $\mathcal{H}$) and then re-expands. The second kind of congruence is typically generated by all the timelike sections of $\mathcal{H}$ containing the r-axis; LTs in (t, x) are the corresponding isometries. This universe starts with a 'big bang' and then expands indefinitely. The third kind of congruence results typically from all the timelike sections of $\mathcal{H}$ containing the *null* line $l : r = 0$ and $t + x = 0$. This universe has a big bang in the infinite past and then expands indefinitely. We have not previously discussed the LTs that serve as the required isometries here; they are called *null rotations* about l. Their characteristic is that they leave exactly *one* null direction invariant, whereas standard LTs leave *two* such directions invariant. Show (geometrically, not algebraically!) how a null rotation about l can be compounded of a rotation about the t-axis, which takes l into some other null direction l' on the central null cone, followed by an LT in (t, r) that brings l' back to l. Show how such a compound transformation can take the 'vertical' plane through l into *any* other timelike plane through l. (For a picture of the three congruences, see Fig. 18.1 of Chapter 18 below.) Also prove that there is essentially only one such congruence on anti-de Sitter space. Describe the motion of the corresponding universe.

15

Linearized general relativity

15.1 The basic equations

Although it seems that one cannot construct a satisfactory full theory of gravitation within special relativity (one that gives consistent results no matter how strong the field) we shall nevertheless see in this chapter how to construct a *special-relativistic linear approximation to general relativity* that is valid when the fields are weak. It is referred to as 'linearized general relativity'. Although its results are approximations, these are often very good approximations, and the method has many applications in cases where the full equations of GR are too difficult to solve.

Suppose, then, that we have a weak gravitational field, in the sense that for its spacetime we can find quasi-Minkowskian coordinates $x^\mu = \{x, y, z, t\}$ (*we shall work in units making $c = 1$*) such that the metric differs from its Minkowskian form

$$\eta_{\mu\nu} := \mathrm{diag}(-1, -1, -1, 1) = \eta^{\mu\nu} \tag{15.1}$$

only by *small* quantities $h_{\mu\nu}$:

$$g_{\mu\nu} = \eta_{\mu\nu} + h_{\mu\nu}, \quad h_{\mu\nu} \ll 1. \tag{15.2}$$

These must clearly be symmetric: $h_{\mu\nu} = h_{\nu\mu}$. We shall assume not only the smallness of the hs themselves, but also that of all their derivatives, and neglect products of any of these. This corresponds to assuming that the hs are small multiples of regular functions $H_{\mu\nu}$,

$$h_{\mu\nu} = \epsilon H_{\mu\nu}, \tag{15.3}$$

and discarding terms involving ϵ^2.

From (15.2) we must have

$$g^{\mu\nu} = \eta^{\mu\nu} - h^{\mu\nu}, \tag{15.4}$$

where $h^{\mu\nu} = \eta^{\mu\alpha}\eta^{\nu\beta}h_{\alpha\beta}$ (which is also of order ϵ); for only this guarantees $g^{\mu\nu}g_{\nu\sigma} = \delta^\mu_\sigma$. It follows that indices on quantities of order ϵ are shifted using $\eta^{\mu\nu}$ and $\eta_{\mu\nu}$, since, for example, $g^{\mu\alpha}h_{\alpha\nu} = \eta^{\mu\alpha}h_{\alpha\nu}$ to first order in ϵ.

Under a Lorentz transformation $p^\mu_{\mu'}$ of the special coordinates, $\{x, y, z, t\} \mapsto \{x', y', z', t'\}$, the ηs transform into themselves, and so we have, from (15.2),

$$g_{\mu'\nu'} = g_{\mu\nu}p^\mu_{\mu'}p^\nu_{\nu'} = \eta_{\mu'\nu'} + h_{\mu\nu}p^\mu_{\mu'}p^\nu_{\nu'}. \tag{15.5}$$

Hence the hs transform as Lorentz tensors; that is, like tensors in special relativity. This allows us to shift our point of view: On the one hand we have the slightly curved spacetime representing the general-relativistic weak field, and a whole family of Lorentz-related special coordinates $\{x, y, z, t\}$ which make the metric manifestly quasi-Minkowskian, as in (15.2). This spacetime we map onto an *exact* Minkowski space (x, y, z, t), where now the Lorentz tensor $h_{\mu\nu}$ represents the gravitational field. (There appears to be an analogy between $h_{\mu\nu}$ and the electromagnetic field $E_{\mu\nu}$—but whereas $E_{\mu\nu}$ is anti-symmetric and represents the field itself, $h_{\mu\nu}$ is symmetric and represents the gravitational potential; it is actually the analog of the electromagnetic 4-potential Φ_μ.)

Both the field equations satisfied by the hs, and the equations of motion of free particles, are impressed on our special-relativistic theory by what happens in the curved spacetime according to general relativity. Consider first the Christoffel symbols, which determine the free orbits according to eqn(10.15). From (10.13) and (10.14) we find, for later reference,

$$2\Gamma^\rho_{\mu\nu} = h^\rho_{\mu,\nu} + h^\rho_{\nu,\mu} - h_{\mu\nu},{}^\rho, \tag{15.6}$$

to first-order. (Recall that indices on quantities of order ϵ are shifted with the ηs.) Next we find, from (10.61), to first order,

$$2R_{\lambda\mu\nu\rho} = h_{\lambda\rho,\mu\nu} + h_{\mu\nu,\lambda\rho} - h_{\lambda\nu,\mu\rho} - h_{\mu\rho,\lambda\nu}. \tag{15.7}$$

And multiplying this equation with $\eta^{\lambda\rho}$ yields [cf. (10.68)]:

$$2R_{\mu\nu} = \Box h_{\mu\nu} + h,{}_{\mu\nu} - h^\lambda{}_{\mu,\nu\lambda} - h^\lambda{}_{\nu,\mu\lambda}, \tag{15.8}$$

where $h := h^\lambda{}_\lambda$ and, as before [cf. (7.37)], $\Box$ denotes the D'Alembertian operator:

$$\Box :=,{}_{\mu\nu}\eta^{\mu\nu} = \frac{\partial^2}{\partial t^2} - \frac{\partial^2}{\partial x^2} - \frac{\partial^2}{\partial y^2} - \frac{\partial^2}{\partial z^2}. \tag{15.9}$$

Since the hs are Lorentz tensors, so are the RHSs of eqns (15.6)–(15.8), and, with them, the component sets $\Gamma^\mu_{\nu\sigma}$ (!), $R_{\lambda\mu\nu\rho}$ and $R_{\mu\nu}$.

It turns out to be surprisingly useful to introduce the trace-reversed version of the hs [cf. before (14.7)]:

$$c_{\mu\nu} := h_{\mu\nu} - \tfrac{1}{2}\eta_{\mu\nu}h. \tag{15.10}$$

Writing $c := c^\lambda{}_\lambda$, we immediately find

$$c = -h, \tag{15.11}$$

and consequently the usual inverse relation

$$h_{\mu\nu} = c_{\mu\nu} - \tfrac{1}{2}\eta_{\mu\nu}c. \tag{15.12}$$

The Ricci tensor (15.8) can then be condensed to

$$2R_{\mu\nu} = \Box h_{\mu\nu} - c^\lambda{}_{\mu,\nu\lambda} - c^\lambda{}_{\nu,\mu\lambda}, \tag{15.13}$$

and a suitable coordinate transformation will even rid us of the last two terms.

To this end consider an *infinitesimal* coordinate transformation

$$x'^{\mu} = x^{\mu} + f^{\mu}(x), \tag{15.14}$$

where the functions f^{μ} of position are of order ϵ, $f^{\mu} = \epsilon F^{\mu}$, just like the hs and cs. (This is one of the few occasions where the old-fashioned 'primed kernel' notation is more convenient than the 'primed index' notation, as one finds very quickly if one tries to use the latter.) Equation (15.14) is not, in general, an infinitesimal Lorentz formation (unless $f_{\mu,\nu} + f_{\nu,\mu} = 0$, see Exercise 15.1), yet in the curved spacetime it nevertheless preserves the quasi-Minkowskian character of the coordinate system, as can be seen from (15.16) below. All that happens in the Minkowski map is a change in the hs, which will be regarded as a *gauge change*, for reasons that will become clear presently. Accordingly, transformations of the form (15.14) are referred to as *gauge transformations*.

Of course, under (15.14) the $g_{\mu\nu}$ transform as tensors, and so we have

$$g_{\mu\nu} = g'_{\alpha\beta} \frac{\partial x'^{\alpha}}{\partial x^{\mu}} \frac{\partial x'^{\beta}}{\partial x^{\nu}} = g'_{\alpha\beta}(\delta^{\alpha}_{\mu} + f^{\alpha}{}_{,\mu})(\delta^{\beta}_{\nu} + f^{\beta}{}_{,\nu}). \tag{15.15}$$

When we substitute from (15.2) for $g_{\mu\nu}$, and analogously for $g'_{\mu\nu}$, (15.15) is seen to imply

$$h'_{\mu\nu} = h_{\mu\nu} - f_{\mu,\nu} - f_{\nu,\mu}, \tag{15.16}$$

in analogy to the gauge transformation $\Phi'_{\mu} = \Phi_{\mu} - \Psi_{,\mu}$ of electromagnetism [cf. (7.44)]. From (15.11) we have the first of the following equations:

$$c' - c = h - h' = 2f^{\lambda}{}_{,\lambda}. \tag{15.17}$$

while the second results on contracting (15.16). Substituting from (15.12) into (15.16) then yields

$$c'_{\mu\nu} = c_{\mu\nu} + \eta_{\mu\nu} f^{\lambda}{}_{,\lambda} - f_{\mu,\nu} - f_{\nu,\mu}. \tag{15.18}$$

Note, however, that all *general-relativistic* tensor components of order ϵ, like $R_{\lambda\mu\nu\rho}$ or $R_{\mu\nu}$ (but not $\Gamma^{\mu}_{\nu\sigma}$!), are left invariant (to first-order) by any such gauge transformation. This is clear from the typical tensor transformation equation (15.15), if we momentarily pretend that $g_{\mu\nu}$ is of order ϵ: it then implies $g'_{\mu\nu} = g_{\mu\nu}$.

As we have remarked already after (7.47), it is known from the theory of differential equations that equations of the form $\Box\Psi = F(x, y, z, t)$ are explicitly solvable for Ψ. In particular, therefore, we can choose f^{μ} so as to satisfy the four equations

$$\Box f^{\mu} = c^{\mu\nu}{}_{,\nu}. \tag{15.19}$$

With that choice, eqn (15.18) yields

$$c'^{\mu\nu}{}_{,\nu} = 0; \quad \text{that is,} \quad h'^{\mu\nu}{}_{,\nu} - \tfrac{1}{2}h'^{,\mu} = 0. \tag{15.20}$$

We are then in what is called the *harmonic gauge*. It is analogous to the Lorenz gauge of electromagnetism characterized by $\Phi^{\mu}{}_{,\mu} = 0$ [cf. (7.45)].

In the gauge (15.20) we find from (15.13) that

$$2R_{\mu\nu} = \Box h_{\mu\nu}, \tag{15.21}$$

so that Einstein's vacuum field equation $R_{\mu\nu} = 0$ reduces to

$$\Box h_{\mu\nu} = 0 \quad (\text{plus } c^{\mu\nu}{}_{,\nu} = 0). \tag{15.22}$$

But then we have

$$\Box h = \Box \eta^{\mu\nu} h_{\mu\nu} = \eta^{\mu\nu} \Box h_{\mu\nu} = 0, \tag{15.23}$$

whence, equivalently to (15.22), we also have

$$\Box c_{\mu\nu} = 0 \quad (\text{plus } c^{\mu\nu}{}_{,\nu} = 0). \tag{15.24}$$

The very form of (15.22) (a 'wave equation') suggests the existence of gravitational waves. However, some wavelike solutions of (15.22) turn out to be mere 'coordinate waves'; that is, solutions that can be reduced to $h_{\mu\nu} = 0$ by a suitable gauge transformation, as we shall see below. But with (15.22), eqn (15.7) implies

$$\Box R_{\lambda\mu\nu\rho} = 0, \tag{15.25}$$

and since $R_{\lambda\mu\nu\rho}$ is a Lorentz tensor *and* gauge invariant [cf. after (15.18)], this *does* show that small disturbances of curvature propagate with the speed of light.

The Einstein tensor $G_{\mu\nu} = R_{\mu\nu} - \frac{1}{2}g_{\mu\nu}R$ [cf. (14.7)] in harmonic gauge becomes

$$G_{\mu\nu} = \frac{1}{2}(\Box h_{\mu\nu} - \frac{1}{2}\eta_{\mu\nu}\Box h) = \frac{1}{2}\Box c_{\mu\nu}, \tag{15.26}$$

so that the full field equations (14.8) now reduce to

$$\Box c_{\mu\nu} = -2\kappa T_{\mu\nu} = -16\pi G T_{\mu\nu} \quad (\text{plus } c^{\mu\nu}{}_{,\nu} = 0). \tag{15.27}$$

Observe that these are *linear* field equations, just like those of the Maxwell field in special relativity [cf. (7.49)]. In terms of the $c_{\mu\nu}$, the eqns (15.27) are also *decoupled*. And, lastly, they are *hyperbolic* partial differential equations, which they would *not* be in terms of the hs; this is of interest for the existence and uniqueness of solutions, and therefore also for *numerical relativity*, the modern field of solving GR problems by use of often very powerful computers.

Since, as can be seen from (15.6), the Γs are of order ϵ, it follows from the definitions (10.28) that covariant derivatives (of all orders) of quantities of order ϵ here reduce to partial derivatives. Consequently the contracted Bianchi identity $G^{\mu\nu}{}_{;\nu} = 0$ [cf. (14.7)] is automatically satisfied because of the gauge condition $c^{\mu\nu}{}_{,\nu} = 0$, via eqn (15.26). But while $G^{\mu\nu}$ is manifestly of order ϵ, we cannot assume the same of $T^{\mu\nu}$, nor thus the effective equality of $T^{\mu\nu}{}_{,\nu}$ and $T^{\mu\nu}{}_{;\nu}$. What is implied by (15.27) is $T^{\mu\nu}{}_{,\nu} = 0$. Yet $T^{\mu\nu}{}_{;\nu} = 0$ is the real equation satisfied by the sources. In practice, this does not invalidate (15.27): because of the exceedingly small numerical value of $\kappa (\sim 2 \times 10^{-48}\,\mathrm{s}^2\,\mathrm{cm}^{-1}\,\mathrm{g}^{-1}$ or $\sim 2 \times 10^{-27}\,\mathrm{cm/g}$ in units where $c = 1$) relative to the other physical quantities occurring in most problems, we can regard κ as of order ϵ, so that in linearized GR $\kappa T^{\mu\nu}{}_{,\nu} = \kappa T^{\mu\nu}{}_{;\nu}$.

One immediate consequence of the linearity of the field equations (15.27)—they are linear also in the $h_{\mu\nu}$—is that solutions can be added. Thus if the tensor pairs $(\overset{a}{h}_{\mu\nu}, \overset{a}{T}_{\mu\nu})$ separately satisfy the field equations for $a = 1, 2, \ldots$, then the metric $g_{\mu\nu} = \eta_{\mu\nu} + \sum \overset{a}{h}_{\mu\nu}$ satisfies the field equations with $\sum \overset{a}{T}_{\mu\nu}$ as the sources. But as a result, just as in the case of Maxwell's theory, as far as the field equations (15.27) are concerned, the sources do not 'feel' each other. Two point masses could sit side by side forever, their separate radial fields simply superimposed. This is tolerable if we are interested, say, in the far field of sources whose motion we know a priori and if we are willing to neglect the 'gravity of gravity'. But if we wish to *find out* how the sources move under their own gravity, we need to solve the equation $T^{\mu\nu}; \nu = 0$ [cf. (14.12)–(14.14)]—not $T^{\mu\nu}, \nu = 0$, which takes into account only the *nongravitational* interactions. One approach—into whose subtleties and limitations we cannot here enter—is to calculate the c's from (15.27), then the Γ's from (15.6), and finally the $T^{\mu\nu}; \nu$ from (10.28).

As in Maxwell's theory [cf. (7.50) and subsequent text], the field equations (15.27) can be solved explicitly (*temporarily in full units*):

$$c_{\mu\nu}(\text{P}) = -\frac{4G}{c^4} \int \frac{[T_{\mu\nu}]\,\mathrm{d}V}{r}, \tag{15.28}$$

where $[\ldots]$ denotes retardation relative to P. And for the same reasons as before, (i) this solution automatically satisfies the harmonic gauge condition $c^{\mu\nu}{}_{,\nu} = 0$ provided $\kappa T^{\mu\nu}{}_{,\nu} = 0$, (ii) the integral represents a Lorentz tensor, and (iii) it is the unique solution in the absence of source-less radiation.

As an example, we shall obtain the linearized metric of a weak stationary field generated by sources in stationary motion (for example, a rotating ball). We shall assume that the stress components T_{ij} within the sources are negligible compared to the other components of $T_{\mu\nu}$, and that unique velocities $\mathbf{v}$ can be associated with various parts of the sources, small enough to allow us to neglect v^2/c^2. Then (*once again in units making $c = 1$*) the energy tensor takes the form

$$T_{\mu\nu} = \begin{pmatrix} \mathbf{0}_3 & -\rho\mathbf{v} \\ -\rho\mathbf{v} & \rho \end{pmatrix} \tag{15.29}$$

[cf. (7.80) and Exercise 7.25]. We now define the gravitational scalar and vector potentials Φ and $\mathbf{w}$ as follows:

$$\Phi := -\int \frac{G\rho\,\mathrm{d}V}{r} = -G \int \frac{T_{44}\,\mathrm{d}V}{r} = \frac{1}{4}c_{44}$$

$$w_i := 4\int \frac{G\rho v_i\,\mathrm{d}V}{r} = 4G \int \frac{T_{i4}\,\mathrm{d}V}{r} = -c_{i4}. \tag{15.30}$$

Since the sources are stationary, we need no retardation brackets in these definitions. (The same is true if the sources are merely periodic, like a double-star system, but do not move appreciably in the time it takes light to cross them.) Now $T_{ij} = 0$ implies $c_{ij} = 0$, so that

$$c = \eta^{\mu\nu} c_{\mu\nu} = c_{44}. \tag{15.31}$$

And this, after a short calculation, together with (15.30) is seen to imply

$$h_{\mu\mu} = 2\Phi, \qquad h_{i4} = -w_i, \qquad h_{ij} = 0 \quad (i \ne j). \tag{15.32}$$

So the metric reads

$$ds^2 = (1 + 2\Phi)\,dt^2 - 2w_i\,dx^i\,dt - (1 - 2\Phi)\sum(dx^i)^2. \tag{15.33}$$

Observe that this is the linearized form of the canonical metric (9.13) for a stationary spacetime, with one important (even though only approximate) improvement: the unspecified lattice metric $k_{ij}\,dx^i\,dx^j$ of (9.13) has here become explicit, $(1 - 2\Phi)\sum(dx^i)^2$, by virtue of the field equations. Thus when using (15.33) to calculate orbits of test particles, we no longer need to restrict ourselves to slow orbits, for which the spatial geometry is almost irrelevant; (15.33) can be used even to predict light paths. Its validity also extends to *non-stationary* sources with negligible stress and low velocities. It is then merely necessary to perform the retardation operation [...] indicated in (15.28) on the integrals of (15.30).

The remarkable similarity of the definitions (15.30) to the corresponding definitions in electromagnetism will not have escaped the reader's attention. They will be further elaborated in Section 15.5 below.

To end this section, we make a remark that will presently be of great importance to us. Equations (15.19) do not *uniquely* determine the gauge transformation that achieves the harmonic gauge (15.20): any vector g^μ of order ϵ and satisfying the wave equation

$$\Box g^\mu = 0 \tag{15.34}$$

can be added to f^μ without affecting the result (15.19). Gauge transformations generated by vectors satisfying (15.34) can therefore be used freely *within* the harmonic gauge. And so can Lorentz transformations of the coordinates $\{x, y, z, t\}$; for under these, $h_{\mu\nu}$ and $c_{\mu\nu}$ are tensors, and the gauge condition (15.20) is tensorial and thus preserved.

15.2   Gravitational waves; The TT gauge

In this and the following two sections we develop some of the basic facts about gravitational radiation. We have, of course, already met an exact gravitational wave in Chapter 13, but that came somewhat out of the blue, and allowed few general conclusions. The systematic approach we shall follow here is the basis of most of the experimental work on wave detection currently in progress.

Let us consider an arbitrary plane wave propagating in the x-direction,

$$c_{\mu\nu} = c_{\mu\nu}(u), \qquad u = t - x. \tag{15.35}$$

This automatically satisfies the first part of the vacuum field equations, (15.24)(i). In order for it also to satisfy the harmonic gauge conditions (15.24)(ii) we need

$$c^{\mu 1}{}_{,1} + c^{\mu 4}{}_{,4} = -\dot{c}^{\mu 1} + \dot{c}^{\mu 4} = 0, \tag{15.36}$$

where the overdot denotes d/du. Integrating the last equation gives

$$c^{\mu 1} = c^{\mu 4} + \text{const}, \tag{15.37}$$

but the constant can at once be discarded. For, as we have seen, solutions are additive, and any solution corresponding to constant cs corresponds to constant hs and thus, via (15.7), to zero $R_{\mu\nu\rho\sigma}$; so it represents undisturbed Minkowski space. Lowering the indices in (15.37) we thus have, in full,

$$c_{11} = -c_{14}, \qquad c_{21} = -c_{24}, \qquad c_{31} = -c_{34}, \qquad c_{41} = -c_{44}. \tag{15.38}$$

These four relations between the ten cs apparently leave six degrees of freedom for the wave. However, as we already alluded to after (15.24), many apparent wave solutions (15.35) are spurious, being mere wavelike disturbances of the coordinate system ('coordinate waves') rather than of the intrinsic spacetime curvature. The $h_{\mu\nu}$ or $c_{\mu\nu}$ representing such waves can be made to vanish by suitable gauge transformations, which shows that they correspond to zero curvature $R_{\mu\nu\rho\sigma}$, which is gauge invariant. We shall now demonstrate this by applying a gauge transformation of the type mentioned at the end of the last section, which preserves the harmonic gauge. Any four smooth functions

$$g_\mu = g_\mu(u) \tag{15.39}$$

will not only automatically satisfy the requirement (15.34), but also, when we apply (15.18), preserve the wave character (15.35) of the cs. In particular, (15.18) leads to

$$
\begin{aligned}
c'_{41} &= c_{41} + \dot{g}_4 - \dot{g}_1, \\
c'_{42} &= c_{42} - \dot{g}_2, \\
c'_{43} &= c_{43} - \dot{g}_3.
\end{aligned}
\tag{15.40}
$$

Evidently, by a proper choice of g_μ we can thus achieve

$$c'_{41} = c'_{42} = c'_{43} = 0. \tag{15.41}$$

But since the new cs also satisfy the harmonic gauge, it then follows from the analog of (15.38) that

$$c'_{11} = c'_{21} = c'_{31} = c'_{44} = 0. \tag{15.42}$$

So we are left with only c'_{23}, c'_{22}, and c'_{33}. Inspection of (15.40) shows that we still have $\dot{g}_4 + \dot{g}_1$ at our disposal and we can use this freedom to achieve

$$c'_{22} = -c'_{33}; \tag{15.43}$$

that is, $c' = c'^{\mu}{}_{\mu} = 0$. For, by reference to (15.18), this requires

$$0 = c^{\mu}{}_{\mu} + 2(g^{\mu}{}_{,\mu}) = c^{\mu}{}_{\mu} + 2(\dot{g}_1 + \dot{g}_4). \tag{15.44}$$

Now only two degrees of freedom remain: the choice of c_{22} and c_{23} (we now drop the primes!). Since $c = 0$, we finally have

$$h_{\mu\nu} = c_{\mu\nu} = \{h_{22} = -h_{33}, \ h_{23} = h_{32}; \ \text{all others zero}\}. \tag{15.45}$$

The gauge which achieves this reduction is called the *transverse-traceless* (TT) *gauge*, and the corresponding hs are often denoted by $h_{\mu\nu}^{\text{TT}}$ ('traceless' because $h = c = 0$, and 'transverse' because all the components are in the spatial directions perpendicular to the direction of propagation).

Now observe, from (15.16), that the gauge transformation generated by (15.39) leaves the transverse hs (h_{22}, h_{33}, and h_{23}) invariant. It follows that these hs must *already* be in their final form if only the wave satisfies $h_{\mu\nu} = h_{\mu\nu}(u)$ plus the harmonic gauge conditions—except that istead of $h_{22} + h_{33} = 0$ we could have

$$h_{22} + h_{33} = \text{const.} \tag{15.46}$$

The possibility of this extra constant arises from our having discarded the constants in (15.37). So the reduction to TT gauge of such a wave is achieved by simply *discarding* all the non-transverse hs and the constant in (15.46). Every non-transverse $h_{\mu\nu}$ by itself thus represents a pure 'coordinate wave' in flat spacetime. In practice we need never solve for the gs in eqns (15.40) and (15.44); the *existence* of the necessary gauge transformation is all that needed to be established.

15.3 Some physics of plane waves

By the result (15.45) of the preceding section, in linearized GR the metric of every plane gravitational wave propagating in the x-direction can be reduced to the following 'canonical' form:

$$\mathbf{ds}^2 = dt^2 - dx^2 - (1 - \lambda)\,dy^2 - (1 + \lambda)\,dz^2 + 2\mu\,dy\,dz, \tag{15.47}$$

where all that is required of λ and μ is that they be of order ϵ but otherwise quite arbitrary smooth functions of $t - x$:

$$\lambda = \lambda(u), \quad \mu = \mu(u), \quad u = t - x. \tag{15.48}$$

We are simply writing $h_{22} = \lambda$ and $h_{23} = \mu$. In this section we shall find it preferable to revert to the 'slightly-perturbed-Minkowski-space' picture rather than using the 'exact-Minkowski-space' map.

It is still possible that (15.47) represents a mere coordinate wave, for which the spacetime is flat. Let us therefore consider the curvature tensor for the metric (15.47). Referring to (15.7), we can quickly convince ourselves that the only not identically vanishing independent components of $R_{\lambda\mu\nu\rho}$ are the following:

$$R_{\alpha11\beta} = R_{\alpha44\beta} = \tfrac{1}{2}\ddot{h}_{\alpha\beta}$$
$$R_{\alpha14\beta} = R_{\alpha41\beta} = -\tfrac{1}{2}\ddot{h}_{\alpha\beta},$$
(15.49)

where α, $\beta = 2$ or 3 and dots denote d/du. So unless $\ddot{\lambda} = \ddot{\mu} = 0$, the metric (15.47) represents a real wave with non-vanishing curvature and $\Box R_{\lambda\mu\nu\rho} = 0$ [cf. (15.25)].

The reader will no doubt have noticed the similarity of the canonical metric (15.47) (when $\mu = 0$) to our earlier exact wave metric (13.1), whose conditions (13.3) and (13.4) are here parallelled. As in Chapter 13, we shall again be interested in what happens when the wave meets a stationary plane of dust or an oncoming plane of photons head-on. And once again we have the following two relevant lemmas:

Lemma I: *Particles satisfying the equations*

$$x, y, z = \text{const}$$
(15.50)

follow timelike geodesics of the metric (15.50);

Lemma II: '*Particles*' *satisfying the equations*

$$x = \pm t; \qquad y, z = \text{const}$$
(15.51)

follow null geodesics of the metric (15.47). And again, in both cases, t is an affine parameter. For proof, we cannot here use the Appendix, since the $dy\,dz$ term spoils the diagonality of the metric. Nevertheless we have, as before,

$$g_{tt} = -g_{xx} = 1, \qquad g_{\mu t} = 0 \quad (\mu \neq t), \qquad g_{\mu x} = 0 \quad (\mu \neq x), \quad (15.52)$$

which trivially suffices to ensure

$$\Gamma_{\mu xx} = \Gamma_{\mu tt} = \Gamma_{\mu xt} = 0.$$
(15.53)

And this, in turn, implies the previous eqns (13.7), from which the lemmas follow.

In linearized GR we can superimpose solutions; this allows us to regard (15.47) as a superposition of a 'pure λ' wave and a 'pure μ' wave, and to discuss these separately, beginning, say, with the λ wave:

$$\mathbf{ds}^2 = dt^2 - dx^2 - (1 - \lambda)\,dy^2 - (1 + \lambda)\,dz^2.$$
(15.54)

As in Chapter 13, we shall concentrate on a 'sandwich' wave; that is, one contained between two parallel planes traveling at the speed of light, with flat spacetime before and aft. Without loss of generality we can suppose $\lambda = 0$ before the wave has passed and $\lambda = $ some linear function of u [cf. (15.49)] thereafter. Within the sandwich, linearized GR imposes no restrictions on λ except $\Box\lambda = 0$, which is satisfied by any smooth function of $u = t - x$. Let us put a plane of static test dust, orthogonal to the x-axis, ahead of the wave, say at $x = 0$. Any two of its free particles separated only by an infinitesimal y-difference dy remain so separated permanently, by Lemma I. But, according to (15.54), the *ruler* distance dY between these particles at any instant $t = $ const is given by $dY = \left(1 - \frac{1}{2}\lambda\right)dy$ (to first-order in ϵ), and so *it* varies as the wave passes. Similarly, two free dust particles originally separated only by an infinitesimal z-difference dz also move relative to each other; as measured by rulers, *their* distance apart dZ at any instant $t = $ const is given by $dZ = \left(1 + \frac{1}{2}\lambda\right)dz$. (Ruler distances between free particles in the x-direction remain invariant.) If we imagine special sets of dust particles colored to mark the y, z coordinate net, forming 'vertical' lines $y = $ const and 'horizontal' lines $z = $ const, then, as the wave passes, whenever λ increases the vertical lines all bunch up and the horizontal lines all spread out, and vice versa when λ decreases.

The diagram representing an active Lorentz transformation, Fig. 2.9, can serve to illustrate this state of affairs, if we replace ξ and η by the 'ruler-distance' coordinates dY and dZ relative to any one dust particle P at the center of the diagram. In Chapter 2 the crucial relation between ξ and η was the constancy of their product for the motion of a point: $\xi\eta = $ const; here it is

$$dY \, dZ = dy \, dz = \text{const} \tag{15.55}$$

(to first-order in ϵ). Relative to P, all neighboring particles move along the hyperbolae (15.55). A small circle of dust particles *around* P, originally satisfying $dy^2 + dz^2 = a^2$, becomes the ellipse

$$\frac{dY^2}{(1 - \lambda)} + \frac{dZ^2}{(1 + \lambda)} = a^2 \tag{15.56}$$

inside the wave. And if the wave is sinusoidal, say

$$\lambda = \epsilon \cos \omega(t - x), \tag{15.57}$$

then the ellipse oscillates between being elongated in the y-direction and being elongated in the z-direction. Its total area, like that of all closed curves made up of specific dust particles, remains constant, since for each area element eqn (15.55) holds.

It is helpful to imagine a small freely floating rigid plate behind the plane of dust, having its center permanently coincident with the particle P, and having a ruler-distance coordinate net dY, $dZ = $ const etched into it. It is relative to these coordinates that eqn (15.56) is to be understood. The particles of the plate do *not* follow geodesics (except, by supposition, its center) and their ruler separations are permanent except for the minute elastic deformations caused by the passing wave. The forces exerted by the wave on the plate are gravitational *tidal* forces, tending to push neighboring points apart or to pull them together. For example, the tidal field f_{PQ} between the point P and another, Q, 'horizontally' beside it at rigid distance dY equals the acceleration relative to P of the free dust particle just passing Q. As we have seen, for that particle we have $dY = \left(1 - \frac{1}{2}\lambda\right) dy$, so that $d^2(dY)/dt^2 = -\frac{1}{2}\ddot{\lambda} dy$. If the wave is sinusoidal, as in (15.57), the tidal field in the y-direction is therefore given by

$$f_{PQ} = \tfrac{1}{2}\epsilon\omega^2 (\cos \omega t)\, dy, \tag{15.58}$$

positive values corresponding to repulsion. The response of the plate to these tidal forces depends on its elastic properties.

It is of interest to note that a plane of orthogonally incoming test photons is affected by the wave very much like a plane of particles, as follows from Lemma II, eqn (15.51). If at the back end of the sandwich λ is increasing, we cannot conclude that it will increase linearly to unity and so focus all the photons onto a vertical line; for our analysis presupposes $\lambda \ll 1$. We can nevertheless legitimately reach the same conclusion from the fact that neighboring photons on the same horizontal leave the wave zone on converging paths, which proceed rectilinearly since we are in flat spacetime, and thus must meet. By iteration along the horizontal, *all* photons on it must meet at the same point, and all the photons of the incoming plane meet on a vertical line. It is not a priori clear how seriously such exact results are to be taken in an approximative theory. But in this particular case we can fall back on the exact wave of Chapter 13 for confirmation. However, for a weak wave of limited extent we can ignore the topological ramifications of Chapter 13.

Let us next consider the 'pure μ' component of the metric (15.47),

$$\mathbf{ds}^2 = dt^2 - dx^2 - dy^2 - dz^2 + 2\mu\, dy\, dz. \tag{15.59}$$

A $45°$ rotation of the axes,

$$y = (y' - z')/\sqrt{2}, \qquad z = (y' + z')/\sqrt{2}, \tag{15.60}$$

(and immediate dropping of the primes) preserves $dy^2 + dz^2$ and converts $2\, dy\, dz$ into $dy^2 - dz^2$, so that the metric (15.59) takes on the form of a λ wave:

$$ds^2 = dt^2 - dx^2 - (1 - \mu)\, dy^2 - (1 + \mu)\, dz^2. \tag{15.61}$$

It follows that a μ wave is nothing but a λ wave rotated by $-45°$.

Now in a pure sinusoidal λ wave, each free particle of the original dust plane in the neighborhood of an arbitrary particle P (through which we might as well draw the x-axis) executes a small-amplitude simple-harmonic motion (s.h.m.) in a direction determined by the hyperbolae of Fig. 2.9 (repeated in Fig. 15.1). In a pure sinusoidal μ wave, each such particle executes s.h.m. in the *orthogonal* direction, namely along the hyperbolae of the λ wave rotated through $-45°$—and these are everywhere orthogonal to those of the original λ wave. (See Fig. 15.1 and cf. Exercise 15.5.) It follows that in a general monochromatic λ, μ wave, each free particle in a y, z plane executes a motion, relative to any nearby center P, which is the superposition of two s.h.m.s of

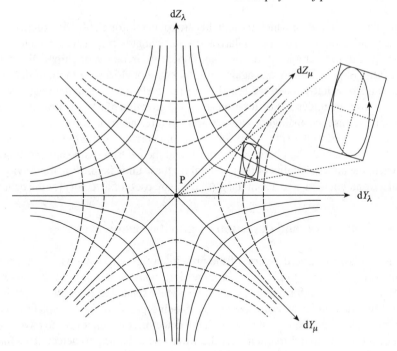

Fig. 15.1

the same frequency in orthogonal directions. In the special case when these are in phase, the resulting motions are all s.h.m.s along small straight-line segments; if the components are exactly out of phase and of the same amplitude, the motions are all little circles; and in the general case, the particles move along little ellipses, whose axes remain fixed in space, and whose dimensions obviously increase linearly with their distance from P. The wave is then said to be linearly, circularly, or elliptically polarized, respectively. Also, the little circles or ellipses can be described in an anti-clockwise or a clockwise sense (looking at the oncoming wave), depending on the phase difference of the components. One accordingly speaks of right-handed or left-handed polarization, or positive or negative *helicity*, respectively. And because the two distinct displacement patterns shown in Fig. 15.1 correspond to a $+$ sign and a $\times$ sign, the respective λ and μ polarization states are often referred to as the $+$ and $\times$ ('plus' and 'cross') states, and λ and μ are written as h_+ and $h_\times$.

It may appear from our discussion and from Fig. 15.1 that a general (λ, μ) wave determines a preferred orientation for the y and z coordinate axes. But this is not the case: both TT conditions, $h_{22} = -h_{33}(h^\mu{}_\mu = 0)$ and $h_{23} = h_{32}(h_{\mu\nu} = h_{\nu\mu})$ are clearly invariant under yz rotations. (Cf. Exercise 15.4.)

In electromagnetic theory, too, every monochromatic plane wave can be compounded of two basic 'linearly polarized' types: one in which the **e** and **b** fields are permanently, let us say, in the y- and z-directions, respectively, if x is the direction of propagation; and one in which these fields are rotated through 90° (in contrast to the 45° separation between the two polarization states in gravity). In electromagnetism it is the tip of the vector **e** itself that oscillates linearly, circularly, or elliptically in these three types of polarization, and again anti-clockwise or clockwise motions determine right-handed or left-handed polarization, respectively. In the gravitational case, the wave is invariant under a 180° rotation about the direction of propagation, in the electromagnetic case the analogous angle is 360°, and for neutrino waves it is 720°. This is closely related to the spin values of the quanta corresponding to these waves: the graviton with spin 2, the photon with spin 1, and the neutrino with spin 1/2. It is also related to the dimensionality of the potential for these three fields: $h_{\mu\nu}$ for gravity, Φ_μ for electromagnetism, and a spinor ϕ_A (the 'square root' of a vector) for the neutrino.

15.4 Generation and detection of gravitational waves

In eqn (15.58) we have the germ of the principle of gravitational wave detection. It is 'merely' necessary to put either a large rigid body in the path of the wave and measure its elastic deformations, or to have widely separated free test masses in the path of the wave and to measure the wiggle in their separation (for example, by optical interferometry). But before one can design a detector, it is imperative to have a good idea of the strength and frequency of the wave one is hoping to detect. This means one must study the possible generation of gravitational waves. It turns out that there is no conceivable way to generate such waves of detectable strength in the laboratory. Only astronomical sources can generate enough power.

Let us see what our formulae tell us. Suppose we have some spatially limited source, like a double star system, at large distance from us. Equation (15.28) then gives us the metric here and now. *Returning to units that make $c = 1$, we have*

$$c^{\mu\nu} = -\frac{4G}{r} \int T^{\mu\nu} \mathrm{d}V, \tag{15.62}$$

where we must remember eventually to evaluate the integral at the 'retarded' time (if that makes a difference); and since the distance r to the various parts of the source is essentially the same, we have taken r outside the integral. Let the center of mass of the source define the spatial origin and imagine our detector far away on the x-axis. Evidently, by the time the wave gets to us, it is essentially plane. As we have seen in Section 15.2, only c^{22}, c^{33}, and c^{23} will then have physical significance, and so it will suffice to study $\int T^{ij} \mathrm{d}V$.

Consider therefore the identity

$$\int (T^{ik} x^j)_{,k} \, \mathrm{d}V = \int T^{ik}_{,k} \, x^j \, \mathrm{d}V + \int T^{ij} \mathrm{d}V, \tag{15.63}$$

which arises from differentiating $T^{ik} x^j$ and integrating the resulting equation, let us say over a volume that extends slightly *beyond* the source region. By Gauss's theorem, which converts the volume integral of a divergence into a surface integral over the boundary, the integral on the LHS is seen to vanish. The first integrand on the RHS can be converted by use of the special-relativistic conservation equation $T^{\mu\nu}{}_{,\nu} = 0$ which the sources will satisfy if we neglect their mutual gravitation—for example, if we think of a double-star system as held together by a string. Thus we find, successively,

$$\int T^{ij}\, dV = -\int T^{ik}{}_{,k}\, x^j\, dV = \int T^{i4}{}_{,4}\, x^j\, dV$$

$$= \frac{d}{dt}\int T^{i4}x^j\, dV = \frac{1}{2}\frac{d}{dt}\int (T^{i4}x^j + T^{j4}x^i)\, dV, \qquad (15.64)$$

where for the last equation we have used the fact that all integrals in this sequence of equations must share the i, j symmetry of the first. We can next play the same trick as in (15.63) once more:

$$0 = \int (T^{4k}x^ix^j){}_{,k}\, dV = \int T^{4k}{}_{,k}\, x^ix^j\, dV + \int (T^{4i}x^j + T^{4j}x^i)\, dV, \qquad (15.65)$$

and we again use the conservation equation, this time to convert $T^{4k}{}_{,k}$ into $-T^{44}{}_{,4}$. Then eqns (15.64) and (15.65) together give

$$\int T^{ij}\, dV = \frac{1}{2}\frac{d^2}{dt^2}\int T^{44}x^ix^j\, dV. \qquad (15.66)$$

This is the well-known *Laue theorem* of SR, which holds for any conserved symmetric tensor $T^{\mu\nu}$. (See Exercise 15.6 for a physical interpretation.) In linearized GR its use lies in facilitating the evaluation of the spatial $c^{\mu\nu}$ corresponding to distant moving sources. We know that in the rest-LIF of any material (in full units) we have

$$T^{44} = c^2\rho, \qquad (15.67)$$

ρ being the mass density. In LIFs moving with low velocity v relative to the rest-LIF, T^{44} differs from $c^2\rho$ only by quantities of order v^2/c^2 (cf. Exercise 7.26). Consequently, in LIFs where the material moves 'non-relativistically', eqn (15.67) applies in good approximation. The integral on the RHS of (15.66) is thus recognized as the quadrupole moment of the source. And so, by (15.62), the second time-derivative of the quadrupole moment determines the wave metric:

$$c^{ij} = -\frac{2G}{c^4 r}\frac{d^2}{dt^2}\int \rho x^ix^j\, dV, \qquad (15.68)$$

in full units. The smallness of the numerical factor should give us pause!

There is, of course, also a non-radiative field due to the source; but that is determined mainly by the coefficients $c^{4\mu}$ [cf. (15.30)–(15.33)].

Note that, both in electromagnetic and gravitational theory, the monopole moment $\int \rho\, dV$ is constant in time and generates no radiation; in electromagnetic theory the dipole moment $\int \rho x^i \, dV$ has non-vanishing second time-derivative and thereby provides the main contribution to electromagnetic radiation; in gravitational theory, on the other hand, the first derivative of the dipole moment is the momentum $\int \rho \dot{x}^i \, dV$, which remains constant. So here the first and main contribution to the radiation comes from the quadrupole, and that, by dimensional necessity, carries the unfortunately high power of c in the denominator.

As an example, let us apply (15.68) to the simplest imaginable source—two stars of equal mass M (here to be treated as mass points) going round the same circle of radius a at opposite ends of a diameter. Let their common angular orbit velocity ω point along the x axis. Then the 2-dimensional quadrupole tensor is given by

$$\int \rho x^i x^j \, dV = 2Ma^2 \begin{pmatrix} \cos^2 \omega t & \cos \omega t \sin \omega t \\ \cos \omega t \sin \omega t & \sin^2 \omega t \end{pmatrix} \qquad (15.69)$$

with $i, j = 2, 3$. Recalling the identities $2\cos^2 A = 1 + \cos 2A$ and $2 \sin A \cos A = \sin 2A$, and substituting (15.69) into (15.68), we find

$$c^{ij} = \frac{8GMa^2\omega^2}{c^4 r} \begin{pmatrix} \cos 2\omega t & \sin 2\omega t \\ \sin 2\omega t & -\cos 2\omega t \end{pmatrix}. \qquad (15.70)$$

Observe that this result is already in TT gauge. It represents the sum of a λ wave of the form $\lambda = \epsilon \cos 2\omega t$ (at fixed x) and a μ wave of the form $\mu = \epsilon \sin 2\omega t$. [To 'retard' the RHS of (15.70), we replace t by $t - x/c$; but at fixed x we can shift the time origin, $t \mapsto t + x/c$, to get back to (15.70).] Not surprisingly, the wave has frequency 2ω, the frequency with which the source returns to identical configurations. And according to our findings in Section 15.3, it is circularly polarized; interestingly, though, the little circles described by test dust relative to an arbitrary center are described at twice the angular speed of the source. If the rotation axis of the source were not along the line of sight, this would manifest itself in a less symmetric polarization state of the wave we observe. (Cf. Exercise 15.9.) In fact, the polarization state serves as an indicator of the orientation of the source.

Results like (15.70), based on eqn (15.68), allow us to estimate the amplitude of waves coming from various likely sources. That amplitude is then the ϵ one uses in eqn (15.58) to predict the response of the detector. A typical value for ϵ from nearby binary star systems is of the order of $\sim 10^{-20}$. A wave of such an amplitude wiggles two free particles 10 km apart by $\sim 10^{-14}$ cm—about 1/30 of the classical radius of the electron. This conveys some inkling of the difficulties of detection. Add to that the fact that the earth is bathed in periodic gravity waves from thousands of binaries with thousands of different frequencies and that, unlike optical telescopes, gravitational 'telescopes' cannot be aimed at specific sources; so some very subtle mathematics will be needed to unravel whatever signal might be detected. 'Burst' sources, on the other hand, like nearby supernovae, would stand out from this background and, moreover, would yield larger amplitudes, of the order of $\sim 10^{-18}$.

The first attempts to observe cosmic gravity waves were made by Weber in the early 1960s. It was he who invented the *bar detector*, which, in modified form, is still being used by a majority of research groups around the world. It consists of a large metal cylinder, weighing up to 5 tonnes, whose mechanical oscillations are driven by gravity waves. Transducers on its surface convert the bar's oscillations into electrical signals. Weber's aluminum cylinders operated at room temperature and eventually reached a sensitivity of $\epsilon \approx 3 \times 10^{-16}$. Modern bars are made of various alloys with 10 times more favorable damping factors, and run at liquid helium temperatures to reduce noise. They are getting close to a sensitivity of $\epsilon \approx 10^{-18}$. Higher sensitivities are expected from various *beam detectors* presently under construction. These consist in principle of two masses separated by a few kilometers, suspended so as to move freely relative to each other, and with a laser beam measuring their relative displacement. (In practice, the arrangement is that of a Michelson interferometer—cf. Section 1.7— which measures the *difference* in tidal displacements along two orthogonal arms.) Ultimate sensitivities of $\epsilon \approx 10^{-22}$ are envisioned for such instruments. There are even plans for a triangular detector in space, consisting of three free-flying spacecraft about 5×10^6 km apart. This LISA project ('Large Interferometric Space Antenna') might fly by ~ 2010.

We end this section by reporting, without proof, two important results. As we stressed in Chapter 14, no proper energy tensor can be assigned to the gravitational field, because of its elusiveness. (By the equivalence principle, a change of reference frame can always eliminate the field at one point.) Also, no such tensor is needed on the RHS of the full non-linear field equations to take care of the 'gravity of gravity'. But in *linearized* GR, it is both possible and useful to construct a *pseudo-energy tensor*, which plays much the same role as a proper energy tensor, provided one sticks to the preferred coordinates.

The linear field equations of linearized GR cannot automatically take care of the gravity of gravity; for example, of the fact that gravity waves carry energy. That they *do* carry energy is clear, for example, from the consideration that sand on a rigid plate would move in response to the tidal forces of a passing wave, and thus do work against (some very small) friction.

If, for the wave, we expand the Einstein tensor $G_{\mu\nu}$ *beyond* the first power of ϵ, say $G_{\mu\nu} = G_{\mu\nu}^{(1)} + G_{\mu\nu}^{(2)} + \cdots$, successive terms being $O(\epsilon)$, $O(\epsilon^2)$, etc., then the vacuum field equation $G_{\mu\nu} = 0$, which the wave satisfies in the full theory, is approximated by the equation

$$G_{\mu\nu}^{(1)} = -G_{\mu\nu}^{(2)} =: -\frac{8\pi G}{c^4} t_{\mu\nu}. \tag{15.71}$$

The symmetric components $t_{\mu\nu}$ defined by this equation in analogy to the full field equations (14.8), (14.11), evidently constitute a Lorentz tensor (the so-called pseudo-energy tensor) and satisfy the conservation equation $t^{\mu\nu}{}_{,\nu} = 0$. If there is matter in the path of the wave, *its* energy tensor $T_{\mu\nu}$ must be added to $t_{\mu\nu}$ in eqn (15.71), and its interaction with the wave will be governed by the joint conservation equation $(T^{\mu\nu} + t^{\mu\nu})_{,\nu} = 0$. So the components of $t_{\mu\nu}$ must have the same physical significance as those of any other energy tensor.

The straightforward calculation of $G_{\mu\nu}^{(2)}$ yields, after averaging over a spacetime region of several wavelengths,

$$t_{\mu\nu} = \frac{c^2}{16\pi G}\left\langle\left(\frac{\partial\lambda}{\partial t}\right)^2 + \left(\frac{\partial\mu}{\partial t}\right)^2\right\rangle\begin{pmatrix} 0 & 0 & 0 & -1 \\ 0 & 0 & 0 & 0 \\ 0 & 0 & 0 & 0 \\ -1 & 0 & 0 & 1 \end{pmatrix} \quad (15.72)$$

for a wave in TT gauge propagating in the x-direction, as in eqn (15.47). The averaging $\langle\ldots\rangle$ is necessary in order to get a physically useful energy density t_{44} and energy current density $-ct_{14}$—whose fluctuations at a single point are meaningless. Perhaps not surprisingly, this energy tensor corresponds to that of a swarm of zero-rest-mass particles (gravitons) traveling parallelly at the speed of light. It is therefore also identical in form to the energy tensor of an electromagnetic wave.

Since calculations like that leading to (15.70) allow us to obtain the wave metric anywhere at some distance from a source, and the pseudo-energy tensor then allows us to calculate the energy current represented by the waves, it is possible (for example, by integrating over a sphere) to calculate the rate $-dE/dt$ at which energy is lost by the source due to the gravitational radiation it emits. In such a manner one can obtain the formula

$$-\frac{dE}{dt} = \frac{G}{5c^5}\left\langle\left(\frac{d^3 J_{ij}}{dt^3}\right)^2\left(\frac{d^3 J^{ij}}{dt^3}\right)^2\right\rangle, \quad (15.73)$$

where J_{ij} is the 'reduced' (that is, trace-free) quadrupole moment of the source,

$$J_{ij} = \int \rho\left(x^i x^j - \frac{1}{3}\delta^{ij}\sum x^k x^k\right)dV. \quad (15.74)$$

This result, in essence, was found by Einstein in 1916! Of course, it is only an approximation, valid as long as the source matter moves non-relativistically. In the simple case of the binary system described by eqn (15.69), formula (15.73) can be shown to imply

$$-\frac{dE}{dt} = \frac{128}{5}\frac{G}{c^5}\left(Ma^2\omega^3\right)^2 = \frac{32G^{7/3}}{5\sqrt[3]{4}c^5}(M\omega)^{10/3}, \quad (15.75)$$

where for the second equation we utilized the Keplerian relation

$$\omega^2 = GM/4a^3, \quad (15.76)$$

expressing the balance of centrifugal and gravitational force. It is not difficult to generalize these results to binaries with unequal masses and non-circular orbits.

A binary system which loses energy will spin ever faster while its orbit shrinks. Consider, for example, the simple circular system discussed above. Its kinetic energy $M(a\omega)^2$ and its potential energy $-GM^2/2a$ make a total of $-GM^2/4a$ when account is taken of the Keplerian relation (15.76). This total energy decreases for decreasing a, and thus, via (15.76), for increasing ω. To date, the only experimental proof we have of the reality of gravitational radiation is the minutely documented observation over the last 25 years of just such a spin-up. The object in question is the famous 'binary pulsar' PSR 1913 + 16, discovered by Hulse and Taylor in 1974, and the object of intense observational scrutiny and theoretical analysis ever since. It consists of two neutron stars of almost equal mass $\sim 1.4 M_\odot$, one live and one dead, in a highly excentric orbit ($e \approx 0.6$) of period ~ 8 h and radius $\sim 1 R_\odot$. The live component is a pulsar of period 59 ms. Much general-relativistic mechanics has gone into the analysis of this system (which has been called a 'veritable laboratory' for GR), and all conceivable non-radiative causes of orbital spin-up (such as tidal dissipation) have been allowed for. What remains is an impressive validation of the general-relativistic prediction that the orbital period should decrease owing to radiation by 7.2×10^{-5} s per year. The agreement between observation and prediction by now lies within 0.5 per cent.

15.5 The electromagnetic analogy in linearized GR

We have already in Section 9.6 seen features of weak stationary gravitational fields that are very reminiscent of electromagnetic fields. In particular, the law of motion (9.18) for slow orbits mimics the Lorentz force law of Maxwell's theory. And in eqns (15.30) above we saw a remarkable resemblance to the electromagnetic expressions for the potentials in terms of the sources. In fact, those results together establish an essentially Maxwellian theory of slow orbits in weak stationary gravitational fields generated by low-stress, low-velocity sources.

In eqns (15.77)–(15.81) below we collect the details of this Maxwellian analogy, writing the electromagnetic formulae on the left and the gravitational formulae on the right. We have already chosen the units of length and time so as to make $c = 1$. Now we additionally choose the unit of mass so as to make $G = 1$, which corresponds to choosing Gaussian units in electromagnetism, in terms of which the Coulomb force becomes simply $q_1 q_2 / r^2$. We write $\mathfrak{E}$ and $\mathfrak{B}$ for the 'gravitoelectric' and the 'gravitomagnetic' fields, respectively:

$$\phi = + \int \frac{\rho \, dV}{r} \qquad \longleftrightarrow \qquad \Phi = - \int \frac{\rho \, dV}{r} \tag{15.77}$$

$$\mathbf{w} = - \int \frac{\rho \mathbf{u} \, dV}{r} \qquad \longleftrightarrow \qquad \mathbf{w} = +4 \int \frac{\rho \mathbf{u} \, dV}{r} \tag{15.78}$$

$$\mathbf{e} = -\operatorname{grad} \phi \qquad \longleftrightarrow \qquad \mathfrak{E} = -\operatorname{grad} \Phi \tag{15.79}$$

$$\mathbf{b} = \operatorname{curl} \mathbf{w} \qquad \longleftrightarrow \qquad \mathfrak{B} = \operatorname{curl} \mathbf{w} \tag{15.80}$$

$$\mathbf{a} = \frac{q}{m}(\mathbf{e} + \mathbf{v} \times \mathbf{b}) \qquad \longleftrightarrow \qquad \mathbf{a} = \mathfrak{E} + \mathbf{v} \times \mathfrak{B}. \tag{15.81}$$

Note, above all, the sign differences in the first two lines: electric force repels, while gravitational force attracts, like charges. Also there is the noteworthy factor 4 in the integral for the gravitational vector potential $\mathbf{w}$: moving masses accordingly generate a four times stronger gravitomagnetic field than direct analogy with electromagnetism would imply. And finally, the absence of a multiplier in the 'gravitational Lorentz force law' (15.81)(ii) stems from the equality of gravitational and inertial mass.

In Chapter 9 we used a very indirect argument to establish the 'force law' (9.18). Now that we are in possession of the explicit form of the law of motion,

$$\frac{d^2 x^\rho}{ds^2} + \Gamma^\rho_{\mu\nu} \frac{dx^\mu}{ds} \frac{dx^\nu}{ds} = 0, \tag{15.82}$$

we can prove it directly. The orbits are assumed to be slow, such that, if they are described at velocity v, we can neglect v^2. Since $ds^2 \approx (1 - v^2) dt^2$, we have $dt/ds \approx \gamma = (1 - v^2)^{-1/2}$. This entails (with $\dot{} = d/dt$)

$$\frac{dx^\rho}{ds} = \dot{x}^\rho \gamma, \qquad \frac{d^2 x^\rho}{ds^2} = \ddot{x}^\rho \gamma^2 + \dot{x}^\rho \dot\gamma \gamma. \tag{15.83}$$

But from (2.10) we know that $\dot\gamma = O(v)$, so that, to first-order in v, we have

$$\frac{dx^\rho}{ds} = \dot{x}^\rho =: (v^i, 1), \qquad \frac{d^2 x^i}{ds^2} = \ddot{x}^i, \qquad \frac{d^2 x^4}{ds^2} = \dot\gamma \gamma. \tag{15.84}$$

As we have seen [in the paragraph following (10.44)], when three of the geodesic equations are satisfied, the last is satisfied automatically. So here we take the first three, and when we substitute from (15.6) and (15.84) into (15.82), these become

$$\ddot{x}^i = -\frac{1}{2}\left(h^i_{\mu,\nu} + h^i_{\nu,\mu} - h_{\mu\nu,}{}^i\right)\dot{x}^\mu \dot{x}^\nu. \tag{15.85}$$

Referring to (15.32); remembering that raising or lowering an i introduces a negative sign; remembering also that $h_{\mu\nu,4} = 0$ by the assumed stationarity of the field; and, for ease of recognition, writing out a few of the summations in full, we discover that (15.85) expresses precisely the law

$$\mathbf{a} = -\text{grad } \Phi + \mathbf{v} \times \text{curl } \mathbf{w}, \tag{15.86}$$

which is thus re-established.

The electromagnetic analogy has many uses. To begin with, it supplies us with a familiar model for the *extra* gravitational force created by the *motion* of the sources, and felt only by *moving* particles (namely, the $\mathfrak{B}$ field). Also occasionally the analogy allows us simply to 'translate' familiar results from electromagnetism to weak GR. For example, recall the electromagnetic potentials created at position $\mathbf{r}$ from its center by a uniformly rotating ball of homogeneous charge density:

$$\phi = \frac{q}{r}, \qquad \mathbf{w} = \frac{1}{5}\frac{R^2 q \boldsymbol{\omega} \times \mathbf{r}}{r^3}. \tag{15.87}$$

Here q is the total charge, R the radius, and ω the angular velocity of the ball. We can translate this into a gravitational result. The charge q becomes the mass m of a rotating ball. It is usual in GR to introduce a quantity a defined by

$$L = I\omega =: ma, \tag{15.88}$$

where L is the angular momentum and I the moment of inertia; so a is the 'angular momentum per unit mass'. For a homogeneous ball, we have $I = (2/5)mR^2$. Thus, translating (15.87) by use of (15.77) and (15.78), we obtain

$$\Phi = -\frac{m}{r}, \qquad \mathbf{w} = -\frac{2I\boldsymbol{\omega} \times \mathbf{r}}{r^3}. \tag{15.89}$$

In the linearized metric (15.33) we require the expression $w_i \, dx^i = \mathbf{w} \cdot d\mathbf{r}$. We find this from (15.89)(ii):

$$\mathbf{w} \cdot d\mathbf{r} = -\frac{2I}{r^3}\boldsymbol{\omega} \cdot (\mathbf{r} \times d\mathbf{r}) = -\frac{2I\omega}{r^3}(x \, dy - y \, dx), \tag{15.90}$$

where we have used the well-known 'interchangeability of dot and cross' in the triple product. So the metric (15.33) around the ball takes the form

$$ds^2 = \left(1 - \frac{2m}{r}\right)dt^2 - \left(1 + \frac{2m}{r}\right)\sum\left(dx^i\right)^2$$
$$+ \frac{4ma}{r^3}(x \, dy - y \, dx) \, dt. \tag{15.91}$$

In terms of polar coordinates r, θ, ϕ it becomes

$$ds^2 = \left(1 - \frac{2m}{r}\right)dt^2 - \left(1 + \frac{2m}{r}\right)[dr^2 + r^2(d\theta^2 + \sin^2\theta \, d\phi^2)]$$
$$+ \frac{4ma}{r}\sin^2\theta \, d\phi \, dt. \tag{15.92}$$

This is, in fact, the linearized version of the famous *Kerr metric*. [And without the cross term that makes it non-static it is clearly the linearized version of the Schwarzschild metric in its isotropic form (11.26).] The Kerr metric is known to correspond exactly and uniquely to a steadily rotating black hole, and as such it plays an important role in theoretical GR. But it is not known exactly which types of *regular* rotating mass distributions might also serve as its source. For most practical purposes the linearized metric (15.92) satisfactorily describes the gravitational field around a rotating star or planet.

As an example, let us work out the periods of free circular orbits in the equatorial plane of a rotating star, and, in particular, let us calculate the difference in periods for orbits described in the positive and negative senses at the same radius. We *could* do this via the geodesics of the metric (15.92), but an alternative method will be instructive.

Consider the magnetic field **b** corresponding to the electromagnetic potentials (15.87). It is the well-known dipole field of a rotating charged ball,

$$\mathbf{b} = \frac{3}{5}R^2 q\left(\boldsymbol{\omega}\cdot\mathbf{r}\frac{\mathbf{r}}{r^5} - \frac{1}{3}\frac{\boldsymbol{\omega}}{r^3}\right). \tag{15.93}$$

Its gravitational analog [taking into account the factor -4 between eqns (15.78)] is

$$\boldsymbol{\mathfrak{B}} = -\frac{12}{5}R^2 m\left(\boldsymbol{\omega}\cdot\mathbf{r}\frac{\mathbf{r}}{r^5} - \frac{1}{3}\frac{\boldsymbol{\omega}}{r^3}\right), \tag{15.94}$$

and in the equatorial plane this reduces to

$$\boldsymbol{\mathfrak{B}} = +\frac{4}{5}\frac{R^2 m\omega}{r^3}. \tag{15.95}$$

We can now find the orbit by equating its radial acceleration $-r\dot{\phi}^2$ to the gravitational acceleration (15.86). To first order the latter is $-m/r^2$, so that $v = r\dot{\phi} = \pm(m/r)^{1/2}$. With that and (15.95), the contribution $\mathbf{v}\times\boldsymbol{\mathfrak{B}}$ to **a** can be written down and added to $-m/r^2$; after canceling a factor $-r$ we then obtain

$$\dot{\phi}^2 = \frac{m}{r^3} \mp \frac{2L}{r^4}\sqrt{\frac{m}{r}}, \tag{15.96}$$

where L is the angular momentum of the ball [cf. after (15.88)] and the top and bottom signs correspond to the positive and negative senses of describing the orbit, respectively. We leave the calculation of the period difference to the Exercises (cf. Exercise 15.11); but we see immediately that the *retrograde* orbit is the faster— simply because the gravitomagnetic force increases the effective pull towards the center, which must be balanced by a greater centrifugal force.

It has long been customary to refer to gravitomagnetic phenomena like the one discussed here in terms of 'frame dragging' or 'space dragging'. This dates back to Einstein ('Mitführung') and his early fascination with Mach's principle, according to which the 'rotating' universe 'drags' the inertial frame relative to the non-rotating earth. But that metaphor can be very misleading. In the present case, for example, if the rotating ball really dragged the space around with it, surely the retrograde orbit would be the slower. Our recommendation is to use the vague concept of dragging at most in a tentative way, and for definite results to rely instead on the solidly established Maxwellian analogy.[1]

We have already discussed in Chapter 9 [cf. after (9.19)] the intimate connection between the rotation rate of the local inertial frame, $\boldsymbol{\Omega}_{\text{LIF}}$, and the gravitomagnetic field. In fact, according to eqn (9.21) and the subsequent text, that rotation rate is given in first approximation by

$$\boldsymbol{\Omega}_{\text{LIF}} = -\tfrac{1}{2}\boldsymbol{\mathfrak{B}}. \tag{15.97}$$

---

[1] Cf. W. Rindler, The case against space dragging, *Phys. Lett.* **A233**, 25 (1997).

We shall now use this formula to derive an interesting result first discussed by Lense and Thirring as early as 1918. The earth's rotation induces a precession in *all* the gyroscopes we can imagine affixed to the stationary lattice surrounding the earth and at rest relative to infinity. The pattern of the rotation directions is given by the familiar magnetic field lines issuing from a rotating ball of charge, as shown in Fig. 15.2, where $\mathbf{\Omega}_{LT}$ denotes the *Lense–Thirring precession* angular velocity, namely $-\frac{1}{2}$ times the RHS of eqn (15.94). (There is a direction reversal between **b** and $\mathfrak{B}$, and another between $\mathfrak{B}$ and $\mathbf{\Omega}_{LT}$.) Thus, for example, the gyroscopes at the poles precess in the sense of earth's rotation, while those in the equatorial plane precess in the opposite sense. For these latter we find, from (15.97) and (15.95)

$$\Omega_{LT} = -\frac{2}{5}\frac{R^2 m\omega}{r^3} = -\frac{J}{r^3} = -\frac{ma}{r^3}. \tag{15.98}$$

For a gyroscope in free circular orbit in the equatorial plane this precession must be added to the de Sitter precession (11.74) discussed in Chapter 11, which applies to orbits around a *non-rotating* earth. Since the time Δt for one orbital revolution is given to sufficient accuracy by the Keplerian formula $2\pi (r^3/m)^{1/2}$, we can find from (15.98) the angle precessed per revolution, $\alpha_{LT} = \Omega_{LT}\Delta t$, due to the Lense–Thirring effect; adding this to the de Sitter angle (11.74) gives us the total precession:

$$\alpha = \pm\frac{3\pi m}{r} - 2\pi a\sqrt{\frac{m}{r^3}}, \tag{15.99}$$

the top and bottom signs referring to positively and negatively described orbits, respectively. For near-earth orbits the first term adds up to $\sim$8 seconds of arc per year, the second to only about 1 per cent of that. The Stanford Gyroscope Experiment (also

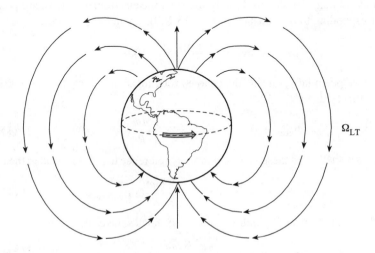

Fig. 15.2

known as Gravity Probe-B)—in the planning since the early 1970s—aims to test these effects early in the new century to high accuracy by flying shielded satellites containing gyrosopes in low *polar* orbits (to avoid large Newtonian effects due to the earth's non-sphericity). As we mentioned before, the de Sitter part of the precession has already been validated satisfactorily (cf. penultimate paragraph of Chapter 11). The Stanford Experiment would be the first to validate gravitomagnetism (in its Lense–Thirring manifestation), in addition to improving the de Sitter data even further. However, somewhat rough (20 per cent) evidence for a different gravitomagnetic effect—the precession of the orbital plane itself, if inclined to the equator (cf. Exercise 15.16)—has already been reported by Ciufolini *et al.* in 1998; this resulted from observations of the procession of the line of nodes of a pair of laser-ranged geodynamics satellites (LAGEOS I and II).

After this digression to the present day, we return to the work of Thirring. Already before his collaboration with Lense he had in 1918 attacked a problem of great Machian interest to Einstein: the gravitomagnetic field inside a uniformly rotating shell, say of mass m and radius R. Now the *magnetic* field $\mathbf{b}$ inside a shell of charge q and radius R rotating at angular velocity ω is constant and given by

$$\mathbf{b} = \frac{2}{3}\frac{q\omega}{R}. \tag{15.100}$$

By our translation from $\mathbf{b}$ to $\mathfrak{B}$ (effected by a factor -4) and from $\mathfrak{B}$ to $\mathbf{\Omega}_{\mathrm{LIF}}$ (effected by a factor $-\frac{1}{2}$), we immediately deduce

$$\mathbf{\Omega}_{\mathrm{LIF}} = \frac{4}{3}\frac{m\omega}{R}. \tag{15.101}$$

Thus all stationary gyroscopes inside the rotating mass shell precess in unison. In fact, the spacetime inside the shell is (in linear approximation) Minkowskian but uniformly rotating. To see this directly, consider the electromagnetic vector potential $\mathbf{w}_{\mathrm{el}}$ corresponding to the constant $\mathbf{b}$ field (15.100), say in the z-direction:

$$\mathbf{w}_{\mathrm{el}} = \frac{1}{3}\frac{q\omega}{R}(-y, x, 0). \tag{15.102}$$

The electric potential ϕ_{el} inside the charged rotating shell is zero. By our translation scheme, this leads to

$$\Phi_{\mathrm{grav}} = 0, \qquad \mathbf{w}_{\mathrm{grav}} = -\frac{4}{3}\frac{m\omega}{R}(-y, x, 0) \tag{15.103}$$

for the mass shell. And these values in turn, by reference to (15.33), lead to the metric

$$\mathbf{ds}^2 = \mathrm{d}t^2 - \sum(\mathrm{d}x^i)^2 + \frac{8}{3}\frac{m\omega}{R}(x\,\mathrm{d}y - y\,\mathrm{d}x), \tag{15.104}$$

which finally, in terms of cylindrical polar coordinates, takes the form

$$\mathbf{ds}^2 = \mathrm{d}t^2 - \mathrm{d}r^2 - r^2\,\mathrm{d}\phi^2 + \frac{8}{3}\frac{m\omega}{R}r^2\,\mathrm{d}\phi\,\mathrm{d}t - \mathrm{d}z^2. \tag{15.105}$$

Note from (15.103) that $w = O(r\omega)$. But we see from (15.32) that in linearized GR w must be considered to be of the same order of smallness as the hs. Consequently its square is neglected. In the present case we thus neglect $r^2\omega^2$. With that, the metric (15.105) becomes equivalent to the metric (9.26) of a lattice uniformly rotating in Minkowski space with angular velocity

$$\omega_{\text{lattice}} = -\frac{4}{3}\frac{m\omega}{R} \tag{15.106}$$

about the z-axis. So the stationary lattice inside the rotating shell, fixed relative to infinity, rotates *negatively* relative to an underlying Minkowski space. By transforming to positively rotating coordinates with angular velocity $4m\omega/3R$ relative to infinity, we would arrive at the Minkowski metric, and our assertion is proved.

Of course, the factor in front of ω in (15.106) and (15.101) (in full units: $4Gm/3c^2R$) is so minute that terrestrial tests are out of the question. For a shell of mass 10^x tonnes and radius 10^y meters, that factor is $\sim 10^{-24+x-y}$.

Nevertheless, it is easy to see why Thirring's result was a victory for the Machians. The distant universe, considered as a rotating mass shell relative to a static earth (never mind its necessarily superluminal transverse velocity!) might similarly, but this time *fully*, drag the local inertial frame at the earth along with it. Already in 1913 Einstein had preliminarily obtained the above results on the rotating shell and reported them enthusiastically to an aging and unresponsive Mach.

Exercises 15

15.1. Prove that the transformation (15.14) is an infinitesimal Lorentz transformation (that is, it transforms $\eta_{\mu\nu}$ into itself) if and only if $f_{\mu,\nu} + f_{\nu,\mu} = 0$.

15.2. Prove that the harmonic gauge condition $c^{\rho\nu}{}_{,\nu} = 0$ is equivalent to $\Gamma^\rho := g^{\mu\nu}\Gamma^\rho_{\mu\nu} = 0$. In the full theory, prove that $\Gamma^\rho = 0$ implies that the coordinates x^ρ themselves satisfy the wave equation, $\Box x^\rho = 0$. Such coordinates are called *harmonic coordinates* and $\Gamma^\rho = 0$ is called the *harmonic coordinate condition*. [*Hint*: for any scalar function ϕ establish the indentity $\Box\phi = g^{\mu\nu}\phi_{;\mu\nu} = g^{\mu\nu}\phi_{,\mu\nu} - \Gamma^\rho\phi_{,\rho}$.]

15.3. Prove that in linearized GR the wave-vector k^μ of a plane wave is an eigenvector of the curvature tensor, in the sense that $R_{\lambda\mu\nu\rho}k^\rho = 0$, independently of any gauge. [*Hint*: establish the following relations in the TT gauge: $\bar{h}_{\mu\nu}k^\nu = 0$, $h_{\mu\nu,\rho\sigma} = k_\rho k_\sigma \bar{h}_{\mu\nu}$. Then by reference to (15.7) prove the eigenvector property in the TT gauge. Finally show that it is gauge-invariant.]

15.4. Prove that under a rotation of the y and z coordinate axes through an angle ψ (that is, $y' = y\cos\psi + z\sin\psi$, $z' = -y\sin\psi + z\cos\psi$), the (λ, μ) wave (15.47) becomes a (λ', μ') wave, such that

$$\lambda' = \lambda\cos 2\psi + \mu\sin 2\psi$$

$$\mu' = -\lambda\sin 2\psi + \mu\cos 2\psi.$$

In particular, note that if $\psi = 45°$, $\lambda' = \mu$, $\mu' = -\lambda$; and if $\psi = 180°$, the wave is unchanged. [*Hint*: the $h^{\alpha\beta}$ are Lorentz tensor components.]

15.5. Prove that the two sets of co-asymptotic hyperbolae of Fig. 15.1 intersect each other everywhere orthogonally. Let the λ-hyperbolae have equation $xy = $ const [cf. (15.55)] in an x, y plane, while the μ-hyperbolae are the λ-hyperbola rotated through $45°$, $x \mapsto \frac{1}{\sqrt{2}}(x - y)$, $y \mapsto \frac{1}{\sqrt{2}}(x + y)$. [*Hint*: if (dx, dy) is a displacement along a λ-hyperbola and $(\delta x, \delta y)$ one along a μ-hyperbola, prove $dx\,\delta x + dy\,\delta y = 0$.]

15.6. Perform a calculation analogous to that which led us from (15.63) to Laue's theorem (15.66), for the electromagnetic four-current density J^μ in place of $T^{\mu\nu}$, thus obtaining

$$\int J^i \, dV = \frac{d}{dt} \int \rho x^i \, dV.$$

In the special case of a *stationary* current distribution, the RHS of this equation vanishes. All currents must then be closed loops, which explains the vanishing of the LHS. In the general case, the RHS measures the changing accumulations of charge due to the variable currents. [*Hint*: (7.38), (7.39).]

15.7. Consider a system of four identical stars equally spaced and orbiting on a common circle under their mutual gravity. Prove that this system does not radiate at all. [*Hint*: the quadrupole tensor of this system is the sum of the quadrupole tensors of two binaries out of phase by $90°$.]

15.8. From the fact that in TT gauge the only non-vanishing components of the curvature tensor are those shown in (15.49), and from the result of the final paragraph of Section 15.2, deduce that the curvature components (15.49) are the only non-vanishing ones for *any* wave metric $h_{\mu\nu} = h_{\mu\nu}(t - x)$ satisfying the harmonic gauge.

15.9. Consider the gravitational radiation generated by the binary system whose quadrupole tensor is (15.69), but now far away in the plane of rotation. To conform to our convention that the wave propagates in the x-direction, let i and j in (15.69) and (15.70) now take the values 1, 2. Since the corresponding cs will not be in the TT gauge, use the result of the final paragraph of Section 15.2 to impose that gauge. Hence prove that the radiation in the plane of the orbit is *linearly* polarized and that its amplitude is only half of that of the frontal radiation at equal distance.

15.10. Using eqn (15.84), verify that (as asserted in the text) the $\rho = 4$ member of the set of geodesic equations (15.82) is satisfied automatically when the other three equations are satisfied, implying (15.86).

15.11. Derive the time difference ΔT between the periods of the negative and positive equatorial orbits at constant distance from the rotating ball discussed in the paragraph containing eqn (15.93). [*Answer*: $\Delta T \approx 4\pi a$ (cf. (15.88)) independently of r.]

15.12. Consider a light-signal guided by mirrors round a circular orbit of radius r in the equatorial plane of the rotating ball of the preceding exercise. Find the time difference ΔT between the periods of the two orbits described in opposite senses and observe that it is now the *positive* orbit that is the faster. [*Hint*: use (15.92). *Answer*: $\Delta T \approx 8\pi ma/r$.]

15.13. A particle is dropped from rest at infinity in the equatorial plane of a ball of mass m and radius R rotating uniformly at angular velocity ω. Prove that the particle experiences a transverse acceleration

$$\sim \frac{4}{5} \frac{R^2 \omega}{c^2} \sqrt{\frac{2G^3 m^3}{r^7}}$$

in the direction of the ball's rotation when at distance r from its center, thus lending credence to the dragging metaphor. [*Hint*: (15.86), (15.95).]

15.14. A Foucault pendulum suspended at the earth's north pole swings freely in a vertical plane Π. Prove that the earth's rotation causes Π to precess relative to the distant universe by $\sim 7.13 \times 10^{-4}$ seconds of arc per day in the sense of its own rotation. [*Hint*: recall from the end of Section 11.2C that for the earth $GM/c^2 = 0.44$ cm.]

15.15. The magnetic field **b** inside an infinite solenoid is strictly parallel to the axis, pointing in the positive sense relative to the current, and everywhere of magnitude $4\pi n I/c$, where I is the current and n is the number of wire turns per unit length. Use this information to prove that the gravitomagnetic field inside a long cylindrical shell of radius R and mass μ per unit length, rotating uniformly at angular velocity $\boldsymbol{\omega}$, is given by $\mathfrak{B} = -8G\mu\boldsymbol{\omega}/c$ in full units. Deduce that the spacetime inside the cylinder, to linear approximation at least, is Minkowskian and rotating at angular velocity $4G\mu\boldsymbol{\omega}/c^2$ relative to infinity. [*Hint*: cf. the argument following (15.103).]

15.16. Give a heuristic Maxwell-type argument for the precession of the axis of a circular non-equatorial orbit (for example, that of an artificial satellite) around the earth's angular velocity vector $\boldsymbol{\omega}$. [*Hint*: Replace the earth by a rotating charged ball and the orbit by a current loop, whose magnetic field will tend to align itself with $\boldsymbol{\omega}$, thus producing a torque. In the gravitational analog, that torque will make the angular momentum vector of a rotating massive ring (roughly representing the satellite in orbit) precess around $\boldsymbol{\omega}$. By following the analogy in detail, determine the direction of the precession.

Part III
Cosmology

16

Cosmological spacetimes

16.1 The basic facts

A. Introduction

Cosmology—the inquiry into the largest-scale features of the universe—has obviously excited Man's curiosity since time immemorial. But in spite of the spectacular advances in astronomy since the days of Galileo and Kepler, cosmology remained more speculation than science until well into the twentieth century. Relevant observations were sparse, and so theories, however fanciful, could not be falsified. All this changed drastically with the advent of the giant telescopes early in the century, which suddenly brought a significant portion of the whole universe into view. And other technologies soon followed. In fact, today one can hope that within a decade or so we may have the answers to such fundamental questions as whether the universe is finite or infinite, and whether it will expand forever or some day halt and recollapse.

B. Regularity of the universe

One of the main features uncovered about the universe is its large-scale regularity. All directions around us appear to be equivalent, and the limited region we have been able to survey in detail appears to be typical of all the rest. In other words, the entire universe appears to be isotropic and homogeneous. Add to this the assumption that the laws of physics, as discovered here on earth, are valid in all regions and at all times, and one can begin to construct cosmological models. These basic assumptions then turn out to be well supported by the agreement of their consequences with the observations, and by the general harmony of the resulting picture. It should also be noted that if the universe is really homogeneous, then that is a good indication that the laws of physics are at least spatially universal; for if they were not, different regions might well have evolved differently.

The apparent regularity of the universe is very fortunate for cosmologists. It is hard to see how without it one could extrapolate from local observations, and how one could even begin to construct useful cosmological models. It is also very mysterious. Whereas the nice roundness of astronomical objects like sun or earth is understandable in terms of energy and gravity, the homogeneity and isotropy of the entire universe, if real, has never been satisfactorily explained. One may eventually be forced to accept that this is simply the way the universe came into being.

C. History

Modern theoretical cosmology found its greatest inspiration in Einstein's general relativity, which provided it with a consistent dynamics and optics for the entire universe, and with such exciting geometrical possibilities as finite and yet unbounded universes. Perhaps most importantly, general-relativistic mechanics had already suggested non-static model universes by the time astronomers found to their great surprise that they were needed.

By contrast, Newton's mechanics had proved sterile as a source of cosmological models. Its laws are inapplicable to infinite mass distributions. Newton's original idea of a *static* universe (cf. Section 1.11), based not on dynamics but rather on symmetry and the existence of absolute space, was beset with contradictions. Suppose, for example, we remove a finite ball of matter from it. What will be the field inside the cavity? If the inside field is zero, then when we put the matter back, it will collapse under its own gravity. It seems the outside masses must provide a centrifugal field inside. But how? Each successive concentric shell yields a zero field inside. Attempts were made in the late nineteenth century to tamper with the inverse square law, just to make a static universe possible. In retrospect it is hard to understand why nineteenth-century astronomers were so bent on the idea that the universe must be static. But even Einstein initially resisted what his own field equations clearly told him: he tampered with them so that his first cosmological model (of 1917) could be static!

However, before we discuss cosmological models, we must look at some of the facts. Evidently the first concern of cosmologists must be with the spatial distribution of stars and galaxies. For Copernicus in the early sixteenth century, just as for Ptolemy and Plato centuries before, all the stars were still fixed to a 'crystalline' sphere, though now centered on the sun rather than on the earth. But as early as 1576, Thomas Digges boldly replaced that sphere by an infinity of stars extending uniformly through all space, the dimmer ones being farther away. The same extension was also made by Giordano Bruno (who for his ideas was burned at the stake in 1600), and mystically foreshadowed a century earlier by Nicholas of Cusa. But, whereas to Digges the sun was still king of the heavens, one who 'raigneth and geeveth lawes of motion to ye rest', Bruno recognized it for what it is: just a star among many. The infinite view was later supported by Newton, who believed that a finite universe would 'fall down into the middle of the whole space, and there compose one great spherical mass. But if the matter was evenly disposed throughout an infinite space ... some of it would convene into one mass and some into another. ... And thus might the sun and the fixed stars be formed' (1692). The really revolutionary content of this passage is the idea of an *evolving* universe, and of gravity as the mechanism causing condensation.

Both Newton and his contemporary Huyghens knew that the stars were immensely far apart. Huyghens had let light from the sun fall onto a pinhole after expanding it optically until its apparent diameter was 30 000 times that of the pinhole, whereupon the pinhole seemed as bright as he remembered (!) Sirius to be at night. He concluded that Sirius is 30 000 times as distant as the sun. Newton gave that multiple as 100 000, basing *his* calculation on the fact that Sirius is about as bright as the planet Saturn. Both Newton and Huyghens assumed that Sirius (the brightest star in the sky) was intrinsically as bright as the sun. But, in fact, it is much brighter, and so it is even farther away than they thought; the correct multiple is about 540 000.

The next revolutionary idea was born sometime around 1750 and has been variously ascribed to Swedenborg, Lambert, Wright, and Kant. They all wrote on the subject of 'island universes', recognizing the finiteness of our galaxy and conjecturing the existence of similar stellar systems far out in space. Various 'nebulae' seen by the astronomers were candidates for this new role. Kant well understood the shape of our own galaxy ('stars gathered together in a common plane'—as indicated by the Milky Way) and so explained the observed elliptical appearance of some of the nebulae as disks seen obliquely. In 1783 Messier cataloged 103 such nebulae, and William Herschel with his powerful 48-inch reflector telescope located no fewer than 2500 before his death in 1822. He became the great observational supporter of the multi-island universe theory, though, ironically, many of his arguments turned out to be quite false. Still, he came to foresee the important division of nebulae into two main classes—galactic and extragalactic. Herschel's theory had its ups and downs in favor, but essentially its verification had to wait for the slow development of observational capacity to match its huge demands. The waiting period culminated in a historic wrangle, continued at one astronomers' conference after another from 1917 to 1924—until it suddenly ended on January 1, 1925: Hubble, with the help of the new (1917) 100-inch telescope at Mount Wilson, had resolved star images in three of the nebulae, and, as some of these were Cepheids, he was able to establish beyond all doubt their extragalactic distances. Only one main feature of the universe as we know it today was still missing: its expansion. From 1912 onwards, Slipher had observed the spectra of some of the brighter spiral nebulae and found many of them redshifted, which presumably meant that these nebulae were receding. But distance criteria were still lacking. Hubble now applied his 'brightest star' measure of distance and, together with Humason, extended the redshift studies to ever fainter nebulae. Finally, in 1929, he was able to announce his famous law: all galaxies recede from us (apart from small random motions) at velocities proportional to their distance from us. The modern era of cosmology had begun.

For another 20 years or so the big 100- and 200-inch telescopes were chiefly respon-sible for enriching our knowledge of the universe. But radar developments during World War II led to radio-astronomy after the war, and with it came a second consid-erable enlargement of the observable universe, as well as the discovery of important new phenomena: quasars in 1962, the thermal background radiation in 1965, and pulsars in 1967. Balloon, rocket, and satellite experiments escaped the obscuring atmosphere of the earth; electronic computers led to new methods of data processing; and solid-state detectors were 50 times more sensitive than photographic plates. In the seventies came infrared, X-ray, and gamma-ray astronomy, which opened up yet further spectral regions, and which located many interesting sources both inside and outside the galaxy emitting such highly energetic radiation. Most recently, the orbiting Hubble space telescope and a new generation of 10-meter earthbound telescopes have collected spectacular new data. Neutrino- and gravitational-radiation astronomy are in a nascent state. And computers and advances in nuclear theory have made possible previously unthinkable numerical investigations into stellar and galactic evolution.

D. Stars and galaxies

However, we must forego a detailed account of the further growth of modern knowledge, and content ourselves with simply listing the main astronomical findings relevant to our purpose. Stars, to begin with, are huge balls of plasma, held together by gravity whose enormous central pressure triggers the generation of energy by nuclear fusion. This occupies most of the star's lifetime and can last ten billion years—or more, or much less, depending mainly on the total mass and chemical composition. In the end, of course, all stars must burn out. Possibly after violent explosions (supernovae) they end up as white dwarfs, neutron stars (pulsars), or black holes. The size of stars can be appreciated by considering that the earth, *with* the moon's orbit, would comfortably fit into the sun, a very average star. About 7000 of them are visible at night to the naked eye; about 10^{11} are contained in a typical galaxy. (Most of us lack mental images for numbers of that size. Here is one possibility: consider a row of books a mile long; the number of *letters* in all those books is about 10^{11}. Or consider a cubical room, $15 \times 15 \times 15$ ft; the number of pinheads needed to fill it is about 10^{11}.)

Stars within a galaxy are very sparsely distributed, being separated by distances of the order of 10 light years. In a scale model in which stars are represented by pinheads, these would be about 50 km apart and the solar system (out to the orbit of Pluto) would have a radius of 5 m. A typical galaxy has a radius of 3×10^4 light years, is 3×10^6 light years from its nearest neighbor, and rotates with an angular velocity that decreases from the center outwards, at an average period of 100 million years. Like a coin, it has a width only about a tenth of its radius, and coins spaced about a meter apart make a good first model of the galactic distribution. About 10^{11} galaxies are within range of the 200-inch Mount Palomar telescope. In the coin model, the farthest of these are 3 km away, but the farthest known quasars are four times as far.

To digress: $\sim 10^{11}$ galaxies are thus *known* to exist, although there may well be infinitely many (if the universe turns out to be infinite). They contain $\sim 10^{11}$ stars each, so a minimum of 10^{22} altogether. As closely packed pinheads these stars would fill a box $20 \times 20 \times 20$ km. One of these pinheads is our sun. Can this possibly be the only center of life in the universe?

At one time it was thought that the galaxies might be more or less evenly distributed throughout space. But this is not so. Galaxies, first of all, join into small groups and large clusters, leaving single galaxies (so-called field galaxies) as the exceptions. Clusters can contain as many as 1000 or even 10 000 and more individual galaxies. They are presumably held together by gravity and have decoupled from the cosmic expansion. Clusters themselves can join into superclusters—but these already expand with the rest of the cosmos. Higher-order clustering seems not to occur. Instead, in the last decade or so, a kind of spongy or filamentary structure has emerged, which has even been matched on computer simulations of structure formation. It seems that clusters form chains, which meet in knots (centers of superclusters) and thus form a kind of irregular lattice. Its cell walls have lesser densities than the edges, and there are huge voids in between. The cells have dimensions of $\sim 10^8$ light-years. Above this scale, however, the homogeneity of the universe seems to set in. After all, the observable universe contains upwards of 10^6 such cells—like little cubic centimeters in a cubic meter.

E. Homogeneity and isotropy

But we must note that homogeneity, in a non-static universe with finite signal speed, is a little tricky to observe. How can we check whether a distant region is similar to our own? We *see* that region as it was millions of years ago. But (i) we do not know exactly *how* many millions, since we do not know the exact expansion history of the universe, which affects the light travel time; (ii) even if we knew the light travel time, we do not know how our *own* region looked that long ago; and (iii) we do not know the exact spatial geometry that would allow us to translate angular measurements into transverse distances.

In a non-isotropic universe, that would be a real problem. (Homogeneity can, of course, exist without isotropy—as in a crystal.) But we are fortunate: we actually observe almost perfect isotropy around ourselves. Since today we strongly believe *not* to be in a special position in the universe, nor to exist at a special time, we conclude that there is isotropy anytime from anywhere. But that guarantees homogeneity! For if any region A went through a different evolution from region B, that difference would be seen as non-isotropy from points located equidistantly between these regions.

Much of the evidence for the regularity of the universe therefore hinges on its isotropy around ourselves. We have several independent ways to observe this. The most important are the following three: (i) optical and radio-astronomical mappings of the sky; (ii) the cosmic background radiation; and (iii) the Hubble expansion. As for the first, suffice it to say that when allowance is made for the obscuring effects of our own Milky Way (though for radio signals it is virtually transparent) the distribution of distant galaxies as seen both in the optical and the radio spectrum is completely, though only crudely, isotropic.

F. Thermal background

Isotropy on an almost unbelievably finer scale is exhibited by the cosmic background radiation. This 2.7 K blackbody microwave radiation was serendipitously discovered by Penzias and Wilson in 1965, unaware that already in 1948 Alpher and Herman had predicted its existence on the basis of Gamow's theory of element production in a hot big bang. It is estimated that about 300 000 years after the big bang (long before galaxies began to condense) the original continuum of particles and photons had cooled down to about 3000 K, at which point the hydrogen ions and electrons could combine into atoms, leaving no charged particles to strongly scatter the photons. The universe became transparent. Henceforth the radiation is decoupled from the matter; it effectively lives in an expanding cage (the material universe), keeping its blackbody profile but cooling, in inverse proportion to the size of the universe, down to its present temperature. Early intensive efforts to study this radiation were superseded by measurements made with the COBE satellite ('Cosmic Background Explorer') launched in 1989, which not only confirmed the perfect blackbody spectral curve of the radiation, but also its incredible isotropy to better than a few parts in 10^4. So perfect is this isotropy that a minute systematic 'dipole' variation of about 5×10^{-3} K has been used to determine the earth's motion through the background as ~ 630 km/s, since by the Doppler effect the observed temperature must be slightly higher fore than aft. In effect, the background radiation provides a local standard of rest in the universe everywhere, relative to which the peculiar velocities of individual galaxies are small.

The isotropy of the thermal radiation reaching us today is a direct indication only of the isotropy of a *thin shell* of continuum at the time of decoupling, namely that shell (the so-called 'surface of last scattering'), centered on us, where the radiation we receive today originated. (With that radiation we are 'seeing back' to within 300 000 years of the big bang!) But the assumed isotropy of *all* such shells implies the homogeneity of the entire universe at decoupling time, and thus, presumably, before and after.

As we shall see later, the density of radiation in the universe follows the law $\rho \propto R^{-4}$ (R = 'radius of the universe'), while that of its matter content obeys $\rho \propto R^{-3}$. At the present time, the latter density greatly outweighs the former. But clearly there must have been a time in the past when the two density curves crossed over. Apparently by coincidence, this seems to have happened around the time of decoupling. For the dynamics of the universe it is in fact an acceptable approximation to piece together, at about $t = 300\,000$ y, an earlier pure-radiation universe with a later pure-matter one.

G. Hubble expansion

Our last argument for the isotropy and homogeneity of the universe comes from its Hubble expansion. The entire universe is found at present to expand by about 1 per cent in 10^8 years. On a human scale, this seems a leisurely pace. But the universe is large; so even a small *proportionate* expansion rate translates into vast relative speeds of widely separated galaxies. A galaxy 10^x light years away from us moves another ($10^x/100$) light years in the next 10^8 years, which translates into a speed of $10^{x-10}c$; and that is c when $x = 10$! (We shall see later that this statement is essentially exact in terms of instantaneous ruler distance and cosmic time.)

The Hubble expansion pattern is best visualized by picturing an infinite regular cubical lattice (in the simplest case of an isotropic flat universe) with knots at all the lattice points representing the galaxies. Suppose at one 'cosmic instant' all the edges of all the lattice cubes have length l. A cosmic time dt later, they *all* have expanded to $l + dl$. Consider now a lattice line of such edges issuing from a given galaxy. Another galaxy n edges away (that is, at distance $x = nl$) has moved with velocity $v = n\,dl/dt$. So we have

$$v = \frac{n\,dl/dt}{nl}x = \frac{dl/dt}{l}x =: Hx, \tag{16.1}$$

the exact Hubble law, with $H = (\mathrm{d}l/\mathrm{d}t)/l$. This H is called *Hubble's parameter*. It evidently need not be constant in time. It can even become negative, when expansion changes into contraction. Its present value, H_0, is called *Hubble's constant*.

The lattice motion pattern here described preserves the spatial homogeneity of the universe. It is also the *only* motion pattern that does that, and the only one consistent with Hubble's law. If Hubble's law were, for example, quadratic in x, a homogeneous lattice around ourselves would quickly lose its homogeneity: distant cells would expand faster than nearby ones. So the linearity of Hubble's law is needed for maintaining the spatial homogeneity of the universe. The fact that H_0 is also direction-independent fits in with the isotropy of the universe.

No reliable figures seem to exist for the degree of that direction-independence, but it is clearly satisfactory. The actual value of Hubble's constant is still uncertain. Hubble originally (in 1929) set it at 540 (km/s)/megaparsec. [The parsec (pc) is a distance unit favored by astronomers and is equivalent to 3.087×10^{18} cm, or 3.26 light years; a megaparsec (Mpc) equals 10^6 pc.] But this figure has undergone several drastic revisions, mainly downwards, and mainly caused by refinements of the various steps leading to a determination of the cosmic distances. The best present estimates seem to be in the range 65 ± 10 (km/s)/Mpc.

H. The big bang

One further aspect of the cosmic expansion needs to be addressed, namely its explanation. Recall Newton's universe: a homogeneous static distribution of stars throughout absolute space. Think of the stars as the knots of our lattice, at rest in AS. Symmetry *relative* to AS then forbids any motion. But if we scrap the ideas of AS *and* of extended inertial frames (cf. Section 1.11), and only consider the lattice *per se*, symmetry *does* permit its Hubble expansion *or* contraction. If the stars were initially mutually at rest, gravity would make them Hubble-contract. But we see the universe expand. Astronomers were at first surprised at that. Was there some mysterious expansion of space itself? However, a quite simple and 'obvious' (though never previously contemplated!) solution was found: the *big bang*. Extrapolating the present expansion of the universe backwards in time, one sees that in the most straightforward scenario it must get ever denser and ever hotter, until a singularity of infinite density and infinite temperature is reached some 10^{10} years ago (on the crude approximation of linear expansion). The time-inverse of this sequence, a symmetric cosmic explosion from infinite density and temperature, is referred to as the big bang. *Why* it happened remains unexplained. But *if* it happened, the observed expansion is no more mysterious than the flying apart of shrapnel from a grenade that explodes in mid-air. And this image also answers the question whether *everything* must expand. If two shrapnel pieces could briefly reach out and hold hands to halt their relative motion, they would henceforth be quite unaffected by the motion of the rest. It is much the same in the universe: the forces holding atoms and molecules together have decoupled their constituents from the general expansion; the gravity that holds the stars in a galaxy together has decoupled *them* from the expansion. We have already seen (in Birkhoff's theorem) that the Schwarzschild metric (and with it the planetary orbits) are unaffected by the existence of expanding surrounding mass shells. The local situation in the universe is quite analogous.

The hypothesis of the big bang (and with it the existence of progressively denser and hotter phases in the past) was strongly boosted by the discovery of the cosmic background radiation which the big-bang theory had predicted. And much further evidence comes from the theory of nucleosynthesis. It so happens that there is about one helium atom for every ten hydrogen atoms practically wherever we look in the universe. This helium could not all have been produced by the fusion of hydrogen in stellar furnaces—there just are not enough of them. The only alternative is the *cosmic* furnace a few minutes after the big bang when the temperature was $\sim 10^9$ K, as had been suggested in a seminal paper by Alpher, Bethe, and Gamow in 1948. Indeed, to have eventually accounted for the observed relative abundances of the light nuclei ^{1}H, ^{2}H, ^{3}He, ^{4}He, and ^{7}Li is one of the great achievements of the big-bang theory. (The heavier nuclei all had to be created in stars, where at comparable temperatures there is a very much higher density of matter and lower density of radiation. Through supernova explosions of massive short-lived stars these nuclei were then released into space.)

But the most straightforward way to justify the big bang is simply to look at the general-relativistic dynamics of presently expanding model universes. Given the known parameters of the universe we live in, and the Penrose-Hawking singularity theorems, it would take the most exotic and unlikely circumstances in the past to prevent its backward extrapolation from terminating in a big bang.

I. Age of the universe

Nevertheless it is important that we have independent estimates for the age of the universe. Radioactive dating methods applied to terrestrial and even lunar rocks have yielded an age of $\sim 4.5 \times 10^9$ years for our solar system (and the sun is expected to live at least that long again). Theories of stellar evolution (especially the Hertzsprung–Russell diagram relating the luminosity of stars to their surface temperature), when applied to various globular clusters (of stars) in our galaxy, yield ages of $(17 \pm 4) \times 10^9$ years for the oldest among them. This is indicative of the age of the galaxy itself. Quite independent estimates for this age can be made from the relative abundances of certain radioactive elements. For example, the two isotopes of uranium, ^{235}U and ^{238}U, though created in roughly equal amounts in stellar interiors and then released into space through explosive events, are actually found to occur in a ratio of about $7 \times 10^{-3} : 1$. This is due to the known much faster rate of decay of ^{235}U. Taking account of the fact that the production is spread out over time, one arrives at estimates of $(15 \pm 5) \times 10^9$ years for the length of time that this process has been going on— presumably for most of the life of the galaxy. Quite similar estimates result from the relative abundances of some other radioactive elements.

What is striking is that these astronomical estimates agree (even if only in a rough way) with the *dynamical* age of the universe; that is, the age of a general-relativistic big-bang model that matches today's observed state of the universe. If every galaxy were to move at *constant* speed away from us, and t_0 denotes the time since the big bang, we would have $x = vt_0$ for a galaxy at distance x having constant velocity v; but we also have, from (16.1), $x = H_0^{-1}v$, so that $t_0 = H_0^{-1}$. This is the so-called *Hubble age* of the universe, which ignores the real dynamics. But even so, it should at least give an order-of-magnitude estimate of the real age of a big-bang universe.

For a present expansion rate of 1 per cent in 10^8 years, the Hubble age would be 10^{10} years, which is not out of line. But there is still a ± 20 per cent uncertainty about the correct value of H_0, which is entirely due to the uncertainty in the cosmological distance scale. Since that same uncertainty enters several other cosmological quantities, it has been found convenient to condense it into a dimensionless fudge factor h of order unity, by writing

$$H_0 = h \cdot 100\,(\text{km/s})/\text{Mpc} = h(3.25 \times 10^{-18})\,\text{s}^{-1}. \tag{16.2}$$

We then find

$$H_0^{-1} = h^{-1}(9.78 \times 10^9)\,y = h^{-1}(3.08 \times 10^{17})\,\text{s}. \tag{16.3}$$

As we have seen, present estimates of H_0 place h between 0.55 and 0.75 which implies, for the Hubble age,

$$13.0 \times 10^9 y < H_0^{-1} < 17.8 \times 10^9 y. \tag{16.4}$$

But gravity must have the effect of decelerating an expanding model universe after the big bang. Conversely, if we run the model backwards in time (GR is time-symmetric) gravity must hasten its collapse. So the collapse time is less than if gravity is switched off and the velocities are uniform. In a universe *with* gravity, therefore, H_0^{-1} overestimates the age [cf. Fig. 16.1(a)]. But by how much? That depends on the density. At zero density the collapse is uniform and H_0^{-1} is the true age; for 'reasonably' low cosmic densities the true age may perhaps lie between $0.8 H_0^{-1}$ and $0.9 H_0^{-1}$. If the density were as high as ρ_{critical}, at which the expanding universe *just* escapes recollapsing, we would have $t_0 = \frac{2}{3} H_0^{-1}$. So, as reference to (16.4) shows, with fairly low cosmic density, fairly low Hubble constant, and fairly low estimates for the age of the galaxy, we could have consistency all around.

How would the existence of a repulsive cosmic Λ-force affect the situation? It would not change any of the local physics responsible for estimating the age of the galaxy or the density of the cosmos. It could, however, reverse the inequality $t_0 < H_0^{-1}$, as Fig. 16.1(b) shows. If the Λ-force is sufficiently big so as eventually to overcome gravity, then the initial deceleration of the universe would eventually change into acceleration, and the tangent to the expansion curve in Fig. 16.1(b) could well intersect the time axis to the right of the origin. Even high values of H_0 and high cosmic densities might then be compatible with the age of the galaxy.

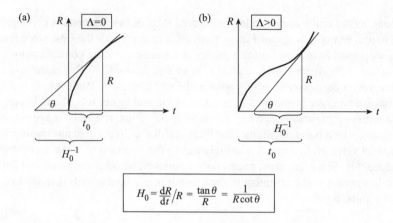

Fig. 16.1

J. Density of universe

The last important datum characterizing the present state of the universe is its average density ρ_0. If we know that (as well as H_0 and Λ and homogeneity-isotropy) we can construct the appropriate dynamical model. But ρ_0 is also the hardest datum to determine, as well as the most controversial, and the numbers are still in flux. In the end, we may have to use alternative data, such as the age of the universe and its present relative acceleration, to determine the dynamics. This would yield ρ_0 and Λ indirectly.

The masses of some of the nearer spiral galaxies can be approximately obtained by analyzing their '*rotation curves*'. These are plots of the velocities v of stars (or gas clouds far beyond the visible edge) versus the radii r of their orbits. If we neglect the non-sphericity of the mass distribution, we can assume the Keplerian relation

$$V^2 = GM(r)/r \qquad (16.5)$$

to find the mass $M(r)$ within the orbit. One would have expected $M(r)$ to level off beyond the visible edge of the galaxy. But, surprisingly, it is v that levels off: $v \approx$ const. This implies $M(r) \propto r$ beyond the edge. Thus was born the concept of a spherical 'halo' of unknown *dark matter*, enveloping the galaxy, with density proportional to $1/r^2$, and in its totality at least ten times as massive as the luminous matter.

But rotation curves cannot be continued sufficiently far out to give reliable estimates of the total masses of galaxies. These can be independently estimated by measuring the velocities of the component galaxies in binary systems, or by applying the *virial theorem* to the velocities and separations of galaxies in gravitationally bound clusters. Such methods have led to an estimate for the density of galactic matter in the universe:

$$\rho_{0(\text{galactic})} \approx 2 \times 10^{-30} h^2 \text{ g/cm}^3. \qquad (16.6)$$

This and several other density estimates involve the Hubble uncertainty factor h. To see why, consider that in the determination of $H_0 = v/x$ only x is in doubt, since v

can be measured directly by observing the redshift and using the Doppler formula. Conversely, if we determine x via the redshift and the Hubble relation $x = v/H_0$, then x carries the uncertainty factor h^{-1}. So, for example, to assign a density to a spiral galaxy via equation (16.5), we measure v directly from the redshift, whence the uncertainty in M is proportional to that in r, while the uncertainty in the density is proportional to r/r^3 and so to h^2—since transverse distances are measured angularly as $x\theta$.

An entirely different approach to determining the density of *baryonic* matter (protons and neutrons) in the universe today comes from the big-bang theory of the formation of the light elements. This highly successful theory is very sensitive to the density of baryons at formation time, and thereby sets an upper limit to that density today,

$$\rho_{0(\text{baryonic})} \lesssim 2 \times 10^{-31} \text{ g/cm}^3, \tag{16.7}$$

without h. That seems to indicate the preponderance of *non*-baryonic matter in the galaxies.

Of course, there is no way, so far, to exclude the possible existence of non-baryonic matter even in intergalactic space. Massive neutrinos have been considered in this context among many other possibilities, but we cannot pursue these questions here. However, we must draw attention to the fact that much of the density debate is conducted in terms of a dimensionless *density parameter*, Ω_0, defined by the first of the following equations:

$$\Omega_0 = \frac{8\pi G \rho_0}{3 H_0^2} = 0.53 \times 10^{29} h^{-2} \rho_0, \tag{16.8}$$

where the last expression holds in units of grams and centimeters. Alternatively, we have

$$\rho_0 = 1.88 \times 10^{-29} \Omega_0 h^2 \text{ g/cm}^3. \tag{16.9}$$

The density parameter Ω_0 is chosen so as to be unity when the density is *critical*; that is, when it is *just* still small enough so as not to make the expanding universe recollapse *in the absence of a Λ-force*. A value $\Omega_0 > 1$ would, then, via the GR field equations, lead not only to a recollapsing but also to a closed (and therefore finite) universe.

Most cosmologists seem to believe that there is enough matter in the galaxies to make Ω_0 as large as 0.3 ± 0.1, while baryonic matter can account for at most a few per cent of the critical value. 'Ordinary' matter thus appears to be in a minority compared to the bulk of the galactic material, whose nature has yet to be determined.

K. Cosmogony

In the rest of this book we shall content ourselves with a study of the smoothed-out motion and geometry of the universe since shortly after the big bang. This is where GR makes its contribution. The fascinating field of *cosmogony* (the study of the formation of the elements and of structures like stars and galaxies out of a primordial mixture of elementary particles in thermal equilibrium) depends on nuclear, atomic and plasma physics, as well as gas dynamics, and is well beyond our present scope. It has, of course, great philosophical interest since, in a sense, it takes over where Darwin left off. If Darwin and modern biology can explain the rise of man from an originally lifeless earth, cosmogony is close to explaining the rise of earths, suns, and galaxies out of amorphous matter, perhaps even the origin of matter itself, given only the immutable laws of nature, and an energizing big bang.

16.2   Beginning to construct the model

As we have seen, the distribution and motion of matter in the universe is by no means random. Though the matter is obviously lumpy, the lumpiness itself appears to be the same in all sufficiently large regions, which are still small relative to all we see. And though, for example, the individual galaxies in clusters have individual velocities, the mass centers of such clusters seem to follow the Hubble expansion pattern quite closely. Accordingly we idealize the actual universe, crudely speaking, by grinding up its matter and redistributing it uniformly so as to match the actual average density and average motion everywhere. We assume that this smoothing out results in a perfectly homogeneous and isotropic mass and velocity distribution. And then we make the very reasonable additional assumption that the motion and geometry of this ideally regular model universe under its own gravity parallels the average motion pattern and geometry of the actual universe.

The material particles of the model (when regarded as mere geometric points) constitute its kinematic *substratum*. We think of it as a space-filling set of moving particles (the *fundamental particles* of cosmology) each of which is a potential center of mass of a cluster of galaxies in the real world. Moreover, each is imagined to carry an observer, called a *fundamental observer*. When in the sequel we loosely speak of galaxies in a model, we shall really mean the fundamental particles. And we ourselves correspond to a fundamental observer.

The a priori demand for the homogeneity of a cosmological model is called the *cosmological principle*—though a better name would be 'cosmological axiom'. It is sometimes loosely formulated by saying that every galaxy is equivalent to every other. It eliminates such in themselves reasonable models as island universes, in which boundary galaxies are atypical; or 'hierarchical' universes where galaxies form clusters, clusters form superclusters, and so on ad infinitum, since then no region is large enough to be typical. Homogeneity is a simplifying hypothesis of great power. Whereas non-homogeneous model universes involve us in global questions, the beauty of homogeneous models is that they can be studied mainly locally: any part of them is representative of the whole.

The assumption of isotropy everywhere is even stronger. As we have seen in the preceding section, it *implies* homogeneity. We accept it as a working hypothesis, which is very strongly supported by the evidence.

In 1948 Bondi and Gold proposed what they called the 'perfect cosmological principle', and based their *steady state cosmology* on it. This strongest of cosmological principles claims that, in addition to being spatially homogeneous and isotropic, the universe is also temporally homogeneous; that is, it presents the same average aspect at all times. It has no beginning or end—a very attractive feature, philosophically. As the universe expands, sufficient new matter must be created to fill the gaps. This constitutes a deliberate violation of energy conservation (and thus of GR), but not by 'much': the spontaneous creation of about one hydrogen atom per 10 cubic kilometers of space per year is all that is needed. The steady state theory has the further advantage of leading to a unique model, which, as such, is highly vulnerable to empirical disproof (cf. Section 2.1). It had many adherents and enjoyed great popularity for almost two decades. But the observational evidence against it (radio source counts, the distribution of quasars, the thermal background radiation, etc.) gradually mounted and eventually overwhelmed it.

So far we have talked rather loosely about homogeneity. We have already (in the preceding section) alluded to the trickiness of defining it in an evolving universe with only finite-speed signals at our disposal. The following definition is due to Walker: Homogeneity means that the totality of observations that any fundamental observer can make on the universe is identical to the totality of observations that any other fundamental observer can make. In other words, if throughout all time we here, as well as observers on all other galaxies, could keep a log of all our observations (for example, the local density of the universe, its expansion rate, the brightness of nearby galaxies, etc.) together with the times at which the observations are made (as measured, say, by standard cesium clocks), then homogeneity is equivalent to the coincidence of all these logs—up to a possible time translation, of course.

A most important corollary of homogeneity (at least in evolving universes) is the existence of a preferred *cosmic time*. This refers to the synchronization of the standard clocks on all the fundamental particles ('fundamental clocks') brought about by simply aligning the logs. A slice through cosmological spacetime at one cosmic instant then finds conditions everywhere identical; and conversely, identical conditions define a cosmic instant: the universe acts as its own synchronization agent.

In the case of the steady state theory (as in all other expanding or contracting homogeneous universes), any two fundamental observers can synchronize their clocks by aligning their records of bouncing light signals off each other. If a homogeneous universe is static, the usual signaling method achieves a 'good' time coordinate, and by homogeneity this agrees with proper time at each fundamental particle. Only when the fundamental particles constitute a stationary but non-static lattice (which would necessarily be anisotropic) is there no kinematically preferred time coordinate, even with homogeneity (cf. Section 9.1).

The assumption of isotropy can be dropped rather more easily than that of homogeneity, without leading to inordinate difficulties. While this seems not called for in modeling the present universe, homogeneous non-isotropic models *have* been considered for various reasons; for example, to investigate whether isotropy could develop out of non-isotropy in the early universe. Some such models also yield interesting examples of what is possible in GR—for instance, universes with closed timelike lines (traveling into one's own past) and 'anti-Machian' universes that rotate relative to each LIF. We mention all this just to make the reader aware of the alternatives. But we ourselves shall stay with homogeneity and isotropy.

16.3   Milne's model

The first cosmological model that we shall now discuss is not meant as a representation of the actual universe. For one thing, it has gravity 'switched off', so it is not even a dynamical model. On the other hand, it is so simple and surveyable and exhibits so many of the kinematic features of the more realistic but also more complicated dynamical models, that it serves as a useful pedagogical introduction to the subject. The model in question was found by Milne in 1932. It is, in fact, one of the 'Friedman models' of general relativity (corresponding to the limit $\rho \to 0$ with $\Lambda = 0$, cf. Section 18.3 below), but it was Milne who reformulated it in more elementary terms.

Since gravity is switched off, the model lives in Minkowski space and can be treated by special relativity. Milne considered an infinite number of *test particles* (no mass, no volume) shot out (for reasons unknown), in all directions and with all possible speeds, at a unique creation event $\mathscr{C}$. Let us look at this situation in some particular inertial frame $S(x, y, z, ct)$, and suppose $\mathscr{C}$ occurred at its origin O at $t = 0$. All the particles, being free, will move uniformly and radially away from O, with all possible speeds short of c. Hence the picture in S will be that of a ball of dust whose unattained boundary expands at the speed of light. At each instant $t = $ const in S, Hubble's velocity–distance proportionality is accurately satisfied relative to O: a particle at distance r has velocity r/t. Still, at first sight, this seems an unlikely candidate for a modern model universe, since (i) it appears to have a unique center, and (ii) it appears to be an 'island' universe. Leaving aside the second objection for the moment, let us dispose of the first: The boundary of the ball behaves kinematically like a spherical light front emitted at $\mathscr{C}$, and thus each particle, having been present at $\mathscr{C}$, will consider *itself* to be at the center of this front! Moreover, since all particles coincided at $\mathscr{C}$, and since all move uniformly, *each* particle will consider the whole motion pattern to be radially away from itself, and of course uniform. There remains the question whether we can have an isotropic density distribution around each particle.

To study this, let τ denote the proper time elapsed at each particle since creation. Then n_0, the proper particle density at any given particle P, is of the form

$$n_0 = N/\tau^3 \quad (N = \text{const}), \tag{16.10}$$

because the expansion is radial relative to P, and because a small comoving sphere centered on P has radius proportional to τ. This τ is clearly the 'cosmic time' of the preceding section (penultimate paragraph), since it figures as time from the big bang in every log. So, for homogeneity, N must be the *same* constant at every particle. This also guarantees that the global density pattern is isotropic and the same around each fundamental particle. For let Q be such a particle and S its inertial rest-frame with Q at the origin. At each moment $t = $ const in S, the particles on a sphere $r = $ const satisfy the equation

$$c^2\tau^2 = c^2t^2 - r^2, \tag{16.11}$$

and thus have τ and with it n_0 constant.

To determine the entire density pattern at some constant t in S, we transform eqn (16.10) from the rest-frame of the general particle P into S. Since P moves, the volume of a small comoving sphere at P is diminished by a γ-factor in S; but the number of particles inside that sphere is the same in S, so the particle density n in S is given by $n = \gamma n_0$. We also have $t = \gamma\tau$. Thus, from (16.10) and (16.11),

$$n = \frac{t}{\tau}n_0 = \frac{t}{\tau^4}N = \frac{Nt}{(t^2 - r^2/c^2)^2}. \tag{16.12}$$

Note how the density approaches infinity at the 'edge' $r = ct$. Beyond every galaxy there are others, and no galaxy is even *near* the edge by its own reckoning. Relativistic kinematics thus gets around the classical objection to island universes—that they must contain atypical edge galaxies. Of course, it must have been an incredibly finely tuned big bang to produce the required density pattern (16.12) and thus global homogeneity!

But though the island nature of Milne's model does not conflict with the cosmological principle, it does offend against another criterion that can be required of model universes: *maximality*. Obviously there is no spacetime singularity at the edge of the ball; spacetime itself is continuous across the edge. So there is more spacetime than substratum, spacetime from which light and particles could enter the substratum and disturb it, spacetime, also, that contains whatever light was emitted at the big bang. A spacetime diagram, Figure 16.2, illustrates this. The steady state theory suffered from the same defect (see end of subsection containing Fig. 18.1 below).

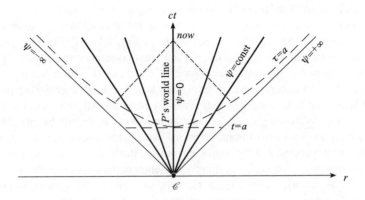

Fig. 16.2

We shall next give an alternative description of Milne's model that brings it into line with the standard general-relativistic description of all homogeneous–isotropic model universes. It is mainly the result of switching from the 'private time' t of some preferred fundamental observer's inertial frame to the 'public time' τ (in Milne's terminology) shared by all. Milne's description (an expanding ball) corresponds to a sequence of cuts $t = $ const through the worldtube of the substratum in M^4. In the standard description, the model is foliated instead by the cuts $\tau = $ const (see Fig. 16.2). Moreover, one uses 'comoving' coordinates, which means that the fundamental particles, though in relative motion, have fixed space coordinates—like the lattice points in a permanently labeled but expanding Cartesian lattice. One such system might be $\{u, \theta, \phi\}$, where u is the velocity of a receding galaxy and θ and ϕ are its angular coordinates. The standard description is then encoded in a special form of the metric, called the *Friedman metric*.

Let us begin by writing the metric of M^4 in the polar form

$$\mathbf{ds}^2 = c^2 \, dt^2 - dr^2 - r^2(d\theta^2 + \sin^2\!\theta \, d\phi^2), \tag{16.13}$$

and assume one of the fundamental particles to be at rest at the spatial origin. The coordinates θ and ϕ are already comoving, and instead of u, for later convenience, we choose ψ, the rapidity, as our radial coordinate. We then have (cf. Exercise 2.13), since $t = \gamma\tau$ and $r = vt$,

$$t = \tau \cosh\psi, \qquad r = c\tau \sinh\psi, \tag{16.14}$$

and this transforms (16.13) into the desired Friedman form,

$$\mathbf{ds}^2 = c^2 \, d\tau^2 - c^2\tau^2\{d\psi^2 + \sinh^2\!\psi \, (d\theta^2 + \sin^2\!\theta \, d\phi^2)\}. \tag{16.15}$$

This metric, plus the information that the spatial coordinates are comoving, tells us a great deal about the model, in addition to characterizing its spacetime background. [By contrast, (16.13) tells us no more than that.] *First*, we can read off that τ is proper time on any fundamental particle (just put $\psi, \theta, \phi = $ const). *Secondly*, that any section $\tau = $ const through the model is a space of constant curvature $-1/c^2\tau^2$ [it is one sheet of the 2-sheeted hyperboloid of Fig. 5.1—cf. Exercise 14.10(ii)]. This would establish τ as cosmic time (a time that connects equal states), if we did not already know it. All cosmic sections $\tau = $ const are of infinite volume, as is to be expected; for a cosmic moment finds the particle density everywhere the same, and we know that there must be infinitely many particles. Milne's model beautifully illustrates how, by the magic of relativity, the universe can already be infinite (in one sense) a mere instant after its point-creation. *Thirdly*, the proper rate of expansion of the substratum is encoded in the coefficient outside the brace: clearly the infinitesimal distance between any two neighboring fundamental particles is proportional to τ (put $d\psi, d\theta, d\phi = $ const). Hence the metric suggest a picture of the substratum as an infinite 3-dimensional lattice of constant negative curvature, expanding at constant rate while remaining similar to itself. Each 'public space' or cosmic moment $\tau = a$

is tangent at each of its points to the flat 'private space' $t = a$ of the fundamental observer present, as Fig. 16.2 illustrates. Thus each public space can be regarded as a composite of identical origin-neighborhoods from the ball model.

There are, however, at least three important facts that are *not* directly visible from the metric (16.15): (i) that the model lives in M^4; (ii) that the model is not maximal; and (iii) that the big bang is a point event; that is, that there exist spacelike sections ($t = $ const) through the substratum containing all the matter in a finite and arbitrarily small volume.

The simple Milne model serves well to illustrate the kinematics of the 2.7 K background radiation. Let the cosmic 'instant' of decoupling be represented by the dashed line $\tau = a$ ($\sim 300\,000$ years) in Fig. 16.2. All the fundamental particles at this instant are the effective sources of the radiation seen today. Ahead of the section $\tau = a$ the universe is transparent. As time goes on, each fundamental observer like P (in whose rest-frame the diagram is drawn) receives the radiation (dotted lines in Fig. 16.2) from ever farther fundamental particles, thus ever more redshifted and therefore ever more cooled. Note incidentally how P could 'see' (not by photons because of the opaqueness before $\tau = a$, but, for example, by neutrinos) fundamental particles at all proper ages $\tau > 0$, but not the big bang itself. (In realistic models with $\rho \neq 0$, as we shall see, even the big bang is in principle visible.)

16.4 The Friedman–Robertson–Walker metric

A. Introduction

To Friedman belongs the great distinction of having been the first to contemplate and analyze a dynamic universe that *moves* under its own gravity. This was one of the few momentous leaps forward against received opinion that was in the air and yet was missed by Einstein. It was also completely ignored for almost a decade. Although Friedman's derivation was to some extent heuristic and not quite rigorous, the metric he found lives on and well deserves to be known by his name. It is, however, often called the *Robertson–Walker metric* these days, for reasons we shall discuss below.

B. Three-metrics of constant curvature

As a preliminary to the derivation of Friedman's metric and to much of our subsequent discussion, we need the various forms of the metric of 3-dimensional spaces of constant curvature. In Chapter 14 we already recognized the metrics of the static lattices of de Sitter space ($\Lambda > 0$) and anti-de Sitter space ($\Lambda < 0$) [cf. (14.26)],

$$d\sigma^2 = \left(1 - \tfrac{1}{3}\Lambda r^2\right)^{-1} dr^2 + r^2(d\theta^2 + \sin^2\!\theta \, d\phi^2), \qquad (16.16)$$

to be 3-spaces of constant curvature $\tfrac{1}{3}\Lambda$. Let us introduce a standard notation for the metric of the unit 2-sphere referred to the usual polar angles θ and ϕ:

$$d\omega^2 := d\theta^2 + \sin^2\!\theta \, d\phi^2 \qquad (16.17)$$

[cf. (8.12)]. Evidently this $d\omega$ also measures the angle between neighboring radii separated by coordinate differences $d\theta$, $d\phi$ (since angle = arc/radius). Now the metric (16.16), as we have seen, has positive or negative curvature according as Λ is positive or negative, and it has zero curvature if $\Lambda = 0$, when it reduces to the usual polar metric of E^3. Writing an η for r, and k/a^2 for the curvature $K = \frac{1}{3}\Lambda$ [with $k = \text{sign}(K)$], we can rewrite (16.16) as

$$d\sigma^2 = a^2 \left\{ \frac{d\eta^2}{1 - k\eta^2} + \eta^2 \, d\omega^2 \right\}, \qquad (16.18)$$

where $k (= 0, 1, \text{ or } -1)$ is called the *curvature index*. The metric in braces corresponds to $a = 1$ and thus to a 3-sphere (or hyperbolic 3-sphere) of *unit* radius, unless $k = 0$. Note how an overall factor a^2 increases the radius of curvature (K^{-2}) in the ratio $1 : a$, as is particularly clear in the case of a sphere where all distances are increased in ratio $1 : a$.

A second standard form of the metric of a 3-space of constant curvature is obtained from (16.18) by setting $\eta = \sin\psi$, ψ, or $\sinh\psi$ according as $k = 1, 0, \text{ or } -1$:

$$d\sigma^2 = a^2 \left\{ d\psi^2 + \begin{pmatrix} \sin^2\psi \\ \psi^2 \\ \sinh^2\psi \end{pmatrix} d\omega^2 \right\}. \qquad (16.19)$$

Here $a\psi$ measures radial distance away from the origin, and the functions $\sin\psi$, etc., measure the lateral spread of neighboring radii. Radii, of course, are geodesics (a typical one being the intersection of the symmetry surfaces $\theta = \pi/2$, $\phi = 0$) and so the metric (16.19) is seen to be consistent with our earlier formulae (8.1) and (8.7).

A third metric form is obtained from (16.18) by introducing a new radial coordinate r [not to be confused with that of eqn (16.16)], via the relation

$$\eta = \frac{r}{1 + \frac{1}{4}kr^2}, \qquad (16.20)$$

namely:

$$d\sigma^2 = a^2 \left\{ \frac{dr^2 + r^2 \, d\omega^2}{(1 + \frac{1}{4}kr^2)^2} \right\}. \qquad (16.21)$$

This exemplifies, in the case of three dimensions, the well-known theorem that any space of constant curvature is *conformally flat*; that is, has a metric that can be expressed as a multiple of the Euclidean metric.

The geometric relations between the three alternative radial coordinates ψ, η, and r *in the case $k = 1$*, are illustrated in Fig. 16.3, where the circle represents a typical geodesic plane of the *unit* 3-sphere; for example, the 2-sphere $\theta = \pi/2$ through the origin $\psi = 0$, here denoted by O. The figure, in conjunction with the following line of equations, should be self-explanatory:

$$\eta = \sin\psi = \frac{2\tan(\psi/2)}{1 + \tan^2(\psi/2)}, \qquad r = 2\tan\frac{\psi}{2}. \qquad (16.22)$$

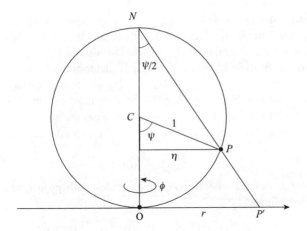

Fig. 16.3

There is, of course, an analogous geometrical picture involving hyperboloids in the case $k = -1$, but it is less enlightening. When $k = 0$, η, ψ and r are all the same.

C. The Friedman–Robertson–Walker metric

We are now in a position to derive the characteristic metric of the most general smoothed-out model universe that satisfies the assumptions of homogeneity and isotropy. In fact, with very little extra effort we can and shall derive the required metric from an apparently much weaker assumption, namely that of *local isotropy* everywhere, by which is meant spatial isotropy in a finite neighborhood around the origin of every comoving LIF.[1] We have already shown in a heuristic way (in Section 16.1) why isotropy everywhere implies homogeneity; so it is intuitively to be expected that local isotropy implies local homogeneity and thus, by iteration, global homogeneity. Global isotropy, on the other hand, is a topological concept quite separate from (and not encodable in) the metric. For example, a cylinder has the same Euclidean metric as the plane, is locally isotropic and globally homogeneous (intrinsically), but nevertheless it is not globally isotropic.

Let $U^\mu = dx^\mu/d\tau$ denote the 4-velocity of the fundamental particles and thus the tangent vector to their worldlines (*'fundamental worldlines'*). An immediate consequence of local isotropy is that all these worldlines must be geodesics. For suppose the 4-acceleration

$$A^\mu = U^\mu{}_{;\nu}U^\nu \tag{16.23}$$

at some event were *not* zero. Being a spacelike 4-vector, it would then have a spatial projection in the local rest-LIF, thus violating local isotropy.

---

[1] Our derivation largely follows one due to J. Ehlers.

Next, consider the tensor $U_{\mu;\nu} - U_{\nu;\mu}$. We can show that in the present case this has *no* components in the direction of U^μ. From (16.23) we already know that $U_{\mu;\nu}U^\nu = 0$; and by differentiating $U_\mu U^\mu = c^2$ we also find $U_{\mu;\nu}U^\mu = 0$. Together these equations justify our assertion. In the LIF determined by U^μ the only non-zero components of the tensor $U_{\mu;\nu} - U_{\nu;\mu} = U_{\mu,\nu} - U_{\nu,\mu}$ (cf. Exercise 10.6) are thus $U_{i,j} - U_{j,i} \equiv \text{curl}\, U_i (i, j = 1, 2, 3)$ and they must vanish by local isotropy. Consequently we have $U_{\mu,\nu} - U_{\nu,\mu} = 0$, which, as we know [cf. after (7.42)], implies that U_μ is the gradient of some function, say $c^2 t$ (for later convenience):

$$U_\mu = c^2 t_{,\mu}. \tag{16.24}$$

This t we take as our time coordinate. It coincides with proper time along each fundamental worldline:

$$dt = c^{-2} U_\mu \, dx^\mu = c^{-2} g_{\mu\nu} dx^\mu \, dx^\nu / d\tau = d\tau,$$

and each section $t = $ const is orthogonal to those worldlines, since

$$U_\mu \, dx^\mu = c^2 t_{,\mu} \, dx^\mu = c^2 \, dt = 0$$

for all dx^μ in the section.

Let us choose arbitrary coordinates x^i in any one hypersurface $t = $ const and declare them 'comoving'; that is, they are invariant along the fundamental worldlines. Then we have

$$\mathbf{ds}^2 = c^2 \, dt^2 - g_{ij} \, dx^i \, dx^j, \tag{16.25}$$

where the g_{ij} can depend on t as well as on the x^i. But consider an infinitesimal triangle of fundamental particles, ABC. Each of its angles must remain constant as time goes on; for if the angle at A varied, so would all such angles at A, by isotropy, and thus the full solid angle at A would have to vary, which is impossible. Hence the triangle remains similar to itself. This implies that the ratios of the g_{ij} at any fundamental particle must remain constant, and so they can involve time only through a common factor, say $R(t, x^i)$:

$$\mathbf{ds}^2 = c^2 \, dt^2 - R^2(t, x^i) \, d\sigma^2, \tag{16.26}$$

where $d\sigma^2$ is a purely 3-dimensional metric.

Now consider the relative expansion rate of neighboring fundamental particles separated by a distance dl, namely (with $\dot{} = d/dt$)

$$H = \frac{(dl)^{\cdot}}{dl} = \frac{(R\, d\sigma)^{\cdot}}{R\, d\sigma} = \frac{\partial R/\partial t}{R}. \tag{16.27}$$

If this were spatially variable, it would have a gradient, which would violate isotropy. So it must be purely a function of time. [The reader will recognize H as Hubble's parameter, cf. (16.1).] But then R itself splits into two factors, one dependent on time, one on space only, as we see on integrating (16.27):

$$\frac{\partial R/\partial t}{R} = H(t) \Rightarrow \ln R = \int H \, dt + g(x^i) \Rightarrow R = R(t)S(x^i).$$

The factor $S(x^i)$ can be absorbed into $d\sigma^2$ so that the metric now simplifies to

$$\mathbf{ds}^2 = c^2 \, dt^2 - R^2(t) \, d\sigma^2. \tag{16.28}$$

Its sections $t = $ const have intrinsic significance, being orthogonal to all the fundamental worldlines. [Cf. (10.4).] By the assumption of local isotropy, each of their points must then be an isotropic point, and hence, by Schur's theorem, the entire section must be a space of constant curvature. The metric $d\sigma^2$ can thus be transformed into some constant multiple of, say, the brace in (16.23), and that multiple can be absorbed into $R^2(t)$. The full metric now at last assumes its standard ('Friedman–Robertson–Walker') form:

$$\mathbf{ds}^2 = c^2 \, dt^2 - R^2(t)\left\{\frac{d\eta^2}{1 - k\eta^2} + \eta^2(d\theta^2 + \sin^2\theta \, d\phi^2)\right\}, \tag{16.29}$$

with possible replacements of the brace from (16.19) or (16.21). Note how t is 'cosmic' time in the sense of connecting identical states of the universe (here: curvature states); it is also the proper time at each fundamental particle.

In Chapter 17 we shall discuss various methods of assigning a distance to far-away galaxies, all based on optical observations. But at this point we define, for future reference, a purely theoretical distance measure, the so-called *proper distance* (or *metric distance*) l between galaxies. It is what we would get if at a given cosmic moment we could lay little rulers end to end between two galaxies. The proper distance from the origin, at time t, to a galaxy at 'comoving' distance ψ [if we use the brace from (16.19) in (16.29)] is evidently given by

$$l = R(t)\psi. \tag{16.30}$$

It is usual to call universes with $k = 1$ 'closed' and those with $k = 0$ or -1 'open', and to regard the former as finite ('compact') and the latter as infinite. This accords with the tacit assumption that the cosmic sections $t = $ const are 3-spheres S^3, Euclidean 3-spaces E^3, or hyperbolic 3-spheres H^3, respectively. But, as we have mentioned before, the global topology is not implicit in the metric. There are all sorts of 'topological identifications' which preserve homogeneity and local isotropy (though most spoil *global* isotropy) and which can produce *finite* 3-spaces even for $k = 0$ or $k = -1$ (cf. Exercise 8.7). The simplest example is the hypertorus, which results from identifying opposite faces of a rectangular cell in E^3. If we lived in a flat universe with this topology, our view would be reminiscent of that in a cabinet of mirrors. We would see infinitely far in all directions, though we see nothing but replicas of the basic cell at ever earlier times. In the seventies, Ellis suggested that we may live in such a 'small universe', which could explain the apparent large-scale homogeneity even if the basic cell were chaotic. In fact, for $k = 0$ there are 18 topologically different space forms, and for both $k = 1$ and $k = -1$ there are infinitely many; for $k = 1$ all forms are finite, while for $k = 0$ or -1 some are finite, some are not.

One cannot help feeling that all but the basic space forms are somewhat contrived. Nevertheless the alternative of an *infinite* universe also has its uneasy aspects, such as the lack of economy: for example, because of the finiteness of the genetic code, there would presumably be infinitely many Einsteins, Shakespeares, etc.

16.5   Robertson and Walker's theorem

In two independent and important papers, Robertson and Walker almost simultaneously discovered (by group-theoretical methods, around 1935) that Friedman's metric applies (in a limited way) to *all* locally isotropic cosmological models, quite independently of GR; that is, without *any* of the assumptions of GR. Their effort was kindled by Milne's 'kinematic relativity' (a theory of relativity starting from cosmology and pointedly independent of GR) which had its optimistic followers for a brief period after about 1935.[2]

What Robertson and Walker proved was that the mere assumption of local isotropy for the substratum and for light propagation leads to certain properties of the model that are conveniently summarized in a Riemannian map with Friedman's metric, having the following features: (i) The motion of the substratum corresponds to fixed spatial coordinates; (ii) light rays follow null geodesics of the metric; (iii) t is a cosmic time coordinate in Walker's sense and corresponds to proper time on the fundamental particles; and (iv) the spatial part of the metric corresponds to radar distance between neighboring fundamental particles.

The two free elements of the metric, k and $R(t)$, can only be determined by additional assumptions (if we approach the task theoretically) or possibly by observation of the actual universe. Observation without further theory, however, is of very limited potential, especially for $R(t)$. (For k, see Exercise 16.9.)

Conforming to current usage and acknowledging Robertson and Walker's theorem, we refer to the metric in question as the Friedman–Robertson–Walker (FRW) metric. Though irrelevant for GR, this theorem found a second important application in another non-general-relativistic theory, namely Bondi and Gold's *Steady State Cosmology*. (See, for example, H. Bondi, *loc. cit.*) As we mentioned earlier (in Section 16.2), this was based on the 'perfect cosmological principle', according to which the universe is not only homogeneous and isotropic, but also presents the same large-scale aspect at all times. (In Bondi's tongue-in-check formulation: 'Geography does not matter, and history does not matter either'.) Since, in particular, Hubble's parameter $H(t)$ must then be constant, eqn (16.27) implies $R = a\exp(Ht)$ for some constant a, which we can absorb into the exponential by a time translation. And since the curvature of the cosmic sections, k/R^2, must likewise be constant in time, k must necessarily vanish. Hence the relevant FRW metric is

$$\mathbf{ds}^2 = c^2\,\mathrm{d}t^2 - e^{2Ht}\{\mathrm{d}r^2 + r^2(\mathrm{d}\theta^2 + \sin^2\!\theta\,\mathrm{d}\phi^2)\},\qquad (16.31)$$

---

[2] A good review of this theory can be found in H. Bondi, *Cosmology*, Cambridge University Press, 1961.

and now the model is fully specified, except for its constant density. Since H is fairly well known, there was no wiggle room against adverse observations, which indeed eventually ruled out this very attractive idea.

Exercises 16

16.1. Prove that, in spite of length contraction, the total proper volume of all the moving matter inside Milne's expanding sphere $r = ct$ in finite, namely maximally $\pi^2 c^3 t^3$ at time t. Hence Milne's fundamental particles must be strictly ideal *point-particles*.

16.2. As an example of a non-isotropic but homogeneous substratum, consider the motion of (non-gravitating) 'test-dust' in the interior (that is, the neck) of Kruskal space, with T, θ, and ϕ as comoving coordinates (cf. quadrants II and IV of Fig. 12.5). Prove that this substratum is indeed homogeneous [*Hint*: Lorentz transformations], and note the analogy to Milne's model. What is cosmic time here? Describe the evolution of this 'universe' by considering the succession of its cosmic sections. [*Hint*: they are 3-cylinders $R \times S^2$.]

16.3. Prove that, in terms of proper distance l, Hubble's law $\dot{l} = Hl$ is strictly satisfied to all distances, with $H = \dot{R}/R$. (This is a manifestation of the *homogeneous* expansion of the cosmic lattice.)

16.4. Consider timelike geodesics (free-particle worldlines) in FRW spaces. Since the radii $\theta, \phi = $ const are totally geodesic, all geodesics issuing from the spatial origin are purely radial; but any fundamental particle can serve as spatial origin, so *all* geodesics are radial relative to *some* fundamental particle. With ψ as radial coordinate, derive the geodesic equations

$$R^2 \dot{\psi} = A = \text{const}, \qquad \dot{t}^2 = 1 + A^2/R^2,$$

where $\dot{} \equiv d/d\tau$ and $c = 1$. (Evidently the fundamental worldlines themselves satisfy these equations.) Suppose a freely moving particle has velocity v relative to the local fundamental observer. Since t is that observer's inertial time, we have $\dot{t} = \gamma(v)$; prove $\gamma v = A/R$. Consequently the motion satisfies $Rp = $ const, where p is the particle's relativistic momentum relative to the substratum. In the case $k = 1$ this equation has a spurious 'explanation': Consider the motion of the particle on a geodesic plane of the substratum, a 2-sphere of radius R: its angular momentum relative to the center is conserved! As a consequence of the constancy of Rp, the random proper motion (that is, motion relative to the substratum) of a *field* galaxy (that is, one not bound gravitationally to others) must have been faster in the past, and will be slower in the future.

16.5. Consider the de Broglie wave associated with the particle of the preceding exercise. Show that its wavelength λ is proportional to R, and thus partakes of the expansion of the universe. We shall find exactly the same behavior for light waves in the next chapter. This suggests the metaphor of space itself expanding. But note that λ is here measured by the fundamental observer present. In Milne's model, in its SR form, the λ of an outgoing wave is progressively lengthened simply because of the progressively greater speed of the fundamental observers that measure it.

16.6. Prove that it is in general impossible for two free particles in FRW space to remain at constant proper distance from each other. [*Hint*: let one of the particles be at the origin and for the other show that $R\psi = $ const implies $dR/d\tau = $ const.] But bear in mind that proper distance is measured instantaneously within a cosmic section. By reference to Fig. 16.2 (careful: here τ is cosmic time), verify that in Milne's model the proper distance between a particle P and a particle at rest in P's private space is *not* constant. [*Hint*: (16.14), (16.15).]

16.7. In an infinitely expanding FRW universe, a particle is projected from the origin at time t_0 with (locally measured) velocity v_0. Prove that it ultimately comes to rest in the substratum at ψ-coordinate

$$\psi = \int_{t_0}^{\infty} \frac{A\,dt}{R(R^2 + A^2)^{1/2}}, \quad A = R(t_0)v_0\gamma(v_0),$$

provided the integral converges. (If, for example, $R \propto t^{1/2}$, the integral does *not* converge.) Illustrate this result graphically for Milne's universe, using Fig. 16.2. [*Hint*: use the equations of Exercise 16.4.]

16.8. In the real lumpy universe the curvature near the lumps vastly outweighs the cosmic curvature k/R^2. Verify this statement by comparing the curvature due to sun and earth with the cosmic curvature of a Milne universe of age 10^{10} years. [*Hint*: see after (11.18).]

16.9. If we lived in a Milne universe, we would, in fact, live in flat Minkowski space, and whatever local experiments we performed to determine the local space curvature, the result would always be zero. Yet that universe has curvature $-1/c^2t_0^2$. What is the connection? To answer this question in general, we define the *private space*[3] of any fundamental observer in any (perfectly smoothed-out) FRW universe as that generated by all the geodesics orthogonal to the observer's worldline at one instant. If local curvature measurements are made (for example, comparing radii and surface areas of little spheres), the curvature found would always be that of the private space. And, of course, by the assumed local isotropy, that private space is locally isotropic *at* the fundamental observer. Fill in the details of the proof of the relation

$$\tilde{K} = K + \frac{H^2}{c^2}$$

between the *cosmic curvature* $K = k/R^2$, the *private curvature* $\tilde{K}$, and Hubble's parameter H, as follows: Write the FRW metric with the brace from (16.21) and replace the numerator by $dx^2 + dy^2 + dz^2$, with $x^2 + y^2 + z^2 = r^2$. By use of the Appendix, calculate $R_{12\,12}/g_{11}g_{22} = -(k/R^2 + \dot{R}^2/c^2R^2)$. But the LHS of this equation is the curvature of the geodesic plane of FRW space determined by the x

[3] Cf. W. Rindler, *Gen. Rel. and Grav.* **13**, 457 (1981).

and y directions [cf. (10.73)]. However, if we regard private space as a 3-space in its own right, its curvature is the negative of the above, since every component of the Riemann tensor changes sign when the $g_{\mu\nu}$ change sign [cf. after (14.32)].

Check the validity of the displayed formula for Milne's model. And note that, *in principle*, this formula allows us to determine K observationally, and with it both k and the present value, R_0, of R.

16.10. Consider a rocket moving radially with constant proper acceleration α in FRW space. How do we obtain its motion? If the universe does not expand appreciably during the journey (which might be the case even for journeys lasting 10^8 cosmic years!), the relevant 2-dimensional metric is $\mathbf{ds}^2 = dt^2 - (R\,d\psi)^2$ with $R = $ const, and the relevant formulae are those of special relativity (cf. Exercises 3.20 and 3.22). Nevertheless, as an exercise, consider the strict solution of the problem, with R given as a function of t. The equation of motion is $g^{\mu\nu}A_\mu A_\nu = -\alpha^2$, where A_μ is the 4-acceleration of the rocket. Using the formula $A_\mu = \frac{1}{2}\mathbb{L}_\mu$ [cf. (10.46)] write out this equation, and note that it involves $\dot{\psi}$ ($\dot{\ } \equiv d/d\tau$) but not ψ itself. Use the metric to replace $\dot{\psi}$ by $(\dot{t}^2 - 1)^{1/2}R^{-1}$ thus obtaining a complicated second-order differential equation for $t(\tau)$. Once that is solved (for example, numerically), we can get $\psi(\tau)$ by quadrature from the metric.

16.11. By considering a small comoving sphere in the steady-state model, prove that a mass $dM = 3VH\rho\,dt$ must be created per volume V in time dt, if a density ρ is to be maintained. Assuming, in cgs units, that $H = 2 \times 10^{-18}$ and $\rho = 10^{-30}$, and given that one year is $\sim 3.2 \times 10^7$ and that the mass of a hydrogen atom is $\sim 1.7 \times 10^{-24}$, prove that one new hydrogen atom would have to be created in a volume of 1 km^3 about every 9 years.

16.12. Since the volume of FRW universes with $k = 0$ or -1 is infinite at each cosmic time $t > 0$, it might be thought that the big bang, too, occurred all over infinite space. But that would be an erroneous view. The general situation is similar to what happens in the Milne model. In spite of the infinite volume of all the *cosmic* sections, the entire substratum can also be enclosed in a finite volume which gets ever smaller towards $t = 0$, thus establishing the big bang as a point event.[4] As Fig. 16.2 makes clear, the volume of a section through the substratum depends on how one cuts it. Even the most general (gravitating) universe permits cuts of finite volume. For simplicity, consider the so-called Einstein–de Sitter model which has zero cosmological constant and $k = 0$, and for which the field equations yield $R \propto t^{2/3}$ (as we shall see later). Using (16.19) in the FRW metric, consider the cut $t = t_0 e^{-\psi}$, centered on a given fundamental particle at time $t = t_0$. [For comparison, note from (16.14)(i) that the cuts $t = a$ in Milne's model, Fig. 16.2, correspond to (cosmic time) $= a/\cosh\psi$.] Choose t_0 small enough to ensure that the cut is spacelike. For larger t_0 the cuts have to be somewhat modified; for example, by having an initial portion with $dt = -\frac{1}{2}R\,d\psi$, until a sufficiently small t value is reached, whereupon the cut proceeds as before. Prove that Δs along each radius $\theta, \phi = $ const from $\psi = 0$

---

[4] Cf. W. Rindler, *Physics Letters* A**276**, 52 (2000).

to $\psi = \infty$ is finite. Also prove that the product $R\psi$ along each generator remains finite. Hence prove that the total volume $\int 4\pi R^2 \psi^2 \, ds$ of these cuts is finite and tends to zero as $t_0 \to 0$. (For early universes that are 'matter-dominated' we *always* have $R \propto t^{2/3}$, while for 'radiation-dominated' early universes we always have $R \propto t^{1/2}$, and then the argument is similar.)

The reader may question how the infinite universe's infinite mass could fit into these finite-volume sections. But the matter coming out of the big bang is essentially collapsed matter, of the kind that goes *into* a black hole. Early cosmic sections $t = \text{const}$ have arbitrarily large densities. Hence the densities near the 'edge' of each finite cut tend to infinity, and thus make the total mass integral infinite also.

17

Light propagation in FRW universes

17.1 Representation of FRW universes by subuniverses

Almost all the information we have of the cosmos comes to us via electromagnetic waves of one kind or another—which, for brevity, we shall here just call 'light'. Evidently, in order to interpret the observations, we must understand what happens to the light on its long journey to us. We shall make a simplifying assumption about the universe, namely that it is sufficiently rarefied so as not to impede the null-geodetic propagation of light.

For the present and many other purposes we often visualize an FRW universe as a succession of cosmic-time cuts $t = $ const, much as we visualized Kruskal space by the succession of cuts shown in Fig. 12.10. But whereas those cuts were somewhat arbitrary, the set of cosmic cuts has intrinsic significance (connecting identical states) and contains much of the essential kinematic information about the model.

For ease of reference, we write down the FRW metric (16.29) once more, *this time with $c = 1$* and using the brace from (16.19):

$$\mathbf{ds}^2 = dt^2 - R^2(t)\{d\psi^2 + \eta^2(\psi)(d\theta^2 + \sin^2\theta\,d\phi^2)\}, \qquad (17.1)$$

where $\eta = \sin\psi$, ψ, or $\sinh\psi$ according as $k = 1, 0$, or -1. All cosmic sections are 3-spaces of constant curvature, and here we shall assume them to have the simplest space-forms (S^3, E^3, or H^3), without any topological identifications. An FRW universe can thus be pictured as an expanding S^3, E^3, or H^3. But our imagination deals more easily with S^2 etc. Fortunately, within each S^3 universe, for example, there live infinitely many S^2 subuniverses, expanding with it, and complete in themselves, in the sense that their particle and photon worldlines together comprise all the particle and photon worldlines of the full universe. These subuniverses are the *symmetry surfaces* of the full universe. One basic symmetry surface of (17.1) is evidently $\theta = \pi/2$, the metric being invariant under $\theta \mapsto \pi - \theta$. This is a 2-*dimensional* FRW *universe*, with constant-curvature spatial sections, the spatial part of the metric now reading $R^2\{d\psi^2 + \eta^2(\psi)d\phi^2\}$. By the rotational symmetry of (17.1), there is such a symmetry surface in every planar direction through its spatial origin, and, by homogeneity, also through every other point.

Now (still in the case $k = 1$) *any* particle- or photon path in the full metric (17.1) corresponds to a great-circle path traced out in *some* S^2-subuniverse! For the intersection of the obvious symmetry surfaces $\theta = \pi/2$ and $\phi = 0$ of (17.1) is totally geodesic, and thus, by the rotational symmetry of the metric, so is *every* radius $\theta, \phi = $ const.

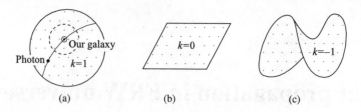

Fig. 17.1

It follows that *any* geodesic issuing from the origin $\psi = 0$ (in particular, a photon- or particle worldline) stays on a radius (as intuition demands). It therefore lies on all the S^2-subuniverses containing that radius, which itself is a spatial geodesic in them.

In the case $k = 1$ we obtain a faithful representation of each such subuniverse by picturing it as a rubber balloon of radius $R(t)$ in Euclidean 3-space with classical time, of which the fundamental clocks partake. Each balloon is blown up at a possibly variable rate prescribed by the Hubble parameter $H(t) = \dot{R}/R$. The substratum corresponds to the material of the balloon, and actual galaxies to ink dots sprinkled more or less uniformly over the surface.

In the cases $k = 0$ and $k = -1$ we replace these balloons by homogeneously expanding 'rubber' planes or saddle surfaces, respectively (see Fig. 17.1). Unfortunately there is no singularity-free way of embedding all of a 2-surface of constant *negative* curvature in Euclidean 3-space; so we must content ourselves with a saddle locally and use our imagination to complete the picture (radii whose spread is proportional to $\sinh \psi$). Since the balloon is the easiest to visualize, we mostly use that in our discussions, it being understood that the other two cases are analogous.

We have already seen (cf. Exercise 16.4) that free particles move through the substratum so as to appear to preserve their relativistic angular momentum Rp with respect to the center of the balloon on which they follow a great circle (or with respect to a center of curvature of the saddle—but this image fails for the plane!) What about photons? For radial light ($\theta, \phi = \text{const}, \mathbf{ds}^2 = 0$) the metric (17.1) yields

$$\mathrm{d}t = \pm R \, \mathrm{d}\psi. \tag{17.2}$$

Recall our definition (16.30) of *proper distance* $l = R\psi$ along a radius, which corresponds to instantaneous ruler distance on the rubber membranes. We have $\mathrm{d}l = R \, \mathrm{d}\psi + \psi \, \mathrm{d}R$; so at the origin $\psi = 0$ a photon satisfies $\mathrm{d}l = \pm \mathrm{d}t$. We conclude that photons move over the balloon as would single-minded little beetles: always along great circles, always at the speed of light locally (that is, with respect to the substratum).

17.2   The cosmological frequency shift

The redshift is the primary piece of information we can gather from the galaxies in the universe. We shall now derive the relevant formula as a first example on the use of the 'rubber' models. The cosmological frequency shift is always denoted by $1 + z$, and it is given by

$$1 + z := \frac{\lambda_0}{\lambda_e} = 1 + \frac{\delta\lambda}{\lambda_e} = \frac{R(t_0)}{R(t_e)}, \qquad (17.3)$$

where λ_e and $\lambda_0 = \lambda_e + \delta\lambda$ are the wavelengths at emission and reception, respectively, of light emitted and received at cosmic times t_e and t_0, respectively. Now for the proof: If two beetles crawl in close succession over a non-expanding track, they arrive as far apart as when they started. But if the track expands or contracts (as does the balloon) proportionately to $R(t)$, then their distance apart at each moment is also proportional to $R(t)$. (This can be seen most readily by imagining equidistant beetles crawling all around a great circle.) Replacing the beetles by two successive wavecrests separated by a wavelength λ, we arrive at (17.3).

What is remarkable about this formula is that the frequency shift depends only on the values of $R(t)$ at emission and reception. What happens in between is irrelevant. Regarded in this way, the cosmological redshift is really an *expansion* effect rather than a velocity effect. Of course, for sufficiently nearby galaxies there is an equivalence with the classical Doppler effect (see Exercise 17.2). But for really distant galaxies, the 'velocity' of the distant substratum at emission, by any reasonable definition (for example, $v = dl/dt$), is quite irrelevant for the ultimately observed redshift.

Note that the same argument that supports the stretching of light waves also supports the stretching of all distances between the photons of the cosmic photon gas (the background radiation). Hence its particle density is proportional to R^{-3}. The energy of each photon, by Planck's relation $E = h\nu$, is proportional to R^{-1}. So altogether the energy density of the photon gas (and therefore also the fourth power of its temperature T, from thermodynamics) is proportional to R^{-4}, whence $T \propto 1/R$.

The 'expansion' aspect of the redshift led Shklovsky in 1967 to suggest an interesting explanation of the puzzling predominance of values $z \approx 2$ among the then-known quasars. It is merely necessary to postulate that the radius of the universe was for a comparatively long time quasi-stationary at approximately one-third of its present value [see Fig. 17.2, where we also introduce the obvious notation $R_1 = R(t_1)$, etc.]. Then one quasar could be several times as far away from us as another (in light-travel time) yet as long as $R(t)$ was the same when each emitted, the observed redshift will be the same. (Present opinion seems to be that there really *was* a maximum of quasar formation at an epoch around $z \approx 2$.) Shklovsky's example shows how unreliable z is as an indicator of cosmic distances, in spite of the classical Doppler relation $z = \delta\lambda/\lambda \propto v$ and Hubble's law $v \propto x$.

To end this section, we show how to derive the frequency-shift formula (17.3) directly, though perhaps less illuminatingly, from the metric. As we have seen, any radial light signal satisfies (17.2); so two successive signals from a galaxy with coordinate ψ to the origin galaxy at $\psi = 0$ will satisfy, respectively,

$$\psi = \int_{t_e}^{t_0} \frac{dt}{R(t)} = \int_{t_e + \Delta t_e}^{t_0 + \Delta t_0} \frac{dt}{R(t)} = \int_{t_e}^{t_0} \frac{dt}{R(t)} + \frac{\Delta t_0}{R(t_0)} - \frac{\Delta t_e}{R(t_e)}, \qquad (17.4)$$

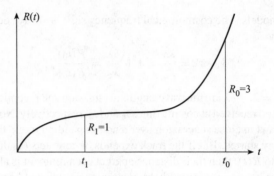

Fig. 17.2

where for the last expression we used the fact that an integral over a short range Δt equals the integrand times Δt. Equation (17.4) implies

$$\frac{\Delta t_0}{\Delta t_e} = \frac{R(t_0)}{R(t_e)}.$$

If we now identify the two signals with successive wavecrests, eqn (17.3) results.

17.3   Cosmological horizons

The 'rubber' subuniverses of the preceding section can also be used to advantage in illustrating the basic ideas behind the concept of cosmological horizons. For definiteness we consider a universe with positive curvature, though most of the arguments apply equally in all three cases. In Fig. 17.1(a) we have marked our own galaxy and a photon on its way toward us. Now it can happen that the universe expands at such a rate that this photon *never* gets to us. (One can obviously blow up the balloon at will so that the photon's proper distance from 'us' stays constant or even increases.) As Eddington put it, light is then like a runner on an expanding track with the winning post (us) forever receding from him. In such a case there will be two classes of (actual or virtual) inward moving photons on every great circle through us: those that reach us at a finite time (or before the big crunch if the universe recollapses), and those that do not. They are separated by the photon that reaches us at $t = \infty$ (or at the big crunch). All such critical photons are shown in the diagram as a dashed circle. In the full universe they constitute a spherical light front moving towards us. This light front is our *event horizon*, and its existence and motion (relative to us) depend on the form of $R(t)$. Events occurring behind it are forever beyond our possible powers of observation (unless we travel away from our galaxy).

It is sometimes said that at the horizon galaxies stream away from us at the speed of light, in violation of SR. Certainly such an event horizon *can* be stationary relative to us ($l_{EH} = \text{const}$), and galaxies must cross it (since it crosses them) at the speed of light, measured locally. Also at the horizon in indefinitely expanding models the redshift becomes infinite [since $R_0 \to \infty$ in eqn (17.3)]—as it would in SR if the source reached the speed of light. But the SR speed limit applies only to objects in an observer's inertial rest-frame. Cosmological observers have *local* inertial frames in which the speed limit applies—but obviously an observer and his horizon can never coexist in a single inertial frame. In open universes, distant galaxies routinely recede from us at superluminary proper speeds, $\dot{l} > c$, since $\dot{l} = Hl$ (cf. Exercise 16.3) and there is no limit to l.

In positively curved universes there is a complication which we shall address below: the 'last' photon to reach us from any direction may already have circled the universe before, and so have been seen by us, possibly more than once.

Figure 17.1 can also be used to illustrate the concept of a *particle horizon*. Suppose the very first photons emitted at our location in the substratum at the big bang are still in the substratum at the present time. (For where else could they be? Unless, of course, there is 'more space than substratum', as in the Milne model.) Let the dashed circle in the diagram now denote the present position of our 'first' photons. In the full universe they constitute a spherical light front moving away from us. As it sweeps outward over more and more galaxies, these galaxies see us for the very first time. By symmetry, however, at the cosmic instant when a galaxy sees *us* for the first time, we see *it* for the first time also. Hence the position of that light front at any cosmic instant (our particle horizon at that instant) divides all galaxies (fundamental '*particles*') into two classes relative to us: those already in our view, and all others. As we shall see below, such horizons can exist even in models with infinite past, though not very realistically.

In order to discuss both types of horizons quantitatively, it is useful to employ a 'conformal diagram' (cf. Fig. 17.3). Let us extract a factor $R^2(t)$ from the FRW metric (17.1) and set $\theta, \phi = $ const (since we are interested only in a typical radius); then we can write

$$\mathbf{ds}^2 = \mathrm{d}t^2 - R^2(t)\,\mathrm{d}\psi^2 = R^2(T)\,(\mathrm{d}T^2 - \mathrm{d}\psi^2) =: R^2(T)\,\widetilde{\mathbf{ds}}^2 \qquad (17.5)$$

with $\widetilde{\mathbf{ds}}^2$ defined by the last equation and

$$T = \int_{t_0}^{t} \frac{\mathrm{d}t}{R(t)}, \qquad (17.6)$$

where t_0 is arbitrary and might as well denote 'now'. The 'conformal time' T is a partly stretched and partly squeezed version of t, which has the advantage of leading to a 'conformally flat' form of the metric. One important fact about conformally related metrics [that is, one metric being a multiple of the other, like $\mathbf{ds}^2$ and $\widetilde{\mathbf{ds}}^2$ in (17.5)] is that they share their null geodesics (as we have seen in Exercise 10.4). But here we have no need of such a general theorem; radii in FRW models are totally geodesic, as we have noted repeatedly, and hence the null geodesics along them are simply the null *lines* (since both are unique). And these are clearly shared. Now the null lines of the metric $\widetilde{\mathbf{ds}}^2$ are the familiar $\pm 45°$ lines of SR, as shown in Fig. 17.3.

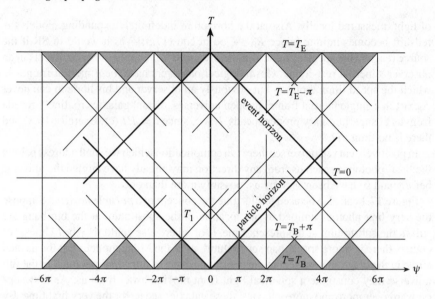

Fig. 17.3

Let $t_0 \le t < t_E$ (E for 'end') be the entire future of the model, where t_E is either finite or infinite, depending on whether or not there is a recollapse. Similarly, let $t_B < t \le t_0$ (B for 'beginning') be the model's entire past, where again t_B can be finite or infinite depending on whether or not there was a big-bang beginning. (One can certainly contemplate models with infinite past, such as the steady state model.) Let us define the conformal times

$$T_E := \int_{t_0}^{t_E} \frac{\mathrm{d}t}{R(T)}, \qquad T_B := -\int_{t_B}^{t_0} \frac{\mathrm{d}t}{R(t)} \tag{17.7}$$

corresponding to t_E, t_B, respectively. Quite independently of t_E and t_B, T_E and T_B can be finite or infinite, depending on the convergence or divergence of the integrals, which in turn depends on the character of $R(t)$. Figure 17.3 shows the spacetime diagram of the metric $\widetilde{\mathbf{ds}}^2$, drawn on the assumption that both T_B and T_E are finite. If not, the upper or lower edge (or both) of the diagram would be at plus or minus infinity, respectively. The vertical lines in the diagram are some of the fundamental worldlines $\psi = \text{const}$, with 'ours' in the center. In the case $k = 1$ the worldlines shown at intervals 2π are actually *all* ours, while those at $\pm\pi$ belong to our antipode. For a more realistic representation we can then imagine the diagram rolled up suitably into a cylinder, representing the history of a great circle through us (cf. Exercise 17.9).

It is now evident that *an event horizon exists whenever T_E is finite* ($\int^{t_E} \mathrm{d}t/R$ *convergent*). For if the diagram has *no* upper edge, then sooner or later our backward light cone sweeps over *all* events, and thus all events become visible. But if T_E is finite, our 'last' backward light cone is our event horizon. It separates events seen from events not seen—unless $k = 1$. In that case, only those events in the shaded triangles above $T_E - \pi$ are never seen. All other events are in at least *one* of the cones going back from $T = T_E$ at $\psi = 0, \pm 2\pi, \pm 4\pi \dots$.

From (17.6) and (17.7), and with reference to Fig. 17.3, we see that the ψ coordinate of the event horizon of the origin observer at time t is given by

$$\psi_{EH} = T_E - T = \int_t^{t_E} \frac{dt}{R(t)}. \tag{17.8}$$

For example, in the steady state theory [cf. (16.31)], where $R = \exp(Ht)$, we find $\psi_{EH} = H^{-1} \exp(-Ht)$, and so the proper distance of the event horizon is given by $l_{EH} = R\psi_{EH} = H^{-1}$. Here this distance is constant, as befits the steady state theory, but in general that is not the case.

The same diagram helps to determine *the condition for a particle horizon to exist:* it is that T_B *be finite* ($\int_{t_B} dt/R$ convergent). For consider an arbitrary cosmic section, say $T = 0$, and one of 'our' earlier *forward* light cones, say that emitted at $T = T_1$. All fundamental observers whose worldlines lie inside that cone at $T = 0$ have seen our T_1-event by then. If there is no lower edge to the diagram, we can slide our T_1-cone back indefinitely, so that at $T = 0$ *no* fundamental worldline is excluded. Then all fundamental observers would already have seen us, and, by symmetry, we them. On the other hand, if we have an 'earliest' forward light cone at T_B, its intersection with $T = 0$ (a spherical light front in the full universe) is our particle horizon at $T = 0$: fundamental observers whose worldlines lie outside of it have not yet seen us, nor we them. Once again, however, the case $k = 1$ is peculiar. The particle horizon then exists precisely until time $T = T_B + \pi$. The diagram shows how *every* fundamental worldline enters *one* of our then equivalent earliest cones at or before that time.

Analogously to (17.8) we can now find the ψ coordinate of the particle horizon of the origin observer at time t:

$$\psi_{PH} = T - T_B = \int_{t_B}^t \frac{dt}{R(t)}. \tag{17.9}$$

It turns out that *all* non-trivial (that is, genuinely gravitating) FRW models with a big bang necessarily have a particle horizon, since for matter-dominated models the field equations imply $R \sim t^{2/3}$ and for radiation-dominated models, $R \sim t^{1/2}$, near the big bang.

And this leads to some problems. First, a speed problem: To be outside particle A's earliest light cone, particle B must have moved away from A faster than the first photon which A emitted in the same direction; that is, faster than light. Here the relativistic speed limit really *was* broken. But recall that the speed limit applies only in the LIF, and that the size of a LIF decreases with increasing spacetime curvature. At the big bang the curvature is infinite, and the LIF has shrunk out of existence. Thus at curvature singularities all the laws of physics have a perfect right to break down. It is believed that GR will actually break down even *close* to the big bang as one enters the regime of quantum fluctuations somewhere near the *Planck time* of $\sqrt{\hbar G/c^5} \approx 10^{-43}$ s.

Then there is the notorious 'smoothness' or 'horizon' problem. Why is the universe so homogeneous? Consider the two most distant regions seen by us in diametrically opposite directions in the sky. Since their light only just reaches us, *we* are on the edge of their particle horizons. Consequently *they* are outside each other's particle horizons, and could never have been in causal contact. So, it is asked, how could the universe have homogenized itself? This is a puzzle *only* if on philosophical grounds one refuses to contemplate a finely tuned Friedman big bang, and prefers to posit an initially chaotic universe that somehow homogenized itself. But then the Friedman particle horizons are irrelevant, since their very existence depends on the finely tuned big bang. In a Milne universe there is no particle horizon, and the diametrically opposite regions could have exchange information at early epochs. So who is to say whether particle horizons exist after a chaotic big bang? By contrast, Penrose has given a powerful argument *against* homogenization.[1] The second law of thermodynamics (entropy increase) is at least as much of a puzzle as is homogeneity. And he points out that only a highly tuned big bang like that of the Friedman models has the necessary low entropy to start off the second law. Homogeneity and isotropy are then simply implicit in the initial conditions.

Event horizons are not quite as ubiquitous as particle horizons among realistic models. But, as we shall see, all indefinitely expanding models with cosmological term as well as all collapsing models have them. On the other hand, the existence of event horizons poses no particular logical or philosophical problems (except the spurious speed problem alluded to in the second paragraph of the present section.)

As can be seen particularly clearly from Fig. 17.3, when a model is run backwards in time, its event- and particle horizons interchange roles. And GR models *can* be run backwards without violating the dynamics, since all the equations are time-symmetric. Just from this it follows that all non-trivial collapsing models have event horizons, since all such big-bang models have particle horizons.

We end this section with a few further remarks concerning the event horizon (EH) and the particle horizon (PH):

1. The reader will no doubt have noticed an analogy between the cosmological event horizon and that of a Schwarzschild black hole. Both hide forever a class of events from the observer, and both are spherical light fronts moving towards (but never reaching) the observer. But the Schwarzschild horizon is the same for *all* external observers and *moves outward* towards all of them, whereas in cosmology each fundamental observer has his own EH that *moves inward* towards only him.

2. Every galaxy but A within A's EH eventually passes out of it. This is immediately clear from Fig. 17.3.

---

[1] See, for example, R. Penrose in *Fourteenth Texas Symposium*, ed. E. Fenyves, New York Academy of Sciences, New York, 1989.

3. Every galaxy B within A's EH remains visible forever at A. For the EH itself carries to A the 'last' image of B, namely that of B crossing the horizon. As B approaches the horizon, its redshift tends to infinity if the model expands indefinitely; for then $R \rightarrow \infty$ at reception [cf. eqn (17.3)]. Also any finite part of B's history up to the crossing event appears infinitely dilated; for example, B's clock apparently never quite reaches noon, if at noon B crosses the horizon. In collapsing models B's light gets infinitely blueshifted (so it appears infinitely bright) near the horizon, since $R \rightarrow 0$ at reception, and B's pre-crossing history appears infinitely accelerated. Although B now hurtles towards A and eventually hits A, A never sees the last part of B's history. B together with all its later emitted photons only meets A at the big crunch.

4. As galaxies are overtaken by A's PH, they come into view at A with infinite redshift (zero luminosity) in big-bang models and with infinite blueshift (infinite luminosity) in models with unlimited expansion in the past (for example, $R \propto \cosh t$). For in the former case R was zero, in the latter, infinite, at emission. (Infinite blueshift is the optical analog of a sonic boom.)

5. In big-bang models with PH (in contrast, for example, to Milne's model), the big bang is visible, in principle, to all observers in all directions at all times. This is evident from Fig. 17.3. (Provided they can 'see' with neutrinos: photons cannot penetrate the earliest opaque 300,000 years.) A useful image: picture a large and expanding balloon, representing the universe; picture identical small circles around all fundamental particles, representing their ever-expanding creation light fronts (PHs); imagine yourself at the center of one of these: as you look down any direction, you see younger and younger galaxies—in principle you even see galaxies just being born *at* your PH. You have the impression of surveying the entire universe. But as time goes on, shell after shell of ever more distant just-born galaxies comes into view. You always *seem* to see the very 'edge of the universe'; but all you see is your *visible universe*, namely that within your particle horizon (that is, the little circle on the large balloon). Its radius, very roughly, is of the order of t_0 light years, if t_0 is the age of the universe in years.

6. Consider non-empty big-bang models. All have a PH. Hence all light signals in such models originate on the lower edge in Fig. 17.3, which is a curvature singularity. Such models either end up on another curvature singularity (coinciding with the upper edge) if they recollapse; or else they ultimately expand into Minkowski or de Sitter spacetime and light can never leave their substratum. Unlike Milne's model, therefore, such models are *maximal*; that is, they have no causal connection to spacetime beyond their substratum.

7. In models without PH, any observer can be present at any event, provided he is willing to travel and provided he starts out early enough. For, in principle, his only travel restriction is to stay within his forward light cone at the start of the trip. And in the absence of a lower edge in Fig. 17.3, that cone can be pushed back to include any event.

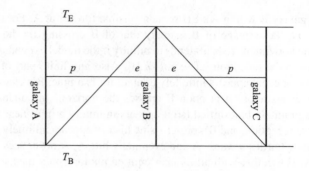

Fig. 17.4

8. If an EH exists, two arbitrary events are, in general, not both knowable to any single observer, even if he travels. For if there is an upper edge in Fig. 17.3, it is easy to specify pairs of events whose forecones do not intersect. Yet to have known both events implies being and remaining in both their forecones. In the case $k = 1$, such unknowable pairs of events can only occur after time $T_E - \pi$.

9. Suppose a model has both an EH and a PH. Let the instantaneous ψ-radii of these horizons be $\psi_{EH} = e$ and $\psi_{PH} = p$, respectively, and note [either from eqns (17.8) and (17.9), or directly from Fig. 17.4] that $e + p = \text{const} = T_E - T_B$. Figure 17.4 shows the entire EH of galaxy B, and half the PHs of galaxies A and C. From this figure we can read off the following facts: (i) galaxies farther than $p + e$ can never be seen at A, nor can they be visited by a traveler from A. That is both A's travel range and maximal PH. (ii) Even a traveler from A can never see events whose instantaneous distance from A exceeds $p + 2e$. That is A's extended EH. (iii) And no traveler from A can ever see a galaxy beyond $2p + 2e$. That is A's extended PH.

17.4   The apparent horizon

The event- and particle-horizons discussed in the last section are conceptually quite unrelated to a Schwarzschild type of horizon that *must* exist in some guise in FRW models: We have seen [cf. after (12.2)] that homogeneous balls even of low density will produce a Schwarzschild horizon provided they are sufficiently large. But in at least the open FRW universes there are at all times homogeneous balls of arbitrarily large size. Surely something special must happen relative to the center of each such ball at the location where its Schwarzschild radius *would* be if the outside were vacuum. Let us imagine a thin shell of vacuum surrounding a ball of such critical size. By Birkhoff's theorem (even with Λ-term), the rest of the universe does not affect the vacuum spacetime in that shell. It must contain the ball's Schwarzschild-horizon light front. If that front points outward (which happens in collapsing universes), that ball would now seem to be doomed to collapse totally in a maximal proper time πm as measured on its surface [cf. (12.13)] or $\frac{1}{2}\pi\tilde{r}$, where $\tilde{r}$ is the 'area coordinate' of the horizon [cf. after (11.1)], since $\tilde{r} = 2m$.

To find the exact location of this horizon, we characterize it as a momentarily stationary light front in terms of the area coordinate (that is, a light front of momentarily stationary area). In the FRW metric (17.1) the area coordinate corresponds to $\tilde{r} = R\eta$, and for its stationarity we need $d\tilde{r} = 0$; that is,

$$R \, d\eta + \eta \, dR = 0 = R \, d\eta + \eta \dot{R} \, dt. \qquad (17.10)$$

For any particle moving radially with this sphere we also have $\theta, \phi = $ const, and then the metric (17.1) gives, when we utilize (17.10),

$$\mathbf{ds}^2 = dt^2 \left(1 - \frac{\eta^2 \dot{R}^2}{1 - k\eta^2} \right). \qquad (17.11)$$

If the particle is a photon, $\mathbf{ds}^2 = 0$. The horizon therefore has $(\ ) = 0$, and, after a little algebra, this yields

$$\tilde{r}_{AH}^2 = (R^2 \eta^2)_{AH} = \frac{R^2}{\dot{R}^2 + k} \qquad (17.12)$$

for its area coordinate; this horizon is called the *apparent horizon*. But here, in contradistinction to the Schwarzschild case, the area coordinate $\tilde{r}$ is now a 'time' *outside* the horizon (cf. Exercise 17.13), which must decrease in contracting universes; no particle can move outward in that region relative to the center.

We shall see [cf. eqns (18.17) and (18.18) below] that the field equations for 'dust' universes with zero cosmological constant Λ directly fix the value of the RHS of (17.12) as $(\frac{8}{3}\pi\rho)^{-1}$ (in units making $c = G = 1$). And so we have

$$\tilde{r}_{AH} = \left(\tfrac{8}{3}\pi\rho\right)^{-1/2}, \qquad (17.13)$$

exactly as in the 'naïve' eqn (12.2) (where we must put $\tilde{r} = r$)!

As an example [for which we again anticipate the dynamics—cf. (18.36) below], we consider the 'oscillating' dust model corresponding to $k = 1$ and $\Lambda = 0$. The graph of its expansion function $R(t)$ is a cycloid, and so its maximal height, R_{max}, is related to its total duration, t_{max}, by $t_{max} = \pi R_{max}$. On the other hand, at the instant of maximal extension ($\dot{R} = 0$), eqn (17.12) gives $\tilde{r}_{AH} = R_{max}$, and so the apparent horizon is then at the equator, the same for each pair of antipodal fundamental particles. The 'Schwarzschild' maximal duration thereafter, as we have seen, is $\frac{1}{2}\pi\tilde{r}$, which here coincides with $\frac{1}{2}\pi R_{max}$, the *actual* proper time elapsed at each fundamental particle before the big crunch.

Just like the Schwarzschild horizon, the cosmological apparent horizon is a limit of 'trapped surfaces' (cf. Section 12.6), which this time lie *outside* of it. (Recall that trapped surfaces have the property that light emitted from them, both outwardly and inwardly, moves inwardly or, more strictly, 'converges'.) And this has relevance for the Penrose-Hawking cosmological singularity theorems. According to them, the existence of an apparent horizon leads under certain very reasonable conditions to a big-crunch singularity, or, in time-reversal, indicates a big-bang singularity in the past. [cf. after (18.10) below.]

17.5   Observables

Astronomers observe galactic redshifts directly, and correlate them with galactic 'magnitude', which, in astronomical usage, means apparent luminosity normalized in a certain way. For these and other observations to be compared with theory, it is necessary for theory to relate observables to the parameters of the model. Several such relations will be obtained in the present section.

But first, let us consider some definitions of distance in cosmology. As in Schwarzschild spacetime (cf. Section 11.2F), various 'reasonable' definitions turn out to be inequivalent, except locally. We have already come across two useful *theoretical* distance measures, namely *proper* distance (or instantaneous ruler distance), $l = R\psi$, and *area* distance, $\tilde{r} = R\eta$. But, of course, these are not directly accessible. More practical methods of obtaining distance estimates in astronomy involve, for example, parallax, radar, apparent size, and apparent luminosity. But for distant galaxies even the first two of *these* methods are impracticable. So we shall here concentrate on 'distance from apparent area', D_A, and 'distance from apparent luminosity', D_L. Since theory must intervene anyway to interpret the observations, one chooses the simplest possible *definitions*, namely those that in a static, unchanging Euclidean universe would actually yield the Euclidean distance. So the D_A of an object of presumed cross-sectional area A, seen subtending a small solid angle Ω, is *defined* by the equation

$$\Omega = \frac{A}{D_A^2}. \tag{17.14}$$

Similarly the D_L of a source that presumably radiates energy isotropically at total proper rate L is defined by the equation

$$S = \frac{L}{4\pi D_L^2}, \tag{17.15}$$

where S is the energy received at the observer crossing unit area in unit time. L stands for (intrinsic, total) *luminosity*, which could be measured in Watts, just as for light bulbs. The *apparent luminosity* S is related to traditional astronomical 'magnitude' m by the formula $m = -2.5 \log_{10} S + \text{const}$; one speaks of 'bolometric' magnitude when S includes energy received over the whole spectrum, not just the visual range. In practice, of course, corrections have to be made for absorption by the atmosphere, etc.

By (17.1), the area of a coordinate sphere $\eta = \text{const}$ at cosmic time t_e is $4\pi R_e^2 \eta^2$, and the solid angle it subtends at the origin is 4π. Consequently, the solid angle subtended by the bundle of radii to a galaxy of cross-sectional area A on that sphere is

$$\Omega = \frac{A}{R_e^2 \eta^2}. \tag{17.16}$$

And this is the solid angle which the galaxy is seen to subtend at whatever time t_0 the light it emits at t_e arrives at the origin. For it is its position at *emission* time that determines the bundle of radii along which the light from it will travel to the origin. (The rubber models should make that clear; the galaxy does *not* expand along with the rest of the universe!) Hence we have, utilizing (17.3),

$$D_A = R_e \eta = (1+z)^{-1} R_0 \eta. \tag{17.17}$$

Suppose, next, that a source of intrinsic luminosity L is placed at the origin and that its light, emitted at t_e, is observed at t_0 on a sphere $\eta = \text{const}$ comoving with the substratum. If the universe were static, the total energy flux through that sphere would, of course, be L. But because the universe expands, the energy $E = h\nu$ of each photon is diminished by a Doppler factor ('Planck effect') and since the time between the arrival of 'successive' photons is also lengthened by a Doppler factor ('number effect'), the total flux through the sphere is $L(1+z)^{-2}$. So the flux per unit area is given by

$$S = \frac{L}{(1+z)^2 4\pi R_0^2 \eta^2}. \tag{17.18}$$

By symmetry, however, the (permanent) η coordinates which two galaxies ascribe to each other are equal. We therefore conclude that (17.18) applies to the flux measured by *us* at time t_0 due to a source at coordinate η relative to us. Eqns (17.15), (17.18), and finally (17.17), then yield

$$D_L = (1+z) R_0 \eta = (1+z)^2 D_A. \tag{17.19}$$

The extremities of this equation constitute an interesting *model-independent* relation between observables:

$$\frac{L}{4\pi S} = (1+z)^4 \frac{A}{\Omega}. \tag{17.20}$$

[It goes back to Etherington (1933) and actually applies without any symmetry assumptions.] The 'disturbing' presence in some of the above formulae, from an observational point of view, is the coordinate η. To eliminate it, consider that to each event along a given incoming ray of light reaching us at time t_0 there correspond a pair of alternative radial coordinates η and ψ, a time t, a redshift z, and an R-value. All these variables are monotonically related (if we accept $\dot{R} > 0$). Now from (17.2) we have

$$\frac{d\psi}{dR} = \frac{d\psi}{dt}\frac{dt}{dR} = -\frac{1}{R\dot{R}}, \tag{17.21}$$

which, after a little computation, yields the following Taylor series for ψ:

$$\psi = \frac{1}{R\dot{R}} \Delta + \frac{1}{2} \frac{R\ddot{R} + \dot{R}^2}{R^2 \dot{R}^3} \Delta^2 + \cdots, \tag{17.22}$$

where $\Delta = R_0 - R_e$ and $R, \dot{R}, \ddot{R}$ are here and for the rest of this chapter understood to be evaluated at t_0. But

$$\eta = (\sin \psi, \psi, \sinh \psi) = \psi + \mathrm{O}(\psi^3), \tag{17.23}$$

so that the RHS of (17.22) also applies to η, as far as it goes. On the other hand, we have

$$R_e = R(1+z)^{-1} = R(1 - z + z^2 + \cdots) \qquad (17.24)$$

as long as $z < 1$ (see below), and so

$$\Delta = R(z - z^2 + \cdots). \qquad (17.25)$$

Substituting this into (17.22), writing η for ψ and substituting *that* into (17.19)(i), we finally find

$$H_0 D_L = z + \tfrac{1}{2}(1 - q_0)z^2 + O(z^3), \qquad (17.26)$$

where, of course, $H_0 = \dot{R}/R$ and we have introduced

$$q_0 := -\frac{R\ddot{R}}{\dot{R}^2} = -\frac{\ddot{R}}{RH_0^2}. \qquad (17.27)$$

This q_0 is a convenient dimensionless parameter that essentially measures $\ddot{R}$. Until recently, cosmologists were so sure that the universe is decelerating (pulled back by gravity), that they introduced the minus sign into the definition and called q_0 the *deceleration parameter*. In retrospect, it might have been better to define an *acceleration* parameter, but we are stuck with q_0.

By use of (17.19) we can convert (17.26) into a formulae for D_A (cf. Exercise 17.15), but that is much less useful. Either of these formulae can serve to determine H_0.

As we noted in connection with (17.24), the above approximations hold only if $z < 1$, and are really useful only when z is not much bigger than 0.5. On the other hand, formulae like (17.26) can be extended to arbitrarily many powers. At the third power the curvature index k begins to appear, because of (17.23). But by now redshifts as big as $z \approx 5$ have been observed, looking back to when the universe was a mere one-sixth of its present size! To utilize such data, the above methods must be replaced by exact model-specific numerical computations, following essentially the same logic. Such more exact methods, applied to supernovae with $0.3 \lesssim z \lesssim 1$, have recently yielded indications that q_0 might well be negative, which would have profound consequences for cosmology.

The main problem with the theory outlined above is the difficulty of knowing the intrinsic luminosity L of the sources, on which a determination of D_L via (17.15) depends. And also, since for high redshift we look far back in time, a theory of how L evolves would be needed—all of which is at present only very imperfectly known. The above-mentioned techniques of observing supernovae in distant galaxies rather than the galaxies themselves, have to some extent overcome this difficulty, since for them one can assume a fairly definite luminosity.

We now turn to another important empirical relation obtained by astronomers (both optical and radio), namely the number of galaxies per unit solid angle of sky whose redshift is less than some given z, or whose apparent luminosity is greater than some given S. Such 'number counts' evidently probe the galactic distribution in depth, or, in other words, radially. Consider a cone of solid angle Ω issuing from the origin at some cosmic instant t_0 ('now') and terminating at coordinate η. Its volume, from (17.1) and (17.23), is given by

$$V = \Omega R^3 \int_0^\eta \eta^2 \, d\psi = \Omega R^3 \int_0^\eta \eta^2 \, d[\eta + O(\eta^3)]$$
$$= \Omega R^3 [\tfrac{1}{3}\eta^3 + O(\eta^5)]. \tag{17.28}$$

Multiplying V by the present particle density n_0, we obtain the total number $N(\eta)$ of galaxies presently in the cone—and this will be the number in it at all times as long as galaxies are neither created nor destroyed. As in converting (17.19), we can replace η^3 in (17.28) by a power series in z; the resulting expression for $N = n_0 V$ is

$$N(z) = n_0 \Omega H_0^{-3} [\tfrac{1}{3}z^3 - \tfrac{1}{2}(1 + q_0)z^4 + \cdots]. \tag{17.29}$$

This formula gives the number of galaxies seen in the solid angle Ω at redshift z or less. But, especially with radio galaxies, it is the energy flux (or apparent luminosity) that is more readily determined than the redshift. So we solve (17.26) for z by setting $z = H_0 D_L + a(H_0 D_L)^2 + \cdots$ and comparing coefficients, which yields $a = -\tfrac{1}{2}(1 - q_0)$. With this expression for z substituted, eqn (17.29) becomes

$$N(D_L) = n_0 \Omega (\tfrac{1}{3}D_L^3 - H_0 D_L^4 + \cdots), \tag{17.30}$$

independently of q_0 up to this order. Above all, this equation serves to determine n_0.

The corresponding formula for the steady state theory, where particles are continuously created, is found to be (cf. Exercise 17.19)

$$N(D_L) = n_0 \Omega (\tfrac{1}{3}D_L^3 - \tfrac{7}{4}H_0 D_L^4 + \cdots). \tag{17.31}$$

The RHS of (17.31) is less than that of (17.30), since at earlier look-back times there were fewer galaxies in the cone than there are now. However, it is not so much this second-order difference between the two equations that lends itself to observational testing. Consider instead the common first-order part of eqns (17.30) and (17.31), and write it in terms of L and S via (17.15):

$$S^{3/2} N(S) = \tfrac{1}{3} n_0 \Omega (L/4\pi)^{3/2}. \tag{17.32}$$

In the steady state theory the luminosity of all galaxies would certainly not be the same. Let us imagine many classes of galaxies with different luminosities L_i and corresponding particle densities n_i. These numbers, however, would have to be permanent. For each class there is an equation like (17.32) with N, L and n_0 replaced by N_i, L_i and n_i. Adding these equations, and writing $N(S)$ for $\sum N_i(S)$ (the total number of galaxies seen at apparent magnitude S and bigger), we find

$$S^{3/2} N(S) = \text{independent of } S. \tag{17.33}$$

This equation is *not* satisfied by the standard model, because as we look farther and farther back in time, there are systematic changes in L_i and n_i. And indeed it was the observed non-constancy of the LHS of (17.33) that contributed in the sixties to the decline of the steady state theory—and to an appreciation of the evolution of intrinsic luminosities.

Exercises 17

17.1. Prove that an affine parameter u along any radial geodesic (also a null geodesic) in the FRW metric is characterized by $du = \text{const} \cdot R^2 \, d\psi$. [*Hint*: Write the radial metric as $dt^2 - R^2 d\psi^2$; then, with $\dot{} \equiv d/du$, we must have $\partial \mathcal{L}/\partial \dot{\psi} = \text{const} = -2R^2 \dot{\psi}$.]

17.2. From (17.3) derive the formula (in full units) $z = l/c + O(\Delta t^2)$, where l is the proper distance of the observed source at the time of observation, and Δt is the look-back time. Compare this with our earlier Doppler formula (4.3). [*Hint*: expand R_e as a power series in Δt.]

17.3. Consider an FRW universe with $R = t$ and $k = 1$. In theory, an observer can see each galaxy by light received from two diametrically opposite directions. Prove that the redshifts in the light arriving simultaneously from the same galaxy but from opposite directions satisfy $(1 + z_1)(1 + z_2) = \exp(2\pi)$.

17.4. According to Planck's law of blackbody radiation, the energy density of photons in the frequency range ν_0 to $\nu_0 + d\nu_0$ is given by

$$du_0 = 8\pi h \nu_0^3 c^{-3} (e^{h\nu_0/kT_0} - 1)^{-1} \, d\nu_0,$$

where k is Boltzmann's constant and T_0 the absolute temperature; and this characterizes the spectrum of blackbody radiation. We have inserted the subscript zero to indicate that we look at the various quantities at some cosmic instant t_0. Now suppose that the radiation in question uniformly fills an FRW universe with expansion factor $R(t)$, and suppose we can neglect its interaction with the cosmic matter. Prove that, as the universe expands or contracts, the radiation maintains its blackbody character. [*Hint*: third paragraph of Section 17.2.]

17.5. By referring to the Kruskal diagram of Fig. 12.5, prove that in the homogeneous Kruskal universe discussed in Exercise 16.2 there exist both particle horizons and event horizons.

17.6. Prove directly from the formulae that when an FRW model is run backwards in time, the particle horizon becomes the event horizon, and vice versa.

17.7. If an event horizon exists, prove that the farthest galaxy in any direction to which an observer can travel if he starts 'now', is the galaxy which lies on his event horizon in that direction 'now'.

17.8. If a model universe has both an event horizon and a particle horizon, prove that the farthest galaxy, on any line of sight, from which a radar echo can be obtained, is that where these horizons cross each other.

17.9. Make a paper model of the cylinder represented by Fig. 17.3 when $k = 1$, by cutting out a suitably drawn diagram and rolling it up so that the lines $\psi = \pm\pi$ coincide. (It will come out best if your vertical extent is not much more than 3π.) Note how the first and last light cones of the origin galaxy re-focus at the antipodal galaxy. Convince yourself that after time $T_B + \pi$ all galaxies have seen the origin galaxy and have therefore been seen by it, and that all events occurring before time $T_E - \pi$ are visible at the origin galaxy.

17.10. Consider the source sphere, around us, of today's 2.7 K radiation, at a look-back time of $\sim 10^{10}$ y. Assume $R \propto t^{1/2}$ for the radiation-dominated universe before recombination (at $\sim 3 \times 10^5$ y after the big bang and at ~ 3000 K), while $R \propto t^{2/3}$ afterwards, assuming the simplest model. Prove that the proper distance of the particle horizon, l_{PH}, at recombination time was $\sim 6 \times 10^5$ lt-y and that our proper distance *then* from the sources we see today was $\sim 3 \times 10^7$ lt-y. Deduce that the angular separation between two points on that source sphere which only *just* came into causal contact at the time of recombination is of the order of one degree. [This is the 'smoothness problem'—cf. after (17.9).]

17.11. In a non-static FRW universe, consider the history of two spheres centered on the origin, one at constant *proper* distance l, the other at constant *area* distance $\tilde{r}$ from it. Show that unless $k = 0$ these two spheres can coincide at most at one cosmic instant.

17.12. Prove that in collapsing universes the apparent horizon is the locus of turn-back points of outgoing light signals in terms of area distance $\tilde{r}$ ($d\tilde{r} = 0$). Also prove that the light-turn-back locus in terms of proper distance l ($dl = 0$) is a different surface (unless $k = 0$), namely $l = -1/H$.

17.13. Prove that in models with $k = 0$ or $k = 1$ the apparent horizon (17.12) always exists (that is, corresponds to a definite locus in the substratum). For the case $k = -1$, look ahead at the field equation (18.9) and deduce that the apparent horizon exists whenever $8\pi\rho + \Lambda > 0$ ($c = G = 1$).

17.14. Re-write equation (17.11) as $\mathbf{ds}^2 = dt^2(1 - \tilde{r}^2/\tilde{r}_{AH}^2)/(1 - k\eta^2)$, and deduce that $\tilde{r} = \text{const} \gtreqless \tilde{r}_{AH}$ implies $\mathbf{ds}^2 \lesseqgtr 0$; interpret this.

17.15. For the expanding (contracting) radiation model with $k = \Lambda = 0$ and $R \propto t^{1/2}$, verify that the apparent horizon and the particle (event) horizon coincide, both being $2t$. What is the corresponding situation for the 'dust' model $k = \Lambda = 0$ and $R \propto t^{2/3}$?

17.16. By reference to the rubber models, describe the most general FRW universe in which it is possible for two galaxies, seen simultaneously by an observer in the same direction at one given instant, to exhibit the same non-zero redshift and the same distance by apparent area, and yet for one to be at twice the proper distance as the other.

17.17. Suppose an ideal big-bang FRW universe is filled uniformly with galaxies, n_0 per unit volume at the present cosmic instant t_0, and all having equal luminosity $L(t)$ at equal cosmic time t. Prove that the present apparent brightness of the sky (energy received per unit solid angle per unit time per unit collection area) is then given by

$$B = (n_0/4\pi R_0) \int_0^{t_0} R(t) L(t) \, dt.$$

In a $k = 1$ universe the same galaxy might be counted repeatedly, but this is as it should be. Note also that for a static and unchanging universe B would be infinite; this constitutes the so-called *Olber's Paradox* for such old-fashioned universes. [*Hint*: Consider contributions to B from a thin collection cone of solid angle Ω, as the light travels down the cone. Its volume element dV at coordinate η and emission time t_e is given by $dV = \Omega R_e^3 \eta^2 \, d\psi$; the number of galaxies in it is $n_0(R_0^3/R_e^3) \, dV$; and each of these galaxies contributes $S = L_e R_e^2/4\pi R_0^4 \eta^2$ to B; in the integration use $d\psi = -dt/R$.]

17.18. From (17.26) and (17.19) derive the following analog of (17.26) for D_A:

$$H_0 D_A = z - \tfrac{1}{2}(3 + q_0)z^2 + O(z^3).$$

17.19. Derive formula (17.31) for the steady state theory. [*Hint*: Recall the relevant FRW metric: $R = \exp(Ht)$, $H = \text{const}$, $k = 0$, and pick $t_0 = 0$; from (17.21) derive $H\psi = \exp(-Ht) - 1 = z$; now the number n_0 of galaxies per unit volume is constant: $N(\psi) = n_0 \int R^3(t)\psi^2 \, d\psi$. Finally, use (17.26) with $q_0 = -1$.]

18

Dynamics of FRW universes

18.1 Applying the field equations

In a famous labor-intensive paper of 1938, Einstein, Infeld, and Hoffmann showed that the field equations of general relativity actually imply the geodesic law of motion—which had originally been regarded as one of the *axioms* of the theory. Of course, that law had been strongly suggested by the equivalence principle, but that could not be considered a 'proof'. Even the rigor of the 1938 paper was soon challenged and numerous improvements followed. But the result itself is not in question: the field equations determine the motion. Nowhere is this more directly apparent than in FRW cosmology, where the field equations determine the expansion factor $R(t)$. (Of course, the assumed symmetries already imply—as we have seen—that the substratum moves geodesically.)

In the present chapter we shall examine the resulting dynamics. We take the field equations in their general form (14.15); that is, *with* the so-called cosmological term $\Lambda g^{\mu\nu}$. Various arguments have at times been given *against* the inclusion of this term: (i) that it was only an afterthought of Einstein's (but: better discovered late than never); (ii) that Einstein himself eventually rejected it (but: authority is no substitute for scientific argument); (iii) that with it the well-established theory of SR is not a special case of GR (but: locally the Λ term is totally unobservable); (iv) that it is *ad hoc* (but: from the formal point of view it belongs to the field equations, much as an additive constant belongs to an indefinite integral); (v) that similar modifications could be made to Poisson's equation in Newton's theory and Maxwell's equations in electrodynamics (but: locally the Λ term is ignored, and cosmologically Poisson's and Maxwell's equations may well need similar modification); (vi) that it represents a space expansion *uncaused* by matter, and thus a field which acts but cannot be acted on (but: in GR, matter and space are intimately related by the field equations, and Newtonian analogies may be misleading); (vii) that one should never envisage a more complicated law until a simpler one proves untenable (but: in cosmology the technical complication is slight, and several recent investigations have suggested that the Λ term indeed may be needed to account for the observations); (viii) (and more technically) that a Λ term in the geometry would destroy the possibility of quantizing gravity (but: an energy tensor $(\Lambda/\kappa)g_{\mu\nu}$ may arise naturally out of quantum fluctuations *in vacuo*, so that the Λ term could be regarded as part of the sources rather than part of the geometry).

Next, what are the sources? We already mentioned that, in effect, our procedure is to grind up all the matter and redistribute it uniformly. The assumed isotropy of the model then restricts the physical characteristics of the sources. In fact, as we shall see, the sources must have an energy tensor [cf. (7.80)] whose components in the rest-LIF are of the form

$$T_{\mu\nu} = \text{diag}(p, p, p, c^2\rho). \tag{18.1}$$

Sources of this kind are called *perfect fluids*. The only stress they can sustain is the isotropic pressure p; ρ is the mass density. It is generally agreed that, except in the early universe, the pressure of the sources can be neglected. A perfect fluid with zero pressure is technically referred to as *dust*. Such dust sits still on the substratum, since any random motion would constitute a pressure. In the early universe, however, uniform radiation (a photon gas) is thought to have predominated. This *does* have pressure; its equation of state (cf. Exercise 7.22) is

$$p = \tfrac{1}{3}c^2\rho. \tag{18.2}$$

It will be advantageous, because of the resulting symmetry, to use the FRW metric (16.29) with the brace from (16.21) in Cartesian coordinates so that $r^2 = x^2 + y^2 + z^2$:

$$\mathbf{ds}^2 = c^2\,dt^2 - R^2(t) \left\{ \frac{dx^2 + dy^2 + dz^2}{\left(1 + \tfrac{1}{4}kr^2\right)^2} \right\}. \tag{18.3}$$

The main labor in applying the field equations

$$G_{\mu\nu} = -8\pi G c^{-4} T_{\mu\nu} \tag{18.4}$$

to this metric lies in the computation of the (modified) Einstein tensor

$$G_{\mu\nu} = R_{\mu\nu} - \tfrac{1}{2}R^\lambda_\lambda g_{\mu\nu} + \Lambda g_{\mu\nu}. \tag{18.5}$$

[We here eschew the usual symbol R for the scalar curvature R^λ_λ so as to avoid notational conflict with the expansion factor $R(t)$.] The computation is simplified if we refer to the expressions listed in the Appendix. Clearly we only need the values of $G_{\mu\nu}$ at the origin $r = 0$, since this is a typical point. Here is what we then find:

$$\frac{G_{11}}{g_{11}} = \frac{G_{22}}{g_{22}} = \frac{G_{33}}{g_{33}} = -\frac{2\ddot{R}}{Rc^2} - \frac{\dot{R}^2}{R^2c^2} - \frac{k}{R^2} + \Lambda, \tag{18.6}$$

$$\frac{G_{44}}{g_{44}} = -\frac{3\dot{R}^2}{R^2c^2} - \frac{3k}{R^2} + \Lambda, \tag{18.7}$$

and $G_{\mu\nu} = 0$ when $\mu \neq \nu$. These expressions, when substituted in the field equations (18.4), imply an energy tensor of the general form (18.1), and thus, as we saw in Chapter 7, of the form (7.86). Writing $U^\mu = (0, 0, 0, c)$ for the 4-velocity of the matter at the origin $r = 0$, and lowering the indices in (7.86), we find

$$T_{\mu\nu} = \text{diag}(-g_{ii}\,p, c^2\rho).$$

With that, the field equations (18.4) read:

$$\frac{2\ddot{R}}{Rc^2} + \frac{\dot{R}^2}{R^2c^2} + \frac{k}{R^2} - \Lambda = -\frac{8\pi Gp}{c^4} \tag{18.8}$$

$$\frac{\dot{R}^2}{R^2c^2} + \frac{k}{R^2} - \frac{\Lambda}{3} = \frac{8\pi G\rho}{3c^2}. \tag{18.9}$$

Note that the high degree of symmetry which we imposed a priori on the metric (just as in the Schwarzschild case) reduces the number of restrictions imposed by the field equations from a potential 10 to just 2.

It is often convenient to replace eqn (18.8) by that which results when we subtract eqn (18.9) from it:

$$\frac{\ddot{R}}{R} = -\frac{4\pi G}{3}\left(\rho + \frac{3p}{c^2}\right) + \frac{1}{3}\Lambda c^2. \tag{18.10}$$

One interesting fact can be read off at once from eqn (18.10). Suppose in a *contracting* universe ($\dot{R} < 0$) we have $\Lambda \leq 0$, and the 'strong energy condition' $c^2\rho+3p \geq 0$ holds. Then total collapse must follow. For with $\ddot{R} < 0$, $\dot{R}$ can only become more negative. Conversely, in such an *expanding* universe, there must have been a big bang in the past. This is an example of the Penrose–Hawking singularity theorem in action. (Cf. end of Section 17.4.)

18.2   What the field equations tell us. The Newtonian analogy

If we multiply the LHS of eqn (18.9) by R^3 and differentiate, we get $\dot{R}R^2$ times the LHS of eqn (18.8); this means

$$\left(\frac{8\pi G\rho}{3c^2}R^3\right)^{\cdot} = -\dot{R}R^2\left(\frac{8\pi Gp}{c^4}\right), \tag{18.11}$$

or, after a little manipulation,

$$(\rho c^2 R^3)^{\cdot} + p(R^3)^{\cdot} = 0. \tag{18.12}$$

Since the volume V of a small comoving ball of the substratum is proportional to R^3, (18.12) can also be written with V in place of R^3. If we then write $\rho c^2 = U$ for the total energy density, A for the surface area of the little ball and r for its radius, we obtain

$$dU = -p\,dV = -pA\,dr, \tag{18.13}$$

the usual equation of continuity (energy balance) in the absence of thermal flow. Of course, isotropy implies that there can be no thermal flow, or, in other words, that the motion must be adiabatic.

If the fluid obeys a simple equation of state,

$$p = wc^2\rho, \tag{18.14}$$

eqn (18.12) becomes

$$\frac{\dot{\rho}}{\rho} = -(1+w)\frac{(R^3)^{\cdot}}{R^3}, \qquad (18.15)$$

which integrates to

$$\rho \propto R^{-3(1+w)}. \qquad (18.16)$$

For pure radiation we have [cf. (18.2)] $w = \frac{1}{3}$, so that $\rho \propto R^{-4}$, while for pure dust ($w = 0$) we have $\rho \propto R^{-3}$. If dust and radiation co-inhabit a universe without interacting, each will satisfy an equation like (18.13) and thus also one like (18.16), so that $(\rho_{\text{dust}}/\rho_{\text{radiation}}) \propto R$. At present, it is estimated that radiation (mainly the 2.7 K background) accounts for only about one-thousandth of the total density. Accordingly, when the universe was about one thousandth as big as today, the radiation- and dust densities crossed over; before that, radiation was dominant. The cross-over time is seen to coincide (for no clear reason) approximately with the decoupling time (cf. Section 16.1), at which the universe was smaller by a factor $\sim 2.7\,\text{K}/3000\,\text{K}$.

As we mentioned, the field equations will specify the motion. But note how they also include an equation of continuity, which, too, is a separate assumption in Newton's theory. This was to be expected, since the GR field equations imply $T^{\mu\nu}{}_{;\nu} = 0$.

Now assume that we have a dust-dominated universe, and accordingly set $p = 0$. Then directly from (18.11) we see $\rho R^3 = \text{const}$ (conservation of mass!). It is convenient to define a constant C by the equation

$$\tfrac{8}{3}\pi G \rho R^3 =: C = \text{const}, \qquad (18.17)$$

so that eqn (18.9) reads

$$\dot{R}^2 = \frac{C}{R} + \frac{\Lambda c^2 R^2}{3} - kc^2. \qquad (18.18)$$

In the case of dust, eqns (18.17) and (18.18) together are fully equivalent to (18.8) and (18.9), *unless* $\dot{R} \equiv 0$. For we have just derived the former pair from the latter; and conversely, (18.17) and (18.18) together imply (18.9), while the result of differentiating R times (18.18) is $\dot{R}R^2c^2$ times (18.8). Equation (18.18) is known as *Friedman's differential equation*, after its discoverer. It is this equation that governs the cosmic motion.

In the cosmological context, the field equations determining the motion allow a Newtonian interpretation, and thus a degree of intuitive understanding that is often unavailable from the formalism alone. Let us consider the Newtonian motion of a small homogeneous ball of dust of radius aR and mass a^3M ($a = \text{const}$) under its own gravity. A particle on its surface (and thus its whole surface) experiences an acceleration

$$\ddot{R} = -\frac{GM}{R^2} \qquad (18.19)$$

towards the center (the as cancel out). Compare this with eqn (18.10), after replacing M by $\frac{4}{3}\pi R^3 \rho$. If $\Lambda = p = 0$, and given the right initial conditions, the ball fits onto the substratum! The comparison also shows that any uniform pressure p contributes an amount $3p/c^2$ to the effective gravitating density and thus (counter-intuitively) hastens the collapse or slows the expansion. There is no simple explanation for this term—it just comes out of the field equations; it does *not* represent the elastic energy stored in each volume element due to the compression, for that is already included in ρ.

But why does pressure have no direct repulsive effect? The reason is symmetry. Consider a uniformly *contracting* Newtonian (or Milnean) universe with gravity switched off. As the stars come into contact, will their mutual pressure slow the collapse? No. Each, being at rest in an inertial frame, finds itself suddenly pressed by stars from all sides. But where is it to move? So it, and all the other stars, must simply continue riding their respective inertial frames unbraked to the bitter end. In much the same way, even realistic universes are unaffected by pressure as such.

The next interpretation suggested by comparing (18.19) and (18.10) is that the cosmological term acts like a distance-proportional *repulsive* force $\frac{1}{3}\Lambda c^2$ per unit distance per unit mass [as, indeed, we have already noted in connection with modified Schwarzschild space, cf. after (14.23).] Equivalently, the cosmological term acts like an all pervasive *negatively* gravitating density of space, $-\Lambda c^2/4\pi G$. However, if we transpose the cosmological term $\Lambda g_{\mu\nu}$ in the field equation (18.4) [with (18.5)] to the RHS, it becomes equivalent to an additional energy tensor

$$T_{\mu\nu(\mathrm{vac})} = (c^4\Lambda/8\pi G)\,\mathrm{diag}(-1, -1, -1, 1) \qquad (18.20)$$

in the rest-LIF, sometimes regarded as the energy tensor of the vacuum. So the vacuum would have a *positive* energy density $c^2\rho = c^4\Lambda/8\pi G$ and *negative* pressure $p = -c^2\rho$, but an *effective* gravitating density $\rho + 3p/c^2 = -\Lambda c^2/4\pi G$ as before.

Now let us assume $p = \Lambda = 0$. We can integrate the Newtonian equation (18.19), after multiplying it by an integrating factor $2\dot{R}$, to find

$$\dot{R}^2 = \frac{2GM}{R} - \tilde{k}c^2, \qquad (18.21)$$

where the constant of integration is written as $-\tilde{k}c^2$ for ease of comparison with the Friedman differential equation (18.18). Since eqn (18.19) applies to balls of arbitrary radius aR and mass a^3M, so does eqn (18.21). Reference to (18.17) and (18.18) then shows that the ball (and all its layers) is interchangeable with a substratum ball if we choose $\tilde{k} = k$ and the instantaneous density compatibly. So the substratum (now identified with the dust) can be pictured as an aggregate of little balls moving Newtonianly!

After multiplying eqn (18.21) by $\frac{1}{2}a^2m$, it actually becomes the Newtonian energy equation for a particle of mass m on the surface of the ball of radius aR:

$$(\text{kinetic energy}) + (\text{potential energy}) = -\tfrac{1}{2}a^2m\tilde{k}c^2. \qquad (18.22)$$

This is the physical interpretation of the Friedman differential equation. If the ball expands and $\tilde{k} = -1$, the particle gets to infinity with energy to spare and constant velocity ac; if $\tilde{k} = 0$ the particle has the exact escape velocity and gets to infinity with zero velocity; and if $\tilde{k} = 1$, the kinetic energy runs out at finite distance and the particle falls back. In terms of the universe, the three cases $k = -1, 0, 1$ correspond, respectively, to substrata that (i) ultimately approximate the Milne substratum ($\dot{R} = 1$), (ii) ultimately come to rest in Minkowski space, and (iii) expand and recollapse. In a quite un-Newtonian way, however, the GR field equations impose negative, zero, or positive curvature on the cosmic sections in these respective cases (in the absence of Λ).

For later use we shall here write down the equations corresponding to (18.17) and (18.18) for radiation-dominated universes—as, indeed, ours is believed to have been in the beginning. Instead of $\rho \propto R^{-3}$ we then have $\rho \propto R^{-4}$ [cf. after (18.16)], and we define a constant D by the equation

$$\tfrac{8}{3}\pi G \rho R^4 =: D = \text{const}, \tag{18.23}$$

so that eqn (18.9) reads

$$\dot{R}^2 = \frac{D}{R^2} + \frac{\Lambda c^2 R^2}{3} - kc^2 \tag{18.24}$$

instead of (18.18). As we trace the model back to the vicinity of the big bang, the last two terms both here and in (18.18) become negligible and we find, respectively,

$$R \propto t^{2/3} \quad \text{(matter)}, \tag{18.25}$$

$$R \propto t^{1/2} \quad \text{(radiation)}. \tag{18.26}$$

18.3   The Friedman models

A. Introduction

We shall now discuss the solutions of the Friedman differential equation (18.18), with a view to obtaining and classifying *all* GR 'dust' universes that are homogeneous and isotropic. These are the *Friedman models*. We shall pretend, although we know this to be false, that the substratum behaves like dust all the way back to the big bang. But it was only for about the first 10^7 years (less than a thousandth of the total) that radiation played a non-negligible gravitating role, and in many contexts we can ignore this.

It will be convenient to employ *units making $c = 1$*, so that Friedman's equation reads

$$\dot{R}^2 = \frac{C}{R} + \frac{\Lambda R^2}{3} - k =: F(R) \quad (C = \tfrac{8}{3}\pi G \rho R^3). \tag{18.27}$$

The symbol $F(R)$ is simply an abbreviation for the three terms preceding it; the parenthesis is a repeat of eqn (18.17), whose sole function is to define C. We can *formally* write down the solution at once by quadrature,

$$t = \int \frac{dR}{\sqrt{F}}, \tag{18.28}$$

and we could proceed to the full solution by using elliptic functions. In special cases the solution can be obtained in terms of elementary functions, as we shall see. However, in the general case it will be enough for us, and more instructive, to give a qualitative rather than an exact analysis. We preface our discussion with some general remarks:

1. A Friedman model is uniquely determined by a choice of the three parameters C, Λ, k, an 'initial' value $R(t_0)$, and the sign of $\dot{R}(t_0)$. For eqn (18.27) then gives $\dot{R}(t_0)$, and thus, in principle, the solution can be iterated uniquely—unless $\dot{R}(t_0) = 0$. In that one case we must fall back on the parent equation (18.8) to get $\ddot{R}(t_0)$, and then again the solution can in principle be iterated uniquely.

2. Since $R = 0$ is a singularity of the Friedman equation, no regular solution $R(t)$ can pass *through* $R = 0$. Regular solutions are therefore entirely positive or entirely negative. Moreover, the solutions occur in matching pairs $\pm R(t)$; this is because, for physical reasons, we must insist on $\rho \geq 0$, which implies that the sign of C must be the same as that of R—but then eqn (18.27) is unaltered by the change $R \mapsto -R$, and this proves our assertion. Since only R^2 occurs in the FRW metric, we therefore exclude no solutions by insisting that $R \geq 0$ and $C \geq 0$.

3. Equation (18.27) also enjoys invariance under the changes $t \mapsto -t$ or $t \mapsto t + \text{const}$. The first implies that to every solution $R(t)$ there corresponds the same solution run backwards, $R(-t)$, as was to be expected from the time-symmetry of the underlying laws; and the second implies that every solution $R(t)$ represents a whole set of solutions $R(t + \text{const})$, differing only in the zero point of time. Bearing these properties in mind, we shall so normalize our solutions that of the pair $R(\pm t)$ we exhibit the expanding one in preference to the collapsing one, and of the set $R(t + \text{const})$ we exhibit, if there is one, that member which passes through the origin $t = 0, R = 0$.

4. As a consequence of time-reversibility and (1) above, every 'oscillating' Friedman model is symmetric about its point of maximal extension, where $\dot{R}_{\max} = 0$; for at that point the model and its time-reversal have identical 'initial' conditions. (The term 'oscillating' here is simply synonimous with 'recollapsing'; genuine oscillations in the sense of repeating the cycle beyond the singularity are not expected.)

B. The static models

It is well to clear out of the way the static models first; that is, those which have $\dot{R} \equiv 0$. As we remarked after (18.18), this is the exceptional case in which Friedman's equation is insufficient and both its parent equations (18.8) and (18.9) must be used. These permit $\dot{R} \equiv 0$ provided

$$\frac{k}{R^2} = \Lambda = 4\pi G\rho \quad (\approx 10^{-57}\,\text{cm}^{-2}). \tag{18.29}$$

(In parentheses we give the value of $4\pi g\rho$ corresponding to $\rho = 10^{-30}$ g/cm^3, which illustrates the typical smallness of Λ and of the curvature k/R^2.) Equations (18.29) of course imply $\rho = $ const, and for a realistic solution we need $\rho > 0$, and thus $k = +1$. This gives the so-called *Einstein universe*, the very first GR model universe to be proposed (by Einstein, in 1917). The spatial part of the model is the static 3-sphere we discussed in Section 8.2. To its inventor at the time it seemed to have every desirable feature. Einstein apparently did not realize that it was unstable: the slightest contraction will make it collapse, the slightest expansion becomes runaway; for the former *increases* the gravitational force on each particle towards any center, while *decreasing* the Λ-repulsion; and the latter does the opposite.

It is of interest to note that, without the Λ term, a non-empty universe ($\rho > 0$) cannot be held in static equilibrium by a positive pressure. Eqn (18.10) rules out this possibility.

The only other way of satisfying (18.29) (less physical but still acceptable as a limiting case) is $k = \Lambda = \rho = 0$, $R = $ any constant. The transformation $Rx \mapsto x$, etc. in (18.3) then leads to the Minkowski metric, and the model consequently represents a static, non-gravitating, substratum filling an infinite inertial frame.

C. The empty models

Models with zero density (or, equivalently, with 'gravity switched off', $G = 0$), like Milne's or the above static Minkowski model, are unrealistic but they provide important limiting cases. We therefore classify them next. Setting $C = 0$ reduces (18.28) to the elementary form (unless $\Lambda = k = 0$)

$$t = \int \left(\tfrac{1}{3}\Lambda R^2 - k\right)^{-1/2} dR, \tag{18.30}$$

which has the following solutions [apart from (a), which we get directly from (18.27)]:

$$
\left.
\begin{aligned}
&\text{(a)} \quad \Lambda = 0, \quad k = 0, \quad R = \text{arbitrary constant} \\
&\text{(b)} \quad \Lambda = 0, \quad k = -1, \quad R = t \\
&\text{(c)} \quad \Lambda > 0, \quad k = 0, \quad R = \exp(t/a) \\
&\text{(d)} \quad \Lambda > 0, \quad k = 1, \quad R = a\cosh(t/a) \\
&\text{(e)} \quad \Lambda > 0, \quad k = -1, \quad R = a\sinh(t/a) \\
&\text{(f)} \quad \Lambda < 0, \quad k = -1, \quad R = a\sin(t/a)
\end{aligned}
\right\} \quad a = |3/\Lambda|^{1/2}.
$$
$$\tag{18.31}$$

Models (a) and (b) we already know: (a) is the empty static model and (b) is Milne's model; these have identical spacetime backgrounds (M^4) but different substrata (that is, motion patterns). The same is true also of the three models (c), (d), and (e), all of which have de Sitter space D^4 (cf. Section 14.5) for their spacetime background. (This situation arises *only* with the empty models; non-empty models with different substrata have different spacetimes.) That models (c), (d), and (e) *must* share de

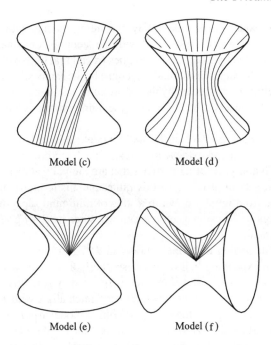

Model (c)   Model (d)

Model (e)   Model (f)

Fig. 18.1

Sitter spacetime is a priori clear, since, as we saw in Section 14.5, D^4 is the unique spacetime that satisfies Einstein's vacuum field equations with $\Lambda > 0$ and is spatially isotropic about every point; and each of the three models in question has these properties. The hyperboloids in the first three diagrams of Fig. 18.1 all represent de Sitter space with two dimensions suppressed, just as in Fig. 14.1. As we mentioned then, geodesic worldlines correspond to timelike central sections of the hyperboloid. In the notation of Fig. 14.1, it can be shown that the substrata of models (c), (d), and (e) correspond, respectively, to plane sections containing one of the following lines: the line $t + x = 0, r = 0$; the t-axis; and the x-axis. Lorentz transformations in the embedding 3-dimensional Minkowski space can carry each of these fundamental worldlines into the 'central' one in models (c) and (e) (cf. Exercise 14.12), which once more shows the equivalence of all these lines.

Model (c) is the well-known *de Sitter universe*. Kinematically it is identical with the steady state universe (16.31). Thus, for example, it has an event horizon. Though unrealistic because of its zero density, it constitutes a limit to which *all* indefinitely expanding models with $\Lambda > 0$ must tend. For indefinite expansion ($R \to \infty$) in a general model causes the RHS of eqn (18.27) to be ultimately dominated by the Λ term, which leads to $R \sim \exp(\frac{1}{3}\Lambda)^{1/2}t$ (a multiplicative constant can be absorbed by a time-translation); and the curvature k/R^2 of the model and its density $3C/8\pi G R^3$ become ultimately small, so that $k = 0$ and $\rho = 0$ are good approximations. As a consequence, for example, *all* indefinitely expanding models with $\Lambda > 0$ possess an event horizon. The horizon associated with the 'central' fundamental observer in diagram (c) is indicated by a pair of dotted lines (recall that light paths correspond to generators), and in the diagrams (d) and (e) it must be the same, since in the absence of density the spacetime is unaffected by the substratum. The significance of the horizon as the light front that reaches the observer in the infinite future is well brought out by the diagram.

Model (d) is the analog in D^4 of the static model (a) in M^4—but it is static only for an instant, corresponding to the waist circle of the hyperboloid. In contrast to M^4, if you fill D^4 with non-gravitating particles that are mutually at rest, this state cannot last: Λ-repulsion sets in at once. In this quite unrealistic model the fundamental particles collapse from infinity down to a state of minimum separation and then go back to infinity again. We call it a *catenary* model, since its $R - t$ graph has the shape of a suspended chain.

Model (e) is the analog of Milne's model in D^4. Here, too, it can alternatively be regarded as an expanding ball of test dust ('horizontal' sections), bounded by a spherical light front. Its expansion, however, is speeded by Λ repulsion. Both models have $k = -1$ and in the beginning they look very much alike, with $R \sim t$.

Model (f), in the same way, is the analog of Milne's model in *anti*-de Sitter space $\tilde{D}^4$ (cf. Section 14.6). Slowed by Λ *attraction*, this test-dust ball finally stops expanding and recollapses. Its substratum corresponds to central sections of the appropriate hyperboloid containing an axis perpendicular to the symmetry axis.

Finally, it is evident from inspection of the diagrams, that models (c), (e), and (f) are all non-maximal (just like Milne's model) in the sense of there being more space than substratum.

D. The three non-empty models with $\Lambda = 0$

For many cosmologists and for many years, these three models were the *only* models ever considered seriously. The vanishing of Λ was taken so much for granted that it was not even stated as an assumption. Putting $\Lambda = 0$ and $C \neq 0$ in Friedman's equation (18.27) makes that equation quite straightforward to solve. For $k = 0$ we immediately find

$$R = \left(\tfrac{9}{4}C\right)^{1/3} t^{2/3} \quad (k = 0). \tag{18.32}$$

This model is called the *Einstein–de Sitter universe*. It is the one that *just* has the escape energy at the big bang and makes it to infinity. Its parameters satisfy some simple relations. For example, from the parenthesized equation in (18.27), it follows that

$$\rho = \frac{1}{6\pi G t^2}. \tag{18.33}$$

And differentiating (18.32) leads to $\dot{R}/R = \tfrac{2}{3}t^{-1}$, so that

$$t_0 = \tfrac{2}{3}H_0^{-1}. \tag{18.34}$$

With $h = 0.7$ this would imply [cf. (16.3)] $t_0 \approx 14 \times 10^9$ years and thus, from (18.33), $\rho \approx 0.9 \times 10^{-29}$ g/cm^3, somewhat on the high side [cf. (16.6)].

When $k \neq 0$, the solutions are as follows:

$$\left.\begin{array}{ll} k = 1: & t = C[\sin^{-1}\sqrt{X} - \sqrt{(X - X^2)}], \\[2mm] k = -1: & t = C[\sqrt{(X + X^2)} - \sinh^{-1}\sqrt{X}], \end{array}\right\} \quad (X = R/C). \quad (18.35)$$

By putting $X = \sin^2(\chi/2)$ and $X = \sinh^2(\chi/2)$, respectively, we can also express these solutions parametrically:

$$k = 1: \quad R = \tfrac{1}{2}C(1 - \cos\chi), \quad t = \tfrac{1}{2}C(\chi - \sin\chi), \quad (18.36)$$

$$k = -1: \quad R = \tfrac{1}{2}C(\cosh\chi - 1), \quad t = \tfrac{1}{2}C(\sinh\chi - \chi). \quad (18.37)$$

The $k = -1$ model is important as a serious contender for the actual universe, but not very remarkable. We see from (18.37), since $\sinh\chi \gg \chi$ for large χ, that $R \sim t$ and so the model ultimately resembles the Milne universe. (Except for one difference: no matter how old this universe gets, each fundamental particle is still surrounded *somewhere* by its 'creation light front' —the particle horizon.)

Though less likely to represent the actual universe (according to present data), the $k = 1$ model is more interesting in itself. Geometers will recognize in eqn (18.36) the parametric equation of a cycloid; that is, the curve traced out by a point on the rim of a rolling circle. Here the radius of that circle is $\tfrac{1}{2}C$, and χ is the angle through which it turns. (The graphs of all three models are shown in Fig. 18.2.)

The full range of χ for the cycloidal universe is 0 to 2π; its maximal radius is given by $R_{\max} = R(\chi = \pi) = C$, and its total duration by $t_{\text{tot}} = t(\chi = 2\pi) = \pi C$. Let us define the total mass M of this universe as the product of the density and the volume at any cosmic instant. The volume is that of a 3-sphere of radius R, which is $2\pi^2 R^3$ (as we have seen in Section 8.2). Consequently we have, using the definition of C in (18.27),

$$M := 2\pi^2 R^3 \rho = 3\pi C/4G. \quad (18.38)$$

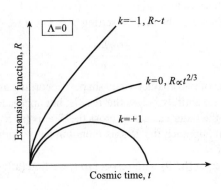

Fig. 18.2

As has been stressed by Lynden-Bell, this relation makes the cycloidal universe 'Machian' in a sense that the $k = 0$ and $k = -1$ universes are not. For if $\Lambda = 0$, and a total finite mass M is decided on, the model is unique: a finite mass always implies $k = +1$ and then M fixes C and thus the entire cycloidal spacetime. In the limit of zero mass, there is *no* spacetime. And even if $\Lambda \neq 0$, the situation is essentially the same for any closed universe. Looking at this slightly differently, we are struck by a powerful constraint that the field equations impose on a closed universe: whereas a Newtonian ball of given mass M can be 'exploded' so as to recollapse at any desired future time, the same is evidently not true for *cycloidal* universes. Here the original kinetic energy is prescribed by the total mass: the duration *must* be $\frac{4}{3}GM$. (The general situation, when $\Lambda \neq 0$, is similar.)

Note that for small R the C term on the RHS of the Friedman equation (18.27) *always* predominates, whence *all* non-empty big-bang models share the behavior $R \sim (\frac{9}{4}C)^{1/3}t^{2/3}$ near $t = 0$. In particular, therefore, it can be seen that they all have a particle horizon. Moreover, because of the symmetry of oscillatory models about the point of maximal extension, and because a particle horizon becomes an event horizon under time-reversal, it is seen that all non-empty oscillatory models have *both* a particle horizon and an event horizon.

E. Non-empty models with $\Lambda \neq 0$

If we allow arbitrary values of Λ, the variety of possible solutions increases substantially. To obtain a qualitative overview of this class of solutions of the Friedman differential equation (18.27), we begin by rewriting it in the form

$$\dot{R}^2 + k = \frac{C}{R} + \frac{\Lambda R^2}{3} =: f(R, \Lambda), \tag{18.39}$$

and, of course, we assume $C > 0$. The function $f(R, \Lambda)$, defined by the last equation, will serve as a kind of potential, playing a role analogous to that of $V(r)$ in the qualitative solution of the Schwarzschild orbit equation (11.34). Figure 18.3 shows some level curves of this function, $f(R, \Lambda) = m(m = -1, 0, \frac{1}{2}, 1, \frac{3}{2}, \dots)$, drawn for some fixed C value. They are obtained by setting $f = m$ in (18.39) and solving for Λ:

$$\Lambda = \frac{3(mR - C)}{R^3}. \tag{18.40}$$

These curves have one of two characteristic shapes, according as $m \leq 0$ or $m > 0$. In the former case, they lie entirely below the R axis, but approach that axis monotonically from below. In the latter case, they cross the R axis (at $R = C/m$), proceed to a maximum, and then approach the R-axis monotonically from above. In all cases, $\Lambda \sim -3C/R^3$ near $R = 0$.

If we differentiate eqn (18.39) with respect to t, we find (after dividing by $\dot{R}$)

$$2\ddot{R} = \frac{\partial f}{\partial R} = -\frac{C}{R^2} + \frac{2\Lambda R}{3}, \tag{18.41}$$

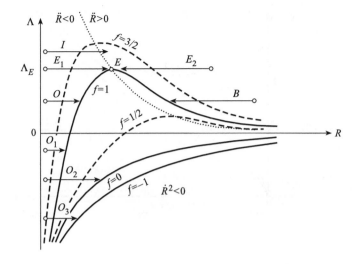

Fig. 18.3

whence the locus of $\ddot{R} = 0$ is given by

$$\Lambda = \frac{3C}{2R^3}. \tag{18.42}$$

This is shown in Fig. 18.3 as the dotted line. Above this locus we have $\ddot{R} > 0$, below it, $\ddot{R} < 0$. It will be noticed that this locus passes through the maxima of the level curves. This is no accident; differentiating $f(R, \Lambda) = m$, we find $d\Lambda/dR = -(\partial f/\partial R)/(\partial f/\partial \Lambda)$, and so the maxima occur where $\partial f/\partial R = 0$, which, by (18.41), is where $\ddot{R} = 0$.

The maximum point of the level curve $f = 1$ (marked E in the diagram) has 'coordinates'

$$R = \frac{3C}{2}, \qquad \Lambda = \frac{4}{9C^2} =: \Lambda_E; \tag{18.43}$$

that is, precisely the values corresponding to the Einstein static universe [cf. (18.29) and (18.27) (ii)]—hence the notation Λ_E.

Now each Friedman model has $\Lambda = $ const. We can therefore obtain solutions by choosing any physically possible starting point ($\dot{R}^2 \geq 0$!) in Fig. 18.3 and proceeding horizontally. The level curves tell us the value of $\dot{R}^2$ (once we have chosen k), and thus the slope of the solution curve, up to sign. The level curves $f = -1, 0, 1$ are of particular importance: for the respective cases $k = -1, 0, 1$, they are the locus of $\dot{R}^2 = 0$ and they constitute the upper boundaries of the 'prohibited' regions where $\dot{R}^2 < 0$.

If we start at $R = 0$ and $\Lambda < 0$, we necessarily get an *oscillating* universe. According as we choose $k = 1, 0,$ or -1, the critical level curves will be $f = 1, 0,$ or -1, respectively; these cannot be crossed, since $\dot{R}^2$ is negative beyond them. Our

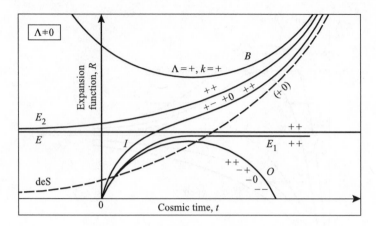

Fig. 18.4

solution curve $R(t)$ therefore starts with infinite slope $\dot{R}$, which gradually decreases and becomes zero at the critical curve. However, at that point in the diagram, $\ddot{R} < 0$, and thus $\dot{R}$ must go on decreasing; since we cannot cross the critical level curve, we get the decreasing half of an oscillating model by going back along the same horizontal $(O_1, O_2, \text{or } O_3)$ to $R = 0$, this time choosing the negative value of $\dot{R}$. All the corresponding solution curves are shown schematically in Fig. 18.4 as O.

If we start at $R = 0$ and $\Lambda > 0$, and choose $k = 0$ or -1, we do not get stopped by a critical level curve; $\dot{R}$ at first decreases from its original infinite value to a minimum at the $\ddot{R} = 0$ locus and then increases again. The result is an *inflexional* universe (see I in Figs 18.3 and 18.4).

The choice $k = 1$ and $\Lambda > 0$ yields a richer variety of solutions. The critical level curve is now $f = 1$. For starting points at $R = 0$ in Fig. 18.3, we get inflexional universes (I) if $\Lambda > \Lambda_E$, and oscillatory universes (O) if $\Lambda < \Lambda_E$. If $\Lambda = \Lambda_E$, we approach the critical curve at its maximum E: there, all derivatives of R vanish, and the solution curve flattens out into a straight line. This corresponds to a big bang universe which approaches the Einstein static universe asymptotically (E_1 in Figs 18.3 and 18.4). By choosing $\Lambda > \Lambda_E$ but *close* to Λ_E, we can construct inflexional models with arbitrarily long quasistationary periods, as in Fig. 17.2. If we start on the other side of the 'hump' of the critical curve (that is, with large R and $\Lambda < \Lambda_E$) and proceed horizontally to the left in Fig. 18.3, $|\dot{R}|$ decreases from large values to zero (and $\ddot{R} > 0$), and we get one half of a *rebounding* universe (B); the other half corresponds to going back along the same horizontal with the opposite sign of $\dot{R}$. Like oscillatory universes, rebounding universes are symmetric about their stationary point, and for the same reason. Observe that for identical C, Λ (and $k = 1$) we can get two different types of model, depending on the initial value of R. Similarly, if for $\Lambda = \Lambda_E$ we approach the critical point E from the right, we get a universe which decreases from infinite extension and approaches the Einstein static universe asymptotically; we prefer to exhibit this model in time-reversed (expanding!) form: E_2 in Figs 18.3 and 18.4. Thus, for $\Lambda = \Lambda_E$ (and $k = 1$) there are *three* different models, including the Einstein static universe.

Figure 18.4 shows the various solution curves schematically; they are labeled according to the signs of Λ and k, in this order. The dashed curve, contrary to our premise $C \neq 0$, represents the empty de Sitter universe; it is included here to show its role as an 'asymptote' to all indefinitely expanding models with $\Lambda > 0$. It can be seen from both Figs 18.3 and 18.4 that the Einstein universe is unstable: a slight perturbation will set it on the course E_1 (in collapsing form), or on the course E_2 (expanding). In fact, the expanding universe E_2 is sometimes regarded as the result of a disturbed Einstein universe, and on this interpretation it is called the *Eddington–Lemaître universe*.

As we have seen, all models with $C = 0$ or with $\Lambda = 0$ allow representation by elementary functions. In fact, the same is true also for models with $k = 0$, and it is sometimes quite useful to have these formulae:

$$R^3 = (3C/2\Lambda)(\cosh(3\Lambda)^{1/2}t - 1) \quad (\Lambda > 0, \ k = 0), \tag{18.44}$$

$$R^3 = (3C/|2\Lambda|)(1 - \cos|3\Lambda|^{1/2}t) \quad (\Lambda < 0, \ k = 0). \tag{18.45}$$

The corresponding models are of type I and O, respectively. (See also Exercise 18.3)

18.4 Once again, comparison with observation

Now that we have subjected the FRW metric to the field equations (assuming 'dust' universes, as we shall continue to do), we still find ourselves left with an embarrass-ingly large choice of possibilities. But whereas feasible observations have little impact on the unrestricted FRW models, they can, in principle, decide between the dynam-ically possible ones. To show this, it will be convenient to work with the following three functions of cosmic time:

$$H = \frac{\dot{R}}{R}, \qquad \Omega = \frac{8\pi G\rho}{3H^2}, \qquad q = -\frac{\ddot{R}}{RH^2}. \tag{18.46}$$

We have met all of these before: H is the Hubble parameter and has dimension $(\text{time})^{-1}$; Ω is the dimensionless density parameter, and q is the dimensionless deceleration parameter. In principle, the present values of these parameters can be determined by observation: H and q from relations such as (17.26), and Ω from estimates of ρ.

In terms of these functions we can now rewrite (i) eqn (18.17), (ii) eqn (18.10) (with $p = 0$), and (iii) eqn (18.8) minus three times eqn (18.9) (*we are still working* *in units that make $c = 1$):*

(i) $\Omega = C/H^2 R^3$; $\qquad\qquad$ $C = \Omega_0 H_0^2 R_0^3$ $\qquad\qquad$ (18.47)

(ii) $\frac{1}{2}\Omega - q = \Lambda/3H^2$; $\qquad$ $\Lambda = 3H_0^2\left(\frac{1}{2}\Omega_0 - q_0\right)$ $\qquad$ (18.48)

(iii) $\frac{3}{2}\Omega - q - 1 = k/H^2 R^2$; $\quad$ $k = H_0^2 R_0^2\left(\frac{3}{2}\Omega_0 - q_0 - 1\right)$.   (18.49)

The second entry in each line solves for one of the constants C, Λ, k; and since these *are* constants, we can evaluate their representations at *any* time, in particular at the present time $t = t_0$; that is the significance of the suffixes zero.

From the second entries in (18.47)–(18.49) it is seen that if we know H_0, Ω_0, and q_0, we can first determine Λ and k/R_0^2 (which yields k *and* R_0 unless $k = 0$, in which case R_0 is arbitrary anyway), and finally C. And, as we have seen, Λ, k, C, and R_0 determine a unique Friedman model. Unfortunately this persuasively simple scheme does not work out in practice: the uncertainties in the current determinations of Ω_0 and q_0, and, to a lesser extent, of H_0, are so great that no direct conclusions are possible from these equations. But Robertson had the idea of coupling these three uncertain data with a fourth: t_0, the age of the universe. Some models can then be eliminated simply because they are too young or too old.

Before we go into this, we digress briefly to discuss conditions for the universe to be closed ($k = 1$). One sufficient but by no means necessary condition would evidently be $q_0 < -1$ [cf. (18.49)]. Next, from (18.48), observe that

$$\Lambda = 0 \quad \Leftrightarrow \quad q_0 = \frac{1}{2}\Omega_0, \qquad\qquad (18.50)$$

and then (18.49) yields

$$k \gtreqless 0 \quad \Leftrightarrow \quad \Omega_0 \gtreqless 1 \quad (\Lambda = 0). \qquad\qquad (18.51)$$

So when $\Lambda = 0$ was assumed as a matter of course, $\Omega_0 = 1$ characterized the critical density beyond which the universe would recollapse, given its present rate of expansion; and via the field equations, this is then equivalent to positive curvature. But suppose $\Lambda \neq 0$. In that case one can define a dimensionless 'density parameter of the vacuum':

$$\Omega(\Lambda) := \frac{\Lambda c^2}{3H^2} = \frac{8\pi G\rho_{\text{vac}}}{3H^2} \qquad\qquad (18.52)$$

(in full units), where ρ_{vac} is the density corresponding to the cosmological term if one regards it as the energy tensor of the vacuum, as in eqn (18.20). From (18.48) we then have

$$\Omega(\Lambda) = \frac{1}{2}\Omega - q, \qquad\qquad (18.53)$$

so that in (18.49) we can eliminate q_0 in favor of $\Omega_0(\Lambda)$:

$$k = H_0^2 R_0^2[\Omega_0 + \Omega_0(\Lambda) - 1]. \qquad\qquad (18.54)$$

The condition for closure is now

$$\Omega_0 + \Omega_0(\Lambda) > 1. \tag{18.55}$$

But this is no longer necessarily coupled with recollapse, nor is $k < 0$ necessarily coupled with indefinite expansion. Reference to Fig. 18.4 shows that with $\Lambda \neq 0$ a closed universe may well expand indefinitely while an open one may recollapse.

We now return to Robertson's idea of using the age of the universe as an additional model parameter. We reject as unrealistic the few non-big-bang models shown in Fig. 18.4, so that we *can* speak of a present age t_0. Let us begin by substituting from (18.47)–(18.49) into Friedman's differential equation (18.27). This yields

$$\dot{y}^2 = H_0^2\{\Omega_0 y^{-1} + (\tfrac{1}{2}\Omega_0 - q_0)y^2 + (1 + q_0 - \tfrac{3}{2}\Omega_0)\}, \quad y = R/R_0. \tag{18.56}$$

Setting $R(0) = 0$, we then have

$$H_0 t_0 = \int_0^{t_0} H_0 \, dt = \int_0^1 \{\ \}^{-1/2} dy =: \phi(\Omega_0, q_0), \tag{18.57}$$

where the empty brace denotes the braced expression of (18.56). The values of this integral for different choices of Ω_0 and q_0 can be machine calculated, and thus the relation between $H_0 t_0$, Ω_0, and q_0 established (cf. Table 18.1).

We knowingly commit a small error in computing t_0 by treating the universe as 'dust' all the way back to the big bang, instead of allowing for the predominance of radiation during the first half-million years or so. But we can afford an error of 0.01 percent.

Since eqn (18.56) can be written as $\dot{R}^2/\dot{R}_0^2 = \{\}$, and since the contents of the brace are dimensionless, that equation is invariant under independent scale changes in R and t. It accordingly determines $R(t)$ only up to such rescalings. But eqn (18.49) determines $\dot{R}_0$ uniquely (unless $k = 0$), and so the scale changes in R and t must be the *same* (unless $k = 0$). Therefore, since $H_0 t_0$, Ω_0 and q_0 are related by (18.57), any two of these parameters determine a Friedman model up to scale; and more cannot

Table 18.1 $H_0 t_0$ as a function of q_0 and Ω_0

q_0 \ Ω_0	0.1	0.2	0.3	0.4	0.5
1.5	0.69	0.67	0.66	0.64	0.63
1.0	0.74	0.72	0.70	0.69	0.67
0.5	0.81	0.78	0.76	0.74	0.72
0.0	0.91	0.87	0.83	0.81	0.79
−0.5	1.07	1.00	0.95	0.91	0.88
−1.0	1.44	1.26	1.16	1.09	1.04
−1.5			1.99	1.58	1.40

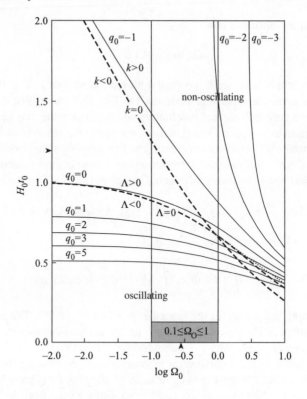

Fig. 18.5

be expected of dimensionless parameters. Any additional dimensional datum like H_0 or t_0 or ρ_0 will then determine the scale and thus the full model.

In the *phase diagram*, Fig. 18.5, where every point corresponds to the present state of a unique Friedman big-bang model (up to scale), we could have chosen any two of the three parameters $H_0 t_0$, Ω_0, q_0 as 'Cartesian coordinates', marking the third in the form of contour lines. The choice $H_0 t_0$ and $\log \Omega_0$ turns out to be convenient. There are three important demarcation lines in this diagram: (i) that which separates $\Lambda > 0$ from $\Lambda < 0$ models, (ii) that which separates $k > 0$ from $k < 0$ models, and (iii) that which separates oscillating from non-oscillating models. As one sees from (18.48), the first of these, the locus $\Lambda = 0$, corresponds to the curve

$$\tfrac{1}{2}\Omega_0 - q_0 = 0. \tag{18.58}$$

The second, the locus $k = 0$, by (18.49) corresponds to the curve

$$\tfrac{3}{2}\Omega_0 - q_0 - 1 = 0. \tag{18.59}$$

As for the demarcation between oscillating and non-oscillating models, we see from Fig. 18.3 that for $k = 0$ or $k = -1$, it is simply $\Lambda = 0$ and thus the curve (18.58).

But for $k = 1$, the required boundary is $\Lambda = \Lambda_E = (4/9)C^{-2}$. Substituting these values for Λ and k into eqns (18.48) and (18.49), and then eliminating R_0 from eqns (18.47)–(18.49), we find the following equation for this locus:

$$27\left(\tfrac{1}{2}\Omega_0 - q_0\right)\Omega_0^2 = 4\left(\tfrac{3}{2}\Omega_0 - q_0 - 1\right)^3. \tag{18.60}$$

It is represented by the dotted line in the diagram. At $k = \Lambda = 0$ *all* the above loci meet; this point is given by $\Omega_0 = 1, q_0 = \tfrac{1}{2}, H_0 t_0 = \tfrac{2}{3}$ and corresponds to the Einstein–de Sitter universe (18.32). So the boundary between oscillating and non-oscillating models runs along the curve $\Lambda = 0$ up to where that curve intersects the locus $k = 0$, and then proceeds independently into the $k, \Lambda > 0$ domain. Models lying on or above this boundary do not oscillate, while those below do. Each point on the dotted line represents an E_1 universe (cf. Figs 18.3 and 18.4), while each point on the other branch of the boundary ($\Lambda = 0$) represents a model of type (18.37). For each oscillating model there always are of course, *two* cosmic times (symmetric with respect to the time of maximum extension) that correspond to the same pair of values Ω_0 and q_0; our phase diagram gives the earlier time, corresponding to the expansive period. For it is based on the *positive* root in (18.57), namely on $H_0 > 0$, the one *certain* empirical fact.

Observations can set limits on the possible values of the parameters $H_0 t_0$, Ω_0, and q_0. Together these limits determine a patch of possible models in the diagram. Unfortunately there is still enough disagreement among observers (and theoreticians) to make that patch rather large and inconclusive. Possibly the values of H_0, t_0, and q_0 may reach reasonable and unchallenged accuracy within the next two decades, so that at least we may know on which side of the various demarcation lines our universe lies. Meanwhile we shall here go out on a limb, and determine (just by way of illustration) a plausible model. We take as 'most likely' for $H_0 t_0$ the value 1.2 (compounded of $h = 0.7$ and $t_0 = 17 \times 10^9$ years) and most likely for Ω_0 the value 0.3. We have marked these values with little black arrows in the diagram. The resulting universe is closed ($k > 0$), accelerating ($q_0 < 0$), and indefinitely expanding. Positive acceleration would be consistent with the recent observations on supernovae that we alluded to earlier, and the finiteness of the universe ($k > 0$) would be reassuring from a philosophical point of view.

However, the mere acceleration ($q_0 < 0$) of the universe—and thus its indefinite future expansion—seems almost assured by the already available H_0, t_0, Ω_0 data, quite apart from the direct supernova evidence. Fig.18.4 (or Table 18.1) shows that, given the almost certain inequality $\Omega_0 > 0.1$, all we need for $q_0 < 0$ is $H_0 t_0 \gtrsim 0.91$— for example, $h > 0.55$, $t_0 > 11.7 \times 10^9 y$. But it takes larger values of H_0, t_0, Ω_0 to bring us into $k > 0$ territory.

18.5 Inflation

In the last two sections of this book we report on two rather speculative topics, which the reader (according to taste) can either accept or reject. But since they form part of the present-day cosmological debate, it will be well to be informed about them. The first of these topics is *inflation*. It affects only the first 10^{-31} s (!!) of the universe. Thereafter it agrees kinematically with the standard FRW cosmology, except for one important detail: if true, the size of the particle horizons (creation light fronts) around each fundamental particle, both now and at all earlier times, would be many, many orders of magnitude larger than without it. And this is one of its greatest attractions. But it also has implications for cosmogony.

Though the basic idea, like most novel ideas, had precursors (Zeldovich, Starobinsky, Sato, etc.), inflation in its modern form was first put forward by Guth in 1980. Since then many cosmologists (especially those coming to cosmology from particle physics) have strongly rallied to this theory, even though it is still very largely based on pure hypothesis. Others, perhaps remembering the equally strong attachment that many people (possibly they themselves) felt for the equally hypothetical steady state theory during the fifties, have remained more skeptical.[1]

Inflationary cosmology is based on not yet fully established 'grand unified theories' (or GUTs). According to these theories, the quantum vacuum is stable at the highest temperatures, but then, some 10^{-37} s after the big bang, when the temperature had dropped to a critical value of $\sim 10^{27}$ K, the vacuum became unstable but still highly energetic. This 'false' vacuum is described by an energy tensor having precisely the form of the cosmological term in Einstein's field equations (but with a huge Λ), this being the only form unaffected by local Lorentz transformations. (The vacuum must look the same in all LIFs.) At this stage, or soon thereafter, the vacuum energy begins to dominate that of any other matter or radiation present, and causes a stupendous expansion of the universe, by a factor of $\sim 10^{43}$, in a mere 10^{-35} s. The expansion is exponential, just like that of the de Sitter model [cf. (18.31)(c)], and for the same reason, namely the (here temporary) presence of a Λ term in the field equations, albeit with a different interpretation. The positive vacuum density corresponding to the Λ term remains constant throughout this expansion [cf. (18.52)]. As a consequence, the matter-energy content of each comoving volume increases in proportion to that volume. In Guth's phrase, this was 'the ultimate free lunch'. In inflationary cosmology the big bang itself was a mere whimper, most of the matter-energy being created during the inflationary phase. When this phase comes to an end, the false vacuum becomes a real vacuum and its energy is converted into radiation. From here on the universe is kinematically and dynamically indistinguishable from one that could have been created by a standard big bang of the right strength. (See Fig. 18.6, which is obviously not drawn to scale, and where BI and EI stand for 'beginning of inflation' and 'end of inflation', respectively.)

We know that the present conditions of our universe (age, density, expansion, acceleration) in principle determine a unique Friedman model—and inflationists have no quarrel with that. We can follow this model back in time. Somewhere around

---

[1] For two recent and sympathetic accounts of inflation, see, for example, A.H.Guth, *Physics Reports* **333–334**,555 (2000) and A. Linde, ibid., p. 575. For a strong critique, see R. Penrose, *loc. cit.* (Footnote 1 on p. 380.)

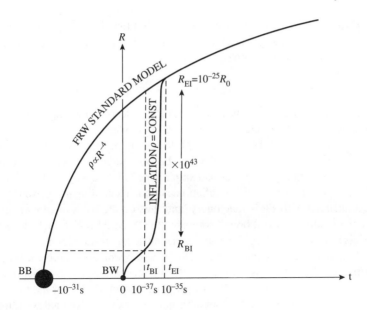

Fig. 18.6

decoupling time, for increased accuracy, we can replace the dust-dominated model by a radiation-dominated model, and still, inflationists have no quarrel with that. Only after we come to a radius of about $10^{-25} R_0$ do the backward continuations diverge: the inflationary R drops precipitously. But since the divergence of the two models in that miniscule earliest stretch of time has essentially no effect on the calculated age of the universe, inflation is irrelevant in the empirical determination of the correct Friedman model along the lines of the preceding section.

Still, as we mentioned earlier, there is one main difference between the universe that emerges from the standard theory at $R = R_{EI}$ and the one that emerges from inflation: the latter has vastly bigger particle horizons. To see why, consider the inflationary epoch from t_{BI} to t_{EI} and during that time let $R = A \exp(Ht)$ with A, H constant. Then we have

$$R_{BI} = A \exp(Ht_{BI}), \qquad R_{EI} = A \exp(Ht_{EI}), \qquad (18.61)$$

two equations which determine A and H. In particular,

$$\log R_{EI} - \log R_{BI} = H(t_{EI} - t_{BI}). \qquad (18.62)$$

Now consider the 'conformal' time T_{BI} reckoned backwards from t_{EI} [cf. (17.7)(ii)]:

$$|T_{BI}| = \int_{BI}^{EI} \frac{dt}{R} = \int_{BI}^{EI} \frac{dR}{R\dot{R}} = \int_{BI}^{EI} \frac{dR}{HR^2}$$

$$= \frac{1}{H}\left[\frac{1}{R_{BI}} - \frac{1}{R_{EI}}\right] \approx \frac{1}{HR_{BI}}. \qquad (18.63)$$

If we regard t_{BI}, t_{EI}, and R_{EI} as given, H becomes a function of R_{BI} through (18.62). Nevertheless we can make the last term in (18.63) arbitrarily large simply by choosing R_{BI} small enough. For, by (18.62),

$$HR_{BI} = R_{BI}(\log R_{EI} - \log R_{BI})/(t_{EI} - t_{BI}), \qquad (18.64)$$

and, of course, $R_{BI} \log R_{BI}$ tends to zero with R_{BI}. (Beetles crawling over the balloon while the balloon is miniscule, cover more 'ψ' than when it is large!)

Thus T_{BI} can certainly be large enough to guarantee that the part of the matter-energy present at t_{BI} which is destined to become today's observable universe (a certain range of ψ in Fig. 17.3) has had ample time to interact by the time inflation is over. But here the logic falters: The inflationary epoch including its horizon stretching has been analized with FRW machinery, and thus on the basis of *already achieved* homogeneity-isotropy! So although compared to standard FRW models the inflationary models have vastly enlarged particle horizons (at any instant far more galaxies are visible), it is not clear how this is related to the homogenization process. What probably helps most is the fact that—owing to the maintenance of constant density during inflation by a linear factor of $\sim 10^{43}$—only one part in $10^{3 \times 43}$ of the final matter-energy was present initially. The observable universe of some 10^{11} galaxies thus grew out of matter-energy equivalent to only 10^{-118} of a galaxy or 10^{-48} of a proton!! The rest was created during inflation. So whereas the big bang in FRW universes must produce all the matter-energy there ever is, inflationary universes only need a weak 'big whimper', possibly a chaotic one, to start things off. Perhaps it can be argued that patches of matter-energy of such low density have a good chance of homogenizing before inflation; or that the homogeneous vacuum that drives inflation is hardly affected by it; or that the matter-energy of the patch that becomes our universe was 'sufficiently' homogenized to make FRW arguments *approximately* applicable, and that the large stretch of conformal time then helped to complete the homogenization. In this connection it may be of some interest to note that in all radiation-dominated FRW models the metric size of the PH near the big bang is proportional to $1/\sqrt{\rho}$, (cf. Exercise 18. 4). So if the big whimper were Friedmannian, the size of the PH at the *start* of inflation would already be as large as that of the standard model at the *end* of inflation (same density.) This may be some indication of the relative ease of homogenization even out of chaos, if the density is sufficiently low.

Another problem that inflation claims to cure is the so-called 'flatness problem'—the alleged improbability of finding the value of Ω_0 even within a factor 10 of unity. But there has been some debate as to the reality both of the problem and of the cure. For Friedman dust-universes we have, from (18.47) and (18.27),

$$\Omega = \frac{C}{\dot{R}^2 R} = \frac{C}{C - kR + \frac{1}{3}\Lambda R^3}. \qquad (18.65)$$

This shows that at the big bang ($R = 0$), Ω always starts at one and then wanders away from that value unless $k = \Lambda = 0$. Consider, for example, the cycloidal universe ($k = 1$, $\Lambda = 0$). It we want $\Omega < 10$, we need $C - R > \frac{1}{10}C$ or $R < \frac{9}{10}C = \frac{9}{10}R_{max}$;

and this is true for fully 60 per cent of the entire time interval. (Even $\Omega < 2$ is true for 18 per cent of that interval.) The situation is similar for all other oscillating models. In the non-oscillating case, a fiduciary value of R might again be C, and a restriction like $0.1 < \Omega < 10$ will again confine us to a non-negligible percentage of the corresponding time range. From this point of view, at least, there is no probability problem. The 'cure' proposed by inflationists was to make $\Omega = 1$, exactly, at the end of inflation and the beginning of the radiative period. For the latter, we have, from (18.47), (18.23), and (18.24),

$$\Omega = \frac{D}{\dot{R}^2 R^2} = \frac{D}{D - kR^2 + \frac{1}{3}\Lambda R^4} \qquad (18.66)$$

instead of (18.62). So if $\Lambda = 0$ (as was assumed for the 'real' cosmological constant) and $\Omega = 1$ for a finite R value, we must have $k = 0$ and $\Omega = 1$ permanently. This led to the need for vast amounts of as yet undetected matter in intergalactic space. However, in light of recent indications that Λ is positive, the claim $\Omega \equiv 1$ seems to have been withdrawn.

One major prediction of inflation, however, has fared well. By analyzing the spectrum of quantum fluctuations during the inflationary phase, it was possible to predict fluctuations that should exist in the temperature distribution of the cosmic microwave background, on scales of $\lesssim 10°$. Accordingly the results of the COBE satellite observations in 1992 were anxiously awaited. And, indeed, they seemed to confirm the predictions as do more recent data, even more strongly. These predictions, moreover, are of great importance in the problem of how inhomogeneities formed in the substratum that eventually led to the formation of structures like the galaxies.

We may also mention the solution of the 'monopole problem'. This has been called an *internal* problem, since to *perceive* it as a problem you must believe in grand unified theory to start with. But then indeed you have a problem explaining where all the magnetic monopoles went, that should have been created at $\sim 10^{27}$ K. According to this theory, monopoles of enormous energy ($\sim 10^{16}$ GeV) should occur at a spacing of about one per particle-horizon volume at the end of inflation. If the particle horizons were as small as in standard cosmology (cf. Exercise 17.10), the mass density of monopoles would exceed that of baryons by a factor 10^{14}! With inflation, on the other hand, there is at most *one* monopole in the universe we see.

18.6 The anthropic principle

It would be reassuring if one could believe that our universe (which seems destined either to recollapse and suffer a fiery death, or else to expand indefinitely and suffer a freezing death) is not the sum-total of all physical existence. A philosophically persuasive line of argumentation has, in fact, led some modern cosmologists to posit the existence of infinitely many alternative universes, all with different physical laws, different physical constants, and different initial conditions. For them that seems to be the only way out of a profound puzzle: why is our universe favorable to human existence? The number of lucky 'coincidences' required to produce an environment in which life as we know it is *possible*, seems to defy the laws of chance.

Let us begin with the big bang. A slightly lower initial expansion rate, or higher density, or higher constant of gravity, would have made the universe recollapse and reheat before it had time to cool sufficiently to make life possible. A slightly higher expansion rate, or lower density, or lower gravitational constant, would have thinned the matter too fast for galaxies to condense. It takes billions of years to cook up and distribute the basic building blocks of life (carbon, oxygen, and nitrogen) in the only suitable furnaces, the interiors of stars. A life-supporting universe must get to be at least that old, and its laws must permit the process. For example, it is 'lucky' that the nuclear force is not quite strong enough to allow the formation of 'diprotons' (proton+proton), a process that would have quickly used up all primordial hydrogen and so deprived the stars of their fuel, and life of one of its bases. Yet that force *is* just strong enough to favor the formation of deuterons (proton + neutron), without which higher nucleosynthesis cannot proceed. Without a 'lucky' energy level in the carbon nucleus the formation of carbon out of helium ($3H^4 \rightarrow C^{12} + 2\gamma$) would have failed. But equally luckily, the oxygen nucleus has an energy level that *prevents* the reaction $C^{12} + H^4 \rightarrow O^{16}$, which would have depleted the carbon as soon as it was formed.

Examples of this kind abound in all the branches of science. Here we shall mention only two others. One from biology: apparently the whole basis of life (DNA) would be in jeopardy if the charge or mass of the electron were only slightly different from what they are. And an example from ecology: water possesses the rare yet vital property that its solid form (ice) is lighter than its liquid form. As a consequence, lakes freeze over in winter and the ice protects the life below, possibly even emerging life. Were it otherwise, more and more ice would grow from the bottom upwards without being melted in the summers, until lakes and oceans were frozen solid.

Can all these 'lucky' coincidences be due to pure chance? The law that multiplies probabilities makes this highly unlikely. One can perhaps hope for the eventual discovery of a 'Theory of Everything'—a theory that fixes all the laws and constants of Nature and shows our universe to be unique. But the mystery would remain: why does the only possible universe permit life? One supposition that cannot be disproved is that a benevolent deity so designed the world. But it is part of the credo of modern science *not* to invoke a deity to explain physical facts. (Newton did not yet feel that so strongly: he believed that God would have to intervene periodically to adjust the planetary orbits, since it seemed to him that the mutual gravity of the planets would lead to instabilities. It took a hundred years before Laplace was able to solve the stability problem the modern way.) Out of all these difficulties grew the *anthropic principle*: if there are infinitely many alternative universes, then there is no mystery in finding ourselves in one that permits our presence.[2]

---

[2] A full and excellent account of this topic can be found in the book by J. D. Barrow and F. J. Tipler, *The Anthropic Cosmological Principle*, Oxford University Press, 1986.

Exercises 18

18.1. Consider the Newtonian ball (part of a Friedman universe) which we discussed in the paragraph containing (18.19). Prove that its total kinetic energy is given by $T = \frac{2}{5}\pi\rho H^2 R^5 a^5$, with $H = \dot{R}/R$, and its total potential energy by $V = -\frac{16}{15}\pi^2\rho^2 G R^5 a^5$ (if the zero-point is at infinity). Deduce that Ω, as defined in (18.46) (ii), is the absolute value of the ratio V/T for every little ball making up the universe. Also show that Friedman's differential equation (18.27), for $\Lambda = 0$, coincides with the Newtonian energy equation of each such ball. Is this implicit in (18.22)?

18.2. Construct a static Einstein-type universe without Λ term, held in equilibrium by *negative* pressure p. Prove that we need $k = 1$ and $p = -\frac{1}{3}c^2\rho$ (in full units). Note again the counter-intuitive effect of pressure.

18.3. Obtain the following three 'purely radiative' solutions of eqn (18.24) without the Λ term:

$$R = (4D)^{1/4}t^{1/2} \quad (k = 0)$$

$$(t - \sqrt{D})^2 + R^2 = D \quad (k = 1)$$

$$(t + \sqrt{D})^2 - R^2 = D \quad (k = -1),$$

and plot the corresponding $R - t$ graphs: half a parabola, a semicircle, a quarter of a hyperbola.

18.4. (i) For the flat dust and radiation models without Λ-term, $R \propto t^{2/3}$ and $R \propto t^{1/2}$, prove that the proper radius of the particle horizon is given by $3t$ and $2t$, or $(\frac{2}{3}\pi G\rho)^{-\frac{1}{2}}$ and $(\frac{8}{3}\pi G\rho)^{-\frac{1}{2}}$ respectively. (ii) For the $t^{2/3}$ universe, find today's proper distance to a source which emitted the light whereby we see it when the radius of the universe was α times what it is today. [*Answer*: $3t_0(1 - \sqrt{\alpha})$.] (iii) For the $t^{2/3}$ universe, what is the proper radius of the particle horizon today (in light years) and what multiple of today's distance to the 'sphere of last scattering' is that? [*Answer*: multiple $\approx (1 - 10^{-3/2})^{-1} = 1.033$.]

18.5. For the cycloidal universe (18.36), prove $dt/R = d\chi$. Deduce that a 'creation photon' circles the universe exactly once between big bang and big crunch. [*Hint*: (17.2).]

18.6. For the cycloidal universe (18.36), prove that the fraction of the total volume visible at parametric time χ is $(\chi - \sin\chi\cos\chi)/\pi$. Thus, at half-time ($\chi = \pi$), the whole universe is already visible. [*Hint*: (8.6) (ii).]

18.7. For the cycloidal universe (18.36), prove $\Omega = 2/(1 + \cos\chi)$. Hence prove that the inequality $\Omega < 2$ holds for 18 per cent of the full time range of this universe. [*Hint*: $\dot{R} = (dR/d\chi)/(dt/d\chi)$.]

18.8. For the hyperbolic universe (18.37), prove $dt/R = d\chi$ and $\Omega = 2/(1 + \cosh\chi)$. Find χ_0 and C if $\Omega_0 = 0.2$ and $t_0 = 2 \times 10^{10}$ years. Hence find the proper distance of the particle horizon today. [*Answer*: $\sim 8 \times 10^{10}$ light-years.]

18.9. Describe how to construct closed oscillating models in which each creation photon circles the universe arbitrarily often between big bang and big crunch. [*Hint*: Fig. 18.3.]

18.10. Prove that if the universe presently accelerates ($q < 0$), we *must* have $\Lambda > 0$. Also prove that the inequality $\Omega \gtreqless \frac{2}{3}(q + 1)$ then determines $k \gtreqless 0$, respectively. Thus, independently of the exact value of q, $\Omega > \frac{2}{3}$ would imply $k = 1$. Prove that also $q < -1$ by itself would guarantee $k > 0$.

18.11. For the 'plausible' model specified at the end of Section 18.4, find the present radius of the universe, R_0, and the necessary value of the cosmological constant Λ. Compare the order of magnitude of your results with (18.29). [*Hint*: use Table 18.1 to estimate q_0. *Answers*: $R_0 \approx 1.9 \times 10^{28}$ cm $\approx 2 \times 10^{10}$ light-years, $\Lambda \approx 2 \times 10^{-56}$ cm^{-2}.]

18.12. With a slight change of emphasis, we can regard Fig. 18.5 as a representation of cosmological phase space, in which every non-vacuous big-bang Friedman model follows a certain trajectory. We have seen that for all such models $\Omega \to 1$ as we go back in time to the big bang [cf. (18.62)]. Verify that similarly $Ht \to \frac{2}{3}$. [*Hint*: use $R \propto t^{2/3}$.] So the trajectory of every such model begins at the 'critical' point $\Omega = 1$, $Ht = \frac{2}{3}$. On a copy of this diagram, plot the trajectories of the three $\Lambda = 0$ models ($k = 0$, $k = 1$, $k = -1$), and also of the three $k = 0$ models ($\Lambda = 0$, $\Lambda > 0$, $\Lambda < 0$). Plot the presumed trajectory of the 'plausible' model specified at the end of Section 18.4.

18.13. Justify the assignation of $\sim 10^{-31}$ s to the stretch so marked along the t-axis of Fig. 18.6, assuming that the universe up to recombination time, $t_{\text{rec}} \approx 10^6$ years, is radiative ($R \propto t^{1/2}$), and that $R_{\text{EI}} \approx 10^{-22} R_{\text{rec}}$.

Appendix

Curvature tensor components for the diagonal metric

One of the most tedious calculations in GR is the determination of the Christoffel symbols $\Gamma_{\nu\sigma}^{\mu}$, the Riemann curvature tensor $R_{\mu\nu\rho\sigma}$, the Ricci tensor $R_{\mu\nu}$, the curvature invariant R, and the Einstein tensor $G_{\mu\nu}$, for a given spacetime metric. Various computational shortcuts exist, but rather than start from scratch each time, it is well to have tables for certain standard forms of the metric. In this appendix we shall deal with the general 4-dimensional *diagonal* metric,

$$\mathbf{ds}^2 = A(dx^1)^2 + B(dx^2)^2 + C(dx^3)^2 + D(dx^4)^2, \tag{A.1}$$

where A, B, C, D are arbitrary functions of all the coordinates, and as a byproduct— without extra labor—we shall get the various curvature components also for the 2- and 3-dimensional diagonal metrics

$$\mathbf{ds}^2 = A(dx^1)^2 + B(dx^2)^2, \qquad \mathbf{ds}^2 = A(dx^1)^2 + B(dx^2)^2 + C(dx^3)^2. \tag{A.2}$$

Rule 1. The components $\Gamma_{\nu\sigma}^{\mu}$, $R_{\mu\nu\rho\sigma}$, $R_{\mu\nu}$, and R for the 2- and 3-dimensional metrics (A.1) are obtained from the 4-dimensional formulae by setting $D = 1$ (not zero!) in the 3-dimensional case, and $C = D = 1$ in the 2-dimensional case, and treating the remaining coefficients as independent of x^4, or of x^3 and x^4, respectively. This can be useful in connection with the result of Exercise 10.13 on composite metrics.

We remind the reader that *all* 2- and 3-dimensional metrics can be 'diagonalized'; that is, brought to the respective forms (A.2) by a suitable transformation of coordinates, and that every *static* spacetime metric can be so diagonalized [cf. (9.5)], as well as many others.

We shall use the following symbols

$$\alpha = \frac{1}{2A}, \qquad \beta = \frac{1}{2B}, \qquad \gamma = \frac{1}{2C}, \qquad \delta = \frac{1}{2D}, \tag{A.3}$$

and the notation typified by

$$A_\mu = \frac{\partial A}{\partial x^\mu}, \qquad B_{12} = \frac{\partial^2 B}{\partial x^1 \partial x^2}, \text{ etc.} \tag{A.4}$$

Then, directly from the definition (10.14), we find (for $\mu = 1, 2, 3, 4$)

$$\Gamma_{23}^1 = 0, \qquad \Gamma_{22}^1 = -\alpha B_1, \qquad \Gamma_{1\mu}^1 = \alpha A_\mu. \tag{A.5}$$

From these three typical Γs *all* others can be obtained by obvious permutations (for example, $\Gamma^2_{33} = -\beta C_2$, $\Gamma^4_{44} = \delta D_4$, etc.).

From the Γs we now obtain the curvature tensor $R_{\mu\nu\rho\sigma}$ as defined by (10.60). We find the following typical components.

$$R_{1234} = 0 \tag{A.6}$$

$$2R_{1213} = -A_{23} + \alpha A_2 A_3 + \beta A_2 B_3 + \gamma A_3 C_2 \tag{A.7}$$

$$2R_{1212} = -A_{22} - B_{11} + \alpha(A_1 B_1 + A_2^2) + \beta(A_2 B_2 + B_1^2)$$
$$- \gamma A_3 B_3 - \delta A_4 B_4. \tag{A.8}$$

Again, all other components can be obtained from these by permutation, and, of course, by making use of the symmetries (10.62)–(10.65). For example, to get R_{3432}, we subject (A.7) to the permutation $1 \to 3, 2 \to 4, 3 \to 2$ (and so, necessarily, $4 \to 1$)—accompanied by $A \to C, B \to D, C \to B, D \to A$. Rule 1 applies for extracting the 2- and 3-dimensional formulae.

Next, we calculate the Ricci tensor $R_{\mu\nu}$, as defined by (10.68). Typically, we find

$$
\begin{aligned}
R_{12} = \quad &\gamma C_{12} \quad &&+\delta D_{12} \\
&-\gamma^2 C_1 C_2 \quad &&-\delta^2 D_1 D_2 \\
&-\alpha\gamma A_2 C_1 \quad &&-\alpha\delta A_2 D_1 \\
&-\beta\gamma B_1 C_2 \quad &&-\beta\delta B_1 D_2
\end{aligned}
\tag{A.9}
$$

$$
\begin{aligned}
R_{11} = \quad\quad\quad &\beta A_{22} &&+\gamma A_{33} &&+\delta A_{44} \\
&+\beta B_{11} &&+\gamma C_{11} &&+\delta D_{11} \\
&-\beta^2 B_1^2 &&-\gamma^2 C_1^2 &&-\delta^2 D_1^2 \\
-\alpha A_1(0 \quad &+\beta B_1 &&+\gamma C_1 &&+\delta D_1) \\
-\beta A_2(\alpha A_2 &+\beta B_2 &&-\gamma C_2 &&-\delta D_2) \\
-\gamma A_3(\alpha A_3 &-\beta B_3 &&+\gamma C_3 &&-\delta D_3) \\
-\delta A_4(\alpha A_4 &-\beta B_4 &&-\gamma C_4 &&+\delta D_4)
\end{aligned}
\tag{A.10}
$$

All other components can be found by making the obvious permutations on these two. The dashed lines indicate the terms that remain in the 2- and 3-dimensional cases, respectively, according to our Rule 1.

For the curvature invariant $R = g^{\mu\nu} R_{\mu\nu}$ we find

$$
\begin{aligned}
\tfrac{1}{4}R = \quad & \alpha\beta(A_{22} + B_{11} - \alpha A_2^2 - \beta B_1^2 - \alpha A_1 B_1 - \beta A_2 B_2 + \gamma A_3 B_3 + \delta A_4 B_4) \\
+ & \alpha\gamma(A_{33} + C_{11} - \alpha A_3^2 - \gamma C_1^2 - \alpha A_1 C_1 + \beta A_2 C_2 - \gamma A_3 C_3 + \delta A_4 C_4) \\
+ & \beta\gamma(B_{33} + C_{22} - \beta B_3^2 - \gamma C_2^2 + \alpha B_1 C_1 - \beta B_2 C_2 - \gamma B_3 C_3 + \delta B_4 C_4) \\
+ & \alpha\delta(A_{44} + D_{11} - \alpha A_4^2 - \delta D_1^2 - \alpha A_1 D_1 + \beta A_2 D_2 + \gamma A_3 D_3 - \delta A_4 D_4) \\
+ & \beta\delta(B_{44} + D_{22} - \beta B_4^2 - \delta D_2^2 + \alpha B_1 D_1 - \beta B_2 D_2 + \gamma B_3 D_3 - \delta B_4 D_4) \\
+ & \gamma\delta(C_{44} + D_{33} - \gamma C_4^2 - \delta D_3^2 + \alpha C_1 D_1 + \beta C_2 D_2 - \gamma C_3 D_3 - \delta C_4 D_4).
\end{aligned}
$$
$$\tag{A.11}$$

Note: the factor $\tfrac{1}{4}$ on the left remains unchanged even in the 2- and 3-dimensional cases obtained by Rule 1.

Lastly, for the Einstein tensor $G_{\mu\nu} = R_{\mu\nu} - \tfrac{1}{2} R g_{\mu\nu}$, we find the following typical components:

$$
G_{12} = R_{12} \tag{A.12}
$$

$$
\alpha G_{11} =
$$
$$
\begin{aligned}
& \beta\gamma(-B_{33} - C_{22} + \beta B_3^2 + \gamma C_2^2 - \alpha B_1 C_1 + \beta B_2 C_2 + \gamma B_3 C_3 - \delta B_4 C_4) \\
+ & \beta\delta(-B_{44} - D_{22} + \beta B_4^2 + \delta D_2^2 - \alpha B_1 D_1 + \beta B_2 D_2 - \gamma B_3 D_3 + \delta B_4 D_4) \\
+ & \gamma\delta(-C_{44} - D_{33} + \gamma C_4^2 + \delta D_3^2 - \alpha C_1 D_1 - \beta C_2 D_2 + \gamma C_3 D_3 + \delta C_4 D_4).
\end{aligned}
$$
$$\tag{A.13}$$

All other components can be obtained from these by permutation. Note, however, that Rule 1 does *not* apply here.

Index

Page numbers in italics refer to footnotes

424   *Index*

   255
gravitational lenses 250–2, 257
gravitational mass 16
gravitational potentials 195, 197
gravitational time dilation 26, 184, 198
gravitational waves 9, 284–95, 321, 323–35
   energy carried by 333–4
   focusing by 289, 328
gravitoelectric field 335
gravitomagnetic field 335
gravitomagnetism 197, 301
graviton 223
Gravity Probe-B 340
group property
   of Lorentz transformation 48
   of Poisson transformation 48
   of tensor transformation 134
Guth 410
gyrocompass 197

Hafele 67
Hall D.B. 66
Hall J.L. *10*
Hamilton's principle 190
harmonic coordinates 341
harmonic gauge 320, 322, 324, 341
Hawking *7*, 279, 280, 383
Hay 80
headlight effect 82, 86
Heaviside 154
helicity 329
Herschel 349
hierarchical universe 358
hindsight effect 86
Hiroshima bomb 113
Hoffmann 391
Holzscheiter 17
homogeneity
   of inertial frame 39
   of universe 347, 351, 358, 359, 412
Hooke 9
horizon
   in accelerated frame 268
   in cosmology 376–83
   in de Sitter space 306, 310–12
   in Schwarzschild space 231, 240, 247, 258–60
horizon problem 380, 410–12
Hubble 349
Hubble age 355
Hubble constant (Hubble parameter) 353, 366, 405
Hubble space telescope 349
Hulse 245
Huyghens 7, 9, 348
hyperbolic motion 54, 71, 75
hyperbolic parameter 53, 59
hypersphere (3-sphere) 170–2, 182

ideal clocks 65, 93, 178
incompressibility 56
index permutation 134
inertial coordinate system 40
inertial force 18
inertial frame 4, 15, 34, 39–41
   local (LIF) 19, 20, 36, 65, 210
inertial mass 16, 21, 110
   conservation of 110
Infeld 391
Inflation 409–13
inflexional universe 375, 376, 404
inner product 135
International Atomic Time (TAI) 10, 26, 186
interval 91
intrinsic geometry 165, 174, 180
invariant 96
invariant speed 15, 47, 57
invariants of electromagnetic fields 147
inverse Compton scattering 122
isochronous four-vectors 106
isometries 191
isotropic point 170
isotropy
   of inertial frames 39
   of universe 347, 351, 358
Israel *7*
Ives 80

Jackson *196*

Kaivolo *80*
Kant 8, 349
Kasner 315
Keating 67
Kepler 347
Kepler's third law 224, 239
Kerr metric 263, 279, 281
   linearized 337
kinematic relativity 368
kinetic energy 112
Kronecker delta 132, 134
Krotkov 17
Kruskal diagram 273, 309
Kruskal metric 273
Kruskal space 267, 272–8, 309

Lambert 349
lambda force 16, 305, 357, 400
Laplace 414
Laue 74, 78, 154
Laue's theorem 331
Leibniz 7
Leinaas *281*
Lemaître 312
Lenard 120
length contraction 39, 62, 151
   paradoxes 63, 74